AF692739

Network Pharmacology of Natural Products

Network Pharmacology of Natural Products

Guest Editor

Alexander George Panossian

MDPI

Basel • Beijing • Wuhan • Barcelona • Belgrade • Novi Sad • Cluj • Manchester

Guest Editor
Alexander George Panossian
Research and Development
PhytoMed AB
Västervik
Sweden

Editorial Office
MDPI AG
Grosspeteranlage 5
4052 Basel, Switzerland

This is a reprint of the Special Issue, published open access by the journal *Pharmaceuticals* (ISSN 1424-8247), freely accessible at: www.mdpi.com/journal/pharmaceuticals/special_issues/2GAFOAG14B.

For citation purposes, cite each article independently as indicated on the article page online and as indicated below:

Lastname, A.A.; Lastname, B.B. Article Title. *Journal Name* **Year**, *Volume Number*, Page Range.

ISBN 978-3-7258-4204-9 (Hbk)
ISBN 978-3-7258-4203-2 (PDF)
https://doi.org/10.3390/books978-3-7258-4203-2

© 2025 by the authors. Articles in this book are Open Access and distributed under the Creative Commons Attribution (CC BY) license. The book as a whole is distributed by MDPI under the terms and conditions of the Creative Commons Attribution-NonCommercial-NoDerivs (CC BY-NC-ND) license (https://creativecommons.org/licenses/by-nc-nd/4.0/).

Contents

About the Editor

Alexander George Panossian

Alexander Panossian, Ph.D., Dr.Sc, is the Founder of Phytomed AB in Sweden and has been the Head of Research & Development of Europharma USA Inc. since 2016. He has multiple advanced degrees in bioorganic chemistry and the chemistry of natural and physiologically active compounds. He completed his doctorate in organic chemistry at the Yerevan State University in Armenia, 1971, and he obtained his scientific degrees from the Moscow Institute of Bioorganic Chemistry in 1975 and the Moscow Institute of Fine Chemical Technology in 1986. Dr. Panossian was made a full professor of bioorganic chemistry and the chemistry of natural and physiologically active compounds in the Russian Federation and later served as the Director of the Laboratory of Quality Control of Drugs of Medical Drug Agency of the Republic of Armenia (1993–2003). In 2003, he moved to Sweden to act as head of research and development at the Swedish Herbal Institute. He has the honor of participating as a guest scientist in the Laboratory of Nobel Laureate Bengt Samuelsson at the Karolinska Institute in Stockholm (1982–1983), at Munich University (1993–1995), and at King College in London, 1996. He was the editor-in-chief of *Phytomedicine, International Journal of Phytopharmacology and Phytotherapy* (Elsevier, Germany) from 2014 to 2017. He has authored or co-authored about 200 articles in peer-reviewed journals. His main research interest is focused on plant adaptogens, anti-stress compounds that are involved in the regulation of the neuroendocrine and immune systems.

Preface

The pharmacology of natural products, particularly botanicals, encompasses pharmacognosy in Galenic preparations and the study of the pharmacodynamics and pharmacokinetics of their active constituents. Pharmacological studies aim to investigate the effect of herbal drugs and other botanicals on physiological systems to provide therapeutic benefits, focusing on the mechanisms related to pharmacological targets and the regulation of physiological processes.

The aim of "network pharmacology" and "network medicine" is to understand the pathophysiology of diseases at the system level. Disease-related interaction networks form the basis for developing novel drugs that target these networks, rather than individual proteins or enzymes.

The network pharmacology of a multi-component drug substance, which comprises several phytochemicals or herbs/botanicals, is more complex due to numerous challenges, including detailed chemical characterization (qualitative and quantitative analysis of phytochemical composition) and standardization, both of which are necessary to ensure reproducible quality and efficacy.

Since botanicals consist of multi-component active compounds and phytochemicals, their interactions lead to novel and unexpected pharmacological activities due to their synergistic and antagonistic effects. Similarly, the pharmacological activity of a combination of several plants or herbal extracts, such as those in botanical hybrid preparations (e.g., TCM formulas), results in a new biologically active substance with unique pharmacological characteristics. However, several challenges exist in network pharmacology for natural product research, which will be addressed in this Special Issue.

The advancement of integrative omics network pharmacology and artificial intelligence in natural products has opened new avenues for the following areas:

- Elucidation of the mechanisms of action of medicinal plants.
- Understanding the synergistic therapeutic actions of complex bioactive components in medicinal plants.
- Providing a rationale for traditional Chinese medicine.
- Enhancing the quality of TCM drug research and the speed and efficiency of developing new TCM products.
- Discovering and developing new botanical hybrid combinations.
- Predicting drug–herb interactions, adverse events, and potential toxic effects.

Alexander George Panossian
Guest Editor

pharmaceuticals

Editorial

Trends and Pitfalls in the Progress of Network Pharmacology Research on Natural Products

Alexander Panossian

Phytomed AB, Sjöstadsvägen 6A, 59344 Västervik, Sweden; ap@phytomed.se

1. Introduction

Herbs, used as food and a source of medicine for centuries, have been extensively studied over time for their chemical and pharmacological properties, with two main aims. One objective has been to identify and purify active compounds, which could then be developed into new, more active pharmaceuticals and analogs targeting specific receptors while being free of adverse effects. The other aim has been to combine several plant extracts or purified ingredients into new, multitarget hybrid combinations with enhanced pharmacological activity.

The first approach has been successfully applied in orthodox Western medicine, resulting in the discovery of compounds such as morphine, salicin, papaverine, and taxol, as well as their synthetic analogs. This approach has shaped the current drug development platform, which is primarily based on the pharmacodynamics and pharmacokinetics of botanicals. It also draws on the accrued knowledge of how biologically active molecules from plant extracts interact with receptor molecules, primarily proteins, involved in cellular and organismal responses to achieve therapeutic effects and maintain organismal homeostasis. Meanwhile, the stimulus–response coupling is often mediated by complex intra- and extracellular communications and feedback regulations within networks of signaling pathways and pharmacological systems, including the neuroendocrine–immune complex, cardiovascular, gastrointestinal, and other systems.

The one-drug-one-target paradigm, commonly explored for Galenic preparations by extracting one or more active constituents of a plant, was based on a reductionistic concept of dissecting biological systems into their constituent parts. However, even purified active compounds are not selective in their action, as they interact with other receptors and molecular networks associated with various functions, inducing adverse events.

In contrast, botanical hybrid preparations (BHP) [1] are designed to target multiple pathways, achieving therapeutic benefits in stress-induced, aging-related, and other chronic diseases [1–5]. When combining botanicals that contain numerous phytochemicals, the most desirable outcome can result in additional therapeutic benefits. However, the intended or expected outcomes can often be unpredictable, as various interactions occur within the target cells of the neuroendocrine–immune, cardiovascular, gastrointestinal, and other regulatory systems involved in the body's response [1–5].

The second approach, primarily used in traditional Chinese medicine (TCM), is based on empirical knowledge and is undergoing modernization by incorporating current understanding of pharmacological systems and network pharmacology, including metabolomics, proteomics, transcriptomics, genomics, and microbiomes (collectively known as OMICS) [6–10].

Complex herbal combinations have a long history in traditional medical systems. Most traditional Chinese medicines contain more than one medicinal plant, suggesting that the concomitant use of additional plants may positively alter the effect of the primary

 https://doi.org/10.3390/ph18040538

herb [6]. Traditional Chinese medicine often provides multitarget (polyvalent) treatment, where ingredients address multiple symptoms and target various organs. In TCM, multi-component herbal preparations are formulated with the belief that each herb included in a prescription targets specific symptoms and that the overall mixture is more than a random assortment of herbs. Instead, the herbs interact harmoniously, each playing a distinct role in the hierarchy, where each herb is figuratively categorized as either "emperor", "minister", "assistant", or "servant". While the rationale of TCM may seem contrived when judged from a scientific perspective, its empirical effects are evident [6].

The resulting impact—whether complementary or antagonistic—is categorized into six basic modes of interaction: reinforcement (where herbs possessing similar medicinal properties are combined enhance efficacy), potentiation, restraint, detoxification, counteraction (where one herb reduces the therapeutic effect of another, representing an antagonistic interaction), and toxicity (though scientific evidence remains limited or lacking) [6].

2. Network Pharmacology of Natural Products and Complex Herbal Mixtures in Drug Discovery Programs

Historically, drug development strategies assumed that a single-target mechanism of action is the best approach for obtaining target-specific therapeutics. This model aims to treat specific conditions and is associated with fewer adverse events. However, many drugs and natural compounds interact with multiple receptors, resulting in polyvalent pharmacological and pleiotropic therapeutic activities through multitarget interactions. This development, in turn, has shifted the paradigm from a "one-target, one-drug" mode to a "network-target, multiple-component therapeutics" mode.

Developing techniques for monitoring the expression of genes, proteins, and metabolites in their entirety in cells, tissues, and organs formed the basis for the new field of systems biology. The aim of "*network pharmacology*" and "*network medicine*" is to understand the pathophysiology of diseases at the systems level. Disease-related interaction networks form the basis for developing novel drugs that target these networks, rather than individual proteins or enzymes [5].

The term "pharmacological target" refers to biomolecules such as DNA, mRNA, and proteins, including transmembrane and nuclear receptors, ion channels, transport proteins, and numerous enzymes to which a drug binds primarily to elicit its pharmacological effect. The complementary binding of drug molecules to a protein with a physiological function can alter the cell's physiological responses, resulting in a pharmacological effect. The efficacy of binding a drug (ligand) depends on the drug's chemical structure and affinity for its specific biomolecule (receptor).

The essence of *network pharmacology* is to evaluate how a study *drug* interacts with therapeutic targets, their associated signaling pathways involved in biological and physiological processes, and the functions linked to diseases, ultimately aiming to achieve a beneficial *therapeutic effect*.

The network pharmacology of a *multi-component drug substance*, which comprises several *phytochemicals* or *herbs/botanicals*, is more complex due to numerous challenges, including detailed chemical characterization (qualitative and quantitative analysis of phytochemical composition) and standardization, both of which are necessary to ensure *reproducible quality* and *efficacy*.

The pharmacology of natural products, particularly botanicals, encompasses pharmacognosy in Galenic preparations and the study of the pharmacodynamics and pharmacokinetics of their active constituents. Pharmacological studies aim to investigate the effect of herbal drugs and other botanicals on physiological systems to provide therapeutic benefits, focusing on the mechanisms related to pharmacological targets and the regulation of physiological processes.

Over the past two decades, we have witnessed a technical revolution, marked by novel methods collectively known as "omics" technologies. The term "omics", derived from the Latin suffix "-ome", refers to the measurement of large quantities to understand a biological question or phenomenon in an integrated manner [5,11,12]. Typical omics technologies include genomics, transcriptomics, proteomics, and metabolomics. More recently, additional "omics" methods have emerged, including epigenomics, lipidomics, and glycomics, as well as microbiomics in the gut, skin, and other organs, which are considered an "extended genome" because microorganisms contribute to the human genome [13].

As outlined above, traditional drug discovery programs focused on developing potent drugs with some specificity for a particular therapeutic target. However, due to the promiscuity of many medicines, which can modulate multiple targets and pathways [14,15], the term "polypharmacology" has been introduced. Bioinformatic tools now have the power to design multi-specific drugs [16–19]. Thus, drug promiscuity has shifted from being seen as the source of unwanted effects to becoming an intentional strategy in drug discovery and development. Network pharmacology can now integrate complex "omics"-based datasets from "multi-omics" and single-cell "omics" technologies to form the foundation for novel next-generation drugs. Notably, medicinal plants and their associated phytochemicals play a pivotal role in this emerging field of drug research.

Network pharmacology has been applied to various traditional medicines, with the most progress being made in the context of traditional Chinese medicine (TCM) [7,8,20]. Network pharmacology aligns perfectly with the holistic philosophy of TCM. While a body of clinical data reports the efficacy of TCM, the mechanistic understanding of herbal TCM formulations is still in its infancy [20–24]. The expectation is that the theory of TCM, which fundamentally differs from the reductionist approach of Western medicine, could eventually be scientifically explained [9,10,25]. Network-based multitarget-multicomponent models could help rationalize the understanding of characteristic TCM syndromes [26]. In TCM, a syndrome describes an imbalance in the body that can lead to various diseases; for example, hot or cold syndromes result in diseases related to excess heat or cold in the body [27].

Network pharmacology, a relatively new discipline, has emerged recently and is rapidly developing [5,28–31]. In 2024 alone, approximately 9000 publications on network pharmacology from around the world were cited and indexed in the PubMed database [32]. In particular, the advancement of integrative omics network pharmacology and artificial intelligence in natural products has opened up new avenues for the following areas:

- elucidation of the mechanisms of action of medicinal plants [3,33–39];
- understanding the synergistic therapeutic actions of complex bioactive components in medicinal plants [1–5,40,41];
- providing a rationale for traditional Chinese medicine, as well as enhancing the quality of TCM drug research and the speed and efficiency of developing new TCM products [7–10,42,43];
- discovering and developing new botanical hybrid combinations [1,2,5];
- predicting drug–herb interactions [44], adverse events [45] and potential toxic effects [12,46–48].

3. Critical Appraisal, Limitations, Challenges, and Future Perspectives

Since botanicals consist of multi-component active compounds and phytochemicals, their interactions lead to novel and unexpected pharmacological activities due to their synergistic and antagonistic effects. Similarly, the pharmacological activity of a combination of several plants or herbal extracts—such as those in botanical hybrid preparations (e.g., TCM formulas)—results in a new biologically active substance with unique pharmacological characteristics [1]. However, there are several challenges in network pharmacology for natural products research:

(I) The reproducibility of the chemical composition (fingerprint) of active compounds and their influence on the pharmacological activity (signature) of the total extract is crucial. This is due to the synergistic, potentiating, and antagonistic interactions between the multiple targets of the numerous active components in the extracts.

(II) Quality and safety issues related to the content of active and potentially toxic compounds should be considered [49]. European drug legislation and pharmacopeia specify active markers and their content in herbal medicines and botanical preparations, ensuring reproducible quality, safety, and efficacy. However, most publications on TCM do not comply with EC standards, particularly when assessing TCM in clinical trials. This is a significant obstacle in determining endpoint outcomes and drawing conclusions regarding the reproducible quality, safety, and efficacy of TCM.

(III) The optimal effective and safe therapeutic dose of *herbal medicines and botanical preparations* should be established, taking into account the "bell-shaped" dose–response relationship. Many in vitro studies have applied supraphysiological concentrations far exceeding the proposed doses in humans. Most studies have overlooked this issue, as pharmacological and toxicological data are often unavailable. The optimal range of doses for botanicals is similar to that of the hormetic zone dose-response pattern. The underlying molecular mechanisms of the reversal of the dose response are not fully understood. The theoretical background of hormesis is related to the hypothesis of interactions between biologically active compounds (ligands, interventions, or drugs) and two target proteins (receptors) that exhibit functionally opposite responses at different concentrations. It has been proposed that a drug acting as a competitive antagonist at either or both receptors changes the relationship between the two opposing concentration–effect curves, resulting in potentiation, antagonism, or reversal of the observed effect. The theoretical model suggests that the total effect in the system can be obtained by the algebraic summation of the two effects produced by the activation of the two opposing receptor populations, generating a classical biphasic hormesis curve. However, in practice, dose-response patterns are significantly complicated due to many other interactions with (i) multiple targets (receptors) with different affinities for the active compound, (ii) other regulatory proteins or mediators within the networks involved in the adaptive stress response, (iii) feedback downregulation within molecular signaling pathways, and/or (iv) the metabolic transformation of active ligands into metabolites, which may act as secondary ligands with varying affinities for different receptors in the adaptive stress response.

(IV) The network analysis model often limits itself to intracellular communications and overlooks extracellular communications and interactions between various physiological regulatory systems, such as the integrative nervous, endocrine, immune, cardiovascular, respiratory, and renal systems. These systems operate across multiple organs and tissues. In this mechanistic model, there is little room for considering the role of physiology and anatomy in the organism's integrative regulatory system, which maintains homeostasis in response to pharmaceutical interventions. The primary challenge in network pharmacology is the lack of sufficient, evidence-based knowledge regarding the molecular biology, pharmacodynamics, and pharmacokinetics of all compounds in botanical preparations, as well as the interactions of mediators that influence their pharmacological activity. While bioinformatics involves building and mechanistically analyzing networks using available in silico tools, artificial intelligence alone can sometimes be insufficient. Valid experimental data, coupled with well-designed preclinical and clinical studies of well-characterized botanical preparations, are needed to ensure the established quality, efficacy, and safety of these compounds.

(V) The conclusions of many network analysis studies go too far and are not supported by evidence in many aspects. In such cases, further studies are often based on misleading backgrounds and can yield inconclusive results. Furthermore, referring only to the latest publications—mainly review articles—and ignoring original studies distorts the scientific history of many inventions.

(VI) Network pharmacology predictions based exclusively on in silico and some in vitro studies have several limitations that must be further validated through animal experiments.

(VII) Scientific information on the direction of correlations between gene expression and physiological function or diseases is used in silico for therapeutic efficacy and toxicity predictions. Additional studies are required where different experimental outcome measures are applied. Furthermore, a strong resistance mechanism can override the effects of a drug. For instance, the drug efflux transporter P-glycoprotein can expel medications from cells before they reach their intended intracellular targets, thereby preventing or reducing most downstream signaling and network effects. Ultimately, clinical studies on predicted diseases in human subjects are essential.

(VIII) One challenge lies in the compatibility of pharmaceutical-grade herbal extracts of reproducible quality and pharmacological activity compared to purified active constituents. The content of a genuine extract is adjusted to a defined content or range of active constituents with a known therapeutic efficacy or content of analytical markers.

(IX) Implementing Lipinsky's so-called "rule of five" is not a relevant tool for screening traditional herbal medicines for new drug discovery.

In conclusion, the network pharmacology approach is highly useful for understanding the mechanisms of action, predicting potential toxic effects, and identifying new therapeutic indications. However, the results of in silico analyses must be validated with experimental findings, including gene and protein expression in isolated cells, and at least confirmed through in vivo animal experiments [49]. A deeper understanding of the molecular mechanism and well-designed preclinical and clinical studies of BHPs would pave the way for advancing the wealth of empirical knowledge in phytomedicine research.

The protein kinase B (AKT) is central in modulating GPCR-mediated PI3K/AKT canonical signaling pathways, which play a role in adaptive stress responses associated with many aging-related diseases and disorders. These pathways also regulate the neuroendocrine-immune and cardiovascular-respiratory systems [2–5,50]. Activating the intracellular PI3K/AKT-mediated signaling pathway triggers a cascade of reactions downstream of proteins, which mediate various cellular functions, including neurogenesis, angiogenesis, neuroinflammatory responses, and other repair mechanisms in many diseases [2,5,30,51–55]. A plethora of plant secondary metabolites target various biased GPCRs, signaling non-specifically to modulate the PI3K/AKT pathway in multiple organs and tissues, thereby regulating compromised cellular and systemic homeostasis and contributing to the resolution of acute and chronic neuroinflammatory disorders and diseases [2,5,30,51–55]. Different types of phytochemicals collectively contribute to various therapeutic actions that target this mediatory intracellular system [2,5,30,51–55]. In this context, further network pharmacology studies of botanicals and BHP are required, with a focus on establishing relationships between distinct phytochemical types of compounds and specific indications for various diseases.

In this Special Issue, we recommend adhering to the commonly accepted terms and definitions below to avoid misuse of terminology (Table 1).

Table 1. Glossary, terminology, and definitions.

Terms	Definitions
Biased ligand	Biased ligands are those that can regulate receptor activity by stabilizing distinct conformations. Since diverse signaling pathways elicit distinct physiological effects, biased ligands that selectively induce beneficial pathways hold promising therapeutic value. Ligand bias refers to the ability of ligands to activate specific receptor signaling responses more effectively than others. It reflects differences in the reactions of a receptor to specific ligands and has implications for the development of highly specific therapeutics.
Biased signaling	Biased signaling is a natural consequence of GPCR or allosteric function and should be expected from both natural agonists and synthetic analogs. Biased signaling of G protein-coupled receptors or receptor tyrosine kinases describes the ability of different ligands to activate an alternative downstream signaling pathway preferentially.
Botanicals	Products that include plant materials, algae, macroscopic fungi, and combinations thereof.
Botanical hybrid preparations/products	The term "botanical hybrid preparation" (BHP) describes the pharmacological activity of a fixed herbal combination with a specific chemical composition.
Herbal substance	All herbal substances are mainly whole, fragmented, or cut plants, plant parts, algae, fungi, and lichen, typically in an unprocessed, often dried form, though some many be used fresh. Herbal substances are precisely defined by the plant part used and the botanical name, following the binomial system (genus, species, variety, and author).
Herbal preparations	Herbal preparations are obtained by subjecting herbal substances to various treatments, including extraction, distillation, expression, fractionation, purification, concentration, and fermentation. These include comminuted or powdered herbal substances, tinctures, extracts, essential oils, expressed juices, and processed exudates.
Herbal medicinal products	Any medicinal product exclusively containing as active substances one or more herbal substances, one or more herbal preparations, or one or more such herbal substances combined with one or more such herbal preparations.
Pharmacognosy	A multidisciplinary subject that encompasses aspects of botany, organic chemistry, biochemistry, and pharmacology, focusing on natural products used in the production and discovery of new drugs.
Pharmacophore	The distinct functional groups or substance classes that possess the biological activity necessary to ensure optimal interaction between their receptor and a specific biological target structure and to trigger (or block) their biological response.
Pharmacodynamics	Pharmacodynamics refers to the effects of drugs on the body, including receptor binding, post-receptor effects, and chemical interactions.
Pharmacokinetics	Pharmacokinetics describes how the body affects a drug's absorption, distribution, metabolism, and excretion after administration.
Pleiotropic action	In pharmacognosy and ethnopharmacology, pleiotropic action refers to all actions of botanical preparation other than those for which it was explicitly intended. It is often used to denote additional beneficial effects but can also indicate detrimental adverse effects.
Polypharmacology	Polypharmacology refers to the binding of a drug to multiple target proteins, with clinical effects being mediated through the modulation of the set of protein targets.

Table 1. *Cont.*

Terms	Definitions
Polyvalent action:	Polyvalence refers to the range of biological activities a botanical preparation may exhibit that contribute to the overall effect observed clinically or in vivo. Polyvalence can result from the following factors: o Various phytochemicals are present, with each type having a different biological effect; o Compounds of a single phytochemical type often exhibit multiple biological effects relevant to treating the disease and improving the patient's health; o Compounds are present that do not affect the cause or symptoms of the disease but modify the absorption, distribution, metabolism, and excretion of active constituents, as well as the adverse effects.
Rule of five	Poor absorption or permeation of a compound is more likely when there are >5 hydrogen bond donors, the molecular mass is >500, cLogP is >5, and the sum of nitrogen and oxygen atoms in a molecule is greater than 10. Many drugs, however, are exceptions to the rule of five, and these are often substrates for biological transporters.
Selective action	o A drug strongly prefers its intended target over other targets; o The degree to which a drug acts on a given site relative to others; o The ability of a drug to affect a particular gene, protein, or signaling pathway in preference to others; o A drug that binds to the same receptor as an agonist but produces an effect opposite to that of the agonist.
Specific action	An accurate, exact (free from ambiguity), and distinctive influence specifically adapted to a purpose or use.
Systems pharmacology	It aims to comprehend how drugs interact with the human body as a complex, integrated biological system. Instead of considering the effect of a drug to be the result of a single specific drug–protein interaction, systems pharmacology refers to the impact of a drug as the outcome of the network of interactions it may have. Interaction networks may include chemical–protein, protein–protein, genetic, signaling, and physiological (at the cellular, tissue, organ, and whole-body levels). Systems pharmacology uses bioinformatics and statistical techniques to integrate and interpret these networks.
Network pharmacology	Network pharmacology is a subfield that combines principles from pharmacology, systems biology, and network analysis to study the complex interactions between drugs and their targets, such as receptors or enzymes, within biological systems. The topology of a biochemical reaction network determines the shape of the drug dose–response curve and the type of drug–drug interactions, thus helping design efficient and safe therapeutic strategies. The topology of network pharmacology utilizes computational tools and network analysis algorithms to identify drug targets, predict drug–drug interactions, elucidate signaling pathways, and explore the polypharmacology of drugs.
Synergy	Two or more agents work together to produce a result that cannot be achieved by any of the agents independently, resulting in the generation of new pharmacological activity specific to the combination of two or more agents. Another definition of synergy is broadly interpreted as the combination of agents that is more effective than the sum of their individual effects, without specifying the differences between potentiation and amplification.

Table 1. *Cont.*

Terms	Definitions
Systems biology	Systems biology involves computational and mathematical analysis and modeling of complex biological systems. It is a biology-based interdisciplinary field that focuses on the complex interactions within biological systems and uses a holistic approach to biological research.

Funding: This research received no external funding.

Conflicts of Interest: Author Alexander Panossian is self-employed at the independent R&D company Phytomed AB. The author declares that the research was conducted in the absence of any commercial or financial relationships that could be construed as a potential conflict of interest.

References

1. Panossian, A.; Lemerond, T.; Efferth, T. State-of-the-art review on botanical hybrid preparations in phytomedicine and phytotherapy research: Background and perspectives. *Pharmaceuticals* **2024**, *17*, 483. [CrossRef] [PubMed]
2. Panossian, A.; Lemerond, T.; Efferth, T. Adaptogens in Long-Lasting Brain Fatigue: An Insight from Systems Biology and Network Pharmacology. *Pharmaceuticals* **2025**, *18*, 261. [CrossRef]
3. Panossian, A.; Seo, E.J.; Efferth, T. Novel molecular mechanisms for the adaptogenic effects of herbal extracts on isolated brain cells using systems biology. *Phytomed. Int. J. Phytother. Phytopharm.* **2018**, *50*, 257–284. [CrossRef] [PubMed]
4. Panossian, A.; Seo, E.J.; Wikman, G.; Efferth, T. Synergy assessment of fixed combinations of Herba Andrographidis and Radix Eleutherococci extracts by transcriptome-wide microarray profiling. *Phytomedicine* **2015**, *22*, 981–992. [CrossRef]
5. Panossian, A.; Efferth, T. Network Pharmacology of Adaptogens in the Assessment of Their Pleiotropic Therapeutic Activity. *Pharmaceuticals* **2022**, *15*, 1051. [CrossRef] [PubMed]
6. Che, C.T.; Wang, Z.J.; Chow, M.S.; Lam, C.W. Herb-herb combination for therapeutic enhancement and advancement: Theory, practice and future perspectives. *Molecules* **2013**, *18*, 5125–5141. [CrossRef] [PubMed]
7. Li, S.; Xiao, W. General expert consensus on the application of network pharmacology in the research and development of new traditional Chinese medicine drugs. *Chin. J. Nat. Med.* **2025**, *23*, 129–142. [CrossRef]
8. Li, S.; Zhang, B. Traditional Chinese medicine network pharmacology: Theory, methodology and application. *Chin. J. Nat. Med.* **2013**, *11*, 110–120. [CrossRef]
9. Han, S.; Lv, A.P.; Li, J.; Jiang, M. Application review of network pharmacology in the study of properties theory of traditional Chinese medicine. *J. Basic Chin. Med.* **2019**, *25*, 127–130.
10. Zhou, Z.; Chen, B.; Chen, S.; Lin, M.; Chen, Y.; Jin, S.; Chen, W.; Zhang, Y. Applications of Network Pharmacology in Traditional Chinese Medicine Research. *Evid. Based Complement. Altern. Med.* **2020**, *2020*, 1646905. [CrossRef]
11. Lay, J.O.; Borgmann, S.; Liyanage, R.; Wilkins, C.L. Problems with the "omics". *Trends Anal. Chem.* **2006**, *25*, 1046–1056.
12. Ouedraogo, M.; Baudoux, T.; Stévigny, C.; Nortier, J.; Colet, J.M.; Efferth, T.; Qu, F.; Zhou, J.; Chan, K.; Shaw, D.; et al. Review of current and "omics" methods for assessing the toxicity (genotoxicity, teratogenicity and nephrotoxicity) of herbal medicines and mushrooms. *J. Ethnopharmacol.* **2012**, *140*, 492–512. [CrossRef]
13. Kinross, J.M.; Darzi, A.W.; Nicholson, J.K. Gut microbiome-host interactions in health and disease. *Genome Med.* **2011**, *3*, 14. [CrossRef]
14. Klaeger, S.; Heinzlmeir, S.; Wilhelm, M.; Polzer, H.; Vick, B.; Koenig, P.A.; Reinecke, M.; Ruprecht, B.; Petzoldt, S.; Meng, C.; et al. The target landscape of clinical kinase drugs. *Science* **2017**, *358*, eaan4368. [CrossRef] [PubMed]
15. Chaudhari, R.; Fong, L.W.; Tan, Z.; Huang, B.; Zhang, S. An up-to-date overview of computational polypharmacology in modern drug discovery. *Expert Opin. Drug Discov.* **2020**, *15*, 1025–1044. [CrossRef]
16. Schneider, P.; Röthlisberger, M.; Reker, D.; Schneider, G. Spotting and designing promiscuous ligands for drug discovery. *Chem. Commun.* **2016**, *52*, 1135–1138. [CrossRef]
17. Da, C.; Zhang, D.; Stashko, M.; Vasileiadi, E.; Parker, R.E.; Minson, K.A.; Huey, M.G.; Huelse, J.M.; Hunter, D.; Gilbert, T.S.K.; et al. Data-driven construction of antitumor agents with controlled polypharmacology. *J. Am. Chem. Soc.* **2019**, *141*, 15700–15709. [CrossRef]
18. Li, X.; Li, Z.; Wu, X.; Xiong, Z.; Yang, T.; Fu, Z.; Liu, X.; Tan, X.; Zhong, F.; Wan, X.; et al. Deep learning enhancing kinome-wide polypharmacology profiling: Model construction and experiment validation. *J. Med. Chem.* **2020**, *63*, 8723–8737. [CrossRef]
19. Berger, S.I.; Iyengar, R. Network analyses in systems pharmacology. *Bioinformatics* **2009**, *25*, 2466–2472. [CrossRef]

20. Wang, Y.Y.; Li, X.X.; Liu, J.P.; Luo, H.; Ma, L.X.; Alraek, T. Traditional Chinese medicine for chronic fatigue syndrome: A systematic review of randomized clinical trials. *Complement. Ther. Med.* **2014**, *22*, 826–833. [CrossRef]
21. Zhang, L.; Yang, L.; Shergis, J.; Zhang, L.; Zhang, A.L.; Guo, X.; Qin, X.; Johnson, D.; Liu, X.; Lu, C.; et al. Chinese herbal medicine for diabetic kidney disease: A systematic review and meta-analysis of randomised placebo-controlled trials. *BMJ Open* **2019**, *9*, e025653. [CrossRef]
22. Yang, J.; Zhu, X.; Yuan, P.; Liu, J.; Wang, B.; Wang, G. Efficacy of traditional Chinese Medicine combined with chemotherapy in patients with non-small cell lung cancer (NSCLC): A meta-analysis of randomized clinical trials. *Support Care Cancer* **2020**, *28*, 3571–3579. [CrossRef] [PubMed]
23. Liang, S.B.; Fang, M.; Liang, C.H.; Lan, H.D.; Shen, C.; Yan, L.J.; Hu, X.Y.; Han, M.; Robinson, N.; Liu, J.P. Therapeutic effects and safety of oral Chinese patent medicine for COVID-19: A rapid systematic review and meta-analysis of randomized controlled trials. *Complement. Ther. Med.* **2021**, *60*, 102744. [CrossRef] [PubMed]
24. Jiao, X.; Jin, X.; Ma, Y.; Yang, Y.; Li, J.; Liang, L.; Liu, R.; Li, Z. A comprehensive application: Molecular docking and network pharmacology for the prediction of bioactive constituents and elucidation of mechanisms of action in component-based Chinese medicine. *Comput. Biol. Chem.* **2021**, *90*, 107402. [CrossRef]
25. Lee, D.Y.W.; Li, Q.Y.; Liu, J.; Efferth, T. Traditional Chinese herbal medicine at the forefront battle against COVID-19: Clinical experience and scientific basis. *Phytomedicine* **2021**, *80*, 153337. [CrossRef]
26. Chen, K.X.; Jiang, H.L.; Luo, X.M.; Shen, J.H. Drug discovery in postgenome era: Trend and practice. *Chin. J. Nat. Med.* **2004**, *2*, 257–260.
27. Li, S.; Zhang, Z.; Wu, L.; Zhang, X.; Li, Y.; Wang, Y. Understanding ZHENG in traditional Chinese medicine in the context of neuro-endocrine-immune network. *IET Syst. Biol.* **2007**, *1*, 51–60. [CrossRef] [PubMed]
28. Noor, F.; Tahir Ul Qamar, M.; Ashfaq, U.A.; Albutti, A.; Alwashmi, A.S.S.; Aljasir, M.A. Network Pharmacology Approach for Medicinal Plants: Review and Assessment. *Pharmaceuticals* **2022**, *15*, 572. [CrossRef]
29. Efferth, T.; Koch, E. Complex interactions between phytochemicals. The multitarget therapeutic concept of phytotherapy. *Curr. Drug Targets* **2011**, *12*, 122–132. [CrossRef]
30. Zhou, M.; Varol, A.; Efferth, T. Multi-omics approaches to improve malaria therapy. *Pharmacol. Res.* **2021**, *167*, 105570. [CrossRef]
31. Ulrich-Merzenich, G.; Zeitler, H.; Jobst, D.; Panek, D.; Vetter, H.; Wagner, H. Application of the "-Omic-" technologies in phytomedicine. *Phytomedicine* **2007**, *14*, 70–82. [CrossRef]
32. Network Pharmacology. Available online: https://pubmed.ncbi.nlm.nih.gov/?term=network%20pharmacology&sort=date&size=200&timeline=expanded (accessed on 2 April 2025).
33. Wu, J.; Huang, H.; Chen, G.; Mao, J.; Tian, X.; Zhan, S.; Peng, Z.; Zhu, Y.; Wang, W. Integration of network pharmacology, bioinformatics and experimental verification strategy to discover the pharmacological mechanisms of mogroside acts against pharyngitis. *J. Ethnopharmacol.* **2025**, *344*, 119499. [CrossRef] [PubMed]
34. Panossian, A.; Hamm, R.; Wikman, G.; Efferth, T. Mechanism of action of Rhodiola, salidroside, tyrosol and triandrin in isolated neuroglial cells: An interactive pathway analysis of the downstream effects using RNA microarray data. *Phytomedicine Int. J. Phytother. Phytopharm.* **2014**, *21*, 1325–1348. [CrossRef] [PubMed]
35. Panossian, A.; Abdelfatah, S.; Efferth, T. Network Pharmacology of Ginseng (Part II): The Differential Effects of Red Ginseng and Ginsenoside Rg5 in Cancer and Heart Diseases as Determined by Transcriptomics. *Pharmaceuticals* **2021**, *14*, 1010. [CrossRef] [PubMed]
36. Panossian, A.; Abdelfatah, S.; Efferth, T. Network Pharmacology of Red Ginseng (Part I): Effects of Ginsenoside Rg5 at Physiological and Sub-Physiological Concentrations. *Pharmaceuticals* **2021**, *14*, 999. [CrossRef]
37. Gao, X.F.; Zhang, J.J.; Gong, X.J.; Li, K.K.; Zhang, L.X.; Li, W. Ginsenoside Rg5: A Review of Anticancer and Neuroprotection with Network Pharmacology Approach. *Am. J. Chin. Med.* **2022**, *50*, 2033–2056. [CrossRef]
38. He, S.; Shi, J.; Chai, H.; Ma, L.; Pei, H.; Zhang, P.; Shi, D.; Li, H. Mechanisms with network pharmacology approach of Ginsenosides in Alzheimer's disease. *Heliyon* **2024**, *10*, e26642. [CrossRef]
39. Liu, C.; Du, L.; Zhang, S.; Wang, H.; Kong, L.; Du, G. Network pharmacology and experimental study of phenolic acids in salvia miltiorrhiza bung in preventing ischemic stroke. *Front. Pharmacol.* **2023**, *14*, 1108518. [CrossRef]
40. Panossian, A.; Hamm, R.; Kadioglu, O.; Wikman, G.; Efferth, T. Synergy and Antagonism of Active Constituents of ADAPT-232 on Transcriptional Level of Metabolic Regulation of Isolated Neuroglial Cells. *Front. Neurosci.* **2013**, *7*, 16. [CrossRef]
41. Panossian, A.; Seo, E.J.; Efferth, T. Synergy assessments of plant extracts used in the treatment of stress and aging related disorders. *Synerg. Res.* **2018**, *7*, 39–49. [CrossRef]
42. Zhang, P.; Zhang, D.; Zhou, W.; Wang, L.; Wang, B.; Zhang, T.; Li, S. Network pharmacology: Towards the artificial intelli-gence-based precision traditional Chinese medicine. *Brief. Bioinform.* **2023**, *25*, bbad518. [CrossRef]
43. Zhao, L.; Zhang, H.; Li, N.; Chen, J.; Xu, H.; Wang, Y.; Liang, Q. Network pharmacology, a promising approach to reveal the pharmacology mechanism of Chinese medicine formula. *J. Ethnopharmacol.* **2023**, *309*, 116306. [CrossRef] [PubMed]

44. Spanakis, M.; Tzamali, E.; Tzedakis, G.; Koumpouzi, C.; Pediaditis, M.; Tsatsakis, A.; Sakkalis, V. Artificial Intelligence Models and Tools for the Assessment of Drug–Herb Interactions. *Pharmaceuticals* **2025**, *18*, 282. [CrossRef]
45. Berger, S.I.; Iyengar, R. Role of systems pharmacology in understanding drug adverse events. *Wiley Interdiscip. Rev. Syst. Biol. Med.* **2011**, *3*, 129–135. [CrossRef]
46. Young, M.; Hoheisel, J.D.; Efferth, T. Toxicogenomics for the prediction of toxicity related to herbs from traditional Chinese medicine. *Planta Med.* **2010**, *76*, 2019–2025. [CrossRef]
47. Börner, F.U.; Schütz, H.; Wiedemann, P. The fragility of omics risk and benefit perceptions. *Toxicol. Lett.* **2011**, *201*, 249–257. [CrossRef] [PubMed]
48. Aardema, M.J.; MacGregor, J.T. Toxicology and genetic toxicology in the new era of "toxicogenomics": Impact of "-omics" technologies. *Mutat. Res.* **2002**, *499*, 13–25. [CrossRef]
49. Panossian, A. Challenges in phytotherapy research. *Front. Pharmacol.* **2023**, *14*, 1199516. [CrossRef]
50. Panossian, A.G.; Efferth, T.; Shikov, A.N.; Pozharitskaya, O.N.; Kuchta, K.; Mukherjee, P.K.; Banerjee, S.; Heinrich, M.; Wu, W.; Guo, D.A.; et al. Evolution of the adaptogenic concept from traditional use to medical systems: Pharmacology of stress- and aging-related diseases. *Med. Res. Rev.* **2021**, *41*, 630–703.
51. Gu, C.; Zhang, Q.; Li, Y.; Li, R.; Feng, J.; Chen, W.; Ahmed, W.; Soufiany, I.; Huang, S.; Long, J.; et al. The PI3K/AKT Path-way-The Potential Key Mechanisms of Traditional Chinese Medicine for Stroke. *Front. Med.* **2022**, *9*, 900809. [CrossRef]
52. Guo, J.; Jiang, X.; Lian, J.; Li, H.; Zhang, F.; Xie, J.; Deng, J.; Hou, X.; Du, Z.; Hao, E. Evaluation of the effect of GSK-3β on liver cancer based on the PI3K/AKT pathway. *Front. Cell Dev. Biol.* **2024**, *12*, 1431423. [CrossRef]
53. Liu, J.; Xu, P.; Liu, D.; Wang, R.; Cui, S.; Zhang, Q.; Li, Y.; Yang, W.; Zhang, D. TCM Regulates PI3K/Akt Signal Pathway to Intervene Atherosclerotic Cardiovascular Disease. *Evid. Based Complement. Altern. Med. ECAM* **2021**, *2021*, 4854755. [CrossRef] [PubMed]
54. Liu, Y.; Jin, B. Mechanism of Traditional Chinese Medicine extract in the treatment of diabetic erectile dysfunction. *J. Ethnopharmacol.* **2025**, *341*, 119332. [CrossRef] [PubMed]
55. Chen, B.; Jin, W. A comprehensive review of stroke-related signaling pathways and treatment in western medicine and traditional Chinese medicine. *Front. Neurosci.* **2023**, *17*, 1200061. [CrossRef] [PubMed]

Disclaimer/Publisher's Note: The statements, opinions and data contained in all publications are solely those of the individual author(s) and contributor(s) and not of MDPI and/or the editor(s). MDPI and/or the editor(s) disclaim responsibility for any injury to people or property resulting from any ideas, methods, instructions or products referred to in the content.

pharmaceuticals

MDPI

Review

State-of-the-Art Review on Botanical Hybrid Preparations in Phytomedicine and Phytotherapy Research: Background and Perspectives

Alexander Panossian [1,*], Terry Lemerond [2] and Thomas Efferth [3,*]

[1] Phytomed AB, Sjöstadsvägen 6A, Lgh 1004, 59344 Västervik, Sweden
[2] EuroPharma USA Inc., Green Bay, WI 54311, USA; terryl@europharmausa.com
[3] Department of Pharmaceutical Biology, Institute of Pharmaceutical and Biomedical Sciences, Johannes Gutenberg University, 55099 Mainz, Germany
* Correspondence: ap@phytomed.se (A.P.); efferth@uni-mainz.de (T.E.)

Abstract: Background: Despite some evidence supporting the synergy concept, the commonly known assumption that combinations of several herbs in one formulation can have better efficacy due to additive or synergistic effects has yet to be unambiguously and explicitly studied. Study aim: The study aimed to reveal the molecular interactions in situ of host cells in response to botanical hybrid preparations (BHP) intervention and justify the benefits of implementing BHP in clinical practice. Results: This prospective literature review provides the results of recent clinical and network pharmacology studies of BHP of *Rhodiola rosea* L. (Arctic root) with other plants, including *Withania somnifera* (L.) Dunal (ashwagandha), (*Camellia sinensis* (L.) Kuntze (green tea), *Eleutherococcus senticosus* (Rupr. and Maxim.) Maxim. (eleuthero), *Schisandra chinensis* (Turcz.) Baill. (schisandra), *Leuzea carthamoides* (Willd.) DC., caffeine, *Cordyceps militaris* L., *Ginkgo biloba* L. (ginkgo), *Actaea racemosa* L. (black cohosh), *Crocus sativus* L. (saffron), and L-carnosine. Conclusions: The most important finding from network pharmacology studies of BHP was the evidence supporting the synergistic interaction of BHP ingredients, revealing unexpected new pharmacological activities unique and specific to the new BHP. Some studies show the superior efficacy of BHP compared to mono-drugs. At the same time, some a priori-designed combinations can fail, presumably due to antagonistic interactions and crosstalk between molecular targets within the molecular networks involved in the cellular and overall response of organisms to the intervention. Network pharmacology studies help predict the results of studies aimed at discovering new indications and unpredicted adverse events.

Keywords: network pharmacology; gene expression; botanical hybrid preparations; *Rhodiola rosea* clinical trials; synergy

1. Introduction

The use of complex herbal formulations comprising fixed combinations of several plant extracts has a long history in TCM, Kampo, Ayurveda, and other traditional medical systems [1–3]. The potential health benefits of consuming combined nutrients or dietary supplements have gained considerable attention due to their impact on overall well-being due to the synergy concept, recently defined as nutrient synergy [4]. The term synergy is differently defined in the various scientific disciplines and comes from the Attic Greek word συνεργία synergia [1] from *synergos*, συνεργός, meaning "working together".

Combining two or more plants assumes that a hybrid botanical preparation (BHP) is more active due to synergistic effects brought about by different mechanisms. Figure 1 illustrates an allegoric analogy with two or more kinds of hybrids from ancient mythology.

Figure 1. Images of two, three, and more kinds of hybrids from ancient mythology. https://en.wikipedia.org/wiki/List_of_hybrid_creatures_in_folklore, accessed on 3 April 2024.

- *Lama* an Assyrian protective hybrid deity with a human head, a bull's body or lion's body, and wings symbolizing the synergistic interaction between these elements.
- The centaur Chiron, from Greek mythology, raised Achilles at the request of Achilles' mother (reproduced from Peter-Paul Rubens in 1630–1635).
- *Navagunjara*—A Hindu creature with the head of a rooster, the neck of a peacock, the back of a bull, a snake-headed tail, three legs of an elephant, tiger, and deer or horse, with the fourth limb being a human hand holding a lotus.
- *Nureonna*—a creature with the head of a woman and the body of a snake (Japanese mythology).
- *Kotobuki*—a Japanese Chimera with the head of a rat, the ears of a rabbit, the horns of an ox, the comb of a rooster, the beard of a sheep, the neck of a Japanese dragon, the mane of a horse, the back of a wild boar, the shoulders and belly of a South China tiger, the arms of a monkey, the hindquarters of a dog, and the tail of a snake.

Based on the assumption of synergistic interaction of several components, researchers propose that combinations of several active ingredients in one formulation can have superior effectiveness and better efficacy due to their multiple effects on various targets (Figure 2) [5–10].

The term botanical "Hybrid" preparation (BHP) is coined to describe the biological/pharmacological activity (conditional pharmacological "signature") of a fixed herbal combination with a specific chemical composition (e.g., TLC of HPLC conditional chemical "fingerprint") (Figure 2) [7]. BHP is like a "newborn" unique biologically active substance, a hybrid of the "parent" ingredients [7].

Similarly, phytochemical hybrid preparation (PHP) is used to determine the pharmacological signature of a combination of phytochemicals comprising specific plant species, e.g., *R. crenulata* (Hook. f. and Thomson) H. Ohba, *R. sacra* (Prain ex Raym.-Hamet) S.H. Fu, *R. kirilowii* (Regel) Maxim., *R. quadrifida* (Pall.) Fisch. & C. A. Mey., and *R. dumulosa* (Franch.) S. H. Fu, etc., characterized by specific chemical fingerprints (Figure 2).

This distinction emphasizes that any new fixed combination exhibits unique biological characteristics and effects different from the ingredients' natural characteristics. That is due to their multitarget effects on various mediators, which interact within various regulatory systems of the host cells and organism [7,8].

Modern technologies in biomedical research and bioinformatics provide potent tools in phytotherapy research and implement a concept of systems biology and network pharmacology, uncovering numerous molecular targets and new mechanisms of action of botanical (Figure 3) [11]. In these studies, the experimental protocol included mRNA microarray hybridization, ingenuity pathway analysis (IPA), and statistical analysis to uncover the mechanism of action of purified compounds, the plant extract, and their combinations comprising BHP. This was achieved by assessing their effects on gene expression in isolated neuronal cells and their potential therapeutic action.

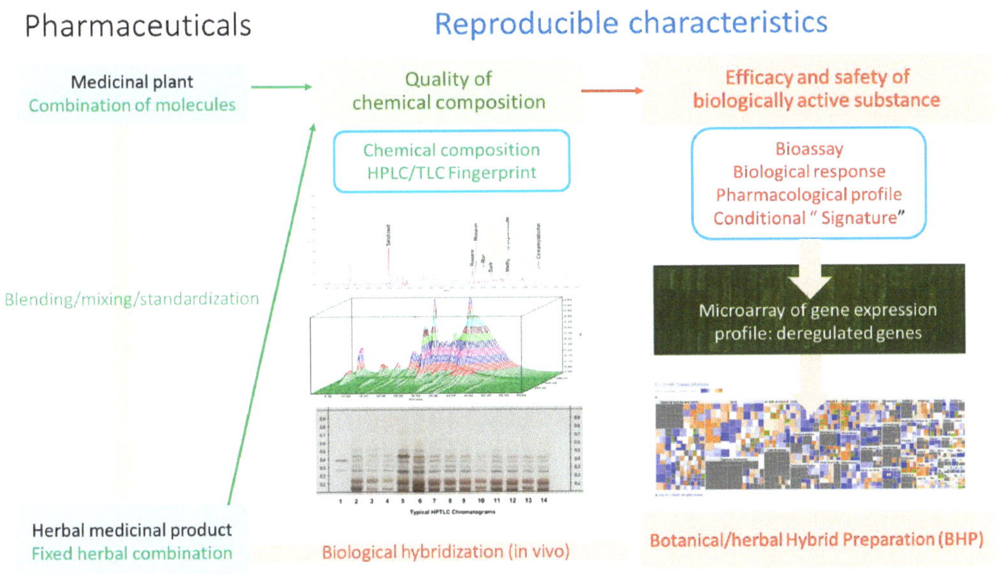

Figure 2. Schematic representation of quality, efficacy, and safety characteristics of herbal medicinal products comprising the mixtures of fixed combinations of molecules from herbal extracts. Reproducible qualitative and quantitative chemical composition by HPLC and TLC fingerprint ensures the reproducible quality of a fixed combination. Reproducible efficacy and safety of a botanical/herbal hybrid preparation (BHP) is characterized by pharmacological profile—conditional signature, e.g., microarray dataset of deregulated genes in response to exposure of BHP in a bioassay providing further information on the effect on physiological functions and diseases in the form of heatmaps.

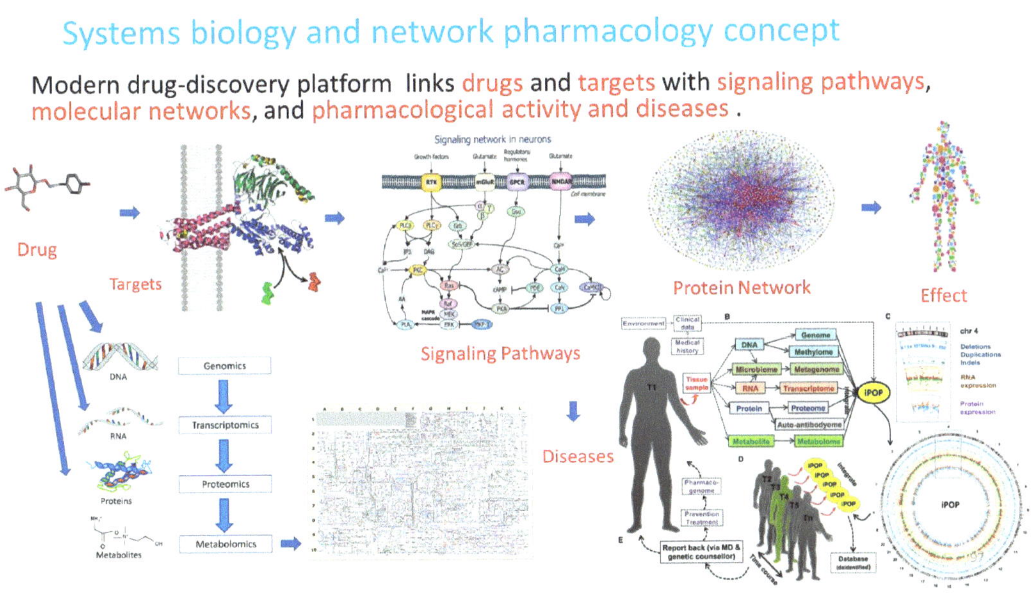

Figure 3. Molecular network-based cause–effect relationships concept. The current drug-discovery platform links drugs and targets with signaling pathways, molecular networks, and organism effects.

Molecular targets of adaptogens, their networks, and signaling pathways are associated with chronic inflammation, atherosclerosis, neurodegenerative cognitive impairment, metabolic disorders, and cancer. Predicting the response of the human body to medication requires an understanding of drug–effect relationships at the organism, organ, tissue, cellular, and molecular levels based on integrative personal OMICS (DNA-genomics, RNA-transcriptomics, microbiomes, proteomics, and metabolomics) profiling and their changes in health and disease, as well as after pharmacological intervention.

Recently, the Herbal Medicinal Products Platform Austria (HMPPA) committee chose *Rhodiola rosea* L. as a medicinal plant in 2023 in Austria [12]. What is behind this choice of HMPPA?

Rhodiola rosea L. (Crassulaceae, syn. *Sedum rhodiola* DC., *Sedum rosea* (L.), Scop cop, known as Roseroot, Rosenroot, Golden Root, Arctic Root, Orpin Rose, Rhodiole, and Rougeâtre) has an extensive history as a treasured medicinal plant and has appeared in the Materia Medica of several European countries [13,14]. In Europe, Rosenroot was formally adopted in Sweden as a natural remedy (national legislation) from 1987 to 2008, and since 2008, as a traditional herbal medicinal product (THMP) and registered as an adaptogen in decreased performance, such as fatigue and weakness [15].

During the last two decades, more than 1200 studies, including 33 clinical and 910 preclinical studies, were conducted in Europe, America, and China, providing results of preclinical and clinical efficacy, safety, and quality of *R. rosea* preparations in various stress-induced disorders, including fatigue syndrome, cognitive deficiencies, mild/moderate depression, anxiety, and burnout symptoms, as well as in healthy subjects under stress (Figure 4) [15–48].

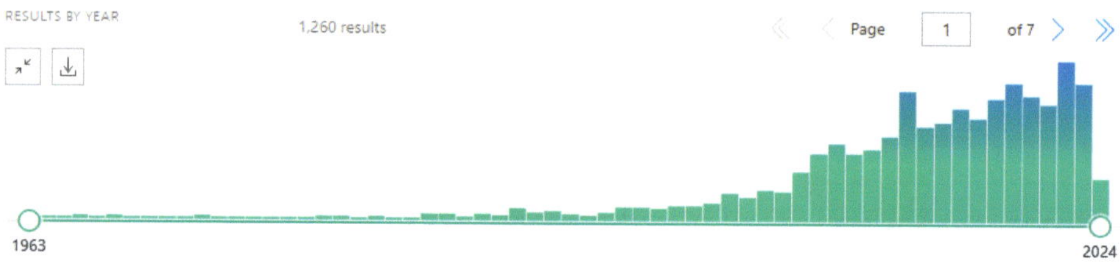

Figure 4. The number of publications (1200 results) worldwide on *R. rosea* from 1963 to 2024; the author's chart extracted from Pubmed: rhodiola rosea—search results—PubMed (nih.gov, accessed on 3 April 2024).

Eighteen clinical studies were conducted on the fixed combinations of *R. rosea* with green tea [49–53], Eleutherococcus, and Schisandra [54–57], Eleutherococcus, Schisandra, and Leuzea [56], caffeine [58,59], Cordyceps [60–63], Ginkgo [64], black cohosh [65], Saffron [66], L-carnosine [67], Eleutherococcus, and Glycyrrhiza [68].

This narrative state-of-the-art review provides the results of recent clinical [49–68] and network pharmacology studies of the BHP of *R. rosea* with other plants [5,6,8–10]. The studies aim to reveal the molecular interactions in situ of host cells in response to the intervention of BHP and justify the benefits of implementing BHP in clinical practice.

2. Synergy and Antagonism of Active Ingredients of *Rhodiola rosea* and Other Plant Extracts

Concomitant treatment of disease by two herbal drugs aiming to achieve better effect suggests their additive (1 + 1 = 2), amplifying (1 + 1 > 2), potentiation (0 + 1 > 1), or synergistic (0 + 0 > 1) result. Hypothetically, their interaction can result in beneficial antagonistic interactions (1 + 1 < 2, or 1 + 1 = 0), e.g., decreased toxicity and detoxifying effects [3,6].

This hypothesis was verified in a set of in vitro studies, where the effects of several BHP and their ingredients on the number of deregulated genes in brain cell cultures were analyzed [5–10]. The composition of genes (signature) deregulated by BHP was quantitatively and qualitatively different from the signature (composition of genes) deregulated by each plant separately, suggesting that the impact of the BHP on the target cells was qualitatively different from the effects of individual ingredients [5] (Figure 5). This implies that the BHP exhibits quite different pharmacological activities when the ingredients are combined. These findings are essential for understanding the unpredictable results of clinical studies of multi-component drugs and dietary supplements [49–68].

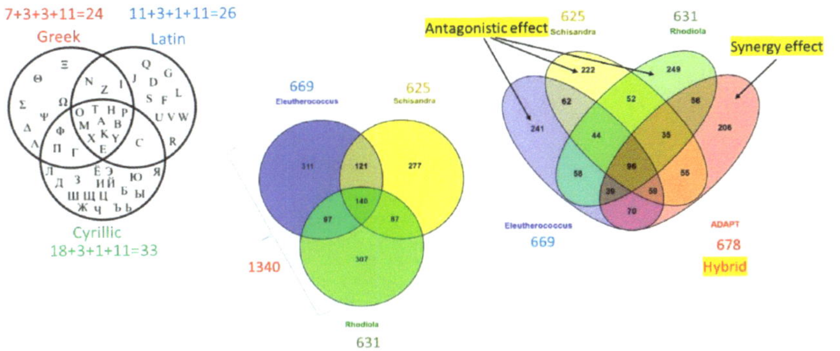

Herbal extracts	Chemical compositions "Fingerprint" Number of compounds identified in extracts, n = A+B+C=ABC	Pharmacological profile, Conditional "Signatures", Number of deregulated genes neuroglia cells, N = A+B+C=D
A - Rhodiola	140	631
B - Eleutherococcus	35	669
C - Schisandra	32	625
Combination A+B+C	207	678
E – Salidroside isolated from Rhodiola	1	640

Figure 5. The upper panel shows a Venn diagram with Greek, Latin, and Cyrillic letter symbols, where overlapped sections contain the same symbols. Eleven symbols are the same in all alphabets. Similarly, *Rhodiola*, *Eleutherococcus*, and *Schisandra* separately deregulated 140 out of 1340 genes in experiments with neuroglia cells; however, only 96 out of 640 genes were deregulated by the hybrid combination (ADAPT) of these three plant extracts. External sections of the Venn diagram show the number of deregulated genes specific to distinct extracts (antagonistic interactions in blue, yellow, and green) and synergy-derived 206 deregulated genes characteristic to ADAPT-232 (in red). The lower panel shows the number of compounds in extracts and 3D-HPLC fingerprints of *Rhodiola*, *Eleutherococcus Schisandra*, and their combination ADAPT-232, and Venn diagrams showing intersections of deregulated genes in neuroglia cells after exposure to these herbal extracts. Authors' drawings adapted from freely accessible publications [5].

Figures 5 and 6 illustrate the essence of hybridization of ingredients of BHP or PHP, their synergy, and the antagonism in these experiments and interpretations. The point is the biological activity of a single compound, e.g., salidroside, an active compound of *Rhodiola rosea* extract, interacts with many proteins in brain cells, deregulating 640 (!) genes in neuroglia cells, associated with various physiological processes and effects in stress and aging-induced disorders (e.g., neurodegeneration). This is illustrated in Figures 5 and 6 [9].

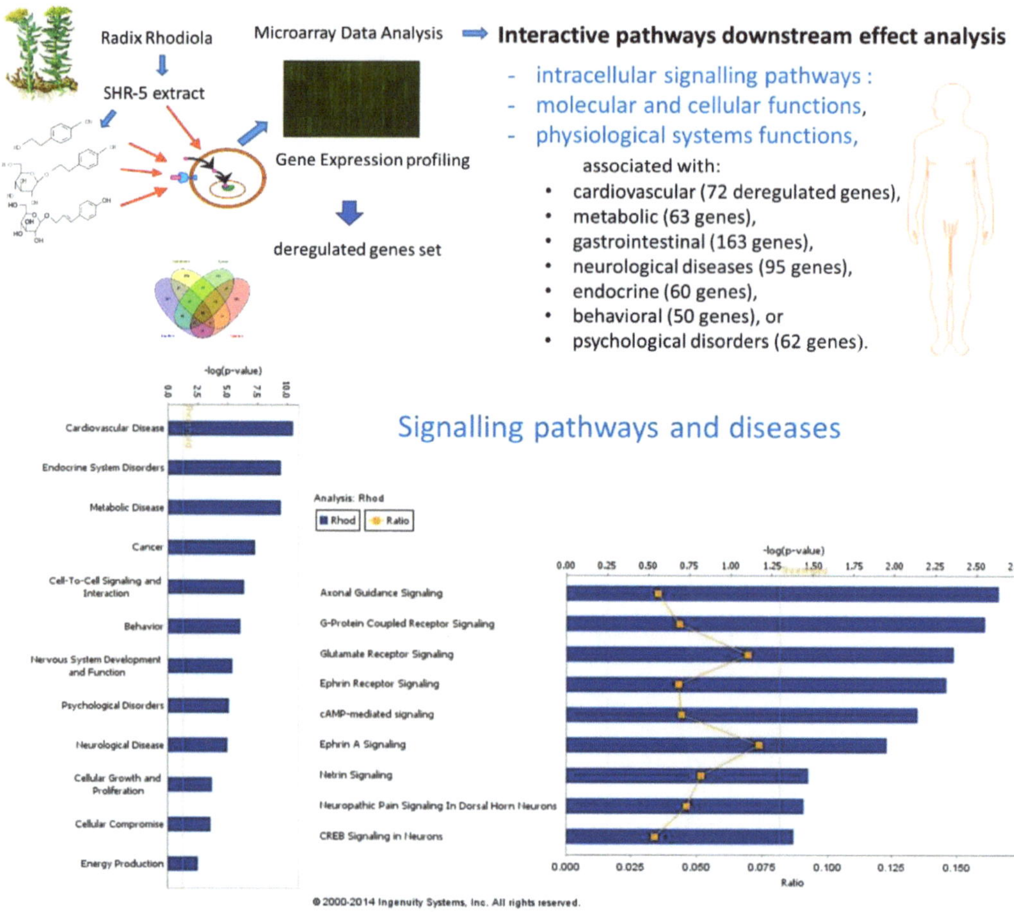

Figure 6. The impact of *Rhodiola* and its active compounds on gene expression in human brain cells suggests potential effects on canonical intracellular signaling pathways, molecular, cellular, and physiological functions, and diseases; updated from the authors' freely accessible publication [9] and authors' drawings.

Meanwhile, the total extract of *Rhodiola*-containing 120 phytochemicals (including salidroside) or the BHP of *Rhodiola, Schisandra*, and *Eleutherococcus* extracts (ADAPT-232) containing 207 phytochemicals deregulate almost the same number of genes (*Schizandra*—625 genes, *Rhodiola*—631 genes, *Eleutherococcus*—669 genes, ADAPT232—678 genes) [6] (Figure 7). Among those deregulated by ADAPT-232, there were 206 genes that were not deregulated by any ingredient of BHP ADAPT-232 due to the synergy effect (Figure 7). The synergy-derived biological effect is characteristic of the BHP (ADAPT-232), which has a distinct pharmacological profile (signature) and typical chemical composition (fingerprint), which are different from *Rhodiola rosea* extracts [5].

Gene expression profiling was conducted on the human neuroglial cell line, T98G, after treatment with either Rhodiola SHR-5 extract or several of its constituents separately, including salidroside, triandrin, and tyrosol. Rhodiola SHR-5 and individual constituents had similar effects on G-protein-coupled receptor (GPCR)-mediated signal transduction through cAMP, phospholipase C, and phosphatidylinositol signaling pathways (Figure 6).

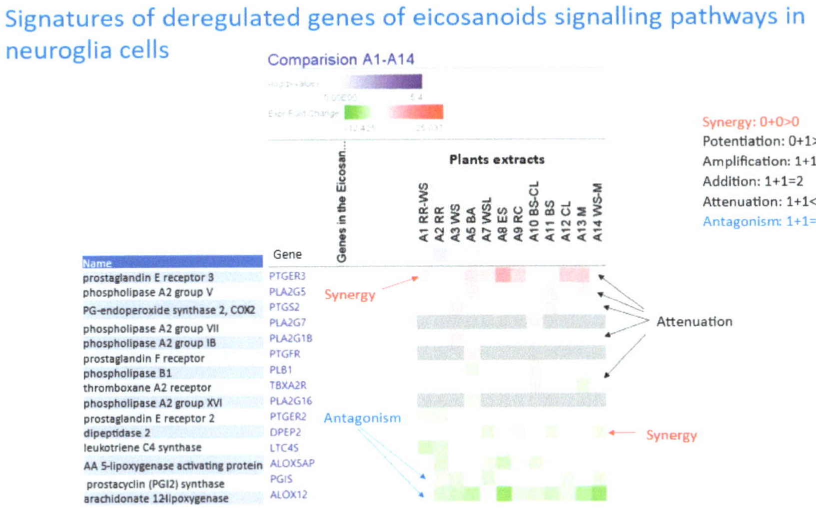

Figure 7. The synergy, potentiation, and antagonistic effects of hybridization of a combination of *Rhodiola* with *Withania*, *Withania* with melatonin, and *Curcuma longa* with *Boswellia* on eicosanoids signaling pathways, which has an essential role in inflammation and neurodegeneration assessed in isolated neuroglia cells. The authors' drawings were adapted from freely accessible publications [65].

The interpretations of microarray data and gene expression changes were conducted using Ingenuity Pathways Analysis (IPA) software (QIAGEN Bioinformatics, Aarhus C, Denmark), which is based on a continuously updated database (the Ingenuity Knowledge Base), gathering research observations with more than 8.1 million findings manually curated from biomedical literature and integrated from 45 third-party databases. The IPA network contains more than 40,000 nodes representing human genes, molecules, and biological functions, linked by 1,480,000 edges representing experimentally observed cause–effect relationships (either inhibiting or activating) associated with gene expression, transcription, molecular metabolism, and receptor binding. Network edges are also linked to activating or inhibiting effects. The IPA core analysis of transcriptomic datasets provides information about the impact of test samples on canonical signaling and metabolic pathways, diseases, and molecular and cellular functions that are activated or inhibited in experiments.

Two statistical methods of analysis of gene expression datasets are used in the IPA: (i) the gene-set-enrichment method, where differentially expressed genes are intersected with sets of genes that are associated with a particular pathway or biological function, providing a so-called "enrichment" score (Fisher's exact test p-value). This score measures the overlap of the observed and predicted regulated gene sets. (ii) The method based on cause–effect relationships related to the direction of effects reported in the literature, which provides the so-called z-score measuring the match of observed and predicted up/down-regulation [10,11]. The predicted effects are based on gene expression changes in the experimental samples relative to the control; z-score > 2, $-\log p$-value > 1.3.

Figure 7 shows the synergy, potentiation, and antagonistic effects of hybridization of a combination of *Rhodiola* with *Withania*, *Withania* with melatonin, and *Curcuma longa* with *Boswellia* on eicosanoids signaling pathways, which play an important role in inflammation and neurodegeneration in neuroglia cells.

In a recent study [8], BHP of *Rhodiola* with *Withania* (Adaptra) positively regulated 22 of 57 genes, which are known to activate the development of neurons (Figure 8), suggesting that Adaptra is potentially helpful in learning and memory, stress, and depression, insomnia, and aging-related neurodegenerative diseases, preventing Alzheimer's and Parkinson's diseases, and aiding recovery from brain injury and stroke.

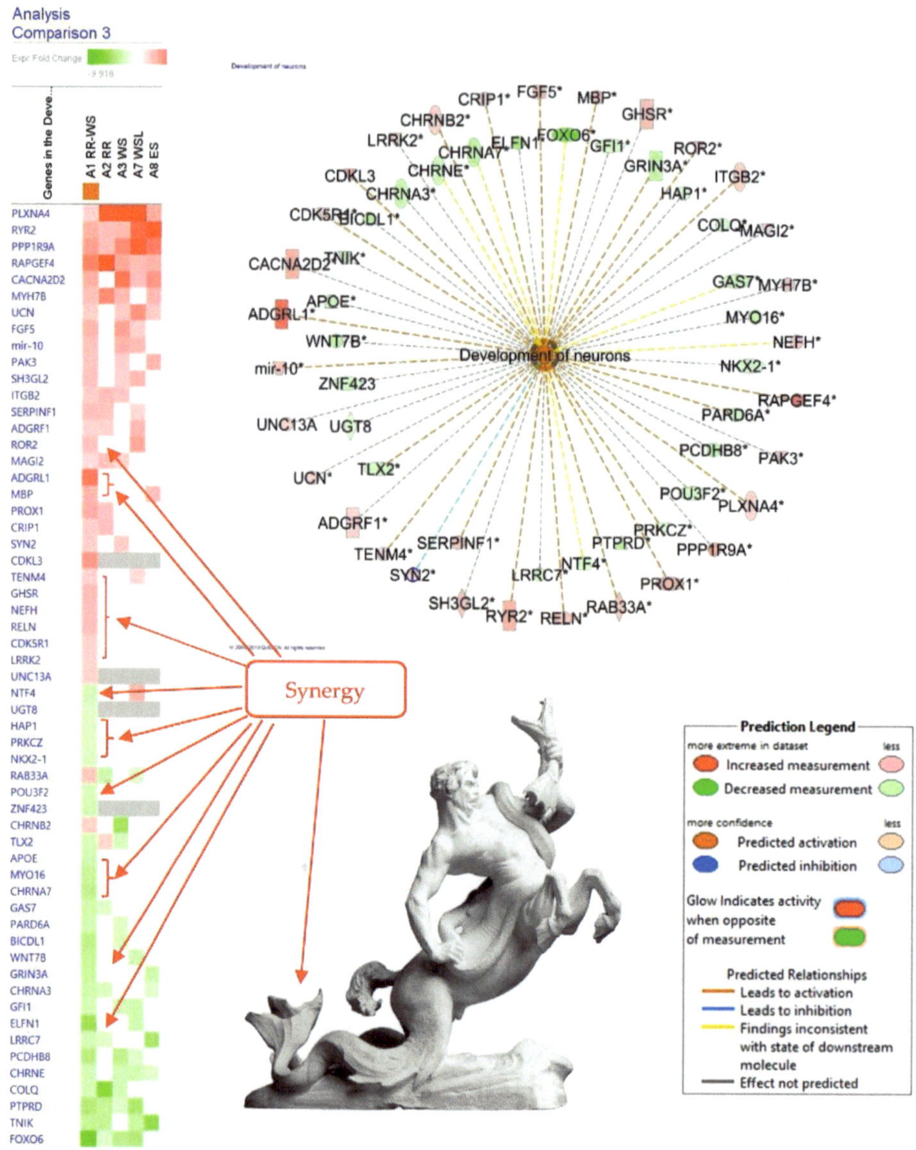

Figure 8. The effects of RR-WS (Adaptra) on gene expression in human T98G neuroglia cells and the predicted activation of the development of neurons. The authors' drawings were adapted from freely accessible publications [8]. The synergy effects (red arrows) of hybridization of *Rhodiola* with *Withania* on neurogenesis signaling pathways in isolated neuroglia cells. The intensity of green and red squares indicates fold-changes compared to control, where green means down- and red means up-regulation. Synergistic or antagonistic effects on gene expression were observed by comparing the impact of the BHP Adaptra combination of RR-WS (sample A1) with the lack of impact of individual extracts of RR (*R. rosea*), WS (*Withania somnifera*), and WSL *Withania somnifera* low dose, corresponding to samples A2, A3, and A7, at a significance level of $p < 0.05$ ($-\log = 1.3$) and a z-score > 2. The symbolic interpretation of synergy and antagonism by the image of hybrid creature from Greek mythology: *ichthyocentaurs* with a human head, a horse's body-derived fish-tail due to their synergistic and antagonistic (e.g., lack of human legs) interactions.

Notably, 25 genes were deregulated due to RR and WS synergistic interactions in the fixed combination Adaptra (Figure 8 and Table 1). This means that in combination, these two ingredients of Adaptra act synergistically. In other words, Adaptra is superior to *Rhodiola* or *Withania* in the activation of neurogenesis and, consequently, has the potential effects mentioned above [8].

Figure 9 shows the Venn diagrams of deregulated genes induced by the treatment of neuroglial cells with *Withania somnifera* (WS), *Rhaponticum cartamoides* L. (RC), and *Eleutherococcus senticosus* (RS) root extracts, as well as their hybrid combination (RC-ES-WS) [6].

Table 1. Effect of *Rhodiola rosea* (RR), *Withania somnifera* (WS), and their combination RR-WS (BHP Adaptra) on genes involved in the regulation of neuronal development *.

Gene Symbol	Entrez Gene Name	Literature Findings	Prediction **	Gene Expression, Fold Change		
				RR-WS	RR	WS
ADGRF1	adhesion G protein-coupled receptor F1	Affects (4)	Affected	2.29	2.28	
ADGRL1 ***	**adhesion G protein-coupled receptor L1**	Increases (2)	Increased	6.93		
APOE	**apolipoprotein E**	Affects (13)	Affected	−2.84		
BICDL1	BICD family like cargo adaptor 1	Affects (2)	Affected	−3.98		−2.34
CACNA2D2	calcium voltage-gated channel auxiliary subunit α2 δ2	Affects (2)	Affected	3.76		6.93
CDK5R1	**cyclin-dependent kinase 5 regulatory subunit 1**	Increases (4)	Increased	2.33		
CDKL3	**cyclin-dependent kinase like 3**	Increases (3)	Increased	4.82		
CHRNA3	cholinergic receptor nicotinic α 3 subunit	Affects (2)	Affected	−3.09	−2.45	
CHRNA7	**cholinergic receptor nicotinic α 7 subunit**	Increases (1)	Decreased	−3.74		
CHRNB2	cholinergic receptor nicotinic β 2 subunit	Increases (8)	Increased	2.45		−5.20
CHRNE	cholinergic receptor nicotinic epsilon subunit	Increases (1)	Decreased	−2.65		−2.59
COLQ	collagen-like tail subunit of acetylcholinesterase	Affects (2)	Affected	−2.65	−6.30	−2.69
CRIP1	cysteine rich protein 1	Increases (1)	Increased	2.41	3.01	
ELFN1	**extracellular leucine-rich repeat and fibronectin type III domain-containing 1**	Affects (1)	Affected	−5.31		
FGF5	fibroblast growth factor 5	Increases (1)	Increased	3.52		4.23
FOXO6	forkhead box O6	Increases (3)	Decreased	−7.93	−2.10	−3.89
GAS7	growth arrest specific 7	Increases (3)	Decreased	−2.85	−2.26	
GFI1	growth factor independent 1 transcriptional repressor	Affects (1)	Affected	−2.65		−2.59
GHSR	**growth hormone secretagogue receptor**	Affects (3)	Affected	3.15		
GRIN3A	**glutamate ionotropic receptor NMDA type subunit 3A**	Decreases (4)	Increased	−3.33		
HAP1	**huntingtin-associated protein 1**	Affects (1)	Affected	−2.21		
ITGB2	integrin subunit β 2	Increases (1)	Increased	2.45	3.05	2.66
LRRC7	leucine-rich repeat containing 7	Affects (1)	Affected	−2.66	−2.11	
LRRK2	**leucine-rich repeat kinase 2**	Affects (4)	Affected	2.26		
MAGI2	membrane-associated guanylate kinase,	Affects (10)	Affected	2.01	3.01	2.18
MBP	**myelin basic protein**	Increases (1)	Increased	3.48		
mir-10	microRNA 100	Increases (1)	Increased	2.89		3.53
MYH7B	myosin heavy chain 7B	Affects (1)	Affected	3.01	5.70	3.08
MYO16	**myosin XVI**	Affects (1)	Affected	−3.32		
NEFH	**neurofilament heavy**	Decreases (18)	Decreased	3.02		
NKX2-1	**NK2 homeobox 1**	Affects (4)	Affected	−2.38		
NTF4	**neurotrophin 4**	Increases (5)	Decreased	−2.61		
PAK3	p21 (RAC1) activated kinase 3	Affects (4)	Affected	2.86		2.36
PARD6A	par-6 family cell polarity regulator α	Decreases (2)	Increased	−2.84		−2.39
PCDHB8	protocadherin β 8	Affects (1)	Affected	−3.97		−3.89
PLXNA4	plexin A4	Increases (5)	Increased	2.25	9.49	10.97
POU3F2	**POU class 3 homeobox 2**	Affects (4)	Affected	−2.65		
PPP1R9A	protein phosphatase 1 regulatory subunit 9A	Affects (6)	Affected	4.51	2.85	5.38
PRKCZ	**protein kinase C ζ**	Decreases (2)	Increased	−2.23		
PROX1	prospero homeobox 1	Increases (1)	Increased	3.76	2.85	
PTPRD	protein tyrosine phosphatase, receptor type D	Increases (3)	Decreased	−4.32	−3.42	−2.11
RAB33A	RAB33A, member RAS oncogene family	Increases (1)	Increased	2.81	−3.11	
RAPGEF4	Rap guanine nucleotide exchange factor 4	Increases (2)	Increased	6.31	11.27	3.92
RELN	**reelin**	Increases (9)	Increased	3.01		
ROR2	**receptor tyrosine kinase-like orphan receptor 2**	Increases (5)	Increased	3.01		
RYR2	ryanodine receptor 2	Increases (2)	Increased	3.75	2.85	3.07
SERPINF1	serpin family F member 1	Increases (1)	Increased	3.04	2.88	
SH3GL2	SH3 domain containing GRB2-like 2, endophilin A1	Affects (2)	Affected	3.02		2.16
SYN2	synapsin II	Affects (3)	Affected	2.41		2.56
TENM4	**teneurin transmembrane protein 4**	Increases (3)	Increased	2.25		

Table 1. *Cont.*

Gene Symbol	Entrez Gene Name	Literature Findings	Prediction **	Gene Expression, Fold Change		
				RR-WS	RR	WS
TLX2	T cell leukemia homeobox 2	Decreases (2)	Increased	−2.64	2.39	−2.59
TNIK	TRAF2 and NCK interacting kinase	Affects (1)	Affected	−3.53	−3.60	
UCN	urocortin	Affects (1)	Affected	2.35		2.38
UGT8	**UDP glycosyltransferase 8**	Affects (2)	Affected	−2.21		
UNC13A	**unc-13 homolog A**	Affects (2)	Affected	2.25		
WNT7B	**Wnt family member 7B**	Affects (2)	Affected	−3.54		
ZNF423	**zinc finger protein 423**	Affects (4)	Affected	−2.66		

* Development of neurons predicted to be increased (z-score −2.87). Overlap *p*-value 7.29 × 10^{-3}. ** Prediction is based on measurement direction and literature data: 22 of 57 genes deregulated by RR-WS have measurement direction consistent with an increase in the development of neurons. *** Twenty-five genes (in **bold** text) deregulated due to synergistic interactions of RR and WS in the fixed combination Adaptra are highlighted in red text.

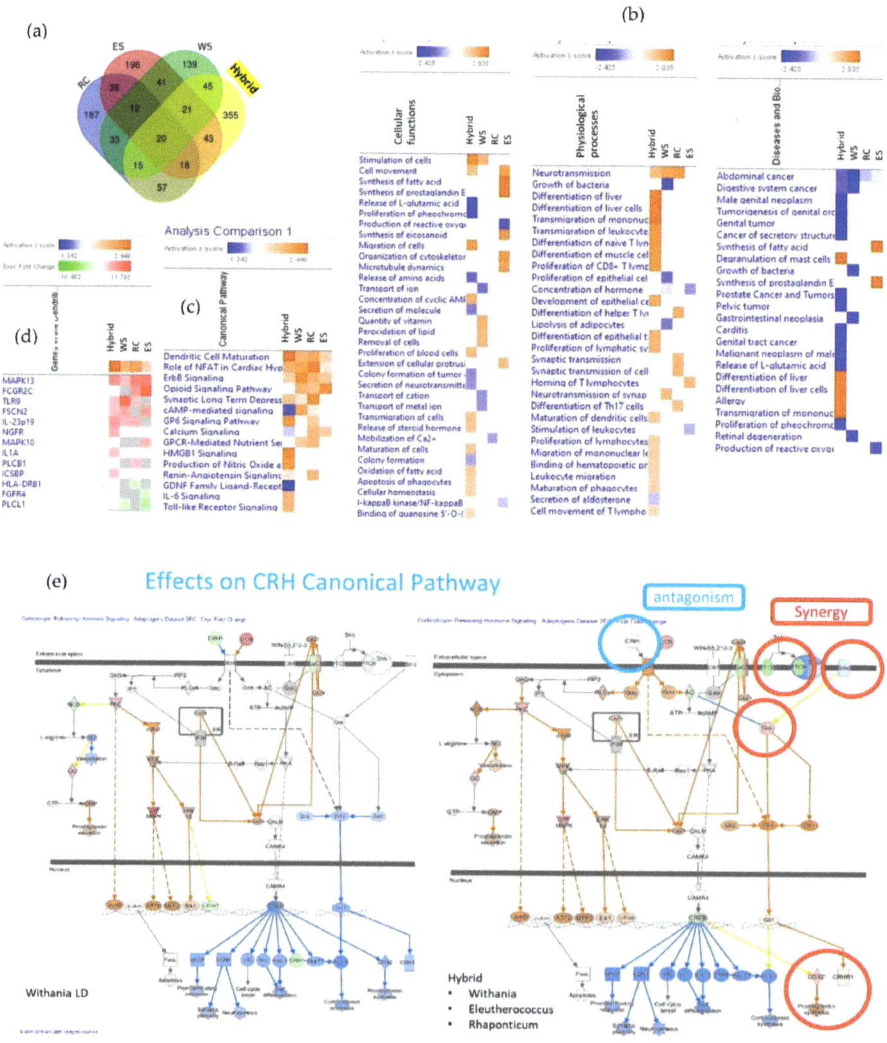

Figure 9. (**a**) Venn diagrams of deregulated genes caused by treatment of neuroglial cells with *Rhaponticum cartamoides* L. (RC), *Eleutherococcus senticosus* (RS), and *Withania somnifera* (WS) root

extracts, as well as their hybrid combination (RC-ES-WS). Values show the number of unique genes up- or downregulated by each extract alone and the number of deregulated genes overlapping extracts. (**b**,**c**) Heatmaps of canonical pathways, cellular functions, physiological processes, and diseases activated (brown) and inhibited (blue) by treatment of neuroglial cells with WS, RC, ES, and the hybrid combination RC-ES-WS. (**d**) Heatmap of gene expression showing synergistic or antagonistic effects of the ingredients of BHP RC-ES-WS on dendritic cell maturation pathway. Upregulated genes are shown in red, while downregulated genes are highlighted in green. (**e**) Effects on CFH Canonical Pathway. Authors' drawings adapted from freely accessible publications [6].

3. Clinical Studies in Human Subjects

3.1. BHP of Rhodiola with Green Tea (Mg-Teadiola®) in Psychological and Social Stress

Green tea contains catechins, tannins, phenolic acids, flavanol glycosides, the alkaloid caffeine, and the amino acid L-theanine [69,70], which are known to be capable of significant effects on the general state of mental alertness or arousal [71], activating adaptive cellular stress responses, inducing the production of cytoprotective proteins, and protecting neurons in animal models of Parkinson's disease, Huntington's disease, Alzheimer disease, and ischemia–reperfusion injury [72,73]. Green tea components, such as epigallocatechin gallate (EGCG), flavonoids kaempferol, and genistein activate protective mechanisms, including antioxidant and detoxifying enzymes via activation of Nrf2 signaling pathway, and upstream PKC, PI3K, and MAPKs modulation [70,72,73].

Both *Rhodiola rosea* and green tea (*Camelia sinensis*) supplementation were known to improve subjective stress perception and mood responses to acute stress. Their combinations with magnesium, vitamins B6, B9, B12, and L-theanine in a BHS Mg-Teadiola® was developed by Sanofi-Aventis Group, France, and studied in two clinical trials conducted in France and the UK to assess stress-protective effects compared to the efficacy of *Rhodiola rosea* and green tea [49–53], as shown in Table 2. The authors hypothesized that the efficacy of BHS Mg-Teadiola® is superior to that of the ingredients and/or placebo. In a DB-R-PC-PG clinical trial (NCT03262376), the single dose effect of Mg-Teadiola® tablets, *Rhodiola*, and green tea extracts was studied in four parallel groups of 100 moderately stressed, otherwise healthy volunteers (DASS score: 13–25) after acute psychological and social stress experimentally induced by The Trier Social Stress Test (TSST, speech, and mental mathematics tasks). The outcome measures were as follows: (i) spectral theta brain activity associated with cognitive task performance, (ii) subjective stress (stress and arousal), (iii) mood (profile of mood states), (iv) salivary cortisol, and (v) cardiovascular parameters (BP, HRV) [49–53].

BHS Mg-Teadiola® significantly alleviated subjective stress and mood responses to acute stress; analyses supported the superiority of the BHS Mg-Teadiola vs. placebo and the ingredients—*Rhodiola* and green tea. The BHS Mg-Teadiola® significantly attenuated subjective stress, tension, and total mood disturbance ratings after acute stress exposure. These effects were found both during the peak stress response and recovery. The salivary cortisol response was unaffected by treatment [49]. The BHS Mg-Teadiola® treatment significantly increased EEG resting state theta activity—considered indicative of a relaxed, alert state, attenuated subjective stress, anxiety, and mood disturbance, and heightened emotional and autonomic arousal; Mg-Teadiola may enhance coping capacity and offer protection from the harmful effects of stress exposure [50].

The BHS Mg-Teadiola® increased spectral theta brain activity during the execution of two attentional tasks, suggesting a potential to increase attentional capacity under stress conditions [51].

Table 2. Clinical studies of BHP of Rhodiola with other plants.

Reference/ Year	BHP Name, Ingredients	Condition	Population (n)/ Country	Dosage and Active Markers	Daily Dose and Duration of Treatment	Study Design * and Comparator	Result and Outcomes
Dye et al., 2020 [49]	**Mg-Teadiola®**: *Rhodiola rosea* L. + *Camelia chinensis* [L.] Kuntze + Mg + vitamins B6, B9, B12+L-theanine	Acute social stress	100 (25 + 25 + 25 + 25) Healthy, moderately stressed (DASS score: 13–25)	125 mg of IC dry extracts of *Camellia sinensis* L. leaf containing 50 mg L-theanine, and 222 mg of IC *Rhodiola rosea* L. root extract (corresponding to 1887 mg plant), and Mg (150 mg elemental) + vitamins B6 (0.7 mg), B9 (0.1 mg), B12 (0.00125 mg) One tablet of Mg- Teadiola® contains 150 mg of Mg, 0.7 mg of vitamin B6, 0.1 mg of vitamin B9, and 1.25 g of vitamin B12, and 222 mg of *Rhodiola rosea* rhizome dry extract, as well as 125 mg of green tea extract, including 50 mg of L-theanine	Single dose One tablet	DB-R-PC-PG, Placebo Capsules Tablets	Subjective stress (stress and arousal), Mood (profile of mood states) TSST
Boyle et al., 2021 [50]	**Mg-Teadiola®**	Acute social stress	25 + 25 + 25 + 25 Healthy, moderately stressed (DASS score: 13–25)			DB-R-PC-PG, Placebo Capsules tablets	TSST Spectral theta brain activity associated with cognitive task performance Salivary cortisol, cardiovascular parameters (BP, HRV)
Boyle et al., 2022 [51]	**Mg-Teadiola®**	Acute social stress	25 + 25 + 25 + 25 Healthy, moderately stressed (DASS score: 13–25)			DB-R-PC-PG, Placebo Capsules tablets	TSST Spectral theta brain activity, attentional capacity
Noah et al., 2022 [52]	**Mg-Teadiola®**:	Chronic negative emotional states	49 + 51 Healthy, moderately stressed (DASS score: >14)		One tablet daily for 28 days	R-PC-PG, Placebo Tablets	Stress, anxiety, depression, sleep, cortisol
Pickering et al., 2022 [53]	**Mg-Teadiola®**	Thermal stimulation	20 + 20 Healthy, moderately stressed (DASS score: >14)			R-PC-PG, Placebo tablets	blood-oxygen-level-dependent (BOLD) signal, stress, anxiety, depression, and sleep, cortisol
Bangratz et al., 2018 [66]	*Rhodiola rosea* + *Crocus sativus* L.	Depression	45	308 mg *Rhodiola* and 30 mg *Crocus*	42 days	OL	
Al-Kuraishy, 2015 [64]	*Rhodiola rosea* + *Ginkgo biloba*	cognitive function	112 (27 + 25 + 30 + 30)	*R. rosea* capsule 500 mg/day, *G. biloba* capsule 60 mg/day (standardized to contain 24% Ginkgo flavone glycosides)	10 days	R-PC-PG, Placebo capsules	Short-term working memory accuracy test (computerized N-back test) psychomotor vigilance task
Liu et al., 2023 [58]	*Rhodiola rosea* + caffeine	physical performance in resistance exercise	48 (12 + 12 + 12 + 12) 24 exercise-trained and 24 untrained healthy subjects	Rhodiola (1.2 g/capsule, 12 mg rhodioloside/salidroside) caffeine (200 mg/capsule; 3 mg/kg)	30 days 2 capsules/day	R-DB-PC-CO (?) Placebo	Significant improvements in muscle strength and muscular endurance compared to the placebo
Yun et al., 2024 [59]	*Rhodiola rosea* + caffeine	muscle endurance and explosiveness in humans	72 (6 groups × 12) 24 exercise-trained and 48 untrained healthy subjects	Rhodiola (1.2 g/capsule) caffeine (200 mg/capsule; 3 mg/kg)	30 days 2 capsules/day	R-OL-PC Placebo	Significant improvements in muscle strength and muscular endurance (5 km run, countermovement jump, standing long jump, 30 m sprint), oxygen consumption (VO2max) compared to the placebo

Table 2. *Cont.*

Reference/Year	BHP Name, Ingredients	Condition	Population (n)/Country	Dosage and Active Markers	Daily Dose and Duration of Treatment	Study Design * and Comparator	Result and Outcomes
Earnst et al., 2004 [60]	*Rhodiola rosea* + *Cordyceps sinensis*	Exercise performance	17 healthy subjects	1000 mg *Cordyceps sinensis* + 300 mg RR 3.0% rosavin and 2.5% salidroside	6 capsules/day 4 days, then a maintenance dose of 3 capsules/day for 11 day	R-DB-PC placebo	No significant difference between or within groups
Coulson et al., 2005 [61]	*Rhodiola rosea* + *Cordyceps sinensis*	Exercise performance	8	1000 mg *Cordyceps sinensis* + 300 mg RR 3.0% rosavin and 2.5% salidroside	6 capsules/day 4 days, then a maintenance dose of 3 capsules/day for 7 day	R-DB-PC placebo	After the pre-post endurance test, no significant difference between intervention and placebo in muscle tissue oxygen saturation; no significant (p </= 0.05) differences in ventilatory threshold (V(T)) or time to exhaustion (T(E)) between or within the treatment or control group
Kriepke et al., 2020 [63]	*Rhodiola rosea* + *Cordyceps sinensis* + blend of other 11 adaptogens	Exercise performance	10 + 11	NS	14-week	R-DB-PC placebo	No significant difference between or within groups
Dieamant et al., 2008 [67]	*Rhodiola rosea* + L-carnosine	Aging skin	62 + 62	1% of RCAC topical	28 days	DB-PC placebo	Protective effect of RCAC on skin barrier function and the positive response produced in human subjects with sensitive skin
Pkhaladze et al., 2020 [65]	**Menopause Relief EP®**: *Rhodiola rosea* EPR-7® (RR) + *Actaea racemose* EP40®, (black cohosh, BC)	Menopausal complaints	220 elderly woman (55 + 55 + 55 + 55)	Menopause Relief EP® capsules, 206.5 mg, containing 200 mg *R rosea* rhizome extract EPR-7® and 6.5 mg of *A. racemose* rhizoma dry extract, EP40®	2 capsules/day for 84 days	R-DB-PC-PG Placebo BC 6.5 mg BC 300 mg	BC is more effective in combination with RR in the relief of menopausal symptoms, particularly psychological symptoms Kupperman Menopausal Index (KMI), Menopause Relief Score (MRS), and menopause Utian Quality of Life (UQOL) index
Narimanian et al., 2005 [54]	**ADAPT-232 (Chisan®)**: *Rhodiola rosea* + *Schisandra* + *Eleutherococcus*	Acute nonspecific pneumonia	60 (30 + 30)	BHP of extracts from roots of *R. rosea* L. (27.6%), from berries of *S. chinensis* (51.0%), and from roots of *E. senticosus* (24.4%), standardized to contain 0.068 mg/mL salidroside, 0.141 mg/mL rosavin, 0.177 mg/mL shisandrin, 0.105 mg/mL gamma-shisandrin, and eleutherosides B and E (0.0 11 and 0.027 mg/mL)	40 mL (20 + 20), 10–15 days	R-DB-PC-PG placebo	Adjuvant therapy with ADAPT-232 decreased the duration of patients' recovery time and the acute phase of the illness; it also increased the mental performance of patients in the rehabilitation period and improved their quality of life (QOL) Duration of antibiotic therapy, psychometric tests, and the QOL

Table 2. *Cont.*

Reference/ Year	BHP Name, Ingredients	Condition	Population (n)/ Country	Dosage and Active Markers	Daily Dose and Duration of Treatment	Study Design * and Comparator	Result and Outcomes
Schutgens et.al., 2009 [18]	**ADAPT-232:** *Rhodiola rosea* + *Schisandra* + *Eleutherococcus*	Ultraweak Biophoton Emission	30 (10 + 10 + 10) Healthy subject experienced levels of stress and of fatigue (tiredness)	One tablet (456 mg) including 140 mg of the proprietary blend ADAPT-232) contains 0.5% schisandrin, 0.47% salidroside, and 0.59% rosavin, 0.11% One Rhodiola tablet (456 mg) including 144 mg SHR-5 extract contains 2.3% salidroside, 0.4% *p*-tyrosol, and 2.7% rosavin	Two tablets 7 days	R-DB-PC-PG Placebo Rhodiola rosea	ADAPT-232 and Rhodiola rosea (SHR-5) were able to reduce photon emission; however, only Rhodiola rosea (SHR-5) significantly reduced photon emission compared with the placebo group; Rhodiola, but not ADAPT-232, reduced fatigue
Aslanyan et al., 2010 [55]	**ADAPT-232 (Chisan®):** *Rhodiola rosea* + *Schisandra* + *Eleutherococcus*	Stressful cognitive tasks (Stroop Color-Word test and the d2Test of attention, fatigue	40 (20 + 20) Healthy women felt stressed over a long period of time by virtue of living under psychologically stressful conditions	One capsule of ADAPT-232S contained 0.5 mg of salidroside, 1.0 mg pf Schizandrin, and 0.35 mg of Eleutherosides B and E	Single dose One tablet	R-DB-PC-PG placebo	Significant improvement in attention and increase in speed and accuracy during stressful cognitive tasks in comparison to placebo Mental performance (attention, speed, and accuracy), arterial blood pressure, and heart rate
Karosanidze et al., 2022 [57]	**ADAPT-232 (Chisan®):** *Rhodiola rosea* + *Schisandra* + *Eleutherococcus*	Long COVID-19	100 (50 + 50) patients with Long COVID symptoms	One daily dose (2 × 30 mL oral solution) contains 180 mg extract of *R. rosea* rhizome, 600 mg of *S.chinensis* berry, and 156 mg of *E. senticosus* radix extracts	60 mL 30 days	R-QB-PC-PG Placebo	There was a significant increase in physical performance and recovery in long-term COVID patients; the duration of fatigue and chronic pain decreased; and the severity of all long-term COVID symptoms was relieved Duration of symptoms of long COVID
Hovhannisyan et al., 2015 [56]	**ADAPT-232S:** *Rhodiola rosea* + *Schisandra* + *Eleutherococcus*	exercise performance	215 (92 + 55 + 68) healthy athletes aged 18–35	One capsule of ADAPT-232S contains 0.5 mg of salidroside, 1.0 mg pf Schizandrin, and 0.35 mg of Eleutherosides B and E	2 capsules × 2 times a day, 30 days	R-DB-PC-PG Placebo	ADAPT-S and ADAPT-232S increase physical performance and the recovery of athletes after heavy physical and emotional loads They significantly decrease inattention, impulsivity, and the perception of stress, reduce fatigue, increase the anabolic index, and have excellent tolerability profiles
	ADAPT-S: *Rhodiola rosea* + *Schisandra* + *Eleutherococcus* + *Rhaponticum*			One capsule of ADAPT-S contains 1.5 mg of salidroside, 1.0 mg of Schizandrin, and 0.35 mg of Eleutherosides B and E, as well as 1.5 mg of 20-hydroxyecdisterone			The effects of ADAPT-S were superior with respect to the anabolic index, blood testosterone, and physical performance index; the results of this study suggest that ADAPT-232S and ADAPT-S might be useful for athletes' recovery after exercising and for preventing the symptoms of overtraining; ADAPT-S was most effective in sports disciplines where high coordination during physical fatigue (wrestling and long jump) is essentially required

* R—randomized, OL—open-label, PC—placebo-controlled, DB—double-blind, SA—single-arm, PG—parallel groups, CO—crossover, SC—sufficiently characterized herbal preparations, IC—insufficiently or purely characterized herbal preparations, NS—not specified.

In a placebo-controlled randomized clinical trial, the repeated dose effect of Mg-Teadiola® tablets for four weeks was studied in two parallel groups of 100 moderately chronically stressed, otherwise healthy volunteers (the Depression Anxiety Stress Scale-42 score: >14) and two groups of 40 individuals after acute thermal stimulation. The outcome measures were as follows: (i) blood-oxygen-level-dependent signal, (ii) stress, (iii) depression, (iv) anxiety, (v) salivary cortisol, and (vi) sleep [52].

Mg-Teadiola® was effective in relieving stress on days 14 and 28 in chronic stress and may diminish pain perception, underlining its potential benefits for patients suffering from pain, in whom comorbidities such as stress and sleep disorders are frequent [42]. Supplementation with Mg-Teadiola® reduced stress on D28 in chronically stressed but otherwise healthy individuals and modulated the stress and pain cerebral matrices during stressful thermal stimulus [53].

3.2. BHP of Rhodiola SHR-5 with Schisandra and Eleutherococcus (ADAPT-232/Chisan®) for Relief of Mental and Physical Fatigue Both in Healthy Subjects and in Patients with Pneumonia and COVID-19

ADAP232/Chisan is the combination of extracts of *Rhodiola rosea* L., roots (SHR-5) *Eleutherococcus senticosus* (Rupr. et Maxim) Harms, roots (SHE-3) and *Schisandra chinensis* (Turcz) Baill., fruits, containing 0.5% schizandrin, 0.47% salidroside, 0.59% rosavin, 0.11% eleutherosides B and 0.19% E. ADAPT 232® capsules and Chisan® oral solution have been used to enhance mental and physical capacities in case of tiredness or during convalescence, as a natural remedy in Sweden since 1979 and as an herbal medicinal product in Denmark since 2002. In 2008, it was approved as a traditional herbal medicinal product in 2008 in Sweden as an adaptogen in case of decreased performance, such as fatigue and sensation of weakness.

Early studies of ADAPT-232 showed improved cognitive functions and endurance under the stress of the cosmonauts. It has been used in space and by cosmonauts during the longest flight in space in the 1990s. It is noteworthy that typically, only carefully verified aids, validated for their efficacy and safety, were allowed for use in space. ADAPT-232 showed a good anti-fatigue effect, significantly increasing accuracy and precision in psychometric tests, particularly in complicated tests, decreasing the number of errors and improving short memory in fatigue under space-like conditions. Significant improvements in concentration, oculomotor coordination, and short-term memory of cosmonauts have been demonstrated in the computerized Monotonic 3 test for attention, as well as in two complicated short-term memory tests [74].

Three randomized, placebo-controlled, double-blind clinical trials (RCT) were conducted on healthy subjects, and two RCT studies were conducted to assess the anti-fatigue effect of ADAPT232/Chisan during recovery of patients with infectious respiratory diseases [54–57]. The efficacy of a single dose of ADAPT-232 on cognitive functions in humans has been demonstrated in the participants who experienced stressful cognitive tasks, namely, the Stroop Color-Word test followed by the d2 test of attention (d2), before and two hours after treatment. The results of the d2 test (attention deficit) are shown on the graph below [55], Figure 10a. ADAP232/Chisan increases tolerance to mental fatigue in healthy subjects [55]. ADAP232 significantly improved attention and increased speed and accuracy during stressful cognitive tasks compared to placebo [55]. ADAP232 significantly decreased the number of mistakes in complicated psychometric tests [54–57] and improved patients' quality of life and recovery from acute nonspecific pneumonia [54], as shown in Figure 10b. ADAPT232/Chisan has beneficial effects on stress-induced fatigue [55] in patients during their recovery from pneumonia [54] (Figure 10b).

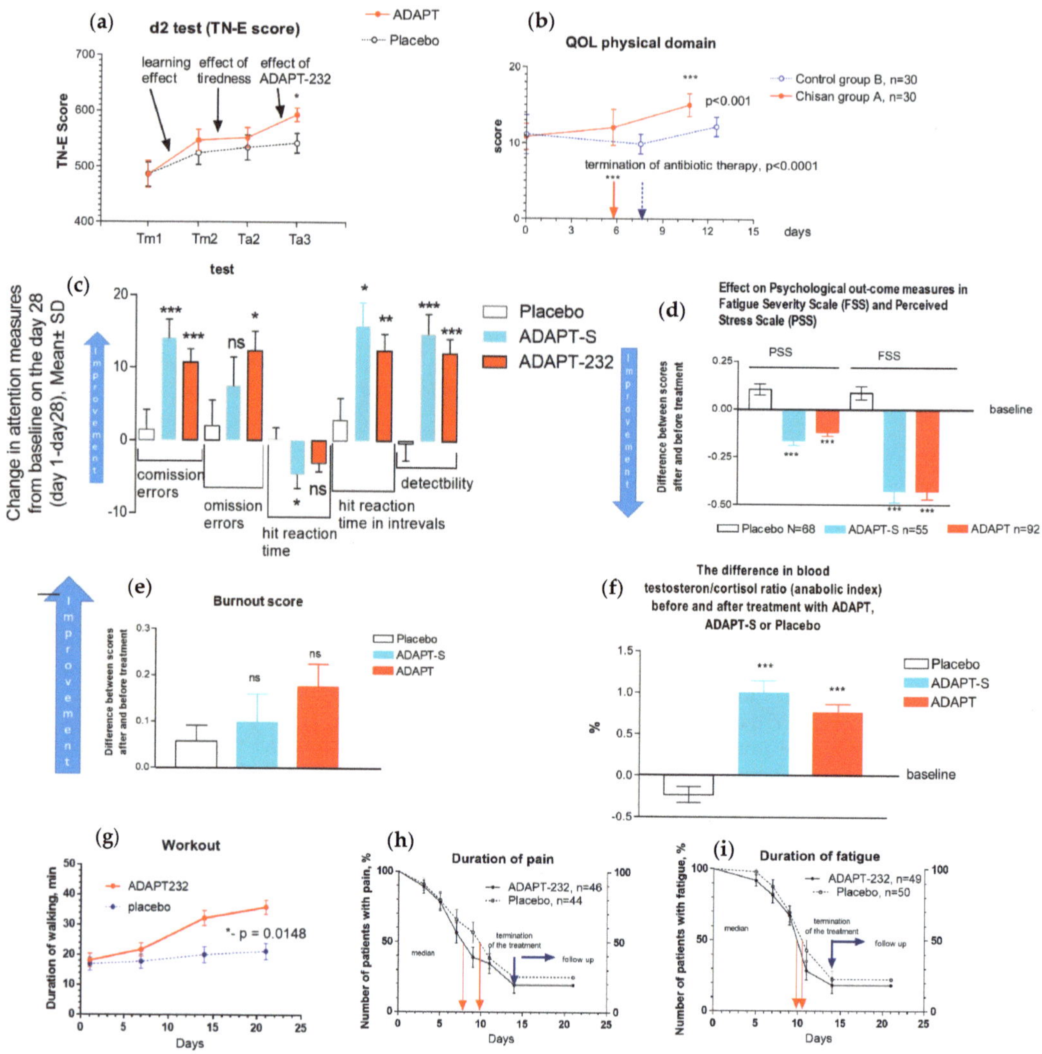

Figure 10. (**a**) Single-dose effects of ADAPT-232 on cognitive functions in participants exposed to stressful cognitive tasks (Stroop CW and d2 tests) four times over three consecutive days (day 1: in the morning, when subjects were not tired; day 2: in the morning, when subjects were not tired and in the afternoon, when subjects were tired; day 3: in the afternoon, two hours after treatment with ADAPT or placebo, when subjects were tired). (**b**)—Effect of ADAPT-232 on physical domain of QOL in patients during their recovery from pneumonia. (**c–f**)—Effect of ADAPT-232 and the BHP, containing Rhodiola, Eleutherococcus, Schisandra, and Leuzea (Rhaponticum cartamoides L.) roots extracts, on decreased inattention, impulsivity, and the perception of stress, fatigue and the anabolic index (testosterone/cortisol ratio) in 200 elite athletes. (**g–i**)—Effect of ADAPT-232 on the duration of pain, fatigue, and physical activity of patients with long COVID symptoms. *—$p < 0.05$; **—$p < 0.01$; ***—$p < 0.001$, *ns*—not significant difference.

ADAP232 and the BHP, containing *Rhodiola*, *Eleutherococcus*, *Schisandra*, and *Leuzea* (*Rhaponticum cartamoides* L.) roots extracts, significantly decreased inattention, impulsivity, and the perception of stress, while also reducing fatigue and increasing the anabolic index (testosterone/cortisol ratio) (Figure 10c–f) in 200 elite athletes. Furthermore, these supplements improved physical performance and the recovery rate of athletes after heavy physical and emotional load, indicating increased adaptation to physical and emotional stress [56].

Recently, the effects of ADAPT-232 on the recovery of patients with long COVID symptoms (fatigue, headache, respiratory insufficiency, cognitive performance, mood disorders, loss of smell, taste, and hair, sweatiness, cough, and pain in joints, muscles, and chest) were studied [59]. One hundred patients confirmed positive with the SARS-CoV-2 test, who were discharged from the intensive care unit and experienced at least three of nine long COVID symptoms in the 30 days before randomization, were included in the study.

ADAPT-232 reduced the duration of pain and fatigue by two days and one day, respectively, in 50% of patients (Figure 10h,i). The rate of cases of fatigue and pain symptoms was significantly lower in the rehabilitation period in the ADAPT-232 treatment group than in the placebo group on days 9 (39% vs. 57%, pain relief, $p = 0.0019$) and 11 (28% vs. 43%, relief of fatigue, * $p = 0.0157$). ADAPT-232 showed a significant increase in physical activity, measured as daily walk time (a workout), compared to placebo (Figure 10g). Clinical testing of C-reactive protein and the blood coagulation marker D-dimer did not show any statistically significant difference over time in the treatment between groups, with the exception of significantly reduced inflammatory cytokine IL-6 in the ADAPT-232 treatment group. In addition, ADAPT-232 significantly decreased blood creatinine compared to the placebo, suggesting prevention of the progression of possible renal failure in patients with long COVID [57].

3.3. BHP of *Rhodiola* with Caffeine for Enhancing Muscle Strength and Muscular Endurance

Alkaloid caffeine is a central nervous system (CNS) stimulant widely consumed in coffee and tea, promoting athletic capacities, including anaerobic exercise capacity, strength, and muscular and aerobic endurance. From 1984 to 2004, the World Anti-Doping Agency (WADA) banned caffeine because it enhances athletic performance in sports such as running, swimming, and cycling, but lifted the ban in 2004, removing caffeine from the list of prohibited substances, despite growing evidence that it is a sports booster. Athletes usually receive 3–6 mg/kg doses of caffeine to enhance resistance exercise performance acutely. However, a dose above 9 mg/kg often fails to cause a positive impact on various performance improvements and may also lead to side effects, such as tachycardia, headache, and anxiety. Long-term caffeine intake leads to an upregulation of adenosine receptor expression, resulting in a progressive reduction of the stimulatory effect on adenosine receptors and the development of caffeine resistance in the human body, which gradually occurs after 28 days of long-term caffeine intake [58]. Furthermore, regular use of caffeine-containing beverages may develop a physical, emotional, and psychological dependence, and one may experience a caffeine withdrawal syndrome characterized by headache, anxiety, irritability, low energy levels, dizziness or light-headedness, mental fogginess, and negative mood after abrupt cessation of caffeine intake. Unlike other psychoactive drugs, caffeine is legal, cheap, and unregulated in almost all parts of the world [75,76].

In two recent studies, the stimulating effects of caffeine, *Rhodiola rosea*, and their concomitant supplementation were compared in healthy human subjects and rats, suggesting the superimposed effects of the two supplements due to their different application mechanisms [58,59]. The impact of caffeine (400 mg/daily), *Rhodiola rosea*, (2.4 g/daily), and their combination for 30 days on muscle strength and muscular endurance physical performance of resistance exercise-untrained volunteers, as well as resistance exercise-trained subjects was assessed in four groups (12 participants in each group) of resistance exercise-trained subjects and two groups of resistance exercise-untrained volunteers, as shown in Table 2 [58]. *Rhodiola*-caffeine supplementation improved the rat model's forelimb grip strength, erythropoietin, dopamine, and oxygen consumption rate. *Rhodiola*-caffeine significantly increased the bench press one-repetition maximum, deep squat 1RM, maximum voluntary isometric contraction, and maximum repetitions at 60% 1RM bench press in resistance exercise-untrained volunteers. BHP *Rhodiola*-caffeine improved resistance exercise performance significantly by increasing the bench press one-repetition maximum, deep squat 1RM, maximum voluntary isometric contraction, and maximum repetitions at 60% 1RM bench press for both resistance exercise-untrained and -trained volunteers. The authors concluded that the stimulating effect of *Rhodiola*-caffeine was superior to *Rhodiola* and caffeine separately, presumably due to their synergistic and different mechanisms of action [58].

In the follow-up open-label placebo-controlled study [59], depicted in Table 2, the participants without prior aerobic training experience in the *Rhodiola*-caffeine group showed significant improvements compared to the placebo group in maximal oxygen consumption (VO2max), 5 km run, countermovement jump (CMJ), standing long jump, and 30 m sprint. For individuals with years of aerobic training experience, the *Rhodiola*-caffeine group exhibited similar results, enhancing performance in the training exercise, except for a 30 m sprint, compared to the placebo group. The authors concluded that the continuous 30-day supplementation of *Rhodiola*-caffeine had superior effects on muscle endurance and explosiveness in both animal and human studies compared to using Rhodiola or caffeine individually [59].

3.4. BHP of Rhodiola with Cordyceps for Boosting Exercise Performance

The aim of enhancing the physical performance of *Rhodiola* by combining it with *Cordyceps* has not been achieved in all three studies; the efficacy of BHP of *Rhodiola* rosea with *Cordyceps* was inactive compared to placebo [60–63]. The fixed combination of Rhodiola rosea with Cordyceps has shown no ergogenic effect on oxygen consumption, cycling time, muscle strength, oxygen uptake, and muscle performance during maximal graded tests following 14 days of supplementation. Meanwhile, both *Rhodiola rosea* and *Cordyceps* alone have demonstrated increased muscle fatigue resistance and strength benefits in active men. *Cordyceps* increased aerobic capacity and *Rhodiola* increased performance and time to exhaustion. The combination of *Rhodiola* with *Cordyceps* negatively affects the efficacy of the intervention due to an unknown antagonistic interaction, probably due to the impact of purine alkaloid Cordycepin [77]. Its chemical structure resembles that of the nucleoside adenosine, which is involved in adenosine triphosphate (ATP) metabolism, providing energy in many processes, including muscle contraction.

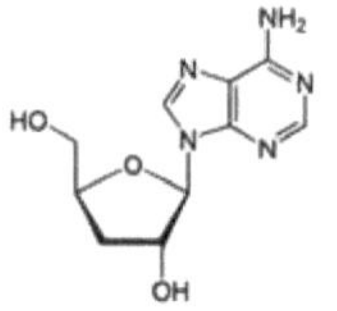
cordycepin

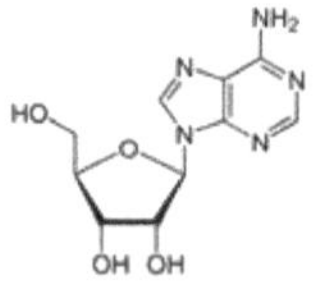
adenosine

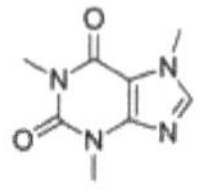
caffeine

3.5. BHP of Rhodiola with Ginkgo biloba for Improvement of Cognitive Function

Ginkgo biloba flavones and polyphenols have a potential neuroprotective effect in Alzheimer's dementia by inhibiting amyloid fibril accumulation [78,79]. *G. biloba* delays age-induced cognitive decline and may have more delicate and subtle therapeutic effects on the speed of cognitive alterations [80]. *G. biloba* is known to improve cognitive impairment and memory, increase cerebral blood circulation in the elderly with dementia syndrome and Alzheimer's disease, and relieve headaches and migraine [79,80]. Several human clinical trials provide preliminary positive evidence of the antidepressant effects and anxiolytic activity of *G. biloba* [81,82].

The effects of BHP of Arctic root with Ginkgo on the cognitive functions of 112 healthy subjects were recruited in a double-blind, randomized, placebo-controlled trial with four parallel groups. The efficacy of *G. biloba*, *R. rosea* extracts, and their fixed combination was assessed in a psychomotor vigilance task (PVT) and short-term working memory accuracy tests, as compared to placebo effects. The critical flicker-fusion frequency, PVT, and computerized N-back test assessed the central cognitive effect. Short-term working memory accuracy and sustained attention were measured in the psychomotor vigilance task, which provided a numerical measure of sleepiness by counting the number of lapses in the attention of the tested subject. *G. biloba* or *R. rosea* improved PVT and exhibited low to moderate working memory accuracy, whereas BHP exhibited superior effects on PVT, short-term working memory accuracy, and critical fusion versus flicker, compared to G. biloba or *R. rosea* when used alone. The author concluded that combining *R. rosea* with *G. biloba* led to a more significant effect on cognitive performance than either *G. biloba* or *R. rosea* when used alone [64].

3.6. BHP of Rhodiola rosea with Actaea racemosa for Relief of Aging-Related Menopausal Symptoms

Aging-related decline in the production of sex hormones is associated with many menopause symptoms, including hot flashes, sweating, sleeplessness, nervousness, irritability, depressive state, palpitation, joint pain, libido changes, vaginal dryness, etc. However, long-term hormonal replacement therapy is known to increase the risk of developing breast cancer. In the meantime, *Rhodiola rosea* is known not only for its stress-protective activity but also for its anti-cancerogenic potential. It was suggested that Rhodiola can counteract menopausal symptoms due to the selective modulation of estrogenic receptors and consequently decrease the risk of cancer in hormone-sensitive tissues during the non-reproductive phase in women, likely due to a decline in the production of sex hormones in menopausal women. Therefore, the BHP (Menopause Relief EP®), consisting of a combination of *Rhodiola rosea* (RR) and *Actaea racemosa* (black cohosh) root extracts, compared with the most effective black cohosh (BC) preparation in women with menopausal complaints [65].

A total of 220 women (mean age 52 years) were randomly assigned to receive two capsules of either BC (6.5 mg), BC500 (500 mg), Menopause Relief EP ® (206.5 mg), or placebo once per day for 12 weeks [65]. The efficacy endpoints were relief of menopausal symptoms, including hot flushes, sweating, heart discomfort, sleep problems, joint and muscular discomfort, depressive mood, irritability, anxiety, physical and mental exhaustion, sexual problems, bladder problems, and vaginal dryness. These were measured using the Kupperman Menopausal Index (KMI), the Menopause Relief Score (MRS), and the menopause Utian Quality of Life (UQOL) index. The symptom relief effects of Menopause Relief EP® were significantly superior in all tests to the effects of BC and placebo after their repeated administration for 6 and 12 weeks. Menopause Relief EP® significantly improved the UQOL index in patients, compared to BC, BC500, and placebo, mainly due to the beneficial effects on the emotional and physical health domains (Figure 11). It was concluded that the impact of this BHP is more effective than the mono-drug in relieving menopausal symptoms, particularly psychological symptoms.

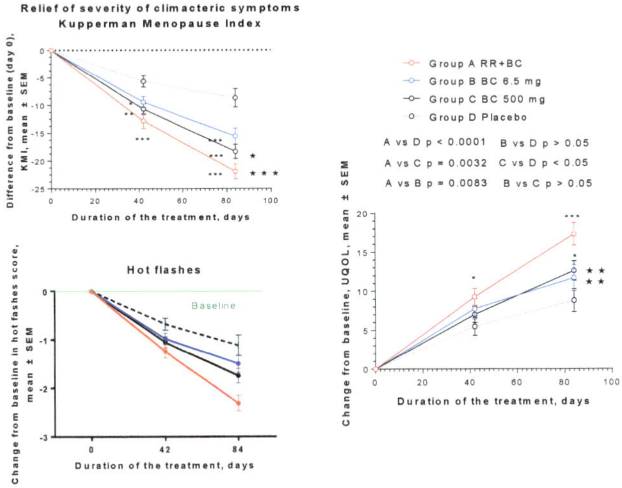

Figure 11. Superior effect of the combination of *Rhodiola* with black cohosh on climacteric symptoms of women and their quality of life. This publication [65] is open access and is distributed under the Creative

Commons Attribution License, which permits unrestricted use, distribution, and reproduction in any medium, provided the original work is properly cited. *—$p < 0.05$; **—$p < 0.01$; ***—$p < 0.001$.

3.7. BHP of Rhodiola with Saffron in Mild and Moderate Depression

Findings from clinical trials conducted to date indicate that saffron supplementation can improve symptoms of depression in patients with MDD. Saffron was found to be as effective as treatment with conventional antidepressants, such as imipramine and fluoxetine, in randomized, double-blind clinical trials [83]. Supplementation of a BHP of *Rhodiola* with saffron extracts in a daily dose of 308 mg *Rhodiola* and 30 mg *Crocus* for six weeks significantly decreased anxiety and depression scores compared to the baseline in 45 patients with mild-moderate depression. The results of this observational study are encouraging, and serious adverse effects were not recorded. However, a double-blind, randomized study with a positive comparator and placebo control design is needed to confirm these results [66].

3.8. BHP of Rhodiola with L-Carnosine in Aging Skin

Carnosine (β-alanine-L-histidine) is an endogenous water-soluble dipeptide. It is mainly distributed in skeletal muscles and its levels depend on age (decreases with age), gender (lower in females), and food (lower in a vegetarian diet). Carnosine plays an essential role in aging-related neurodegeneration, diabetes mellitus, cardiovascular diseases, cancer, and other disorders. The meta-analysis of clinical trial findings indisputably demonstrated a significant beneficial effect of L-carnosine only for diabetes mellitus and cognitive impairment [84]. The pathophysiology of sensitive skin consists of an inflammatory reaction resulting from the abnormal penetration of potentially irritating substances in the skin, which occurs due to skin barrier dysfunction and changes in the production of local neuromodulators. A double-blind comparative study was conducted on 124 volunteers with sensitive skin who received either a formulation containing 1% of *Rhodiola*-L-carnosine BHP or a placebo, applied twice daily for 28 consecutive days [67]. The reduction of transepidermal water loss, positive perceptions of improvements in skin dryness and skin-comfort sensation, and the reduction of discomfort sensation after the stinging test were measured. The *Rhodiola*-L-carnosine BHP treatment produced in vivo protective effects in skin barrier function and a positive subjective response from volunteers with sensitive skin. The authors suggested that the protective effect of *Rhodiola*-L-carnosine BHP on skin barrier function and the positive reaction produced in human subjects with sensitive skin is due to an increase in opioid peptides release, an inhibitory effect on neuropeptides production, and modulation of cytokines production by keratinocytes under ultraviolet stress, as observed in vitro testing [67].

4. Critical Appraisal of the Studies and Challenges

4.1. The Rationale for Including Caffeine or Caffeine-Containing BHP Based on Rhodiola Extracts

There is an essential difference in pharmacological profiles (signatures) between adaptogens and conventional stimulants, such as caffeine. Caffeine temporarily increases physical and cognitive functions, but unlike adaptogens, it does not exhibit neuro-, hepato-, cardioprotective actions nor does it improve survival under stress. Additionally, caffeine does not induce energy replenishment and is associated with poor quality arousal and recovery process after exhaustive physical load. Notably, unlike *Rhodiola* and other adaptogens, prolonged use of caffeine can cause the user to develop both tolerance and addiction [3].

The magnesium and caffeine content in green tea was uncontrolled and not specified [40–43]; therefore, there is no guarantee that the effect of Mg-Teadiola will consistently provide reproducible efficacy.

In this context, stern concerns occur regarding the rationale for including caffeine or green tea containing caffeine in the formulation of the safe adaptogenic BHP supplement

Mg-Teadiole [40–43]. Addiction to caffeine and other undesired effects of regular caffeine supplementation are well known. In this context, the formulation's rationale does not stand up to scrutiny.

Similarly, combining *R. rosea* extracts with caffeine to increase physical performance has no apparent benefits, particularly in the studies [58,59], where the daily dose of caffeine was at the safety threshold—400 mg.

4.2. Issues Regarding Product Quality and the Quality of Clinical Trials

The magnesium and caffeine content in green tea is uncontrolled and not specified; therefore, there is no guarantee that the effect of Mg-Teadiola will provide consistently reproducible efficacy.

The authors declare the labeled amount of active ingredients [40–43] but do not adhere to Extensions of the Consolidated Standards of Reporting Trials Statement for Herbal Medicinal Interventions (CONSORT) regarding the quality of the product. Mg-Teadiole was not adequately characterized for several factors, including extraction solvents, the dry herb: dry native extract ratio (DER), the content of active markers (caffeine, theanine, Mg^{+2}, salidroside, rosavin, etc.), and the analytical methods validated for selectivity, accuracy, and precision. In addition, TLC and HPLC fingerprints were not provided to ensure reproducible quality and reproducible pharmacological activity. The placebo and Mg-Teadiola were distinguishable by appearance. Reporting the masking procedure is not convincing enough to provide an adequate double-blind study design.

The reporting of *Rhodiola*-caffeine studies did not comply with the requirements outlined in CONSORT, particularly regarding characteristics of the herbal product, e.g.,

- The part(s) of the plant used to produce the product or extract;
- The type and concentration of extraction solvent used;
- Dry herb: dry native extract ratio (DER);
- Qualitative testing (product's chemical fingerprint, HPLC fingerprint);
- Doses (number of capsules per day);
- Description and the results of analytical methods validation (selectivity, accuracy, and precision);
- Sample size (12 subjects per group) justification;
- Product randomization (who generated the randomization allocation sequence and where the verum and placebo were encoded);
- Method of randomization (a method used to generate the random allocation sequence, including details of restriction);
- Allocation concealment;
- Masking (how the care provider, investigator, and outcomes assessor were blinded to study preparations);
- Implementation (who enrolled participants and assigned participants to their groups, etc.);
- Procedure for treatment compliance (how measurements of compliance of individual patients with the treatment regimen under study were documented);
- Monitoring;
- Settings and locations where the data were collected;
- Deviations from the protocol;
- Quality assurance and quality control;
- The voucher specimen (i.e., the retention sample was retained and, if so, kept or deposited);
- Statistical analysis of adverse events (e.g., the odds ratio (OR) statistics);
- The other measures to exclude the risk of bias were not specified.

4.3. Reproducibility and Consistency of the Results of Studies

The reproducible quality of herbal interventions is the primary issue in ensuring the reproducible efficacy of herbal medicines. However, along with the problem of a lack of independent replications, there are other concerns related to poor reporting and methodological quality issues. These include unclear risk of bias, optimized doses, and treatment regimens to rectify and observe the reproducible efficacy in various clinical studies [44,85].

Green tea and Rhodiola rosea products are known as botanicals with significant variability that depends on numerous factors, including genetic, environmental (climate, soil, insects, pest, microbiological infection) factors, processing methods, and storage conditions (extraction solvent polarity, temperature, duration, etc.) [13,14,86]. The content of active ingredients in herbal preparations depends on many factors, including the geographic and climate zone where the herbs were grown, the season and conditions under which they were harvested, and how they were dried, extracted, and prepared to form the final dosage.

5. Limitations of the Review

5.1. Literature Search Strategy

The studies were selected using the keywords *Rhodiola rosea, network pharmacology, gene expression, Ingenuity pathways IPA, clinical trials, synergistic effect* in PubMed and in our private library. The results were summarized using a simple analytical framework: Search, Appraisal, Synthesis, and Analysis (SALSA).

5.2. Appraisal

Some clinical studies, e.g., Teadiola (see Table 2), were designed to compare the efficacy of Arctic root with green tea and black cohosh. These studies show the superior efficacy of BHP compared with mono-drugs and placebo, presumably due to the synergy of BHP's ingredients [40–43].

In the clinical study of binary BHP of Rhodiola and black cohosh, BHP was compared only with black cohosh and placebo, but not with Arctic root. BHP's efficacy was superior to that of black cohosh, suggesting that these ingredients act at least additively. The limitation of this study was the lack of inclusion of one more group—Arctic root.

Other clinical studies had no active control as a mono-drug, except for a placebo, e.g., BHP of Arctic root and cordyceps, which were found inactive in contrast to cordyceps, which was found active in similar clinical trials, suggesting a possible antagonistic mechanism of action between these mono drugs.

5.3. Terminology

There are several definitions of synergy [87–91], including two commonly used definitions of synergy in pharmacology and pharmacognosy related to efficacy, effectiveness, and potency of the combinations of two agents with the same pharmacological activity: "A combination of agents that is more effective than is expected from the effectiveness of its constituents is said to show synergy" [88]. In this review, we adhere to the definition of synergy, which is formulated as "two or more agents working together to produce a result not obtainable by any of the agents independently" and can be interpreted as the generation of new pharmacological activity, which is specific only for the combination of two or more agents [87].

6. Future Perspectives

There are many challenges in BHP and PHP research, including:

(I) The impact of the chemical composition of active compounds (conditional fingerprint) on the pharmacological activity (conditional signature) of the total extract is crucial due to synergistic, potentiating, and antagonistic interactions between the multiple targets of numerous active components of the extracts.

(II) The optimal effective and safe therapeutic dose of *Rhodiola rosea* extract should be established, considering the "bell shape" dose–response relationship of *Rhodiola rosea*. Supraphysiological concentrations far exceeding the proposed dose in humans were applied in many in vitro studies. Most studies have ignored this issue; the pharmacological and toxicological data are usually unavailable. The optimal range of doses of adaptogens is like that of the hormetic zone dose–response pattern. The underlying molecular mechanisms of hormesis are not fully understood. The theoretical background of hormesis is related to the hypothesis of interactions of biologically active compounds (ligand/intervention/drug) with two target proteins (receptors) that have functionally opposite responses at different concentrations. It was proposed that a drug acting as a competitive antagonist at either or both of the receptors changes the relationship between the two opposing concentration-effect curves, resulting in potentiation, antagonism, or reversal of the observed effect; the theoretical model suggested that the total effect in the system can be obtained by the algebraic summation of the two effects resulting from the activation of the two opposing receptor populations providing classical hormesis biphasic curve. However, in practice, dose–response patterns are significantly complicated due to many other interactions with:

- multiple targets (receptors) of different affinity to the active compound,
- other regulatory proteins or mediators in the networks involved in the adaptive stress response,
- feedback down regulations in the molecular signaling pathways and/or,
- the metabolic transformation of active ligands into metabolites, a secondary ligand, which have different affinity to various receptors of adaptive stress response.

(III) Quality and safety issues related to the content of potentially toxic compounds should be considered. For example, the cyanogenic nitrile Lotaustralin can be highly toxic and thus is encouraged to be subjected to the quality control process. Although Rhodiola is generally safe, two critical points that need to be considered are possible herb-drug interaction, established in some studies, and potentially toxic cyanogenic nitrile lotaustraline detected in some Rhodiola extracts.

7. Conclusions

Considering botanicals comprising multi-component active compounds, their interactions result in novel, unexpected pharmacological activity due to their **synergistic and antagonistic** effects.

The most important finding from BHP's network pharmacology studies was the evidence supporting the synergistic interaction of its ingredients, revealing unexpected new pharmacological activities unique to and specific to the new BHP.

The results of some studies show the superior efficacy of BHP, such as the combinations of Arctic root with black cohosh, Arctic root with Ashwagandha, Arctic root with green tea, Arctic root with Ginkgo, Arctic root with Eleuthero and Schisandra, compared to the mono-drugs. On the contrary, some a priori designed combinations, such as Rhodiola with Cordyceps, were inactive in several clinical studies, presumably due to antagonistic interactions and crosstalk between molecular targets within the molecular networks involved in the cellular and overall response of organisms to the intervention.

The network pharmacology approach is suitable for understanding their mechanisms of action and predicting possible toxic effects, new indications for use, and lack of activity. However, the results of in silico analysis must be based on experimental findings of gene and protein expression of isolated cells and validated at least through in vivo experiments on rodents.

Author Contributions: Conceptualization, A.P. and T.L.; writing—preparation of original drafts, A.P.; writing—review and editing, A.P. and T.E. All authors have read and agreed to the published version of the manuscript.

Funding: This research received no external funding.

Institutional Review Board Statement: Not applicable.

Informed Consent Statement: Not applicable.

Data Availability Statement: Data sharing is not applicable.

Conflicts of Interest: The authors A.P. and T.E. declare no conflict of interest; T.L. is the founder of EuropharmaUSA Inc., Green Bay, Wisconsin, USA; AP is self-employed at the research and development company Phytomed AB and has no shares or financial interest in any pharmaceutical company.

References

1. Hu, L.Y.; Cai, A.Q.; Li, B.; Li, Z.; Liu, J.P.; Cao, H.J. Chinese herbal medicine for post-viral fatigue: A systematic review of randomized controlled trials. *PLoS ONE* **2024**, *19*, e0300896. [CrossRef] [PubMed]
2. Ibáñez, B.; Melero, A.; Montoro, A.; Merino-Torres, J.F.; Soriano, J.M.; San Onofre, N. A Narrative Review of the Herbal Preparation of Ayurvedic, Traditional Chinese, and Kampō Medicines Applied as Radioprotectors. *Antioxidants* **2023**, *12*, 1437. [CrossRef] [PubMed]
3. Panossian, A.G.; Efferth, T.; Shikov, A.N.; Pozharitskaya, O.N.; Kuchta, K.; Mukherjee, P.K.; Banerjee, S.; Heinrich, M.; Wu, W.; Guo, D.A.; et al. Evolution of the adaptogenic concept from traditional use to medical systems: Pharmacology of stress- and aging-related diseases. *Med. Res. Rev.* **2021**, *41*, 630–703. [CrossRef] [PubMed]
4. Townsend, J.R.; Kirby, T.O.; Sapp, P.A.; Gonzalez, A.M.; Marshall, T.M.; Esposito, R. Nutrient synergy: Definition, evidence, and future directions. *Front. Nutr.* **2023**, *10*, 1279925. [CrossRef] [PubMed]
5. Panossian, A.; Hamm, R.; Kadioglu, O.; Wikman, G.; Efferth, T. Synergy and Antagonism of Active Constituents of ADAPT-232 on Transcriptional Level of Metabolic Regulation of Isolated Neuroglial Cells. *Front. Neurosci.* **2013**, *7*, 16. [CrossRef] [PubMed]
6. Panossian, A.; Seo, E.J.; Efferth, T. Synergy assessments of plant extracts used in the treatment of stress and aging related disorders. *Synerg. Res.* **2018**, *7*, 39–49. Available online: https://www.sciencedirect.com/science/article/pii/S2213713018300312?via=ihub (accessed on 3 April 2024). [CrossRef]
7. Panossian, A.; Seo, E.J.; Wikman, G.; Efferth, T. Synergy assessment of fixed combinations of Herba Andrographidis and Radix Eleutherococci extracts by transcriptome-wide microarray profiling. *Phytomedicine* **2015**, *22*, 981–992. [CrossRef] [PubMed]
8. Panossian, A.; Seo, E.J.; Efferth, T. Novel molecular mechanisms for the adaptogenic effects of herbal extracts on isolated brain cells using systems biology. *Phytomedicine* **2018**, *50*, 257–284. [CrossRef] [PubMed]
9. Panossian, A.; Hamm, R.; Wikman, G.; Efferth, T. Mechanism of action of Rhodiola, salidroside, tyrosol and triandrin in isolated neuroglial cells: An interactive pathway analysis of the downstream effects using RNA microarray data. *Phytomedicine* **2014**, *21*, 1325–1348. [CrossRef]
10. Panossian, A.; Seo, E.J.; Efferth, T. Effects of anti-inflammatory and adaptogenic herbal extracts on gene expression of eicosanoids signaling pathways in isolated brain cells. *Phytomedicine* **2019**, *60*, 152881. [CrossRef]
11. Panossian, A.; Efferth, T. Network Pharmacology of Adaptogens in the Assessment of Their Pleiotropic Therapeutic Activity. *Pharmaceuticals* **2022**, *15*, 1051. [CrossRef] [PubMed]
12. HMPPA. *Rhodiola rosea.* Phytochemical pharmacological and clinical evidence. In Proceedings of the Scientific eSymposium, Herbal Medicinal Products Platform Austria (HMPPA) and Society for Medicinal Plant and Natural Product Research (GA), Virtual, 23 November 2023. Available online: https://www.hmppa.at/rhodiola-rosea-esymposium/ (accessed on 3 April 2024).
13. Tao, H.; Wu, X.; Cao, J.; Peng, Y.; Wang, A.; Pei, J.; Xiao, J.; Wang, S.; Wang, Y. Rhodiola species: A comprehensive review of traditional use, phytochemistry, pharmacology, toxicity, and clinical study. *Med. Res. Rev.* **2019**, *39*, 1779–1850. [CrossRef] [PubMed]
14. Panossian, A.; Wikman, G.; Sarris, J. Rosenroot (*Rhodiola rosea*): Traditional use, chemical composition, pharmacology and clinical efficacy. *Phytomedicine* **2010**, *17*, 481–493. [CrossRef]
15. HMPC. *EMA/HMPC/24186/2023. Assessment Report on Rhodiola rosea L., Rhizoma et Radix. Draft–Revision 1*; Committee on Herbal Medicinal Products (HMPC): Amsterdam, The Netherlands, 2023; pp. 1–50.
16. Darbinyan, V.; Kteyan, A.; Panossian, A.; Gabrielian, E.; Wikman, G.; Wagner, H. *Rhodiola rosea* in stress induced fatigue—A double blind cross-over study of a standardized extract SHR-5 with a repeated low-dose regimen on the mental performance of healthy physicians during night duty. *Phytomedicine* **2000**, *7*, 365–371. [CrossRef] [PubMed]
17. Darbinyan, V.; Aslanyan, G.; Amroyan, E.; Gabrielyan, E.; Malmström, C.; Panossian, A. Clinical trial of *Rhodiola rosea* L. extract SHR-5 in the treatment of mild to moderate depression. *Nord. J. Psychiatry* **2007**, *61*, 343–348. [CrossRef] [PubMed]
18. Schutgens, F.W.; Neogi, P.; Van Wijk, E.P.; Van Wijk, R.; Wikman, G.; Wiegant, F.A. The influence of adaptogens on ultraweak biophoton emission: A pilot experiment. *Phytother. Res.* **2009**, *23*, 1103–1108. [CrossRef]
19. Yu, H.L.; Zhang, P.P.; Zhang, C.; Zhang, X.; Li, Z.Z.; Li, W.Q.; Fu, A.S. Effects of *Rhodiola rosea* on oxidative stress and negative emotional states in patients with obstructive sleep apnea. *Lin. Chuang Er Bi Yan Hou Tou Jing Wai Ke Za Zhi* **2019**, *33*, 954–957. (In Chinese) [CrossRef]
20. Skarpanska-Stejnborn, A.; Pilaczynska-Szczesniak, L.; Basta, P.; Deskur-Smielecka, E. The influence of supplementation with *Rhodiola rosea* L. extract on selected redox parameters in professional rowers. *Int. J. Sport. Nutr. Exerc. Metab.* **2009**, *19*, 186–199. [CrossRef]
21. Olsson, E.M.; Von Schéele, B.; Panossian, A.G. A randomised, double-blind, placebo-controlled, parallel-group study of the standardised extract shr-5 of the roots of *Rhodiola rosea* in the treatment of subjects with stress-related fatigue. *Planta* **2009**, *75*, 105–112. [CrossRef]
22. Parisi, A.; Tranchita, E.; Duranti, G.; Ciminelli, E.; Quaranta, F.; Ceci, R.; Cerulli, C.; Borrione, P.; Sabatini, S. Effects of chronic *Rhodiola rosea* supplementation on sport performance and antioxidant capacity in trained male: Preliminary results. *J. Sports Med.*

Phys. Fit. **2010**, *50*, 57–63. Available online: https://www.minervamedica.it/en/journals/sports-med-physical-fitness/article.php?cod=R40Y2010N01A0057 (accessed on 3 April 2024).
23. Goyvaerts, B.; Bruhn, S. *Rhodiola rosea* special extract SHR-5 in burnout and fatigue syndrome. *Erfahrungsheilkunde* **2012**, *61*, 79–83. [CrossRef]
24. Edwards, D.; Heufelder, A.; Zimmermann, A. Therapeutic effects and safety of *Rhodiola rosea* extract WS® 1375 in subjects with life-stress symptoms—Results of an open-label study. *Phytother. Res.* **2012**, *26*, 1220–1225. [CrossRef] [PubMed]
25. Noreen, E.E.; Buckley, J.G.; Lewis, S.L.; Brandauer, J.; Stuempfle, K.J. The effects of an acute dose of *Rhodiola rosea* on endurance exercise performance. *J. Strength. Cond. Res.* **2013**, *27*, 839–847. [CrossRef] [PubMed]
26. Mao, J.J.; Li, Q.S.; Soeller, I.; Xie, S.X.; Amsterdam, J.D. *Rhodiola rosea* therapy for major depressive disorder: A study protocol for a randomized, double-blind, placebo- controlled trial. *J. Clin. Trials* **2014**, *4*, 170. [CrossRef] [PubMed]
27. Duncan, M.J.; Clarke, N.D. The Effect of Acute *Rhodiola rosea* Ingestion on Exercise Heart Rate, Substrate Utilisation, Mood State, and Perceptions of Exertion, Arousal, and Pleasure/Displeasure in Active Men. *J. Sports Med.* **2014**, *563043*, 1–8. [CrossRef] [PubMed]
28. Punja, S.; Shamseer, L.; Olson, K.; Vohra, S. *Rhodiola rosea* for mental and physical fatigue in nursing students: A randomized controlled trial. *PLoS ONE* **2014**, *9*, e108416. [CrossRef] [PubMed]
29. Shanely, R.A.; Nieman, D.C.; Zwetsloot, K.A.; Knab, A.M.; Imagita, H.; Luo, B.; Davis, B.; Zubeldia, J.M. Evaluation of *Rhodiola rosea* supplementation on skeletal muscle damage and inflammation in runners following a competitive marathon. *Brain Behav. Immun.* **2014**, *39*, 204–210. [CrossRef] [PubMed]
30. Dimpfel, W. Neurophysiological effects of *Rhodiola rosea* extract containing capsules (A double-blind, randomized, placebo-controlled study). *Int. J. Nutr. Food Sci.* **2014**, *3*, 157–165. [CrossRef]
31. Cropley, M.; Banks, A.P.; Boyle, J. The Effects of *Rhodiola rosea* L. Extract on Anxiety, Stress, Cognition and Other Mood Symptoms. *Phytother. Res.* **2015**, *29*, 1934–1939. [CrossRef]
32. Mao, J.J.; Xie, S.X.; Zee, J.; Soeller, I.; Li, Q.S.; Rockwell, K.; Amsterdam, J.D. *Rhodiola rosea* versus sertraline for major depressive disorder: A randomized placebo-controlled trial. *Phytomedicine* **2015**, *22*, 394–399. [CrossRef]
33. Ahmed, M.; Henson, D.A.; Sanderson, M.C.; Nieman, D.C.; Zubeldia, J.M.; Shanely, R.A. *Rhodiola rosea* Exerts Antiviral Activity in Athletes Following a Competitive Marathon Race. *Front. Nutr.* **2015**, *2*, 24. [CrossRef] [PubMed]
34. Thu, O.K.; Spigset, O.; Nilsen, O.G.; Hellum, B. Effect of commercial *Rhodiola rosea* on CYP enzyme activity in humans. *Eur. J. Clin. Pharmacol.* **2016**, *72*, 295–300. [CrossRef]
35. Kasper, S.; Dienel, A. Multicenter, open-label, exploratory clinical trial with *Rhodiola rosea* extract in patients suffering from burnout symptoms. *Neuropsychiatr. Dis. Treat.* **2017**, *13*, 889–898. [CrossRef] [PubMed]
36. Lekomtseva, Y.; Zhukova, I.; Wacker, A. *Rhodiola rosea* in Subjects with Prolonged or Chronic Fatigue Symptoms: Results of an Open-Label Clinical Trial. *Compl. Med. Res.* **2017**, *24*, 46–52. [CrossRef]
37. Concerto, C.; Infortuna, C.; Muscatello, M.R.A.; Bruno, A.; Zoccali, R.; Chusid, E.; Aguglia, E.; Battaglia, F. Exploring the effect of adaptogenic *Rhodiola rosea* extract on neuroplasticity in humans. *Complement. Ther. Med.* **2018**, *41*, 141–146. [CrossRef] [PubMed]
38. Jówko, E.; Sadowski, J.; Długołęcka, B.; Gierczuk, D.; Opaszowski, B.; Cieśliński, I. Effects of *Rhodiola rosea* supplementation on mental performance, physical capacity, and oxidative stress biomarkers in healthy men. *J. Sport Health Sci.* **2018**, *7*, 473–480. [CrossRef]
39. Timpmann, S.; Hackney, A.C.; Tamm, M.; Kreegipuu, K.; Unt, E.; Ööpik, V. Influence of *Rhodiola rosea* on the heat acclimation process in young healthy men. *Appl. Physiol. Nutr. Metab.* **2018**, *43*, 63–70. [CrossRef]
40. Ballmann, C.G.; Maze, S.B.; Wells, A.C.; Marshall, M.M.; Rogers, R.R. Effects of short-term *Rhodiola rosea* (Golden Root Extract) supplementation on anaerobic exercise performance. *J. Sports Sci.* **2019**, *37*, 998–1003. [CrossRef]
41. Gao, L.; Wu, C.; Liao, Y.; Wang, J. Antidepressants effects of Rhodiola capsule combined with sertraline for major depressive disorder: A randomized double-blind placebo-controlled clinical trial. *J. Affect. Disord.* **2020**, *265*, 99–103. [CrossRef]
42. Koop, T.; Dienel, A.; Heldmann, M.; Münte, T.F. Effects of a *Rhodiola rosea* extract on mental resource allocation and attention: An event-related potential dual task study. *Phytother. Res.* **2020**, *34*, 3287–3297. [CrossRef]
43. Williams, T.D.; Langley, H.N.; Roberson, C.C.; Rogers, R.R.; Ballmann, C.G. Effects of Short-Term Golden Root Extract (*Rhodiola rosea*) Supplementation on Resistance Exercise Performance. *Int. J. Environ. Res. Public Health* **2021**, *18*, 6953. [CrossRef] [PubMed]
44. Li, X.; Chen, W.; Xu, Y.; Liang, Z.; Hu, H.; Wang, S.; Wang, Y. Quality Evaluation of Randomized Controlled Trials of Rhodiola Species: A Systematic Review. *Evid. Based Complement. Altern. Med.* **2021**, 9989546. [CrossRef] [PubMed]
45. Lu, Y.; Deng, B.; Xu, L.; Liu, H.; Song, Y.; Lin, F. Effects of *Rhodiola rosea* Supplementation on Exercise and Sport: A Systematic Review. *Front. Nutr.* **2022**, *9*, 856287. [CrossRef] [PubMed]
46. Amsterdam, J.D.; Panossian, A.G. *Rhodiola rosea* L. as a putative botanical antidepressant. *Phytomedicine* **2016**, *23*, 770–783. [CrossRef] [PubMed]
47. Ivanova Stojcheva, E.; Quintela, J.C. The Effectiveness of *Rhodiola rosea* L. Preparations in Alleviating Various Aspects of Life-Stress Symptoms and Stress-Induced Conditions-Encouraging Clinical Evidence. *Molecules* **2022**, *27*, 3902. [CrossRef] [PubMed]
48. Sanz-Barrio, P.M.; Noreen, E.E.; Gilsanz-Estebaranz, L.; Lorenzo-Calvo, J.; Martínez-Ferrán, M.; Pareja-Galeano, H. *Rhodiola rosea* supplementation on sports performance: A systematic review of randomized controlled trials. *Phytother. Res.* **2023**, *37*, 4414–4428. [CrossRef] [PubMed]

49. Dye, L.; Billington, J.; Lawton, C.; Boyle, N. A combination of magnesium, B vitamins, green tea and rhodiola attenuates the negative effects of acute psychosocial stress on subjective state in adults. *Curr. Dev. Nutr.* **2020**, *4*, nzaa067_023. [CrossRef]
50. Boyle, N.B.; Billington, J.; Lawton, C.; Quadt, F.; Dye, L. A combination of green tea, rhodiola, magnesium and B vitamins modulates brain activity and protects against the effects of induced social stress in healthy volunteers. *Nutr. Neurosci.* **2022**, *25*, 1845–1859. [CrossRef] [PubMed]
51. Boyle, N.B.; Dye, L.; Lawton, C.L.; Billington, J. A Combination of Green Tea, Rhodiola, Magnesium, and B Vitamins Increases Electroencephalogram Theta Activity During Attentional Task Performance Under Conditions of Induced Social Stress. *Front. Nutr.* **2022**, *9*, 935001. [CrossRef]
52. Noah, L.; Morel, V.; Bertin, C.; Pouteau, E.; Macian, N.; Dualé, C.; Pereira, B.; Pickering, G. Effect of a Combination of Magnesium, B Vitamins, Rhodiola, and Green Tea (L-Theanine) on Chronically Stressed Healthy Individuals-A Randomized, Placebo-Controlled Study. *Nutrients* **2022**, *14*, 1863. [CrossRef]
53. Pickering, G.; Noah, L.; Pereira, B.; Goubayon, J.; Leray, V.; Touron, A.; Macian, N.; Bernard, L.; Dualé, C.; Roux, V.; et al. Assessing brain function in stressed healthy individuals following the use of a combination of green tea, Rhodiola, magnesium, and B vitamins: An fMRI study. *Front. Nutr.* **2023**, *10*, 1211321. [CrossRef] [PubMed]
54. Narimanian, M.; Badalyan, M.; Panosyan, V.; Gabrielyan, E.; Panossian, A.; Wikman, G.; Wagner, H. Impact of Chisan (ADAPT-232) on the quality-of-life and its efficacy as an adjuvant in the treatment of acute non-specific pneumonia. *Phytomedicine* **2005**, *12*, 723–729. [CrossRef] [PubMed]
55. Aslanyan, G.; Amroyan, E.; Gabrielyan, E.; Nylander, M.; Wikman, G.; Panossian, A. Double-blind, placebo-controlled, randomised study of single dose effects of ADAPT-232 on cognitive functions. *Phytomedicine* **2010**, *17*, 494–499. [CrossRef] [PubMed]
56. Hovhannisyan, A.; Nylander, M.; Wikman, G.; Panossian, A.G. Efficacy of Adaptogenic Supplements on Adapting to Stress: A Randomized, Controlled Trial. *Athl. Enhanc.* **2015**, *4*, 4. [CrossRef]
57. Karosanidze, I.; Kiladze, U.; Kirtadze, N.; Giorgadze, M.; Amashukeli, N.; Parulava, N.; Iluridze, N.; Kikabidze, N.; Gudavadze, N.; Gelashvili, L.; et al. Efficacy of Adaptogens in Patients with Long COVID-19: A Randomized, Quadruple-Blind, Placebo-Controlled Trial. *Pharmaceuticals* **2022**, *15*, 345. [CrossRef] [PubMed]
58. Liu, C.; Zhao, H.; Yan, Y.; Yang, W.; Chen, S.; Song, G.; Li, X.; Gu, Y.; Yun, H.; Li, Y. Synergistic Effect of *Rhodiola rosea* and Caffeine supplementation on the Improvement of Muscle Strength and Muscular Endurance: A Pilot Study for Rats, Resistance Exercise-Untrained and -Trained Volunteers. *Nutrients* **2023**, *15*, 582. [CrossRef]
59. Yun, H.; Lu, B.; Su, W.; Wang, J.; Zheng, J.; Wang, J.; Wang, Z.; Li, Y.; Sun, Y.; Liu, C. Combined effects of *Rhodiola rosea* and caffeine supplementation on aerobic endurance and muscle explosiveness: A synergistic approach. *Front. Nutr.* **2024**, *11*, 1335950. [CrossRef]
60. Earnest, C.P.; Morss, G.M.; Wyatt, F.; Jordan, A.N.; Colson, S.; Church, T.S.; Fitzgerald, Y.; Autrey, L.; Jurca, R.; Lucia, A. Effects of a commercial herbal-based formula on exercise performance in cyclists. *Med. Sci. Sports Exerc.* **2004**, *36*, 504–509. [CrossRef] [PubMed]
61. Colson, S.N.; Wyatt, F.B.; Johnston, D.L.; Autrey, L.D.; FitzGerald, Y.L.; Earnest, C.P. Cordyceps sinensis- and *Rhodiola rosea*-based supplementation in male cyclists and its effect on muscle tissue oxygen saturation. *J. Strength Cond. Res.* **2005**, *19*, 358–363. [CrossRef]
62. Liao, Y.H.; Chao, Y.C.; Sim, B.Y.; Lin, H.M.; Chen, M.T.; Chen, C.Y. Rhodiola/Cordyceps-Based Herbal Supplement Promotes Endurance Training-Improved Body Composition but Not Oxidative Stress and Metabolic Biomarkers: A Preliminary Randomized Controlled Study. *Nutrients* **2019**, *11*, 2357. [CrossRef]
63. Kreipke, V.C.; Moffatt, R.J.; Tanner Ma, C.J.; Ormsbee, M.J. Effects of Concurrent Training and a Multi-Ingredient Performance Supplement Containing *Rhodiola rosea* and Cordyceps sinensis on Body Composition, Performance, and Health in Active Men. *J. Diet. Suppl.* **2021**, *18*, 597–613. [CrossRef]
64. Al-Kuraishy, H.M. Central additive effect of Ginkgo biloba and *Rhodiola rosea* on psychomotor vigilance task and short-term working memory accuracy. *J. Intercult. Ethnopharmacol.* **2015**, *5*, 7–13. [CrossRef]
65. Pkhaladze, L.; Davidova, N.; Khomasuridze, A.; Shengelia, R.; Panossian, A.G. *Actaea racemosa* L. Is More Effective in Combination with *Rhodiola rosea* L. for Relief of Menopausal Symptoms: A Randomized, Double-Blind, Placebo-Controlled Study. *Pharmaceuticals* **2020**, *13*, 102. [CrossRef] [PubMed]
66. Bangratz, M.; Ait Abdellah, S.; Berlin, A.; Blondeau, C.; Guilbot, A.; Dubourdeaux, M.; Lemoine, P. A preliminary assessment of a combination of rhodiola and saffron in the management of mild-moderate depression. *Neuropsychiatr. Dis. Treat.* **2018**, *14*, 1821–1829. [CrossRef]
67. Dieamant, G.d.e.C.; Velazquez Pereda, M.d.e.l.C.; Eberlin, S.; Nogueira, C.; Werka, R.M.; Queiroz, M.L. Neuroimmunomodulatory compound for sensitive skin care: In vitro and clinical assessment. *J. Cosmet. Dermatol.* **2008**, *7*, 112–119. [CrossRef]
68. Dzampaeva, Z.V.; Datieva, F.S.; Esenova, Z.S.; Takoeva, E.A. Efficacy of Complex Phytoadaptogens as an Adjunct to Non-surgical Treatment of Chronic Periodontitis: A Randomized Clinical Trial. *J. Int. Soc. Prev. Community Dent.* **2021**, *11*, 348–356. [CrossRef] [PubMed]
69. Tan, J.; Engelhardt, U.H.; Lin, Z.; Kaiser, N.; Maiwald, B. Flavonoids, phenolic acids, alkaloids and theanine in different types of authentic Chinese white tea samples. *J. Food Compos. Anal.* **2017**, *57*, 8–15. [CrossRef]
70. Phuah, Y.Q.; Chang, S.K.; Ng, W.J.; Lam, M.Q.; Ee, K.Y. A review on matcha: Chemical composition, health benefits, with insights on its quality control by applying chemometrics and multi-omics. *Food Res. Int.* **2023**, *170*, 113007. [CrossRef]

71. Nobre, A.C.; Rao, A.; Owen, G.N. L-theanine, a natural constituent in tea, and its effect on mental state. *Asia Pac. J. Clin. Nutr.* **2008**, *17* (Suppl. S1), 167–168.
72. Grosso, C.; Santos, M.; Barroso, M.F. From Plants to Psycho-Neurology: Unravelling the Therapeutic Benefits of Bioactive Compounds in Brain Disorders. *Antioxidants* **2023**, *12*, 1603. [CrossRef]
73. Son, T.G.; Camandola, S.; Mattson, M.P. Hormetic dietary phytochemicals. *Neuromol. Med.* **2008**, *10*, 236–246. [CrossRef]
74. Panossian, A.; Wagner, H. Stimulating effect of adaptogens: An overview with particular reference to their efficacy following single dose administration. *Phytother. Res.* **2005**, *19*, 819–838. [CrossRef]
75. Sellami, M.; Slimeni, O.; Pokrywka, A.; Kuvačić, G.; Hayes, L.D.; Milic, M.; Padulo, J. Herbal medicine for sports: A review. *J. Int. Soc. Sports Nutr.* **2018**, *15*, 14. [CrossRef]
76. Sajadi-Ernazarova, K.R.; Anderson, J.; Dhakal, A.; Hamilton, R.J. *Caffeine Withdrawal*; StatPearls Publishing: Treasure Island, FL, USA, 2024. Available online: https://www.ncbi.nlm.nih.gov/books/NBK430790/ (accessed on 3 April 2024).
77. Łysakowska, P.; Sobota, A.; Wirkijowska, A. Medicinal Mushrooms: Their Bioactive Components, Nutritional Value and Application in Functional Food Production—A Review. *Molecules* **2023**, *28*, 5393. [CrossRef] [PubMed]
78. Xie, L.; Zhu, Q.; Lu, J. Can We Use *Ginkgo biloba* Extract to Treat Alzheimer's Disease? Lessons from Preclinical and Clinical Studies. *Cells* **2022**, *11*, 479. [CrossRef] [PubMed]
79. Liao, Z.; Cheng, L.; Li, X.; Zhang, M.; Wang, S.; Huo, R. Meta-analysis of Ginkgo biloba Preparation for the Treatment of Alzheimer's Disease. *Clin. Neuropharmacol.* **2020**, *43*, 93–99. [CrossRef] [PubMed]
80. Barbalho, S.M.; Direito, R.; Laurindo, L.F.; Marton, L.T.; Guiguer, E.L.; Goulart, R.d.A.; Tofano, R.J.; Carvalho, A.C.A.; Flato, U.A.P.; Capelluppi Tofano, V.A.; et al. *Ginkgo biloba* in the Aging Process: A Narrative Review. *Antioxidants* **2022**, *11*, 525. [CrossRef]
81. Cave, A.E.; Chang, D.H.; Münch, G.W.; Steiner-Lim, G.Z. A systematic review of the safety and efficacy on cognitive function of herbal and nutritional medicines in older adults with and without subjective cognitive impairment. *Syst. Rev.* **2023**, *12*, 143. [CrossRef]
82. Hort, J.; Duning, T.; Hoerr, R. Ginkgo biloba Extract EGb 761 in the Treatment of Patients with Mild Neurocognitive Impairment: A Systematic Review. *Neuropsychiatr. Dis. Treat.* **2023**, *19*, 647–660. [CrossRef]
83. Hausenblas, H.A.; Saha, D.; Dubyak, P.J.; Anton, S.D. Saffron (*Crocus sativus* L.) and major depressive disorder: A meta-analysis of randomized clinical trials. *J. Integr. Med.* **2013**, *11*, 377–383. [CrossRef]
84. Sureshkumar, K.; Durairaj, M.; Srinivasan, K.; Goh, K.W.; Undela, K.; Mahalingam, V.T.; Ardianto, C.; Ming, L.C.; Ganesan, R.M. Effect of L-Carnosine in Patients with Age-Related Diseases: A Systematic Review and Meta-Analysis. *Front. Biosci.* **2023**, *28*, 18. [CrossRef]
85. Panossian, A. Challenges in phytotherapy research. *Front. Pharmacol.* **2023**, *14*, 1199516. [CrossRef]
86. Deka, H.; Barman, T.; Dutta, J.; Devi, A.; Tamuly, P.; Paul, R.K.; Karak, T. Catechin and caffeine content of tea (*Camellia sinensis* L.) leaf significantly differ with seasonal variation: A study on popular cultivars in North East India. *J. Food Comp. Anal.* **2021**, *96*, 103684. [CrossRef]
87. Skirven, T.M.; Osterman, A.L.; Fedorczyk, J.; Amadio, P.C. Preface. In *Rehabilitation of the Hand and Upper Extremity*; Skirven, T.M., Osterman, A.L., Fedorczyk, J., Amadio, P.C., Eds.; Elsevier: Philadelphia, PA, USA, 2011; Volume 1, pp. 152–162. Available online: http://www.theopedia.com/synergism (accessed on 3 April 2024).
88. Berenbaum, M.C. Synergy, additivism and antagonism in immunosuppression. A critical review. *Clin. Exp. Immunol.* **1977**, *28*, 1–18. [PubMed]
89. Geary, N. Understanding synergy. *Am. J. Physiol. Endocrinol. Metab.* **2013**, *304*, E237–E253. [CrossRef] [PubMed]
90. Lederer, S.; Dijkstra, T.M.H.; Heskes, T. Additive Dose Response Models: Defining Synergy. *Front. Pharmacol.* **2019**, *10*, 1384. [CrossRef]
91. Roell, K.R.; Reif, D.M.; Motsinger-Reif, A.A. An Introduction to Terminology and Methodology of Chemical Synergy-Perspectives from Across Disciplines. *Front. Pharmacol.* **2017**, *8*, 158. [CrossRef]

Disclaimer/Publisher's Note: The statements, opinions and data contained in all publications are solely those of the individual author(s) and contributor(s) and not of MDPI and/or the editor(s). MDPI and/or the editor(s) disclaim responsibility for any injury to people or property resulting from any ideas, methods, instructions or products referred to in the content.

pharmaceuticals

MDPI

Article

Multi-Omics Analysis in Mouse Primary Cortical Neurons Reveals Complex Positive and Negative Biological Interactions Between Constituent Compounds of *Centella asiatica*

Steven R. Chamberlin [1,2,3,*], Jonathan A. Zweig [2,3], Cody J. Neff [2,3], Luke Marney [3,4], Jaewoo Choi [3,4], Liping Yang [3,4], Claudia S. Maier [3,4], Amala Soumyanath [2,3], Shannon McWeeney [1,5] and Nora E. Gray [2,3]

1 Department of Medical Informatics and Clinical Epidemiology, Oregon Health & Science University, Portland, OR 97239, USA; mcweeney@ohsu.edu

2 Department of Neurology, Oregon Health & Science University, Portland, OR 97239, USA; soumyana@ohsu.edu (A.S.); grayn@ohsu.edu (N.E.G.)

3 BENFRA Botanical Dietary Supplements Research Center, Oregon Health & Science University, Portland, OR 97239, USA; marneyl@oregonstate.edu (L.M.); claudia.maier@oregonstate.edu (C.S.M.)

4 Department of Chemistry, Oregon State University, Corvallis, OR 97331, USA

5 Knight Cancer Institute, Oregon Health & Science University, Portland, OR 97239, USA

* Correspondence: chambest@ohsu.edu

Abstract: Background: A water extract of the Ayurvedic plant *Centella asiatica* (L.) Urban, family Apiaceae (CAW), improves cognitive function in mouse models of aging and Alzheimer's disease and affects dendritic arborization, mitochondrial activity, and oxidative stress in mouse primary neurons. Triterpenes (TT) and caffeoylquinic acids (CQA) are constituents associated with these bioactivities of CAW, although little is known about how interactions between these compounds contribute to the plant's therapeutic benefit. **Methods:** Mouse primary cortical neurons were treated with CAW or equivalent concentrations of four TT combined, eight CQA combined, or these twelve compounds combined (TTCQA). Treatment effects on the cell transcriptome (18,491 genes) and metabolome (192 metabolites) relative to vehicle control were evaluated using RNAseq and metabolomic analyses, respectively. **Results:** Extensive differentially expressed genes (DEGs) were seen with all treatments, as well as evidence of interactions between compounds. Notably, many DEGs seen with TT treatment were not observed in the TTCQA condition, possibly suggesting CQA reduced the effects of TT. Moreover, additional gene activity seen with CAW as compared to TTCQA indicates the presence of additional compounds in CAW that further modulate TTCQA interactions. Weighted Gene Correlation Network Analysis (WGCNA) identified 4 gene co-expression modules altered by treatments that were associated with extracellular matrix organization, fatty acid metabolism, cellular response to stress and stimuli, and immune function. Compound interaction patterns were seen at the eigengene level in these modules. Interestingly, in metabolomics analysis, the TTCQA treatment saw the highest number of changes in individual metabolites (20), followed by CQA (15), then TT (8), and finally CAW (3). WGCNA analysis found two metabolomics modules with significant eigenmetabolite differences for TT and CQA and possible compound interactions at this level. **Conclusions:** Four gene expression modules and two metabolite modules were altered by the four treatment types applied. This methodology demonstrated the existence of both negative and positive interactions between TT, CQA, and additional compounds found in CAW on the transcriptome and metabolome of mouse primary cortical neurons.

Keywords: *Centella asiatica*; triterpenes; caffeoylquinic acids; mouse primary cortical neurons; transcriptome; metabolome; co-expression; omics integration

1. Introduction

Traditional and complementary medicines are often sought to improve or maintain health, and it is estimated that 50% of the population in industrialized countries use these approaches [1]. In the elderly, these interventions are primarily used to promote resilience [2] since many of these are botanical extracts with purported anti-aging effects [3–6]. However, claims are controversial, and few clinical trials have been conducted with these treatments [7].

The study of botanical extracts for healthcare is challenging, in part due to the complex mix of compounds found in these plants. Unlike single-chemical drugs or isolated natural products, botanical extracts may contain hundreds of compounds [8]. Many traditional practitioners believe that these botanical mixtures are more clinically effective than isolated compounds due to interactions between the constituents. Interactions can occur between individual compounds, groups or fractions of compounds, constituents from different parts of the plant, or even between compounds derived from different botanical species, as seen in complex traditional herbal formulas [9,10]. These mixtures can interact with biological systems in a way that is additive, synergistic, or antagonistic [9,11–15]. Fraction-based methods have been developed to evaluate the synergy between compounds in whole plant extracts using a combination of mass spectrometry, isolation of natural products, and synergy assays [16]. In a whole organism, the nature of these interactions can vary depending on the compounds and be considered pharmacokinetic or pharmacodynamic. Pharmacokinetic interactions affect the bioavailability of a constituent in an organism through changes in its metabolism or cellular transport induced by another constituent. Pharmacodynamic interactions occur when two compounds affect each other's binding to targets or have similar or opposing biological activity mediated through interactions at different points in molecular pathways [11,13].

While synergy studies have traditionally utilized targeted assays, leveraging untargeted omic analysis methods [14,15] in pre-clinical models offers an effective way to understand multiple bioactivities and mechanisms of action that are modulated by this mix of compounds, including understanding the interactions between the compounds. For the current project, we propose to evaluate two untargeted molecular domains, transcriptomics and metabolomics, in an in vitro model system to broadly study compound interactions in the biological activity of compounds within a water extract (CAW) of the botanical *Centella asiatica* (L.) Urb., family Apiaceae. This plant is native to both Asia and Australia and has a history of use in several systems of traditional medicine, including traditional Chinese medicine and Ayurvedic medicine [17], for the treatment of various conditions, including cognitive impairment [18]. Current in vitro, in vivo, and clinical research also supports the neurological benefits of *Centella asiatica* [19–30]. CAW can evoke particularly potent neuroprotective effects both in vitro and in vivo that may be mediated by effects on dendritic arborization [19,30], increased synaptic density [19,24,30], mitochondrial biogenesis [19,23,30], and activation of endogenous antioxidant mechanisms [19,23,30].

Two classes of compounds have been associated with the health-promoting bioactivity of CAW: pentacyclic triterpenes (TT) and mono- and di-caffeoylquinic acids (CQA) [31], although the contribution of each compound group to the overall effects of CAW and the mechanisms by which the compounds can elicit those effects have not been fully elucidated. The goal of this project is to begin to determine the contributions to bioactivity of these two compound classes separately or combined, as well as in the complete CAW extract.

In this study, mouse primary cortical neurons were treated with either CAW, groups of TT or CQA compounds separately, or a combination of the two (TTCQA). The in vitro CAW concentration chosen (50 μg/mL) was previously found to increase dendritic arborization, synaptic density, mitochondrial biogenesis, and activation of antioxidant response in mouse primary hippocampal neurons [30,32,33]. In vivo, CAW administered in drinking water has also been shown to affect these same pathways in both the hippocampus and cortex of treated mice [24,29,34]. All compounds were applied at concentrations equivalent to their presence in CAW. Transcriptomic and metabolomic data were collected from these treated cells. The primary outcomes in this study, derived from these two untargeted molecular domains, are both the individual gene expression and metabolite abundance levels, as well as gene and metabolite signatures of co-expression and co-abundance. This design allowed us to assess the effects of, and nature of interactions between, the TT and CQA compounds, as well as interactions between these compounds and other unknown compounds in CAW, using these outcomes.

2. Results

2.1. Analysis of Culture Media Post-Incubation

Neuron culture media was collected and frozen after the 48 h treatment period at the time of cell harvest. The cultured media was then extracted with OstroPlate® solid-phase extraction, and concentrations of selected phytochemicals were measured using liquid chromatography coupled to multiple reaction monitoring mass spectrometry (LC-MRM-MS; Supplementary Table S1). In many of the incubations, compound concentrations were somewhat higher than the pre-incubation time zero starting value, possibly due to evaporation of solvent during incubation (Supplementary Table S1). Interestingly, the post-incubation concentration of CQA metabolites was ten-fold lower in CAW incubations compared to incubations containing CQA alone or TTCQA. This suggests a rapid degradation of CQAs in the presence of other CAW constituents. CQA metabolites are known to undergo chemical changes in non-acidic solutions [35,36]. Triterpene-glycoside content (MS and AS) was similar in all cultures containing these compounds. Triterpene aglycones (MA and AA) were present in higher concentrations post-incubation in CAW than the TT or TTCQA cultures, potentially arising from precursors present in CAW, as they were considerably higher than the time 0 values (Supplementary Table S1). Differences in response to CAW compared to the compounds may in part be due to these concentration differences.

2.2. CAW and Its Constituent Compounds Induce Extensive Gene Expression Changes

Of the 40, 39 culture samples were of sufficient quality for RNA-seq processing, displaying a high level of inter-neuronal networking and neuro-spheres (n = 8 per treatment except for TTCQA, which had n = 7). RNA extracted from the samples was of high quality with an average RNA Integrity Number (RIN) of 9.9. For the metabolomics processing, we had 50 cell culture samples of sufficient quality (n = 10 per treatment).

To investigate both differentially expressed genes (DEG) and differentially abundant metabolites (DAM), we compared each of the four treatment groups to the vehicle control (CAW vs. Ctrl, TT vs. Ctrl, CQA vs. Ctrl, and TTCQA vs. Ctrl). RNA-seq genes with expression less than 10 counts in any of the samples were removed, leaving 18,491 genes for analysis. Our in-house library provided 192 metabolites for analysis. A false discovery rate cutoff of 0.05 was used for both DEG and DAM analyses. While there was a high degree of treatment-related gene expression change in the neurons (CAW 946 upregulated, 1721 downregulated; CQA 33 upregulated, 165 downregulated; TT 514 upregulated, 1.246 downregulated; TTCQA 303 upregulated, 678 downregulated), based on fold change (Supplementary Table S2), the same was not true for changes in metabolites. In contrast, the metabolomics analysis found only 3 downregulated metabolites for CAW, 8 upregulated and 7 downregulated metabolites for CQA, 3 upregulated and 5 downregulated metabolites for TT, and 12 upregulated and 8 downregulated metabolites for TTCQA (Supplementary Table S3). This difference in the magnitude of DEGs as compared to DAMs was largely because the full transcriptome (18,491) was evaluated, but only a select number of metabolites (192) was assessed. When compared by the percentage of affected genes or metabolites, relative to the total number tested, the difference is not as dramatic (transcriptome: 21.8% of the 18,491 genes had a change in expression; metabolome: 16.1% of the 192 metabolites had a change in abundance).

A principal component analysis (PCA) was used to assess the effect of each treatment relative to the control and also to compare the treatments to each other, using both the transcriptomic and metabolomic data. This work was performed with the PCA tools R package [37]. Metabolites and genes in the lowest decile of variability across samples were removed. There was not a strong separation of samples seen between treatments for both data types, as would be expected since some of these treatments contain the same compounds (Figure 1A). However, for the independent treatment vs. control analyses, there is greater separation of samples as the number of differentially expressed metabolites between the groups being compared increases (Figure 1C). The same is generally true with the transcriptomic data (Figure 1B). This confirmed that the treatments were having some effect.

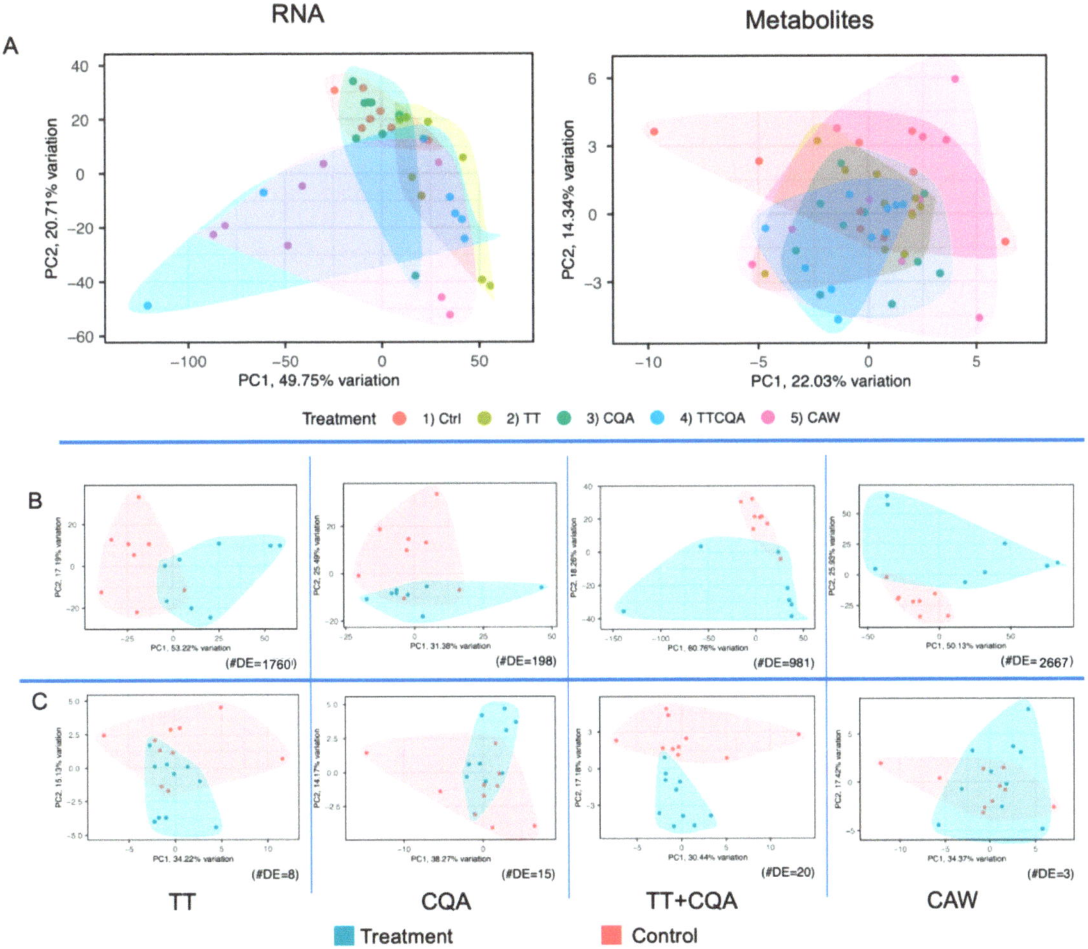

Figure 1. PCA plots for the two untargeted molecular domains of metabolomics and transcriptomics. (**A**) All treatments plotted together, including the vehicle control, for either metabolic or RNA-seq

data. (**B**) RNA-seq data, each treatment compared individually to control. (**C**) Metabolic data, each treatment compared individually to control. For (**B**,**C**), the number of differentially expressed metabolites or genes (#DE) is shown in the lower right corner of each plot. The plots are also ordered left to right by increasing number of CA compounds in each treatment.

2.3. Complex Interactions Were Seen Between CAW Compounds, Both Activating and Inactivating Genes

Transcriptomic analysis revealed many interactions were apparent between constituent compounds in CAW. Only 17% of the DEGs that are seen in the TT and CQA separate treatments are also seen in the combined TTCQA treatment. (Supplementary Figure S1), However, there are 420 new DEGs seen in this combination, of which only 263 are seen in the full CAW treatment, indicating the potential 'silencing' of 157 DEGs by interactions with unknown additional compounds in CAW. Interestingly, 177 of the TT DEGs not seen in the TTCQA treatment return in the full CAW treatment, and there are only 26 DEGs in common to all four treatments (Supplementary Figure S1A). In the metabolomic data, only two metabolites downregulated by a constituent treatment (CQA) were seen in the full CAW treatment (Supplementary Figure S1B).

A more detailed classification of the different types of gene expression interactions (additive, synergistic, antagonistic) from the differential gene expression model between TT and CQA can be seen in Figure 2 (See Section 4.6).

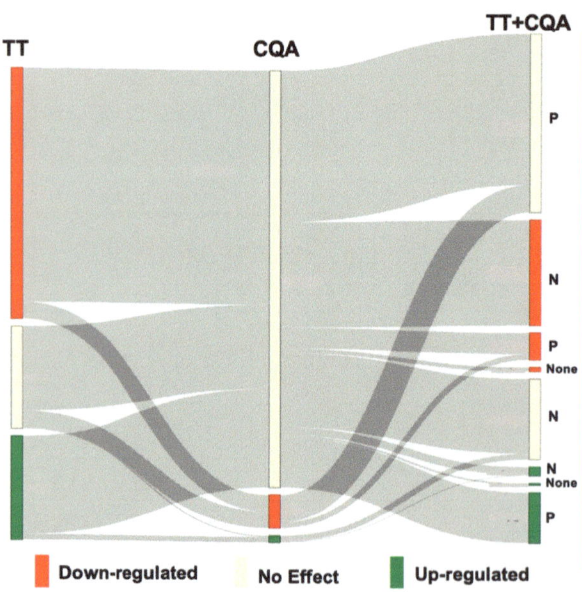

Interaction	TT	CQA	TTCQA	# Genes
Negative	Down	No Effect	Down	286
	No Effect	No Effect	Down	238
	No Effect	Up	No Effect	5
	No Effect	Up	Up	1
	Up	No Effect	No Effect	370
	Up	No Effect	Up	39
	Up	Up	No Effect	24
	Up	Up	Up	3
None	Down	No Effect	Down	21
	Up	No Effect	Up	9
Positive	Down	Down	Down	29
	Down	Down	No Effect	54
	Down	No Effect	Down	104
	Down	No Effect	No Effect	752
	No Effect	Down	No Effect	82
	No Effect	No Effect	Up	182
	Up	No Effect	Up	69

Figure 2. Effect on gene expression of the TT and CQA compounds administered separately and interactions between these groups observed in the TTCQA-treated samples. The figure on the left shows a mapping of the interactions listed in the table on the right. Any gene with significant expression changes, relative to control, seen with any of the TT, CQA, or TTCQA treatments are represented in the figure and table (N = 2268 unique genes). The relative number of genes and the expression status are represented within each of the three columns labeled TT, CQA, or TTCQA (red = significant downregulation, green = significant upregulation, beige = no significant expression changes seen with this treatment). The TT and CQA compound 'Interaction effect' for the combined group is shown next to the TTCQA column ('P' = positive, or synergistic; 'N' = negative, or antagonistic; 'none' = additive, or no interaction). The gray ribbons show the number of genes interacting between

TT and CQA, the interaction status, and the final expression status in the TTCQA treatment. The table on the right shows the number of genes in each category. As an example, a positive interaction has a higher fold change value in the TTCQA treatment than the sum of the two-fold change values from TT and CQA, and a negative interaction has a lower TTCQA fold change than the sum of TT and CQA fold changes. So a positive interaction could still result in a downregulated or unregulated TTCQA gene, and a negative interaction could result in an upregulated or unregulated TTCQA gene.

For this analysis, a little less than half of the genes examined had a negative interaction between TT and CQA (n = 966). A substantial number of these genes had a significant effect from the TT treatment (n = 728), while very few had an effect from CQA (n = 33). There was an increased downregulation in the TTCQA treatment in 524 of these genes, 238 of which did not have a significant change from either TT or CQA (considered a negative interaction for the sake of this analysis). Other interactions reduced or neutralized significant upregulated effects seen in either TT or CQA (n = 443).

We saw a larger number of genes with a positive interaction between the TT and CQA treatments (n = 1272). Most of these were either downregulated by TT, CQA, or both and had their downregulation either diminished (n = 154) or neutralized (n = 888) in the TTCQA treatment. Interestingly, there were 182 genes that were not affected by either the TT or CQA treatment that were significantly upregulated by the TTCQA mixture.

Finally, there were 30 genes that did not have any interaction between the TT and CQA treatment, all of which were either upregulated (n = 21) or downregulated (n = 9) by only TT (Figure 2). A table of the parameter estimates for each treatment can be found in (Supplementary Table S4) for the TT and CQA gene expression interaction classification.

2.4. Multiple Modules Capturing a Diverse Variety of Compound Interactions Point to Distinct Cellular Mechanisms Affected by CAW and Its Constituent Compounds in the Transcriptomic Data

Weighted Correlation Gene Network Analysis (WCGNA) was used to find groups of genes expressed in a correlated manner across the 39 samples, irrespective of DEG status, indicating probable participation in a common biological function. The soft thresholding power was set at 12 to create a scale-free topology in the weighted co-expression network. The 18,491 genes clustered into 14 modules (see Section 4.7). A least squares linear method with post-hoc Tukey multiple comparison tests were used to identify treatment differences, relative to control, within these modules (FDR cutoff of 0.05). Eight of these modules showed significant eigengene differences between at least one of the treatments and the control. To functionally classify these modules, we performed an over-representation analysis using a hypergeometric test to see if the number of genes from a module mapped to a pathway was greater than would be expected by random chance. Genes in each module were used to enrich *Mus musculus* pathways from the Reactome Pathway Database using the ReactomePA package in R version 1.46.0 [38]. Genes in five of the eight modules significantly enriched molecular pathways in Reactome. Sample eigengene distributions for these five modules are shown in Figure 3 by treatment group.

The size of the five modules varied from 256 to 2490 genes. Module 1 (M1) was enriched for extracellular matrix organization and collagen biosynthesis pathways. CAW was the only treatment with a significant effect on this eigengene (70% upregulated DEGs) (Figure 3, Module 1). Fatty acid metabolism pathways were associated with Module 2 (M2), and three treatments show a significant effect (TT, TTCQA, and CAW; 80% downregulated DEGs) (Figure 3, Module 2). Module 3 (M3) enriched pathways included cellular response to stress and stimuli and were significantly impacted by the TT treatment (61% upregulated DEGs) (Figure 3, Module 3). Module 4 (M4) enriched pathways are primarily involved in immune system functions. Here again, only the TT treatment showed a significant impact

(90% downregulated DEGs). (Figure 3, Module 4). Enriched pathways for Module 5 (M5) are mostly associated with Electron Transport and Mitochondrial Biogenesis, only affected by the CQA treatment (91% downregulated DEGs) (Figure 3, Module 5). The pathways mentioned here were the highest level. For a more detailed description of smaller enriched pathways see Supplementary Table S5.

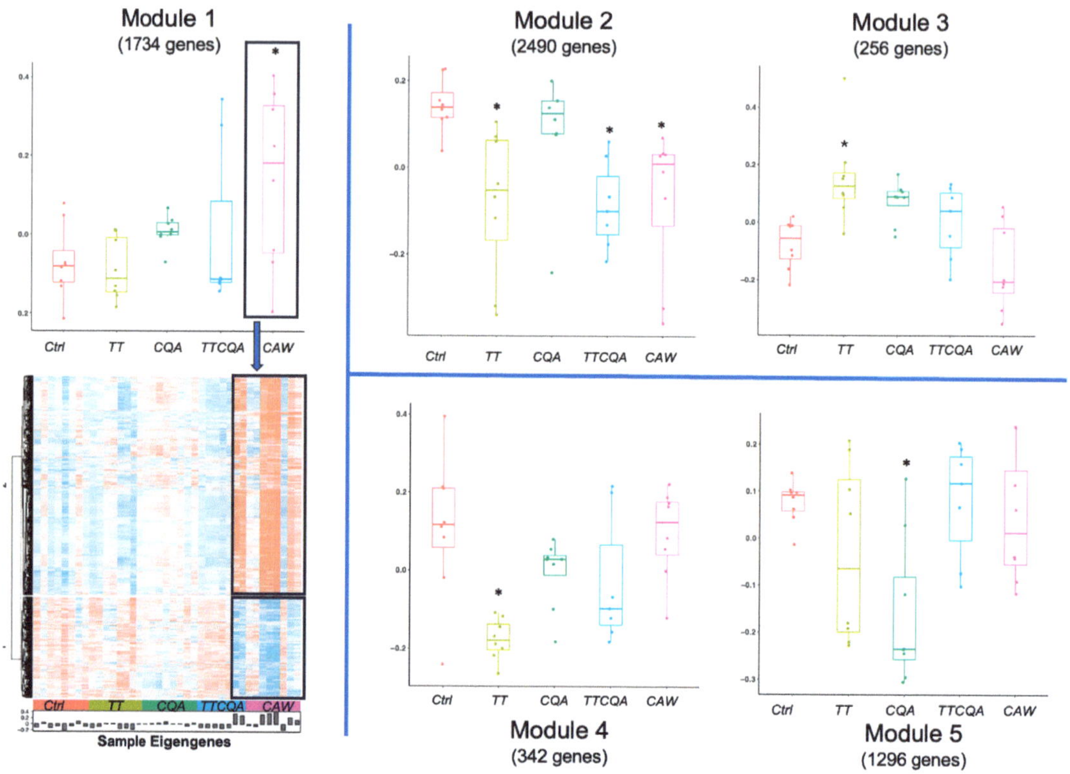

Figure 3. Sample eigengene distributions for the five gene co-expression modules with significant eigengene differences between at least one treatment and control. Boxplots for module eigengene distributions by each treatment group (* p-value $\leq$ 0.05). Module 1 = Extracellular Matrix Organization and Collagen Biosynthesis, Module 2 = Fatty Acid Metabolism, Module 3 = Cellular Response to Stress and Stimuli, Module 4 = Immune System, Module 5 = Electron Transport and Mitochondrial Biogenesis. Module 1 also shows a heat map of individual gene expression levels (blue = low, red = high). The highlighted section represents samples in the significant treatment for this module. See Supplementary Figure S6 for heatmaps for Modules 2–5.

Module 1 of Figure 3 also shows an associated heatmap of individual gene expression levels for this module with groups defined by hierarchical clustering. The two highest-level clusters show a reversal of expression patterns associated with the treatment of significant effect (CAW), with an upregulated cluster for the treatment group showing a mostly downregulated pattern in the control group, and vice versa. For heatmaps associated with Modules 2–5, see Supplementary Figure S6. The number of module genes in the enriched pathways was generally small in our analysis, compared to the total number in the module (M1 = 6%, M2 = 2%, M3 = 11%, M4 = 21%, M5 = 4%). It is not unusual for the fraction of genes related to the primary biological function of the module to be less than 20% of all genes in that module. While not all genes in the module were associated with the enriched function(s), genes that were not in these pathways but had a strong relationship to the module could be (guilt by association) [39]. To evaluate the strength of the relationship of both enriched and non-enriched genes to the entire module, we assessed both module membership (MM) and intramodular connectivity (IC). MM, or kME, is the correlation between individual module gene expression and the eigenvalue of that module. IC, or kWithin, is the gene degree, or number of neighboring connections within the module, in the weighted gene co-expression network that we constructed. The MM of the genes in enriched pathways was evaluated using a *t*-test with Bonferroni correction and found that genes within four out of the five modules (M1, M2, M3, M4) had significantly higher MM scores than genes not found in enriched pathways. Interestingly, the MM of M5 was significantly lower than the non-enriched pathway MM in this module, having a weak negative correlation (r = −0.19 average) with the eigengene.

Next, the treatment DEGs contribution to each module was evaluated. While not all treatment DEGs were found in the five modules, the significant treatments, with respect to the eigengenes, had the highest number of significant DEGs in the module. CAW DEGs comprised 28% of the genes in M1, whereas all other treatment DEGs were less than 1% of this module. For M2, TT DEGs were 30% of the total module genes, TTCQA DEGs were 23%, CAW DEGs were 38%, and CQA DEGs were less than 1%. TT DEGs in M3 were 22% of total genes, and none of the other treatments were more than 10%. In M4, TT DEGs comprised 63% of the module, and none of the other treatment DEGs were more than 7% of the module. Treatment DEG contributions were lower in M5, but the significant eigengene treatment CQA still had the highest percentage of the module at 9%.

The MM measure was significantly higher for the DEGs associated with significant eigengene treatments compared to non-DEGs in the module in two modules, M2 and M4 (M2: TT 0.54 vs. 0.18, TTCQA 0.52 vs. 0.22, and CAW 0.58 vs. 0.12; M4: TT 0.68 vs. 0.31; Bonferroni adjusted *t*-test). No other significant treatment DEGs in the other three modules were found to have significantly higher MM measures.

Each module displays a distinctly different pattern of CAW compound interaction at the eigengene level (Figure 3, all modules). M1 (Extracellular Matrix Organization) shows an increasing effect as the number of compounds increases, reaching significance with the full CAW treatment. M2 (Fatty Acid Metabolism) shows the effect of the TT treatment carried through to the full CAW treatment. CQA does not have an effect in this module and does not seem to diminish the TT effect, nor do the additional unknown compounds in CAW. The significant effect of TT seen in M3 (Cellular Response to Stress) seems to be diminished by CQA and even further diminished by the unknown compounds. The same pattern is seen with the TT treatment in M4 (immune function), but here TT has a downregulating effect, whereas it had an upregulating effect in M3. In M5 (Electron Transport and Mitochondrial Biogenesis), only the CQA treatment had a significant effect, which seems to be negated through interactions with the TT compounds and not further impacted by interactions with unknown compounds in CAW.

Information about the individual genes in these five modules and the pathways they enrich can be found in Supplementary Table S5.

2.5. *TT Treatment Has the Greatest Effect on Functions Associated with Metabolite Co-Abundance, and This Effect Is Diminished by Interactions with Other Compounds*

A modified protocol for WCGNA, with normalized metabolite abundance values, was used to find groups of metabolites that were co-expressed across the 50 samples (10 from each treatment plus control) [40]. Using a soft thresholding power of 12 to construct the co-abundance network, the 192 metabolites clustered into 6 modules (see Section 4.7). Two of the six modules had significant differences between at least one of the treatments and control (FDR cutoff of 0.05). A Tukey multiple comparison test was employed to identify the specific treatment differences in each of these two modules (Figure 4A). Metabolites from both modules significantly enriched 31 Reactome *Mus musculus* pathways for Module 1 and 345 pathways for Module 2 (see Supplementary Table S6 for pathway information), using a hypergeometric test.

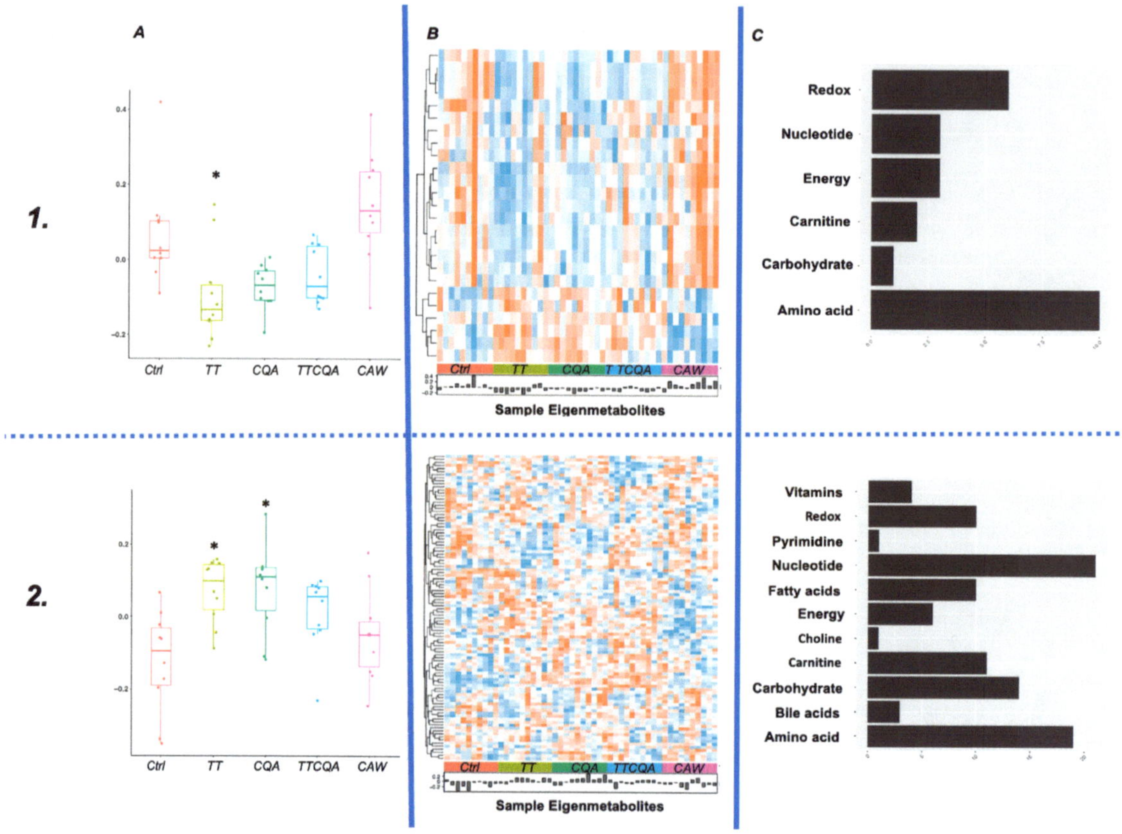

Figure 4. Metabolite co-abundance modules with significant eigenmetabolite differences between at least one treatment and control (**1.** Module 1 n = 25 metabolites, **2.** Module 2 n = 100 metabolites). (**A**) Boxplots for module eigenmetabolite distributions by each treatment group (* adj p-value ≤ 0.05). (**B**) Heatmap of individually scaled metabolite abundance for the module (red = high abundance, blue = low abundance). Samples are ordered by treatment group and labeled on the horizontal axis. Histogram shows eigenmetabolite values for each sample. (**C**) Metabolite category counts for all metabolites in the module.

The majority of metabolites in Module 1 (M1) were amino acids (Figure 4C, M1), which was reflected in several of the 31 enriched pathways ('branched-chain amino acid catabolism', 'phenylalanine and tyrosine metabolism'). Most of the metabolites in Module 2 (M2) were either nucleotides or amino acids (Figure 4C, M2). This was also reflected in several of the 345 enriched pathways for this module ('metabolism of amino acids and derivatives', 'metabolism of nucleotides').

Information about individual metabolites in both modules can be found in Supplementary Table S7.

There were no enriched pathways in common between those associated with metabolite module M1 and the pathways associated with the five transcriptomic modules described in Section 2.4 However, there were common enriched pathways between metabolite module M2 and four of the transcriptomic modules (transcriptomic M1, M3, M4, M5). Some of these pathways include 'signaling by PDGF' for M1, 'cellular responses to stress and stimuli' for M3, 'homeostasis' for M4, and 'respiratory electron transport' for M5. For a complete list of the common pathways, see Supplementary Table S8.

2.6. *Integration of Transcriptome and Metabolome Shows Little Overlap Between TT, CQA, and TTCQA in Significant Gene Activity, but Commonality in Pathways Affected*

In the previous section, it was noted that there were several common biological pathways enriched with both genes and metabolites from the WGCNA modules generated for both omics domains. To focus more specifically on how an individual treatment impacts genes and metabolites that are functionally related to each other within a certain biological function (pathway), we performed an integration analysis as described in Section 4.8. This integration approach considered all DEGs for a specific treatment and then used a computational framework to find metabolites, from both the DAMs and WGCNA modules, that were also impacted by the same treatment and were statistically associated with the DEGs. This framework used network methods and prior knowledge of biological functions (pathways).

The 1760 DEGs and 133 DAMs (8 individual DAMs, 25 from Module 1, and 100 from Module 2) identified following TT treatment were used as seeds to construct the composite network as described in Section 4.8. This network was constructed without the necessity of additional connector genes/proteins to contain the maximum number of seeds possible. However, there were genes in the metabolite-gene network that were not differentially expressed by TT. There were 6 communities detected in the composite network that contained both a gene and metabolite seed, for a total of 37 metabolites and 42 genes in close functional relationship to each other (see Supplementary Table S9). Reactome pathway enrichment, conducted at the individual community level using both seed and non-seed genes, identified a total of 29 pathways containing at least one DEG for the TT treatment overall (see Supplementary Table S10). An example of the network derived from one TT community from the composite network is in Figure 5. Individual network figures for each community containing both seed metabolites and genes are in Supplementary Figure S2.

The 198 DEGs and 115 DAMs (15 individual DAMs and 100 from Module 2) affected by CQA treatment were used as seeds. In this case, connector genes/proteins were used for maximum seed inclusion in the composite network, which yielded five communities, meeting our criteria of containing both a seed metabolite and a gene. These communities contained a total of 23 seed metabolites and 26 seed genes (see Supplementary Table S11), and these DEGs contributed to the enrichment of 74 pathways (see Supplementary Table S10). Module network figures are in Supplementary Figure S4.

The 981 DEGs and 20 individual DAMs (no Module metabolites) altered with TTCQA treatment were used as seeds. Again, connector genes/proteins were used to construct the composite network, yielding two communities of interest, containing a total of 2 seed metabolites and 24 seed genes (see Supplementary Table S12), contributing to the enrichment of 90 pathways (see Supplementary Table S10). Module network figures are in Supplementary Figure S5.

No composite network communities built using CAW seed genes and metabolites met the necessary criteria.

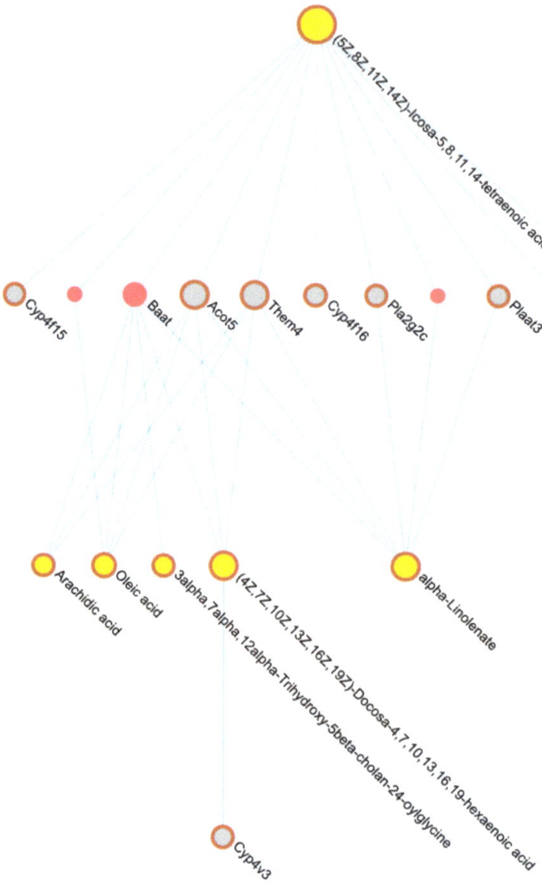

Figure 5. TT Integration Community Network Example. Network is derived from seed genes and metabolites in one community of the TT integrated composite network. Seed genes and metabolites are outlined in red. Genes are indicated in gray, and metabolites in yellow. Connector genes (not seed genes) are the red circles that are not outlined. The size of the circles is related to the number of network connections, or degree. This network enriched mostly fatty acid metabolism pathways.

Next, pathway and gene overlap were evaluated between treatments using the composite network module level data, as described above. Comparisons were performed between TT and CQA, TTCQA and TT, and TTCQA and CQA. There were very few genes in common between any of these treatments (GPT for TT and CQA, CYP4F15 and ACOT5 for TTCQA and TT, and none for TTCQA and CQA). However, while there was only one pathway in common between TT and CQA, there were 6 pathways in common between TTCQA and TT and 34 between TTCQA and CQA.

2.7. Functional Relationships of DEGs for All Treatment Groups Are Not by Random Chance

A network approach was used to determine the overall functional similarity of the genes differentially expressed by each treatment. First, a protein–protein interaction (PPI) network was constructed using the STRING Database [41]. Protein interactions with a physical interaction score ≥ 400 (scale 0–1000), which is a moderate to high confidence in the interaction, were kept, and then the DEGs from each treatment were mapped onto this network, and the average network shortest path between the DEGs within each treatment separately was calculated. Permutation testing (1000 permutations) in the PPI network was then conducted by randomly distributing the same number of genes that were differentially expressed in each treatment. This was performed to see if the average shortest path distance between the DEGs for a specific treatment was closer than would be expected by random distribution. DEGs associated with all four treatments were significantly closer in the STRING PPI than would be expected by chance, as was also the case with the DEGs found in the five co-expression modules.

3. Discussion

Herbal or botanical healthcare products generally consist of complex mixtures in the form of powders or extracts of single or multiple herbs. The effects of the product may be due to the bioactivities of individual compounds present in the mixture but may also result from complex interactions between the different components. Studying interactions that may occur between multiple components of a botanical mixture presents a significant challenge. Several methods exist to understand these interactions, including the use of isobolograms of bioactivity vs. the proportion of the components being tested and biochemometric approaches where the activity of mixtures including or missing specific components is evaluated [16,42–47]. Untargeted transcriptomic analysis has previously been used to study individual and combined effects of herbs and phytochemicals in neuroglial cell lines [14,15].

In the present study, the primary outcomes derived from two untargeted omics domains, as well as their integration, were used to evaluate interactions between components of *Centella asiatica* water extract, CAW, which contains a complex set of components as previously reported [31,48]. This study compared different subsets of CAW constituent compounds, as well as the complete CAW extract, in a nested compound design.

While the earlier studies used targeted analyses of specific effects of the extract, here the effects of the nested subsets of compounds were compared in an untargeted fashion to identify broad patterns of activity and interactions for further research. For example, previous in vitro synergy experiments have generally evaluated a single biological activity in a targeted fashion, such as cell viability or an anti-microbial effect, often in a pairwise fashion between two compounds [44], and previous systems biology in silico methods are reliant on computational predictions [43]. To our knowledge, this is the first in vitro approach using systems biology methods to study compound interaction effects on bioactivity in a complex plant extract with a nested compound design with two untargeted molecular domains.

At the individual gene level, the TT treatment stood out in the gene expression results, with the most activated genes of the three select compound treatments (TT, CQA, and TTCQA) and the most activated genes that were still retained in the full CAW treatment, as compared to those retained from the CQA and TTCQA treatments in CAW. This would be consistent with the body of previous research focusing on TTs as the main bioactive compound in *Centella asiatica*, with activity demonstrated for many health conditions, including neurodegenerative, dermatological, and others [17]. However, in this study, there were many complex interactions reflected in the overlap of activated genes between TTs and other treatments. A large portion of gene activity seen with either the TT or CQA treatments was not seen when the two groups were combined or in the complete CAW extract, with 66% of the TT and CQA gene effects disappearing with TTCQA and 56% of these effects disappearing with CAW. Interestingly, 10% of gene expression lost on combining compounds in TTCQA is restored with the full CAW treatment, which suggests that the additional compounds within CAW are modulating these effects. Based on the post-incubation cultured media analysis, this return of gene activity in CAW could in part be modulated by the degradation of CQA compounds, which appear to inhibit the TT activity in the TTCQA treatment. Since there was no apparent degradation of either TT or CQA compounds seen in the TTCQA treatment (the post-incubation concentrations remained relatively unchanged for CQA in TTCQA), it could be assumed that the TT inactivation was through some other type of compound inhibition. There was also a large amount of new gene activity unique to TTCQA and CAW that was not observed in TT or CQA groups alone, suggesting the presence of interactions between TT and CQA (in the case of TTCQA) and/or the involvement of additional compounds in the case of CAW for interactions with these genes. Further studies are needed to identify whether DEGs unique to CAW could be the result of interactions between TT or CQA and other compounds within the extract.

Continuing with individual genes, the combination TTCQA treatment inactivated many genes whose expression was altered by TT. Most of this interaction is positive on genes downregulated by TT, with no effect by CQA alone. However, there is also a large number of upregulated TT genes neutralized by negative interaction with CQA (Figure 2 and Supplementary Figure S1).

Some of these inactivated genes explain the therapeutic mechanisms of TT. The effect of TT on these genes not only has a therapeutic effect on neurodegenerative diseases but also on endocrine, dermatological, cardiovascular, digestive, respiratory, gynecological, and rheumatoid diseases (Asiatic acid impact on *AKT*, *mTOR*, *NF-κB*, *BDNF*, *CPT-1*, *SOX2*, *BCL2*, *IL18*, *CASP-3*, and *NLRP3* modulates the therapeutic effect for some neurological, digestive, dermatological, endocrine, cardiovascular, and respiratory conditions. Asiaticoside's impact on *NF-κB*, *MAPK*, *BCL2*, *TLR4*, *TRAF6*, *IL18*, and *IL10* modulates the therapeutic effect for some neurological, endocrine, cardiovascular, digestive, and respiratory conditions. Additionally, madecassoside impacts *MAPK*, *TLR2*, and *IL10*, modulating the therapeutic effect for some neurological, endocrine, dermatological, and rheumatoid conditions [17,49]. This finding could suggest a therapeutic rationale for not combining TT and CQA compounds, i.e., using isolated TT compounds. While TT compounds are more widely studied, recent research has also shown health benefits from the CQA compounds, such as ameliorating cognitive impairment in an Alzheimer's mouse model [28].

When evaluating higher-level functions using gene co-expression modules affected by each treatment, TT again showed the greatest effect across both the transcriptomics modules and metabolomics modules. However, there were very different patterns of interaction with TT and other compounds for these different modules. In Module 2, which was enriched for fatty acid metabolism and had a dominant gene downregulation effect from TT, there was virtually no interaction with CQA or unknown compounds. This module effect was still seen with TTCQA and CAW. One of the pathways within fatty acid metabolism was arachidonic acid metabolism. Downregulation of this pathway could explain some of the anti-inflammatory and immune-modulating effects of both TT and CAW [17]. With both Module 3 and Module 4, the significant effect of TT, seen with eigengenes (Figure 3), appears to be diminished by CQA and then further diminished by unknown compounds in CAW. Both of the overall functions associated with these modules through the enrichment analysis (cellular response to stress and stimuli, immune functions) were previously researched for CA. While the pattern of compound interaction is the same in these two modules, the effect of that interaction is different. For Module 3 (cellular response to stress and stimuli), the effect is overall an upregulation of genes, which seems to be neutralized by interactions. The reverse is true for Module 4 (immune functions). However, both triterpene compounds and extracts of CA, water, or ethanol have shown effects in these functions in previous research [17]. However, most of these were in vivo studies. The metabolomic Module 1 had the same pattern of interaction and differential abundance as the transcriptomic Module 4, but the metabolomic Module 2 had a pattern of interaction distinct from any transcriptomic module. For this module, both TT and CQA have relatively equivalent significant effects that are apparently neutralized through their interaction and then further reduced by interactions with unknown compounds.

The significant effect for CAW seen in transcriptomic Module 1 (extracellular matrix organization) appears to be mediated through interactions of all the compounds tested in this experiment since no significant effect is seen until all compounds are together in the full CAW extract, with possibly a strong contribution from unknown compounds (Figure 3, Module 1). Collagen formation sub-pathways, within extracellular matrix organization, are dominant in this module. Interestingly, TT has been found to reduce the deposition of the extracellular matrix and is thus useful in treating liver, pulmonary, and other fibrotic conditions [50,51]. However, a methanol extract of *Centella asiatica* has been found to stimulate collagen and extracellular matrix formation, which is beneficial in certain dermatological conditions, as well as wound healing [17,49,52]. TT compounds alone have shown benefit in treating dermatological conditions, but this effect appears to be modulated through other mechanisms besides collagen formation [17].

The integration analysis conducted for this experiment was metabolomics-centric since we had a much smaller number of metabolites of interest than genes. Finding genes associated with these metabolites helped to focus on specific functions for the compound interaction analysis. This is likely why some of the functionality seen with the gene co-expression analysis is not reflected in this integration. Generally, functions seen here are more closely reflected in the metabolite co-abundance findings, except for fatty acid metabolism.

Distinctly different functions between the TT and CQA treatments were apparent at the pathway level based on this integration analysis. Some of the top pathways associated with DEGs in the TT integration included the metabolism of amino acids, nucleotides, and fatty acids. The top pathways associated with DEGs in the CQA integration included chromatin organization, epigenetic regulation of gene expression, and DNA repair. There were also functional variations across network community groups within a treatment (see Supplementary Tables S9, S11, and S12). Many of the same pathways seen with CQA are also seen in TTCQA, such as chromatin organization, epigenetic regulation of gene expression, and DNA repair. Interestingly, there is almost no overlap between these three treatments at the DEG level, but a substantial amount of overlap at the pathway level, particularly between TTCQA and CQA, suggesting that some functions are retained in the TTCQA combination but mediated through different genes.

The PPI network analysis is another 'guilt by association' approach that considers two genes in a PPI to have a more likely functional relationship the shorter the path between them [53]. There is also some evidence that the network proximity of DEGs in a PPI is associated with the therapeutic synergy of the associated compounds [54]. This analysis revealed that none of the collections of altered genes were distributed by random chance. This would be expected for collections of compounds with structural similarity, such as the four compounds in the TT treatment of the eight compounds in the CQA treatment. However, it is more surprising to see this phenomenon maintained within the combination of these two sets of compounds (TTCQA), especially in the large number of structurally diverse compounds seen with CAW. It is possible that the lack of random distribution seen in CAW is mediated through compound interactions, and even possibly driven by the two therapeutic classes of compounds.

Future analyses could address some of the limitations of this current study. This study sample size was relatively small and may not be able to detect smaller significant effects, particularly with the metabolomics differential analyses. Additionally, the individual experiments were not able to collect sufficient cells from the cultures to perform the two omics analyses on the same batch of neuronal cultures. This particularly impacts our methodological approach to omics integration. Using a knowledge-based approach has limitations due to incomplete annotation of biological functions. The two independent co-expression/abundant analyses were data-driven and likely a more complete picture of functional molecular relationships, although still subject to limitations from the use of known pathway data for the enrichment analysis. Even the co-expression/abundance analyses are based on the assumption of a linear relationship between biological molecules, which is not always the case. Some sample variability can be seen in the eigengenes and eigenmetabolites in Figures 3 and 4A. While the two omics experiments were performed in separate batches, the samples within each omics domain were from the same culture batch, so this would not have a source of variation. This could be related to interactions at the cellular transport level.

Since this was an in vitro experiment, some of the findings may not translate at an organism level. Some compounds present in these treatments may not be absorbed when given orally, and, in general, most pharmacokinetic mechanisms are not accounted for in this type of experiment. Future studies should explore the in vivo absorption of these compounds and also examine metabolomic and transcriptomic changes in the brains of mice treated with TT, CQA, TTCQA, and CAW. In humans, the concentration of TT and CQA compounds and endogenous metabolites could be measured in plasma following treatment with these mixtures, although these may not correlate directly to levels in the brain or other sites of action of these compounds.

It is difficult to know how the concentrations used in this experiment relate to the medicinal use of *Centella asiatica* as both traditional and commercial preparations vary widely in composition and dosage and corresponding plasma and tissue concentrations of TT and CQA are not usually known. However, in a clinical trial, a *Centella asiatica* TT preparation that improved symptoms of diabetic neuropathy resulted in steady state total TT plasma levels of about 0.6 μM [55]. Human pharmacokinetic studies of 2 g and 4 g doses of CAW [56] resulted in plasma C_{max} values for total TT of 0.5 and 1.2 μM, respectively. CQA levels resulting from *Centella asiatica* administration are more difficult to evaluate as they are widely distributed among commonly consumed food plants [57]. Nevertheless, the TT plasma concentrations seen in humans were of a similar order to the in vitro concentrations used in this study, supporting the potential clinical relevance of the data obtained here.

In conclusion, this study revealed some compelling patterns for further investigation. Four gene expression modules and two metabolite modules were altered by the 4 types of treatments applied. This methodology demonstrated the existence of both negative and positive interactions between TT, CQA, and additional compounds found in CAW on the transcriptome and metabolome of mouse primary cortical neurons. It will be interesting to confirm the pathways that emerged from this study in targeted future experiments. Additionally, it will be informative to evaluate how the phenotypic endpoints that previous work has shown to be affected by CAW are impacted by TT, CQA, and TTCQA treatment.

4. Materials and Methods

4.1. Mouse Primary Cortical Neuron Cell Cultures

Mouse primary cortical neurons were isolated and cultured (37 °C) following a previously published protocol [58]. Briefly, six-well plates were coated with poly-L-lysine (PLL; Sigma, Burlington, MA, USA), three days prior to primary neuron harvest. After one day, the PLL was removed, and the plates were washed with double distilled water. Following the wash, plating media (Minimal Essential Media, Fetal Bovine Serum 4.6%,

Glucose 0.55%, Antibiotic-antimycotic: 10,000 U/mL Penicillin 10,000 U/mL Streptomycin and 25 µg of amphotericin B/mL, (Invitrogen, Carlsbad, CA, USA)) was added to each well for the remaining two days. Then, neurons were isolated from the cortices of C57BL6 mouse embryos at embryonic days 16–18, incubated in a mixture of HBSS and trypsin (2.5%) (Gibco, Grand Island, NY, USA) at 37 °C, then dissociated. Neurons were plated onto six-well plates (1 million/well) and incubated (37 °C) for 3 h after which the plating media was replaced with supplemented neurobasal media (with B27 1:50 dilution, Glutamax 1:100 dilution, antibiotic-antimycotic: penicillin 10,000 U/mL, streptomycin 10,000 U/mL and amphotericin B 25 µg/mL, (Invitrogen, Carlsbad, CA, USA)). Neurons were then cultured for five days at 37 °C before receiving experimental treatments.

4.2. Treatments

Centella asiatica dried plant material (aerial parts; batch number X20090016) was purchased from Oregon's Wild Harvest (OWH; Redmond, OR, USA). Voucher samples were deposited in the BENFRA Center laboratories at Oregon Health & Science University (code number BEN-CA-6) and at the Herbarium at Oregon State University (code number OSC-V-265416). A water extract (CAW) was prepared by reflux extraction of the dried plant aerial parts minus the root, filtering, and lyophilization [23–25,59]. The ratio of dried plant material to dried extract was 5:1. The CAW treatment was prepared from this dried extract (0.050 mg/mL in 0.025% *v*/*v* aqueous methanol vehicle in each well). The triterpene (TT) and caffeoylquinic acid (CQA) content in the CAW was determined using liquid chromatography coupled with multiple reaction monitoring mass spectrometry (LC-MRM-MS) as previously described [31]. The triterpene (TT) treatment was prepared (0.025% *v*/*v* aqueous methanol vehicle final concentration in the cell culture) with madecassoside (1.7945 µg/mL; 1.84 µM), asiaticoside (0.7376 µg/mL; 0.769 µM), madecassic acid (0.0377 µg/mL; 0.073 µM), and asiatic acid (0.021 µg/mL; 0.043 µM) concentrations equivalent to that in the CAW 50 µg/mL solution (LC-MRM-MS verified concentrations). The caffeoylquinic acid (CQA) solution was prepared in the same fashion using chlorogenic acid (0.3749 µg/mL; 1.06 µM), neo-chlorogenic acid (0.1695 µg/mL; 0.478 µM), crypto-chlorogenic acid (0.1492 µg/mL; 0.421 µM), 3,4-dicaffeoylqinic acid (0.1146 µg/mL; 0.221 µM), 3,5-dicaffeoylqinic acid (0.0884 µg/mL; 0.171 µM), 4,5-dicaffeoylqinic acid (0.0975 µg/mL; 0.189 µM), 1,3-dicaffeoylqinic acid (0.1291 µg/mL; 0.250 µM), and 1,5-dicaffeoylqinic acid (0.1946 µg/mL; 0.377 µM), (LC-MRM-MS verified concentrations). The triterpene and caffeoylquinic acid (TTCQA) solution was prepared using the 12 compounds that were used in the TT and CQA treatments, prepared in the same fashion. The control vehicle consisted of aqueous methanol (0.025% *v*/*v*). These are the stock treatment 'theoretical values'. These stock treatment solutions were then quantified using LC-MRM-MS as described in Section 4.3. These are the expected time zero values and can be found in the first four lines of Supplementary Table S1. Purified reference *Centella asiatica* TT and CQA were purchased from Chemfaces (Wuhan, Hubei, Peoples Republic of China).

CAW or compound treatments were initiated after five days of incubation and lasted for 48 h. There were four compound treatments (CAW, TT, CQA, TTCQA) and one vehicle control treatment. After the 48 h treatment period, an aliquot of medium from each well was collected for analysis of post-incubation cultured media compound concentrations, and cells were harvested by trypsinization (0.25% trypsin in Hanks' balanced salt solution, 37 °C, 5 min) after media was removed and cells rinsed 1× with phosphate-buffered saline. Detached cells were then centrifuged at 300× *g* for 5 min at room temperature and the resulting cell pellet was flash-frozen and stored at −80 °C until analysis. Eight samples for each experimental condition were prepared for RNA sequencing and ten samples were prepared for each condition for metabolomics analyses.

4.3. Analysis of Compound Concentrations in the Media After 48 h of Treatment

For quantifying 12 CA phytochemical marker compounds (3 mono-CQA, 5 di-CQAs, and 4 TTs) in media, a SPE method was developed that removes phospholipids and proteins (using Ostro® Protein Precipitation and Phospholipid Removal Plate (Waters Corp., Milford, MA, USA)) prior to LC-MRM-MS analysis. Each media sample was thawed and a 100 μL aliquot was transferred into an Eppendorf tube, that contained 300 μL methanol containing 0.1% formic acid and 5 μL digoxin-d3 (10 μg/mL) as an internal standard. Media sample was loaded into an Ostro plate well and 300 μL acetonitrile containing 0.1% formic acid was added and mixed by pipetting up and down 3 times. The plate was placed onto a vacuum manifold and the elution solvent was drawn into glass inserts of the collecting plate. The 1-mL glass inserts containing the eluate were placed into a 2-mL micro-centrifuge tube and dried in Speed Vac for approximately 1 h to obtain solid residues. For reconstitution, 50 μL of 70% methanol containing 0.1% formic acid was added to the glass inserts and resuspended by pipetting up and down 3 times. Samples were centrifuged at 13,000 rpm at 4 °C for 10 min, and each supernatant was transferred to an LC-MS vial and stored at −20 °C until LC-MRM-MS analysis using a Waters Xevo TQXS system connected to a Waters UPLC I-class. The MS method details have been described by us previously [31]. Chromatograms from LC-MRM-MS, targeted to detect only TT and CQA compounds, are provided in Supplementary Figure S7. These show the presence of the required compounds in the stock solutions for TT, CQA, TTCQA, and CAW and that their concentrations are in line with those of the CAW extract. An untargeted LC-qTOF-MS chromatogram of all compounds detected in the CAW extract can be found in Supplementary Figure S8 showing the presence of additional compounds. Previous studies employing LC-qTOF-MS with a slower solvent gradient allowed the detection and annotation of 117 compounds in CAW extracts [48].

Cell pellets were resuspended in RLT-β-mercaptoethanol (QIAGEN) lysis buffer and quickly frozen for RNA extraction at the Oregon Health & Science University (OHSU) Gene Profiling Shared Resource. RNA quality was assessed at the GPSR and assigned an RNA Integrity Number (RIN). Only samples with RIN $\geq$ 8.0 were considered for sequencing, which occurred at the OHSU Massively Parallel Sequencing Shared Resource facility. This was performed on a HiSeq 2500 (Illumina, San Diego, CA, USA) and aligned to the mm38 mouse genome with the Star aligner [60]. Subsequent QA/QC was performed using the MultiQC package version 1.22 [61].

4.4. Metabolomics

Cells for control and treatments (approximately 1 million each) were thawed on ice and transferred to Precellys Lysing Kit bead blender tube with pre-supplied beads (2 mL Tissue Homogenizing Mixed Beads Kit (CKMix), Item No. 10409, Cayman Chemicals, Ann Arbor, MI, USA). Extraction solvents (500 μL, cold 80% methanol in water containing 12-[[(cyclohexylamino)carbonyl]amino]-dodecanoic acid (CUDA, 20 ppm, 10 μL)) were added to each vial, and homogenized. The homogenates were incubated for 1 h at −20 °C, then centrifuged at 10,000 rpm at 4 °C for 10 min. Each supernatant was transferred to a new vial and dried using a Speed Vac. Each residue was reconstituted in a cold 50% aqueous acetonitrile, vortexed for 20 s, and centrifuged at 10,000 rpm at 4 °C for 10 min. Each supernatant was transferred to an LC-MS vial and stored at −80 °C until LC-MS analysis. Quality control (QC) samples were generated by pooling 10-μL aliquots from each sample extract and were analyzed with samples. Blanks contained only extraction solvent.

The LC-MS/MS metabolomics workflow was described by us previously (see [62]) and entails two chromatographic methods: (1) Reverse-phase liquid chromatography was carried out using ACE Excel C18-PFP (1.7 μm, 2.1 mm × 100 mm). The sample injection volume was 5 μL, and the flow rate was 0.5 mL/min. The mobile phases consisted of water (A) and acetonitrile (B), both containing 0.1% formic acid. The gradient was as follows: initial hold at 5% B for 3 min, then increase to 80% B in 13 min, held until 16 min, then shift to 5% B and hold for equilibration from 16.5 min to 20.5 min. The column oven temperature was maintained at 30 °C. (2) Hydrophilic interaction liquid chromatography, the method was identical to that previously reported [62].

For both chromatographic methods ESI-MS/MS data were acquired in positive and negative ionization modes using a quadrupole time-of-flight MS system (AB Sciex tripleTOF 5600; Framingham, MA, USA). Individual metabolites were annotated using an existing in-house library based on retention time, exact mass, and MS/MS of 650 compounds compiled in the Mass Spectrometry Metabolite Library of Standards (MSMLS, IROA Technologies, Bolton, MA, USA). Raw data files were imported and processed using PeakView (ver. 1.2, Sciex) and MultiQuant (ver. 3.0.2, Sciex) software. Metabolites were verified using chromatographic retention time (error < 10%), accurate mass (error < 10 ppm), MS/MS fragmentation (score > 70), and isotopic pattern (error < 20%). The RP-C18 LC method yielded 117 metabolites (58 metabolites in positive ion mode and 59 in negative ion mode), and the HILIC method resulted in 234 metabolites (135 metabolites in positive ion mode and 119 in negative ion mode). The metabolites that were assigned in both methods were evaluated and the metabolites detected with the lower coefficient variation (CV) value in the QC samples were kept. This evaluation resulted in 192 high confident unique metabolites (178 metabolites (<20% QC CV), 10 metabolites (<30% QC CV), and 4 metabolites (>30% QC CV), Supplementary Table S13).

4.5. Differential Analyses for Gene Expression and Metabolite Abundance

For differential gene expression, the limma-voom package was used, a modification of the limma package designed for RNA-seq count data [63]. Genes were removed from the data if they did not have a transcript count of at least 10 in each sample. Next, a $\log_2$ transformation of counts per million (l cpm) was performed, followed by a trimmed mean of M-values (TMM) normalization. Heteroscedasticity was controlled with the voom function to meet the assumptions of a linear method. Differential calculations were conducted between each of the four treatments (TT, CQA, TTCQA, CAW) and the control (FDR cutoff 0.05).

For metabolite abundance, the MetaboAnalyst package version 5.0 was used [64]. The data were median normalized, $\log_2$ transformed, and then Pareto scaled (Supplementary Figure S3). *t*-tests were performed to compare each treatment to the control for all 192 metabolites (FDR cutoff 0.05).

4.6. TT and CQA Gene Expression Interaction Classification

To assess the interaction between the TT and CQA treatments on normalized gene expression data, parameter estimates were derived from a least squares linear method with the control group as a reference, using the limma R package version 3.58.1 [63]. For each group, the effects (parameter estimates) of each treatment were assessed relative to the control. Next, the sum of the effects of TT and CQA treatments was contrasted with the effect of the TTCQA treatment using the contrasts.fit eBayes functions in limma. Any gene found to be differentially expressed by any of these three treatments was examined for interaction (N = 2268). We included genes that might not be differentially expressed by either one or two of the other treatments to see if an interaction might create or inhibit a significant expression.

If the TTCQA effect on the expression of a particular gene was greater than the sum of the TT and CQA effects, the interaction was classified as positive, or synergistic. If the TTCQA effect was less than the sum, the interaction was classified as negative, or antagonistic. If they were equal, the interaction was classified as an additive, or no interaction [65]. Since these coefficients represent $\log_2$ fold changes, negative values represent downregulation and positive values represent upregulation. So, any negative interaction that decreases this coefficient value would potentially increase downregulation, and positive interactions, as well as additive, would increase upregulation.

4.7. Weighted Gene Correlation Network Analysis (WGCNA Gene Co-Expression and Metabolite Co-Abundance)

The WGCNA package in R [66] was used for both RNA-seq and metabolomic data [40]. Both data types were normalized for the creation of this network by variance stabilizing transformation. To create a scale-free network, Pearson correlation was first used to construct an adjacency matrix for all gene or metabolite pairs using a soft thresholding power of 12 for both data types. The matrix was transformed into a topological overlap matrix (TOM) which was then converted to a dissimilarity matrix (1-TOM). Hierarchical clustering was applied to the final matrix to identify clusters of genes or metabolites with similar expression or abundance profiles. The minimum node size was 5 for metabolomics data and 30 for RNA-seq data. Module eigengenes and eigenmetabolites (first principal components) were calculated using the moduleEigengenes of WGCNA. Module membership for each gene or metabolite was calculated as the correlation between the normalized expression or abundance values with the module eigengene or eigenmetabolite and was calculated with the signedKME in WGCNA. Intramodular connectivity was the degree of each gene or metabolite in the adjacency matrix described above, calculated with the intramodular connectivity function in WGCNA.

Following the construction of these modules in each of the two omics domains, we then tested for differences between each treatment and control seen at the eigengene or eigenmetabolite level.

4.8. Integration of Transcriptomics and Metabolomics Data

A modification of a protocol previously used to integrate transcriptomic and metabolomic data for the study of neurodegenerative disease [67] was used to obtain a more comprehensive functional perspective for comparing the four treatments. This protocol was modified to fit the limitations of the data. Since this project was a repeated experiment design, with the transcriptome coming from different samples than the metabolome, rather than a split design with both omics coming from the same samples, a 'data-driven' approach was not appropriate. A 'data-driven' approach would correlate expression and abundance levels (or eigenvalues) across domains to create correlation-based networks for further analyses. Instead, a 'knowledge-based' approach was used, with existing molecular pathway information, to construct networks within and between individual genes and metabolites. These networks and subsequent analyses were conducted using OMICSNET software version 2.0 [68].

Genes and metabolites associated with significant treatment effects, relative to control, were used as seeds to construct the 'knowledge-based' networks in OMICSNET. Analyses were performed separately for each treatment. Any gene differentially expressed by a treatment was used as a seed. Any individual metabolite affected by treatment or member of a metabolite module whose eigenmetabolite was affected by treatment was also used as a seed. Due to the sparse number of individual metabolites affected, we also included co-expressed metabolites. Seed genes were used to construct a primary protein–protein interaction (PPI) network using the STRING database [41]. Genes and proteins are used interchangeably here. Connector proteins (non-seed) were added if necessary to create a more fully connected network. Seed metabolites were used to first create a primary metabolite-protein network using the KEGG pathway database [69]. A secondary PPI was created using the proteins in the primary metabolite-protein network using STRING, adding connector proteins if necessary. Finally, a composite network was created by linking the three independent network layers with common nodes (genes/proteins). To control the network size, the 'Minimum network' function was used in OMICSNET to create the smallest network possible which still contained as many seeds as possible.

Once an integrated network was created, tightly clustered communities of nodes were detected using the Walktrap algorithm [70]. Only communities that contained both seed metabolites and seed genes were further analyzed. These communities identified close functional relationships between genes and metabolites both affected by the same treatment. Subsequent Reactome pathway over-representation analysis was conducted using all genes in a community (seeds and connectors) using ReactomePA [38]. A comparison of pathways impacted by each treatment was then performed to identify overlapping and distinct functions using the integrated omics data (Figure 6).

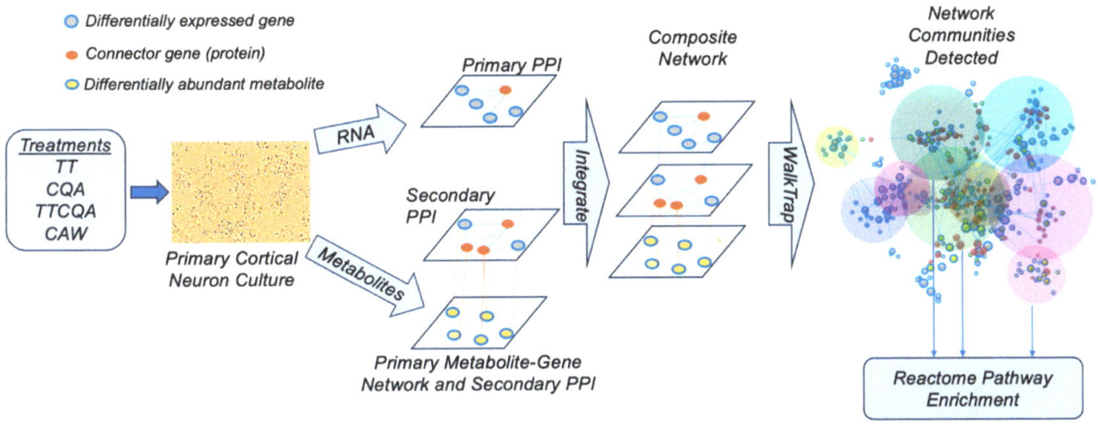

Figure 6. Omics integration methodology. Individual seed networks are shown in the center of the figure. For metabolite seeds two networks are shown, a primary metabolite-gene network and a secondary PPI created from the genes in the primary network. Seed genes are shown in this secondary PPI if present, but are not used to construct this network. For the seed genes, only one primary PPI network is created. The red connector (non-seed) genes (protein) are added to both PPIs, if necessary, to increase connectivity in the overall network for community detection that will better contain seeds of interest.

Supplementary Materials: The following supporting information can be downloaded at: https://www.mdpi.com/article/10.3390/ph18010019/s1, Figure S1 (Molecular effects of 4 treatments), Figure S2 (TT integration Module 1), Figure S3 (Pre and Post normalized metabolite distributions), Figure S4 (CQA integration Module 1), Figure S5 (TTCQA integration Module 8), Figure S6 (Heatmaps for RNA co-expression Modules 2-5), Figure S7 (LC-MRM-MS of stock solutions used for preparation of media), Figure S8 (UPLC-MS analysis of CAW extract and compounds); Table S1 (Concentration of phytochemical compounds in media cultured with neurons), Table S2 (Gene expression fold changes for treatments relative to control), Table S3 (Metabolite abundance fold changes for treatments relative to control), Table S4 (Parameter estimates for TT and CQA interaction analysis), Table S5 (Detailed pathway enrichment information for RNA co-expression modules), Table S6 (Detailed pathway enrichment information for metabolite co-abundance modules), Table S7 (Detailed information about individual metabolites in the co-abundance modules), Table S8 (Pathways in common between metabolite Module 2 and transcriptomic modules), Table S9 (Integrated gene and metabolite information for the TT treatment), Table S10 (Pathway information for the integrated TT treatment analysis), Table S11 (Integrated gene and metabolite information for the CQA treatment), Table S12 (Pathway information for the integrated CQA treatment analysis), Table S13 (Raw metabolite peak intensities for all metabolites).

Author Contributions: Conceptualization, S.R.C., A.S., N.E.G., S.M. and C.S.M.; data curation, S.R.C.; formal analysis, S.R.C. and L.M.; funding acquisition, S.R.C., A.S. and N.E.G.; investigation, S.R.C., J.A.Z., C.J.N., J.C., L.Y. and L.M.; methodology, J.C., L.Y., L.M. and C.S.M.; visualization, S.R.C.; writing—original draft, S.R.C.; writing—review and editing, S.R.C., A.S., N.E.G., S.M., C.S.M. and L.M. All authors have read and agreed to the published version of the manuscript.

Funding: This research was funded by the National Center for Complementary and Integrative Health (NCCIH) of the National Institutes of Health (NIH) (U19 AT010829, A.S.), as well as a supplementary grant for integrative health practitioners from the NCCIH (U19 AT010829-02S1, A.S.) and a K08 grant from NCCIH (1K08AT012477-01, S.R.C.). The authors also wish to acknowledge use of OSU's mass spectrometry center and the following NIH equipment grants, S10OD026922 and S10RR027878 (C.S.M.).

Institutional Review Board Statement: Not applicable.

Informed Consent Statement: Not applicable.

Data Availability Statement: The original contributions presented in this study are included in the article/supplementary material. Further inquiries can be directed to the corresponding author for RNAseq files.

Acknowledgments: We thank Daniel Bottomly for his advice about carrying out differential expression analyses and feedback on a draft of the manuscript.

Conflicts of Interest: The authors declare no conflicts of interest. The funders had no role in the design of this study; in the collection, analyses, or interpretation of data; in the writing of the manuscript; or in the decision to publish the results.

References

1. WHO. Ageing and Health. 2022. Available online: https://www.who.int/news-room/fact-sheets/detail/ageing-and-health (accessed on 15 December 2023).
2. AARP/NCCAM. Complementary and Alternative Medicine: What People Aged 50 and Older Discuss with Their Health Care Providers; Survey Report. 2011. Available online: https://www.aarp.org/pri/topics/health/prevention-wellness/complementary-alternative-medicine/ (accessed on 15 December 2023).
3. Nascimento, A.L.; Lozano, A.; Melo, J.G.; Alves, R.R.; Albuquerque, U.P. Functional aspects of the use of plants and animals in local medical systems and their implications for resilience. *J. Ethnopharmacol.* **2016**, *194*, 348–357. [CrossRef]
4. Ward, L.; Pasinetti, G.M. Recommendations for Development of Botanical Polyphenols as "Natural Drugs" for Promotion of Resilience Against Stress-Induced Depression and Cognitive Impairment. *Neuromol. Med.* **2016**, *18*, 487–495. [CrossRef] [PubMed]
5. Ho, Y.S.; So, K.F.; Chang, R.C. Anti-aging herbal medicine--how and why can they be used in aging-associated neurodegenerative diseases? *Ageing Res. Rev.* **2010**, *9*, 354–362. [CrossRef]
6. Ismail, Z.; Mohamed, R.; Mohd Hassan, M.H.; Wan Su, K. Usage of traditional medicines among elderly and the prevalence of prednisolone contamination. *Malays. J. Med. Sci. MJMS* **2005**, *12*, 50–55. [PubMed]
7. Xu, Q.Q.; Shan, C.S.; Wang, Y.; Shi, Y.H.; Zhang, Q.H.; Zheng, G.Q. Chinese Herbal Medicine for Vascular Dementia: A Systematic Review and Meta-Analysis of High-Quality Randomized Controlled Trials. *J. Alzheimer's Dis.* **2018**, *62*, 429–456. [CrossRef]
8. Enke, C.G.; Nagels, L.J. Undetected components in natural mixtures: How many? What concentrations? Do they account for chemical noise? What is needed to detect them? *Anal. Chem.* **2011**, *83*, 2539–2546. [CrossRef] [PubMed]
9. Van Vuuren, S.; Viljoen, A. Plant-based antimicrobial studies-methods and approaches to study the interaction between natural products. *Planta Med.* **2011**, *77*, 1168–1182. [CrossRef] [PubMed]

10. Panossian, A.; Lemerond, T.; Efferth, T. State-of-the-Art Review on Botanical Hybrid Preparations in Phytomedicine and Phytotherapy Research: Background and Perspectives. *Pharmaceuticals* **2024**, *17*, 483. [CrossRef]
11. Efferth, T.; Koch, E. Complex interactions between phytochemicals. The multi-target therapeutic concept of phytotherapy. *Curr. Drug Targets* **2011**, *12*, 122–132. [CrossRef]
12. Wagner, H.; Ulrich-Merzenich, G. Synergy research: Approaching a new generation of phytopharmaceuticals. *Phytomedicine* **2009**, *16*, 97–110. [CrossRef]
13. Caesar, L.K.; Cech, N.B. Synergy and antagonism in natural product extracts: When 1 + 1 does not equal 2. *Nat. Prod. Rep.* **2019**, *36*, 869–888. [CrossRef]
14. Panossian, A.; Seo, E.J.; Wikman, G.; Efferth, T. Synergy assessment of fixed combinations of Herba Andrographidis and Radix Eleutherococci extracts by transcriptome-wide microarray profiling. *Phytomedicine* **2015**, *22*, 981–992. [CrossRef]
15. Panossian, A.; Hamm, R.; Kadioglu, O.; Wikman, G.; Efferth, T. Synergy and Antagonism of Active Constituents of ADAPT-232 on Transcriptional Level of Metabolic Regulation of Isolated Neuroglial Cells. *Front. Neurosci.* **2013**, *7*, 16. [CrossRef] [PubMed]
16. Junio, H.A.; Sy-Cordero, A.A.; Ettefagh, K.A.; Burns, J.T.; Micko, K.T.; Graf, T.N.; Richter, S.J.; Cannon, R.E.; Oberlies, N.H.; Cech, N.B. Synergy-directed fractionation of botanical medicines: A case study with goldenseal (*Hydrastis canadensis*). *J. Nat. Prod.* **2011**, *74*, 1621–1629. [CrossRef] [PubMed]
17. Sun, B.; Wu, L.; Wu, Y.; Zhang, C.; Qin, L.; Hayashi, M.; Kudo, M.; Gao, M.; Liu, T. Therapeutic Potential of *Centella asiatica* and Its Triterpenes: A Review. *Front. Pharmacol.* **2020**, *11*, 568032. [CrossRef]
18. Shinomol, G.K.; Muralidhara; Bharath, M.M. Exploring the Role of "Brahmi" (Bacopa monnieri and *Centella asiatica*) in Brain Function and Therapy. *Recent. Pat. Endocr. Metab. Immune Drug Discov.* **2011**, *5*, 33–49.
19. Gray, N.E.; Alcazar Magana, A.; Lak, P.; Wright, K.M.; Quinn, J.; Stevens, J.F.; Maier, C.S.; Soumyanath, A. *Centella asiatica*: Phytochemistry and mechanisms of neuroprotection and cognitive enhancement. *Phytochem. Rev.* **2018**, *17*, 161–194. [CrossRef]
20. Veerendra Kumar, M.H.; Gupta, Y.K. Effect of *Centella asiatica* on cognition and oxidative stress in an intracerebroventricular streptozotocin model of Alzheimer's disease in rats. *Clin. Exp. Pharmacol. Physiol.* **2003**, *30*, 336–342. [CrossRef] [PubMed]
21. Veerendra Kumar, M.H.; Gupta, Y.K. Effect of different extracts of *Centella asiatica* on cognition and markers of oxidative stress in rats. *J. Ethnopharmacol.* **2002**, *79*, 253–260. [CrossRef] [PubMed]
22. Gupta, Y.K.; Veerendra Kumar, M.H.; Srivastava, A.K. Effect of *Centella asiatica* on pentylenetetrazole-induced kindling, cognition and oxidative stress in rats. *Pharmacol. Biochem. Behav.* **2003**, *74*, 579–585. [CrossRef] [PubMed]
23. Gray, N.E.; Harris, C.J.; Quinn, J.F.; Soumyanath, A. *Centella asiatica* modulates antioxidant and mitochondrial pathways and improves cognitive function in mice. *J. Ethnopharmacol.* **2016**, *180*, 78–86. [CrossRef] [PubMed]
24. Gray, N.E.; Zweig, J.A.; Caruso, M.; Martin, M.D.; Zhu, J.Y.; Quinn, J.F.; Soumyanath, A. *Centella asiatica* increases hippocampal synaptic density and improves memory and executive function in aged mice. *Brain Behav.* **2018**, *8*, e01024. [CrossRef]
25. Gray, N.E.; Zweig, J.A.; Caruso, M.; Zhu, J.Y.; Wright, K.M.; Quinn, J.F.; Soumyanath, A. *Centella asiatica* attenuates hippocampal mitochondrial dysfunction and improves memory and executive function in β-amyloid overexpressing mice. *Mol. Cell. Neurosci.* **2018**, *93*, 1–9. [CrossRef]
26. Soumyanath, A.; Zhong, Y.-P.; Henson, E.; Wadsworth, T.; Bishop, J.; Gold, B.G.; Quinn, J.F. *Centella asiatica* Extract Improves Behavioral Deficits in a Mouse Model of Alzheimer's Disease: Investigation of a Possible Mechanism of Action. *Int. J. Alzheimer's Dis.* **2012**, *1*, 381974.
27. Han, J.; Miyamae, Y.; Shigemori, H.; Isoda, H. Neuroprotective effect of 3,5-di-O-caffeoylquinic acid on SH-SY5Y cells and senescence-accelerated-prone mice 8 through the up-regulation of phosphoglycerate kinase-1. *Neuroscience* **2010**, *169*, 1039–1045. [CrossRef]
28. Matthews, D.G.; Caruso, M.; Alcazar Magana, A.; Wright, K.M.; Maier, C.S.; Stevens, J.F.; Gray, N.E.; Quinn, J.F.; Soumyanath, A. Caffeoylquinic Acids in *Centella asiatica* Reverse Cognitive Deficits in Male 5XFAD Alzheimer's Disease Model Mice. *Nutrients* **2020**, *12*, 3488. [CrossRef] [PubMed]
29. Matthews, D.G.; Caruso, M.; Murchison, C.F.; Zhu, J.Y.; Wright, K.M.; Harris, C.J.; Gray, N.E.; Quinn, J.F.; Soumyanath, A. *Centella Asiatica* Improves Memory and Promotes Antioxidative Signaling in 5XFAD Mice. *Antioxidants* **2019**, *8*, 630. [CrossRef] [PubMed]
30. Gray, N.E.; Zweig, J.A.; Murchison, C.; Caruso, M.; Matthews, D.G.; Kawamoto, C.; Harris, C.J.; Quinn, J.F.; Soumyanath, A. *Centella asiatica* attenuates Aβ-induced neurodegenerative spine loss and dendritic simplification. *Neurosci. Lett.* **2017**, *646*, 24–29. [CrossRef]
31. Yang, L.; Marney, L.; Magana, A.A.; Choi, J.; Wright, K.; McFerrin, J.; Gray, N.E.; Soumyanath, A.; Stevens, J.F.; Maier, C.S. Quantification of Caffeoylquinic Acids and Triterpenes as Targeted Bioactive Compounds of *Centella asiatica* in Extracts and Formulations by Liquid Chromatography Mass Spectrometry. *J. Chromatogr. Open* **2023**, *4*, 100091. [CrossRef]
32. Rowe, K.; Gray, N.E.; Zweig, J.A.; Law, A.; Techen, N.; Maier, C.S.; Soumyanath, A.; Kretzschmar, D. *Centella asiatica* and its caffeoylquinic acid and triterpene constituents increase dendritic arborization of mouse primary hippocampal neurons and improve age-related locomotion deficits in Drosophila. *Front. Aging* **2024**, *5*, 1374905. [CrossRef]

33. Gray, N.E.; Zweig, J.A.; Matthews, D.G.; Caruso, M.; Quinn, J.F.; Soumyanath, A. *Centella asiatica* Attenuates Mitochondrial Dysfunction and Oxidative Stress in Aβ-Exposed Hippocampal Neurons. *Oxid. Med. Cell. Longev.* **2017**, *2017*, 7023091. [CrossRef]
34. Zweig, J.A.; Brandes, M.S.; Brumbach, B.H.; Caruso, M.; Wright, K.M.; Quinn, J.F.; Soumyanath, A.; Gray, N.E. Loss of NRF2 accelerates cognitive decline, exacerbates mitochondrial dysfunction, and is required for the cognitive enhancing effects of *Centella asiatica* during aging. *Neurobiol. Aging* **2021**, *100*, 48–58. [CrossRef]
35. Wianowska, D.; Gil, M. Recent advances in extraction and analysis procedures of natural chlorogenic acids. *Phytochem. Rev.* **2019**, *18*, 273–302. [CrossRef]
36. Xue, M.; Shi, H.; Zhang, J.; Liu, Q.Q.; Guan, J.; Zhang, J.Y.; Ma, Q. Stability and Degradation of Caffeoylquinic Acids under Different Storage Conditions Studied by High-Performance Liquid Chromatography with Photo Diode Array Detection and High-Performance Liquid Chromatography with Electrospray Ionization Collision-Induced Dissociation Tandem Mass Spectrometry. *Molecules* **2016**, *21*, 948. [CrossRef] [PubMed]
37. Blighe, K.; Lun, A. PCAtools: Everything Principal Components Analysis. 2019. Available online: https://github.com/kevinblighe/PCAtools (accessed on 15 January 2024).
38. Yu, G.; He, Q.-Y. ReactomePA: An R/Bioconductor package for reactome pathway analysis and visualization. *Mol. BioSyst.* **2016**, *12*, 477–479. [CrossRef]
39. Van Dam, S.; Võsa, U.; van der Graaf, A.; Franke, L.; de Magalhães, J.P. Gene co-expression analysis for functional classification and gene-disease predictions. *Brief. Bioinform.* **2018**, *19*, 575–592. [CrossRef] [PubMed]
40. Pei, G.; Chen, L.; Zhang, W. WGCNA Application to Proteomic and Metabolomic Data Analysis. *Methods Enzym.* **2017**, *585*, 135–158.
41. Szklarczyk, D.; Gable, A.L.; Nastou, K.C.; Lyon, D.; Kirsch, R.; Pyysalo, S.; Doncheva, N.T.; Legeay, M.; Fang, T.; Bork, P.; et al. The STRING database in 2021: Customizable protein-protein networks, and functional characterization of user-uploaded gene/measurement sets. *Nucleic Acids Res.* **2021**, *49*, D605–D612. [CrossRef]
42. Gilbert, B.; Alves, L.F. Synergy in plant medicines. *Curr. Med. Chem.* **2003**, *10*, 13–20. [CrossRef]
43. Wang, X.; Xu, X.; Tao, W.; Li, Y.; Wang, Y.; Yang, L. A systems biology approach to uncovering pharmacological synergy in herbal medicines with applications to cardiovascular disease. *Evid.-Based Complement. Altern. Med.* **2012**, *2012*, 519031. [CrossRef]
44. Lin, J.; Jiang, H.; Ding, X. Synergistic combinations of five single drugs from *Centella asiatica* for neuronal differentiation. *Neuroreport* **2017**, *28*, 23–27. [CrossRef]
45. Britton, E.R.; Kellogg, J.J.; Kvalheim, O.M.; Cech, N.B. Biochemometrics to Identify Synergists and Additives from Botanical Medicines: A Case Study with *Hydrastis canadensis* (Goldenseal). *J. Nat. Prod.* **2018**, *81*, 484–493. [CrossRef] [PubMed]
46. Caesar, L.K.; Kellogg, J.J.; Kvalheim, O.M.; Cech, R.A.; Cech, N.B. Integration of Biochemometrics and Molecular Networking to Identify Antimicrobials in *Angelica keiskei*. *Planta Med.* **2018**, *84*, 721–728. [CrossRef] [PubMed]
47. Kellogg, J.J.; Todd, D.A.; Egan, J.M.; Raja, H.A.; Oberlies, N.H.; Kvalheim, O.M.; Cech, N.B. Biochemometrics for Natural Products Research: Comparison of Data Analysis Approaches and Application to Identification of Bioactive Compounds. *J. Nat. Prod.* **2016**, *79*, 376–386. [CrossRef] [PubMed]
48. Alcazar Magana, A.; Wright, K.; Vaswani, A.; Caruso, M.; Reed, R.L.; Bailey, C.F.; Nguyen, T.; Gray, N.E.; Soumyanath, A.; Quinn, J.; et al. Integration of mass spectral fingerprinting analysis with precursor ion (MS1) quantification for the characterisation of botanical extracts: Application to extracts of *Centella asiatica* (L.) Urban. *Phytochem. Anal.* **2020**, *31*, 722–738. [CrossRef] [PubMed]
49. Park, K.S. Pharmacological Effects of *Centella asiatica* on Skin Diseases: Evidence and Possible Mechanisms. *Evid.-Based Complement. Altern. Med.* **2021**, *2021*, 5462633. [CrossRef] [PubMed]
50. Wei, L.; Chen, Q.; Guo, A.; Fan, J.; Wang, R.; Zhang, H. Asiatic acid attenuates CCl(4)-induced liver fibrosis in rats by regulating the PI3K/AKT/mTOR and Bcl-2/Bax signaling pathways. *Int. Immunopharmacol.* **2018**, *60*, 1–8. [CrossRef] [PubMed]
51. Dong, S.-H.; Liu, Y.-W.; Wei, F.; Tan, H.-Z.; Han, Z.-D. Asiatic acid ameliorates pulmonary fibrosis induced by bleomycin (BLM) via suppressing pro-fibrotic and inflammatory signaling pathways. *Biomed. Pharmacother.* **2017**, *89*, 1297–1309. [CrossRef]
52. Yao, C.H.; Yeh, J.Y.; Chen, Y.S.; Li, M.H.; Huang, C.H. Wound-healing effect of electrospun gelatin nanofibres containing *Centella asiatica* extract in a rat model. *J. Tissue Eng. Regen. Med.* **2017**, *11*, 905–915. [CrossRef] [PubMed]
53. Valdeolivas, A.; Tichit, L.; Navarro, C.; Perrin, S.; Odelin, G.; Levy, N.; Cau, P.; Remy, E.; Baudot, A. Random walk with restart on multiplex and heterogeneous biological networks. *Bioinformatics* **2019**, *35*, 497–505. [CrossRef] [PubMed]
54. Tang, Y.-C.; Gottlieb, A. SynPathy: Predicting Drug Synergy through Drug-Associated Pathways Using Deep Learning. *Mol. Cancer Res.* **2022**, *20*, 762–769. [CrossRef] [PubMed]
55. Lou, J.S.; Dimitrova, D.M.; Murchison, C.; Arnold, G.C.; Belding, H.; Seifer, N.; Le, N.; Andrea, S.B.; Gray, N.E.; Wright, K.M.; et al. *Centella asiatica* triterpenes for diabetic neuropathy: A randomized, double-blind, placebo-controlled, pilot clinical study. *Esper. Dermatol.* **2018**, *20*, 12–22. [CrossRef]

56. Wright, K.M.; Bollen, M.; David, J.; Speers, A.B.; Brandes, M.S.; Gray, N.E.; Alcázar Magaña, A.; McClure, C.; Stevens, J.F.; Maier, C.S.; et al. Pharmacokinetics and Pharmacodynamics of Key Components of a Standardized *Centella asiatica* Product in Cognitively Impaired Older Adults: A Phase 1, Double-Blind, Randomized Clinical Trial. *Antioxidants* **2022**, *11*, 215. [CrossRef] [PubMed]
57. Alcázar Magaña, A.; Kamimura, N.; Soumyanath, A.; Stevens, J.F.; Maier, C.S. Caffeoylquinic acids: Chemistry, biosynthesis, occurrence, analytical challenges, and bioactivity. *Plant J.* **2021**, *107*, 1299–1319. [CrossRef] [PubMed]
58. Kaech, S.; Banker, G. Culturing hippocampal neurons. *Nat. Protoc.* **2006**, *1*, 2406–2415. [CrossRef] [PubMed]
59. Houghton, P.J.; Ramanp, A. *Handbook for the Fractionation of Natural Extracts*; Chapman and Hall: London, UK, 1998; pp. 54–65.
60. Dobin, A.; Davis, C.A.; Schlesinger, F.; Drenkow, J.; Zaleski, C.; Jha, S.; Batut, P.; Chaisson, M.; Gingeras, T.R. STAR: Ultrafast universal RNA-seq aligner. *Bioinformatics* **2013**, *29*, 15–21. [CrossRef]
61. Ewels, P.; Magnusson, M.; Lundin, S.; Käller, M. MultiQC: Summarize analysis results for multiple tools and samples in a single report. *Bioinformatics* **2016**, *32*, 3047–3048. [CrossRef] [PubMed]
62. Choi, J.; Shoaib, M.; Yin, T.; Nayyar, G.; Shinozaki, K.; Stevens, J.F.; Becker, L.B.; Kim, J. Tissue-Specific Metabolic Profiles After Prolonged Cardiac Arrest Reveal Brain Metabolome Dysfunction Predominantly After Resuscitation. *J. Am. Heart Assoc.* **2019**, *8*, e012809. [CrossRef] [PubMed]
63. Ritchie, M.E.; Phipson, B.; Wu, D.; Hu, Y.; Law, C.W.; Shi, W.; Smyth, G.K. limma powers differential expression analyses for RNA-sequencing and microarray studies. *Nucleic Acids Res.* **2015**, *43*, e47. [CrossRef]
64. Pang, Z.; Chong, J.; Zhou, G.; de Lima Morais, D.A.; Chang, L.; Barrette, M.; Gauthier, C.; Jacques, P.; Li, S.; Xia, J. MetaboAnalyst 5.0: Narrowing the gap between raw spectra and functional insights. *Nucleic Acids Res.* **2021**, *49*, W388–W396. [CrossRef] [PubMed]
65. Law, C.W.; Zeglinski, K.; Dong, X.; Alhamdoosh, M.; Smyth, G.K.; Ritchie, M.E. A guide to creating design matrices for gene expression experiments. *F1000Res.* **2020**, *9*, 1444. [CrossRef] [PubMed]
66. Langfelder, P.; Horvath, S. WGCNA: An R package for weighted correlation network analysis. *BMC Bioinform.* **2008**, *9*, 559. [CrossRef] [PubMed]
67. Horgusluoglu, E.; Neff, R.; Song, W.M.; Wang, M.; Wang, Q.; Arnold, M.; Krumsiek, J.; Galindo-Prieto, B.; Ming, C.; Nho, K.; et al. Integrative metabolomics-genomics approach reveals key metabolic pathways and regulators of Alzheimer's disease. *Alzheimer's Dement.* **2021**, *18*, 1260–1278. [CrossRef] [PubMed]
68. Zhou, G.; Pang, Z.; Lu, Y.; Ewald, J.; Xia, J. OmicsNet 2.0: A web-based platform for multi-omics integration and network visual analytics. *Nucleic Acids Res.* **2022**, *50*, W527–W533. [CrossRef] [PubMed]
69. Kanehisa, M.; Furumichi, M.; Sato, Y.; Ishiguro-Watanabe, M.; Tanabe, M. KEGG: Integrating viruses and cellular organisms. *Nucleic Acids Res.* **2021**, *49*, D545–D551. [CrossRef] [PubMed]
70. Pons, P.; Latapy, M. Computing Communities in Large Networks Using Random Walks. In *Computer and Information Sciences-ISCIS 2005, Prooceedings of the 20th International Symposium, Istanbul, Turkey, October 26–28, 2005*; Springer: Berlin/Heidelberg, Germany, 2005.

Disclaimer/Publisher's Note: The statements, opinions and data contained in all publications are solely those of the individual author(s) and contributor(s) and not of MDPI and/or the editor(s). MDPI and/or the editor(s) disclaim responsibility for any injury to people or property resulting from any ideas, methods, instructions or products referred to in the content.

pharmaceuticals

MDPI

Article

Efficacy of Kan Jang® in Patients with Mild COVID-19: A Randomized, Quadruple-Blind, Placebo-Controlled Trial

Levan Ratiani [1], Elene Pachkoria [1], Nato Mamageishvili [2], Ramaz Shengelia [2], Areg Hovhannisyan [3] and Alexander Panossian [4,*]

[1] Department of Infectious Diseases, The First University Clinic, Tbilisi State Medical University, Gudamakari St., Tbilisi 0141, Georgia; l.ratiani@tmsu.edu (L.R.); e.pachkoria@tsmu.edu (E.P.)
[2] Department for History of Medicine and Bioethics, Faculty of Medicine, Tbilisi State Medical University, Vazha-Pshavela Ave. 33, Tbilisi 0162, Georgia; nato15sg@yahoo.com (N.M.); r.shengelia@tmsu.edu (R.S.)
[3] Institute of Fine Organic Chemistry of the National Academy of Science, Azatutian Ave. 26, Yerevan 375014, Armenia; dopingareg@gmail.com
[4] Phytomed AB, Sjöstadsvägen 6A, Lgh 1004, 59344 Västervik, Sweden
* Correspondence: ap@phytomed.se; Tel.: +46-733306226

Abstract: ***Background and aim***. This study aimed to assess the efficacy of the treatment of Kan Jang®, a fixed combination of *Andrographis paniculata* (Burm. F.) Wall. ex. Nees and *Eleutherococcus senticosus* (Rupr. & Maxim.) Maxim extracts in patients with mild symptoms of COVID-19. ***Methods***. One hundred and forty patients received six capsules of Kan Jang® (n = 68, daily dose of andrographolides—90 mg) or placebo (n = 72) and supportive treatment (paracetamol) for 14 consecutive days in a randomized, quadruple-blinded, placebo-controlled, two-parallel-group design. The efficacy outcomes were the rate of cases turning to severe, the detection rate of coronavirus SARS-CoV-2 over the time of treatment, the duration, and the severity of symptoms (sore throat, runny nose, cough, headache, fatigue, loss of smell, taste, pain in muscles) in the acute phase of the disease. Other efficacy measures included improving cognitive and physical performance, quality of life, and the levels of inflammatory blood markers—interleukin 6 (IL-6), C-reactive protein, and D-dimer. ***Results***. Kan Jang® significantly ($p < 0.05$) reduced the rate of cases turning to severe (5.36%) compared to the placebo (17.86%) and decreased the detection rate of SARS-CoV-2 virus over the time of the treatment. The statistical difference in the rates of patients with clinical deterioration in the Kan Jang treatment and placebo control groups was significant ($p = 0.0176$) both in the 112 patients in the included-per-protocol (IPP) analysis and in the 140 patients in the intended-to-treat (ITT) analysis ($p = 0.0236$); the absolute risk reduction in cases thanks to the Kan Jang treatment was 12.5%, and the number we needed to treat with Kan Jang was 8. The patient's recovery time (number of sick days at the home/clinic) was shorter in the Kan Jang group compared with the placebo group. The rate of attenuation of inflammatory symptoms in the Kan Jang® group was significantly higher, decreasing the severity of cough, sore throat/pain, runny nose, and muscle soreness compared with the placebo group. Kan Jang® significantly decreased the Wisconsin Upper Respiratory Symptoms scores compared to the placebo in the sample size of 140 patients. However, the relief of fatigue and headache and the decrease in IL-6 in the blood were observed only in a subset of 86 patients infected during the second three waves of the pandemic. Kan Jang® significantly increased physical activity and workout; however, it did not affect cognitive functions (attention and memory), quality of life score, inflammatory marker D-dimer, and C-reactive protein compared with the placebo group. ***Conclusions***. Overall, the results of this study suggest that Kan Jang® is effective in treating mild and moderate COVID-19 irrespective of the SARS-CoV-2 variant of infection.

Keywords: adaptogens; Kan Jang®; *Andrographis paniculata; Eleutherococcus senticosus*; clinical trial; mild COVID-19; IL-6; inflammatory symptoms

1. Introduction

The pathogenesis and progression of coronavirus infection is a multistep process [1,2], characterized by a wide variety of inflammatory symptoms and requires an appropriate therapeutic strategy, including targeting the initiation of defense response to the pathogen [3–6] and numerous extra- and intracellular interactions between components of host-defense and life-cycle regulation systems on all levels of regulation—genomic, transcriptomic, proteomic, metabolomic and macrobiotic [7]. Effective prevention or treatment of COVID-19 and other viral infections requires pharmaceutical intervention, which directly targets the virus and indirectly affects various defense systems mechanisms, including immune and inflammatory response, activation of detoxification processes, and reparation of oxidative stress-induced damages in compromised cells (Figure 1) [8]. This can be achieved with herbal preparations that have polyvalent and pleiotropic actions on host-defense systems by synergistically targeting multiple elements of molecular networks involved in inflammatory defense response; these are presumably more effective than mono-drugs that target only one receptor [9,10].

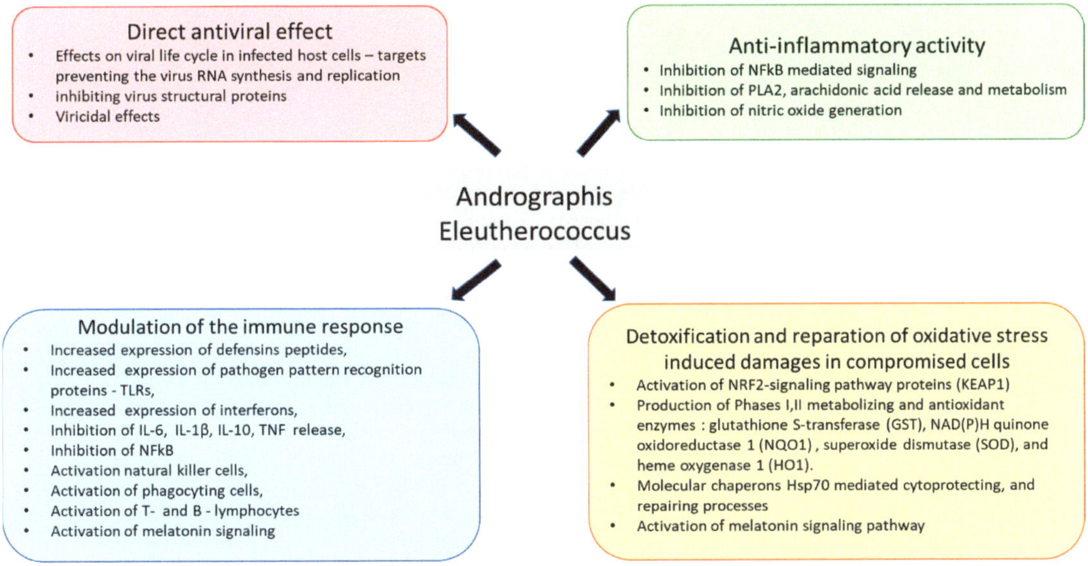

Figure 1. Schematic diagram of reported effects of Andrographis and Eleutherococcus elucidated in animal and cell culture models: (i) modulatory effects on immune response (blue block), (ii) anti-inflammatory activity (green block), (iii) detoxification and repair of oxidative-stress-induced damage in compromised cells (brown block), and (iv) direct antiviral effect via infraction with viral docking or replication (red block), modified from Panossian and Brendler [8].

A blend of extracts of *Andrographis paniculata* (Burm. F.) Wall. Ex. Nees with *Eleutherococcus senticosus* (Rupr. & Maxim.) Maxim, known in Scandinavia under the proprietary name Kan Jang, was demonstrated to be effective in several clinical trials treating various viral upper-respiratory-tract infections (URTIs) [11–18]. It should be emphasized that Kan Jang and *A. paniculata* are two entirely different active substances of two different chemical compositions, with two unique pharmacological and toxicological profiles (biological signatures; Figure 2), which are different from their ingredients and purified compounds, e.g., andrographolides and eleutherosides [9]. The gene expression network and pharmacology network studies suggest a potential antiviral response, virus clearance, and antigen presentation due to the synergistic interaction of *A. paniculata* and *E. senticosus* (Figure 2).

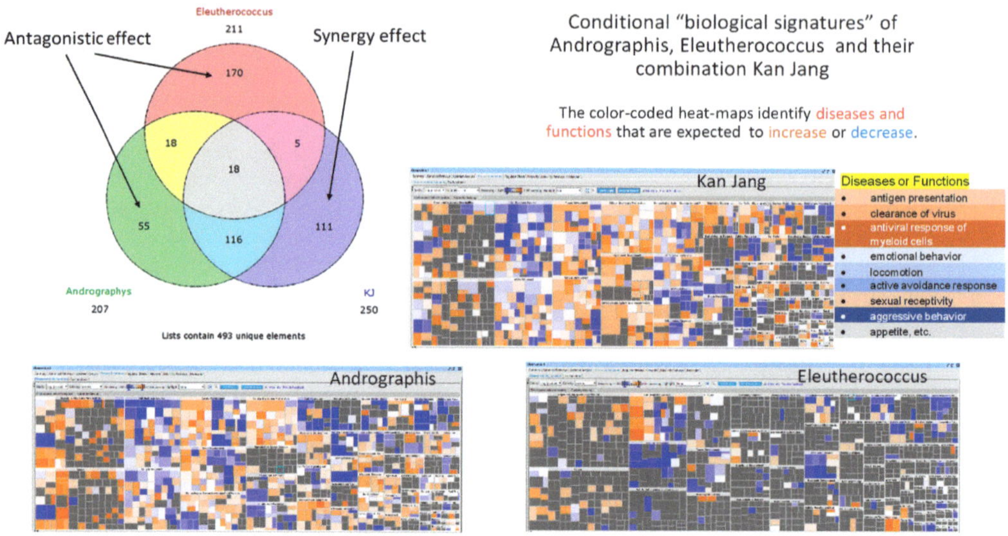

Figure 2. Venn diagrams show the number of deregulated genes in neuroglia cells in response to Andrographis, Eleutherococcus, and their fixed combination Kan Jang (KJ). The diagram includes overlapping sections and sections associated with synergistic interactions of Andrographis and Eleutherococcus in Kan Jang (111 unique genes) [12]. Conditional "biological signatures" are shown in the form of color-coded heatmaps, which identify the diseases and biological functions associated with gene expression in response to Andrographis, Eleutherococcus, and Kan Jang.

The evidence from many preclinical studies with *A. paniculata* and *E. senticosus* and their combination suggests that they can be used in prevention and treatment of viral infections, including coronavirus, at all stages of progression of inflammation, as well as in aiding the recovery of the organism by (i) modulating innate and adaptive immunity, (ii) anti-inflammatory activity, (iii) detoxification and repair of oxidative-stress-induced damage in compromised cells, (iv) direct antiviral effects of inhibiting viral docking or replication, and (v) improving quality of life during convalescence [8].

This study aimed to assess the clinical efficacy and safety of Kan Jang® in 140 patients with mild or moderate COVID-19 inflammatory symptoms in the last three days who were not requiring intensive care unit (ICU) admission as per protocol.

Recently, we published the results of an interim analysis of this clinical trial in 86 patients recruited in Georgia during three waves of COVID-19 (Figure 3) [18].

"The latest wave caused several coronavirus variants of SARS-CoV-2 was characterized by a remarkable speed, lesser severity, comorbidity, lower death rate, and much higher positivity rate, but more cases with breathlessness, and characterized by newer symptoms like gastrointestinal, particularly in a younger population, as compared to the 1st wave" [19]. The further recruiting of patients in the ongoing study was conducted from 6 May 2022 due to the spread of the last wave of the SARS-CoV-2 variants pandemic in Georgia (Figure 3).

This study provides the results of the analysis in the overall sample size of 140 patients included per protocol (IPP) and the intention to treat (ITT), including an additional subset of 54 patients admitted during the last wave of the pandemic, who were affected mainly by known and unknown mutants/variants of SARS-CoV-2.

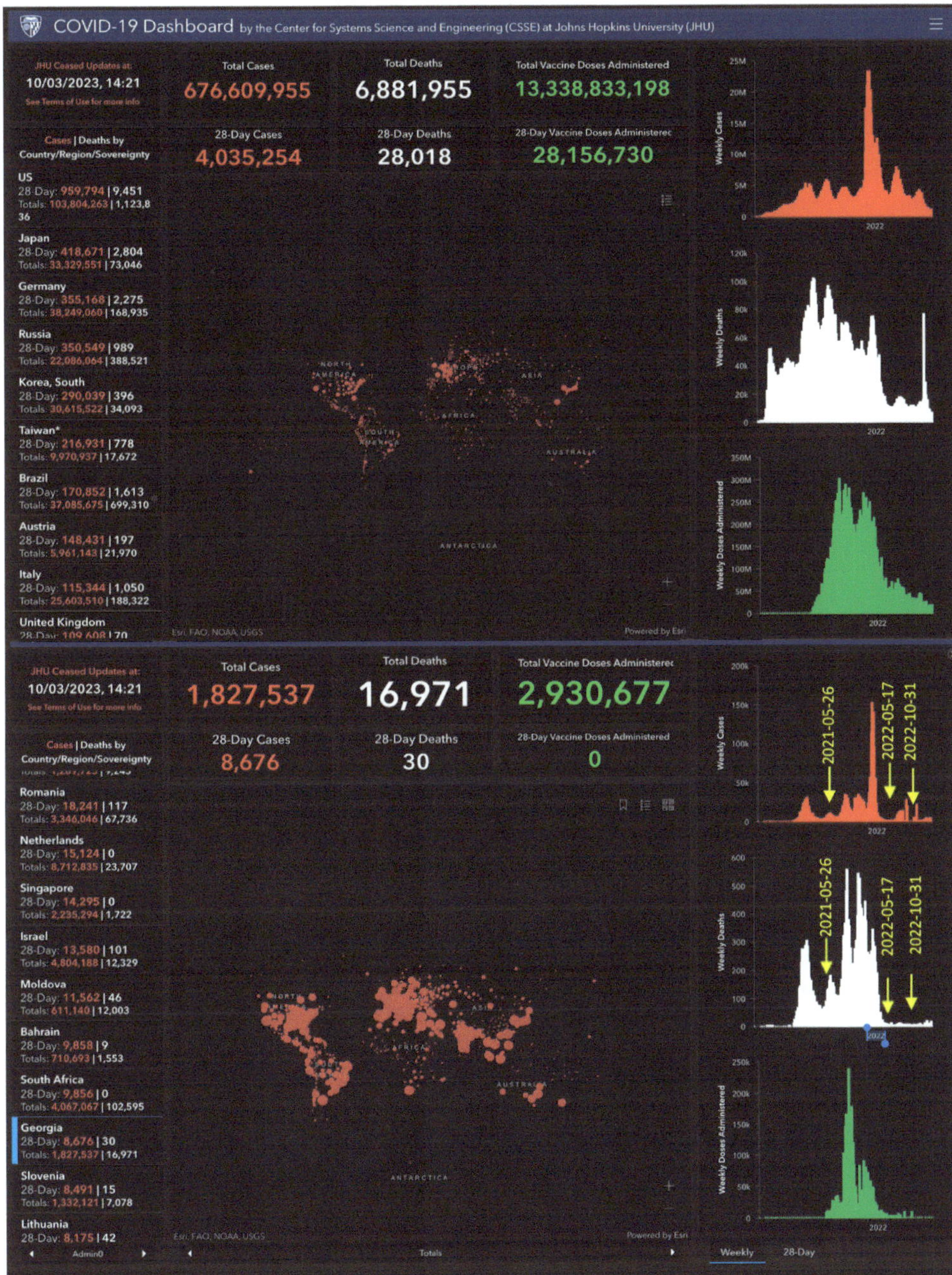

Figure 3. The figure shows six waves of the COVID-19 pandemic, including the number of infected patients recorded over time (in red color charts) in Georgia and worldwide. In white colors are the number of deaths, and the number of vaccinations are in green colors. Eighty-six patients were studied during three waves characterized as the most severe period of the pandemic from 26 May 2021 to 30 March 2022, while a subset of 54 patients was studied on the last wave of less harmful

SARS-CoV-2 variants' infection from 17 May 2022 to 30 October 2022. All 140 patients were diagnosed as having mild or moderate COVID-19 according to the WHO classification [16]. The background of Figure 3 was made using a screenshot of the COVID-19 Dashboard from the Center for Systems Science and Engineering (CSSE) at Johns Hopkins University (JHU). This is available online from https://www.arcgis.com/apps/dashboards/bda7594740fd40299423467b48e9ecf6; (accessed on 28 July 2023), and the details can be seen in high-resolution and real-time online.

2. Results

2.1. Patients

Demographic and Baseline Characteristics

One hundred and fifty-eight patients with confirmed diagnoses based on a positive SARS-CoV-2 test, experiencing mild to moderate COVID-19 symptoms [20], were screened for compliance with the inclusion criteria and assessed for eligibility. One hundred and forty patients with at least 3 to 8 symptoms (fatigue, headache, nasal discharge, loss of smell, taste, cough, muscle pain, and body temperature from 37 to 38 °C) for the last three days before admission to the hospital met the inclusion criteria and were randomly assigned to one of two treatment groups, Kan Jang® (A) or placebo (B) (Figure 4).

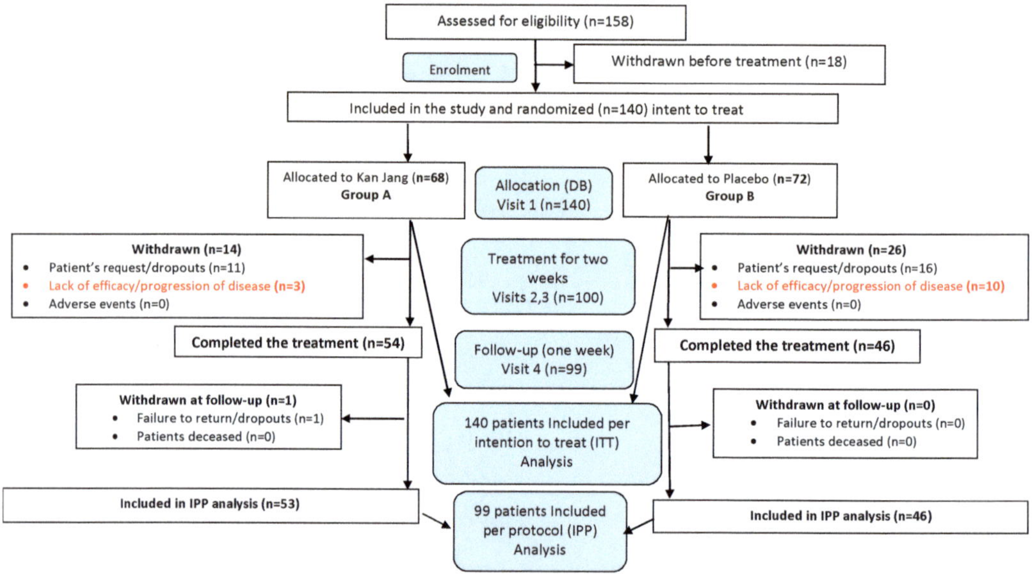

Figure 4. Schematic diagram of the trial. For details of the disposition of patients, see Supplement S1.

The groups did not show differences in baseline demographic, physical, or other critical clinical measurements, except for physical activity, muscle pain, loss of smell, and URTI scores, which were worse in the Kan Jang group than the placebo group (Table 1).

Several blood parameters, specifically blood serum IL-6, D-dimer, C-reactive protein, neutrophils, lymphocytes, monocytes, eosinophils, and basophils counts, were substantially higher than the normal ranges typical for acute viral inflammation. However, no significant difference between groups for these inflammatory markers was observed (Table 1).

Table 1. Baseline demographic characteristics, outcome measures, and laboratory biochemical and hematological measurements.

		Unit		Group A Kan Jang			Group B Placebo		Signif. of Difference
Parameters			**n**	**Mean**	**SD.**	**n**	**Mean**	**SD.**	***p*-Value**
Age		Years	68	50.35	17.82	72	45.74	16.82	0.143 [b]
Gender, Male/Female, %—53/87 = 61%		Male/Female	68	28/40 = 70%			72	25/47 = 53%	0.785
BMI		kg/m^2	68	24.58	3.274	72	24.52	3.376	0.908 [b]
Start of symptoms		days	68	<3		72	<3		
Viral load, SARS-CoV-2		%	68	100		72	100		
Body temperature		°C	68	37.86	0.58	72	37.7	0.59	0.148 [b]
Fatigue	100% patients	A.U.	68	1.794	0.407	72	1.819	0.539	0.755 [b]
Headache	91% of patients	AU.	64	1.859	0.393	63	1.794	0.480	0.443 [b]
Sore throat	62% of patients	AU.	49	1.776	0.422	38	1.632	0.633	0.2573 [b]
Cough	55% of patients	AU.	41	1.902	0.490	36	1.833	0.378	0.606 [b]
Pain in muscles **	51% of patients	AU.	39	1.872 **	0.656	32	1.469 **	0.671	0.013 [a]
Runny nose	27% of patients	AU.	23	1.957	0.475	15	1.733	0.458	0.247 [b]
Loss of smell **	16% of patients	AU.	8	2.375 **	0.744	14	1.714 **	0.469	0.036 [b]
Loss of taste	4% of patients	AU.	4	2.500	0.577	2	3.000	0	0.312 [b]
Physical activity **		AU.	68	12.75 **	2.984	72	13.85 **	3.005	0.019 [b]
Physical activity (daily walk) **		min	68	7.279 **	8.614	72	12.57 **	13.45	0.008 [b]
Decreased attention (d2-test)		%E (errors)	68 68	28.34	21.47	72	26.00	26.83	0.189 [b]
URTI **		WI score	68	17.59 *	6.497	72	14.69 *	5.832	0.006 [a]
QOL		WI score	68	36.29	12.15	72	37.14	12.81	0.163 [b]
Blood serum IL-6 (normal level < 7 pg/mL)		pg/mL	68	12.60	54.29	72	9.39	17.86	0.970 [b]
D-dimer (normal range from 0.1 to 0.5 mg/L)		mg/L	68	0.812	1.528	72	4.431	32.94	0.672 [b]
C-reactive protein (normal level < 5 mg/L)		mg/L	68	12.74	13.82	72	16.86	23.30	0.989
ALT (normal level < 35 U/L)		U/L	68	27.69	19.90	72	26.79	20.46	0.831
AST (normal level < 32 U/L)		U/L	68	25.32	19.07	72	26.83	19.49	0.241
Total WBC count (normal range: 3.6–11.0 × 10^9 cells/L)		10^9/L	68	5.872	1.863	72	5.271	1.936	0.064
Erythrocytes, RBC (normal range: 3.8–5.8 × 10^{12} cells/L)		10^{12}/L	68	4.699	0.479	72	4.770	0.601	0.231
Hemoglobin, Hb (normal range: 13.5–17.0 g/dL)		g/dL	68	12.95	1.647	72	13.51	1.706	0.053
Hematocrit, HCT (normal range: 40–50, L/L)		L/L	68	40.48	4.924	72	41.56	6.179	0.064
Platelet count (normal range: 150–380 × 10^3 cells/µL)		10^3 µL	68	207.1	49.49	72	200.8	51.22	0.565
Neutrophil count (normal range: 1.8–7.5 × 10^9 cells/L)		10^9/L	68	60.090 *	11.80	72	62.49 *	13.43	0.458
Lymphocyte count (normal range: 1.0–4.0 × 10^9 cells/L)		10^9/L	68	28.72 *	11.47	72	27.54 *	12.33	0.560
Monocyte count (normal range: 0.1–1.0 × 10^9 cells/L)		10^9/L	68	9.194 *	15.47	72	6.457 *	3.572	0.132
Eosinophil count (normal range: 0.1–0.4 × 10^9 cells/L)		10^9/L	68	1.615 *	1.391	72	1.285 *	1.136	0.144
Basophil Count (normal range: 0.01–0.1 × 10^9 cells/L)		10^9/L	68	0.471 *	0.229	72	0.477 *	0.296	0.722

*—over the normal range; **—significantly different from a parallel group; [a]—unpaired parametric *t*-test; [b]—unpaired nonparametric Mann–Whitney rank test.

2.2. *Efficacy*

The therapeutic efficacy of Kan Jang® was assessed by comparing (i)—differences in the time to resolve inflammatory symptoms in Kan Jang® and placebo groups of patients and (ii)—differences in the relief severity of inflammatory symptoms from the baseline in the Kan Jang® and placebo groups of patients. Treatment groups were compared for all efficacy outcome measures to assess the primary and secondary endpoints.

2.2.1. Primary Endpoints

The Rate of Patients with Clinical Deterioration and Virus Clearance

In Kan Jang® Group A of 68 patients, 12 were dropouts, and 3 were withdrawn from the study due to lack of efficacy and disease progression; they continued the treatment with steroids and antibiotics (Figure 5).

In Placebo Group B of 72 patients, 16 were dropouts, and 10 were withdrawn from the study due to lack of efficacy and disease progression; they continued the treatment with steroids and antibiotics (Figure 5a).

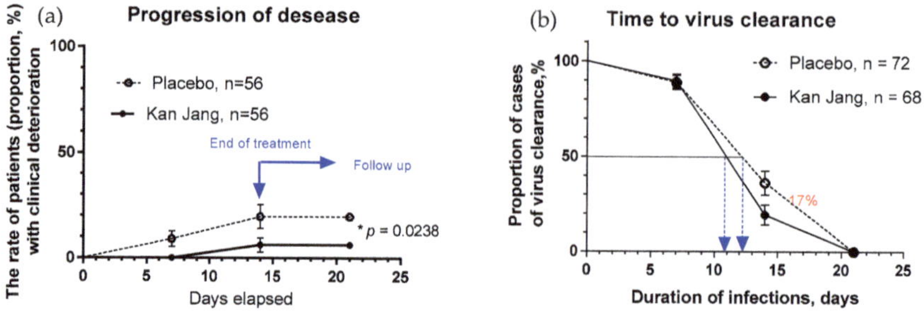

Figure 5. (**a**)—The rates of patients with clinical deterioration in the treatment and control groups; hazard ratio Kan Jang/placebo = 0.2906, 95% CI of ratio from 0.094 to 0.894. (**b**)—The virus clearance in the treatment and control groups: Kaplan–Meier curves show the percent of patients with SARS-CoV-2 virus over the time from randomization (Day 1) to the end of the treatment (Day 14) and the follow-up period for one week (Day 21) in the treatment and control groups; hazard ratio Kan Jang/placebo = 1.686, 95% CI of ratio from 0.8698 to 3.269. * $p < 0.05$.

The disease progression rate in the placebo group (56 patients) was 17.86%, while in the Kan Jang group (56 patients) it was 5.36% (p = 0.0176, significant result at 95% significance level (confidence 95%), power—92.99%, IPP analysis; Figure 5a). The statistical difference (st. error 0.0477) in the rates of patients with clinical deterioration in the Kan Jang treatment (A) and placebo control (B) groups was significant (p = 0.0236 in 140 patients included in ITT analysis, observed power—90.66%).

The overall clinical effectiveness, defined as the ratio of proportions of effective to ineffective cases in the Kan Jang (94.54/5.36 = 17.65) vs. placebo control (82.14/17.86 = 4.6) groups, was 3.84. Absolute risk reduction (ARR%) by Kan Jang treatment was 12.5%, while the relative risk reduction was 214.8%. The number of patients that we needed to treat with Kan Jang to prevent one additional bad outcome (defined as the number we needed to treat, NNT = 1/ARR) was eight.

The percentage of patients with a negative SARS-CoV-2 virus test was 17% lower in the Kan Jang group compared to the placebo group after 14 days of the treatment (19.3% vs. 36.3%, difference—17.0 ± 8.5%, p = 0.819; Figure 5b). The rate of virus clearance was faster at 1.6 days shorter in the Kan Jang group (10.8 days) compared to the placebo group (12.4 days), along with the time to virus clearance in 50% of patients (hazard ratio Kan Jang/placebo = 1.686, 95% CI of ratio from 0.8698 to 3.269; Figure 5b).

Recovery Time and Time to Fever Resolution

The proportion of patients with high body temperature (from >37 °C to <38 °C) was 17.3% lower in the Kan Jang group (29.5%) compared to the placebo group (46.8%) after one week of treatment; however, this difference was statistically insignificant (Figure 6a). The duration of increased body temperature (from >37 °C to <38 °C) was also shorter in the Kan Jang group compared to the placebo group; median recovery in the Kan Jang® group was six days while in the placebo group it was 7 days (hazard ratio Kan Jang/placebo = 1.336, 95% CI of ratio from 0.808 to 2.309; Figure 6a).

The median (50%) number of patients who elapsed at the clinic was reached at three days less for the Kan Jang treatment (11 days) compared to the placebo (14 days) (hazard ratio Kan Jang/placebo = 0.7857, 95% CI of ratio from 0.530 to 1.164; hazard ratio Kan Jang/placebo = 1.232, 95% CI of ratio from 0.778 to 1.951; Figure 6b).

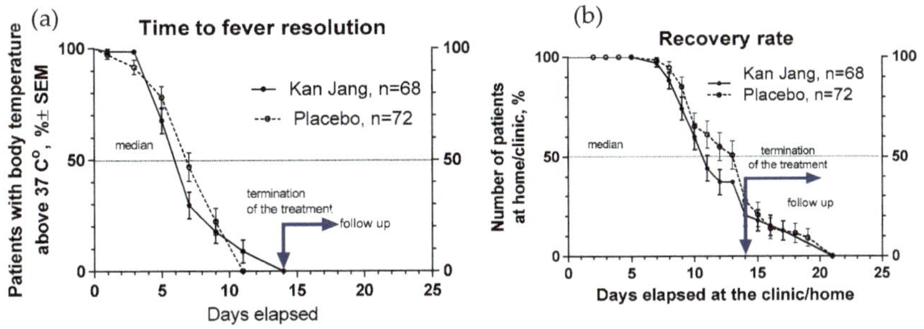

Figure 6. (**a**)—Duration of increased body temperature (from >37 °C to <38 °C) in the treatment and control groups; median recovery: Kan Jang®—6 days, placebo—7 days; hazard ratio Kan Jang/placebo = 1.336, 95% CI of ratio from 0.808 to 2.309. (**b**)—Duration of hospitalization in the treatment group and control group; Kaplan–Meier curves show the percent of patients hospitalized over the time from randomization (Day 1) to the end of the treatment (Day 14) and followed up for one week (Day 21) in the treatment and control groups.

The Severity and Time to Resolution of Inflammatory Symptoms

The occurrence of various inflammatory symptoms differed at the baseline in the cohort of COVID-19 patients recruited in this study (Table 1); the most common symptoms were fatigue (in 100% of patients), headache (in 91% of patients), sore throat (in 62% of patients), cough (in 55% of patients), and muscle pain (51%), while other symptoms were less frequently observed—rhinorrhea (27%) and loss of smell (16%) and taste (4%).

The results of the assessment of upper respiratory symptoms using the Wisconsin URTI survey (including runny and plugged nose, sneezing, sore throat, cough, hoarseness, headache and congestion, and feeling tired) show a beneficial effect of Kan Jang compared to the placebo (Figure 7).

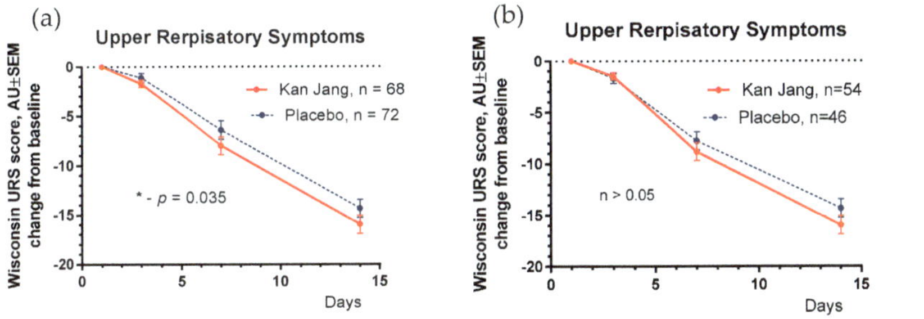

Figure 7. (**a**)—Between-groups comparison of the changes from the baseline of Wisconsin Upper Respiratory Symptoms scores in patients of group A (Kan Jang) and group B (placebo) over the time from Day 1 to Day 21 shows the statistically significant positive effect of Kan Jang treatment vs. placebo both (**a**)—in 140 patients included in ITT analysis ($p = 0.035$), and (**b**)—in 100 patients included in IPP analysis ($p > 0.05$); * $p < 0.05$.

The severity of all inflammatory symptoms gradually decreased from the baseline to the end of therapy (Day 14) and the follow-up period (21 days) in both groups of patients (Figures 7–11). However, the relief from inflammatory symptoms for patients in the Kan Jang group was significantly better compared to placebo effects over the time of the treatment and follow-up, including relief of sore throat (Figure 8), rhinorrhea (nasal discharge/runny nose, Figure 9), myalgia (muscle pain, Figure 10), and cough (Figure 11).

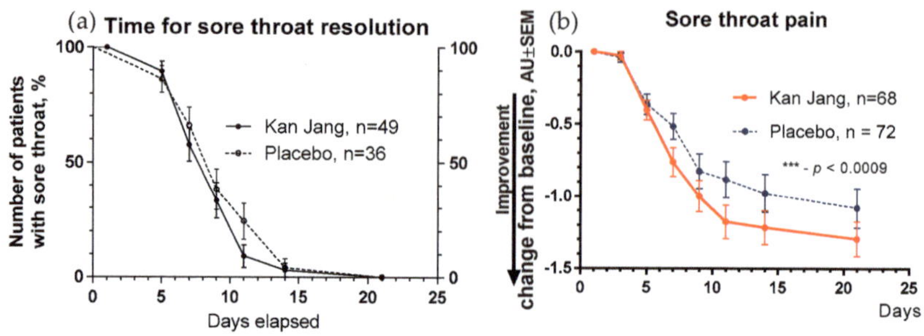

Figure 8. (**a**)—Time to relieve sore throat in the treatment and control groups: Kaplan–Meier curves show the percent of patients with a sore throat over the time from randomization (Day 1) to the end of the treatment (Day 14) and follow-up for one week (Day 21); median recovery, Kan Jang®, —7 days, placebo—11 days; hazard ratio Kan Jang/placebo = 1.36, 95% CI of ratio from 0.7350 to 2.516. (**b**)—Relief of the sore throat; the changes in the severity of the symptom from the baseline of patients in group A (Kan Jang) and group B (placebo) over the time from Day 1 to Day 21. Between-groups comparison of the changes in the symptom severity from the baseline over time shows significant interaction ($p = 0.0009$). The Kan Jang® treatment has a statistically significant effect on the relief of the sore throat compared to the placebo in the ITT analysis. *** $p < 0.0001$.

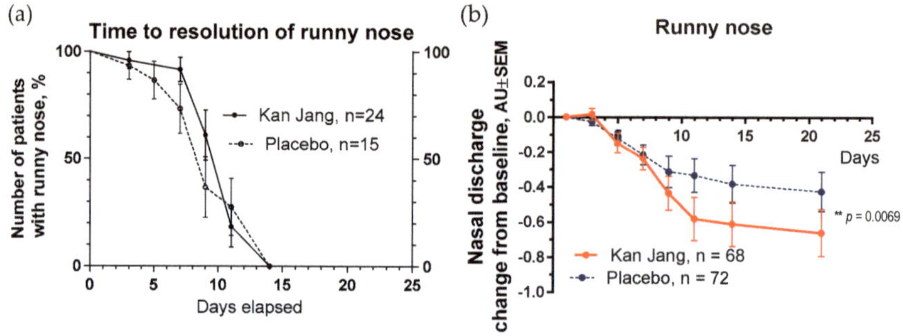

Figure 9. (**a**)—Time to a resolution of runny nose in the treatment and control groups: Kaplan–Meier curves show the percent of patients with a runny nose over the time from randomization (Day 1) to the end of the treatment (Day 14) and follow-up for one week (Day 21) and in the treatment and control groups; median recovery: Kan Jang®—14 days, placebo—14 days; hazard ratio Kan Jang/placebo = 0.672, 95% CI of ratio from 0.2321 to 1.944. (**b**)—Reduction in nasal discharge; the changes in the severity of the symptom from the baseline of patients in group A (Kan Jang) and group B (placebo) over the time from Day 1 to Day 21. Between-groups comparison of the changes in the symptom severity from the baseline over time shows significant interaction ($p = 0.0069$). The Kan Jang® treatment has a statistically significant effect on the reduction in nasal discharge compared to the placebo ITT analysis. ** $p < 0.01$.

The rates of resolution of cough over time were similar in the Kan Jang and placebo groups (Figure 11a); however, the relief from severity of the cough for patients taking Kan Jang over the time from the baseline was significantly more effective compared to the placebo (Figure 11b). Kan Jang® relieved the severity of cough compared to the placebo, as was evident in the ITT analysis in 140 randomized patients ($p = 0.053$), in the IPP analysis in 100 patients who completed the study ($p = 0.018$), in a subset of 54 patients during last wave of the COVID-19 pandemic in Georgia ($p = 0.0049$, ITT analysis), and in a subset of 42 patients who completed the study during the last wave of the COVID-19 pandemic in Georgia ($p = 0.0047$, IPP analysis).

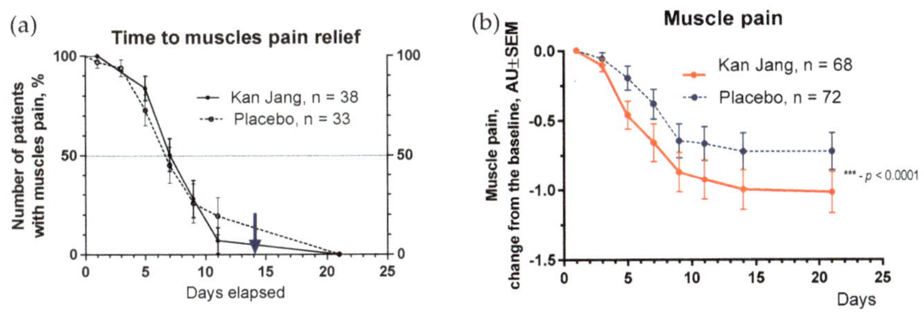

Figure 10. (**a**)—Time to muscle pain relief in the treatment and control groups. Kaplan–Meier curves show the percent of patients with muscle pain over the time from randomization (Day 1) to the end of the treatment (Day 14) and follow-up for one week (Day 21); median recovery, Kan Jang®—9 days, placebo—11 days; hazard ratio Kan Jang/placebo = 0.8965, 95% CI of ratio from 0.4434 to 1.812. (**b**)—Relief of the muscle pain; the changes in the severity of the symptom from the baseline of patients in group A (Kan Jang) and group B (placebo) over the time from Day 1 to Day 21. Between-groups comparison of the changes in the severity of the symptom from the baseline over time in 140 patients included in the ITT analysis shows significant interaction ($p < 0.0001$). Statistical significance of the effects of Kan Jang® treatment vs. placebo on muscle pain relief is shown as: *** $p < 0.0001$.

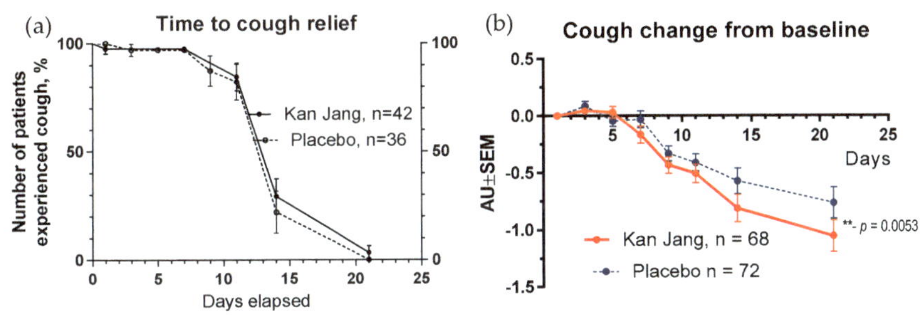

Figure 11. (**a**)—Time to resolution of cough in the treatment and control groups: Kaplan–Meier curves show the percent of patients with a cough over the time from randomization (Day 1) to the end of the treatment (Day 14) and follow-up for one week (Day 21) and in the treatment and control groups; hazard ratio Kan Jang/placebo = 0.7027, 95% CI of ratio from 0.2744 to 1.799. (**b**)—Relief of the cough; the changes in the severity of the symptom from the baseline of patients in group A (Kan Jang) and group B (placebo) over the time from Day 1 to Day 21. Between-groups comparison of the changes in the symptom severity from the baseline over time shows significant interaction ($p = 0.0053$). The Kan Jang® treatment significantly relieved the cough compared to the placebo in the ITT analysis, ** $p < 0.01$.

2.2.2. Secondary Endpoints

Secondary endpoints comprised the measures of (i)—immune response marker IL-6 concentration in the serum, (ii)—blood hypercoagulation marker D-dimer, (iii)—inflammatory marker C-reactive protein, (iv)—physical activity, (v)—physical performance, and (vi)—cognitive performance.

Blood Serum Markers of Immune Response and Inflammation

At the beginning of the study, the baseline levels of all selected markers of immune response and inflammation—IL-6, C-reactive protein, and D-dimer—were considerably higher than typical blood values (Table 1). In 3 days of the treatment with Kan Jang, the level of IL-6 reached the standard limit of 7 pg/mL, while in the placebo group, 14 days after randomization (Figure 12a).

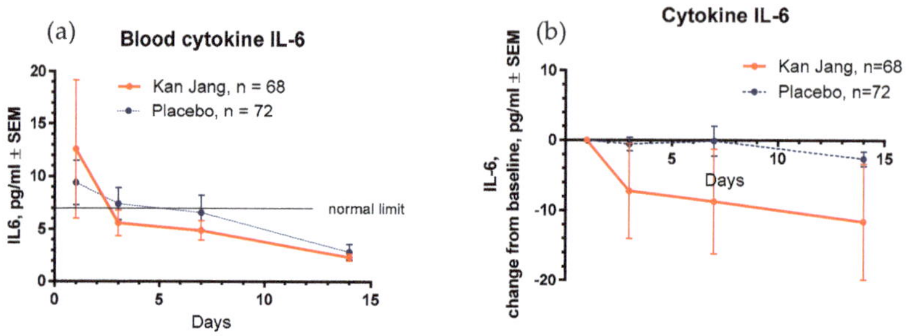

Figure 12. (**a**)—Concentration of IL-6 (mean ± SD) in the blood of patients in group A (Kan Jang) and group B (placebo) over the time from Day 1 to Day 14. (**b**)—The changes from the baseline in the levels (mean ± SD) of cytokine IL-6 in the blood of patients in group A (Kan Jang) and group B (placebo) over the time from Day 1 to Day 14. Between-groups comparison of the changes in the level of cytokine IL-6 in the blood from the baseline over time shows an insignificant difference (p = 0.1619) between groups A and B.

Between-groups comparison of the changes in the level of cytokine IL-6 in the blood did not show significant with treatment time ($p = 0.1619$), despite the Kan Jang® treatment showing a noteworthy decrease in cytokine IL-6 in the blood (−11.575 pg/mL on day 14) compared to the placebo (−2.246 pg/mL on day 14; Figure 12b, ITT analysis).

The levels of C-reactive protein and D-dimer were also normalized in the blood of all patients at the end of treatment; however, the Kan Jang® treatment made no significant difference in the effects on C-reactive protein and D-dimer compared to the placebo ($p > 0.05$, Figures in Supplement S2).

Physical Activity and Physical and Cognitive Performance

The Kan Jang® treatment significantly increased physical activity ($p = 0.0023$) and workout time (in min, $p = 0.020$) in patients compared to the placebo (Figure 13).

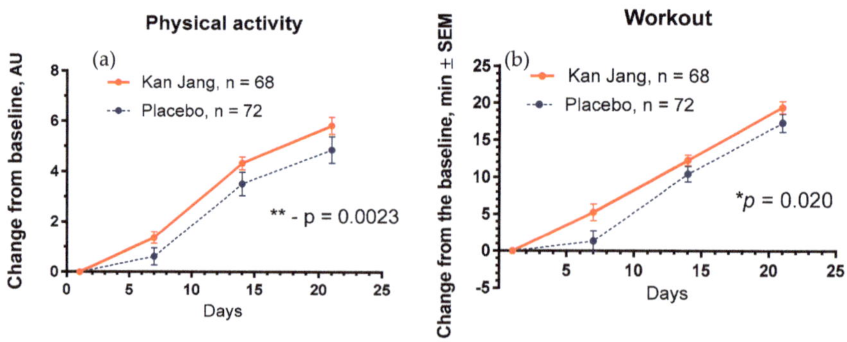

Figure 13. Between-groups comparison of the changes from the baseline in (**a**)—the overall physical activity and (**b**)—physical performance/workout time (in min) of patients in group A (Kan Jang) and group B (placebo) over the time from Day 1 to Day 21 shows the statistically significant positive effect of Kan Jang treatment vs. placebo in the 140 patients included in the ITT analysis. * $p < 0.05$, ** $p < 0.001$.

No significant differences between the effects of Kan Jang® and placebo were observed on cognitive performance, quality of life, and other secondary outcomes.

2.3. Safety

The extent of exposure to Kan Jang was 84 capsules consumed in 14 consecutive days, in a total of 1.26 g andrographolides. The dose was 6 capsules daily in a total of 1672.5 mg of the fixed combination of 1560 mg Herba Andrographis standardized native extract (corresponding to 90 mg of andrographolides) and 112.5 mg Radix Eleutherococci native extract, which is equivalent to 7–12.6 g and 1.9–3.4 g of herbal substances, correspondingly.

Regardless of causality, adverse events were monitored for all patients from the first dose and through the one-week follow-up period. No adverse events of allergic reactions like urticaria, angioedema, paresthesia, anaphylactic reactions, rush, and pruritus (itchy skin) were observed after initiation of study treatments, including events likely to be related to the underlying disease or likely to represent concomitant illness (Supplement S3). Deaths, other serious adverse events, and other significant adverse events that deserve special attention were not recorded in the 140 patients during the treatment in either study group. The results of blood analysis showed the normalizing anti-inflammatory effect of Kan Jang. Vital signs related to safety did not reveal evidence of a drug effect.

3. Discussion

Several antiviral drugs (Paxlovid™ [21], Remdesivir, Bebtelovimab [22], Molnupiravir [23], Andrographis [24–26]) and immune modulators (Olumiant, an interleukin-6 receptor blocker [27]) were approved by drug regulatory authorities for use against COVID-19 for treating non-hospitalized or hospitalized COVID-19 patients. Recently, a systematic review and meta-analysis of 50 clinical trials (involving 11,624 patients) of Chinese herbal medicine in COVID-19 concluded that they are effective and safe in combination with conventional "Western" drugs, but the certainty of the evidence ranged from moderate to very low [23]. Similar conclusions were made in other systematic reviews and meta-analyses of Chinese herbal medicine in COVID-19 [28–32].

Herba Andrographidis preparations are the first herbal medicine for treating COVID-19 and were formally adopted in Thailand [24–26] despite contradictory conclusions from the efficacy and safety results in clinical studies [33–40].

In this clinical trial, the efficacy of Kan Jang in patients with mild or moderate symptoms of COVID-19 was studied. In the first subset of 86 patients, Kan Jang was found to effectively relieve sore throat, muscle pain, and runny nose in patients with mild or moderate symptoms of COVID-19 [18]. Further increase in the sample size to 140 patients improved Kan Jang's efficacy in ameliorating cough, sore throat/pain, runny nose, and muscle soreness and reducing the Wisconsin Upper Respiratory Symptoms score compared to the placebo. Kan Jang® decreased blood inflammatory marker IL-6 over the recovery time in the subgroup of 86 patients with mild or moderate symptoms; however, the effect was statistically insignificant in the sample size of 140 patients, which was enriched with the 54 patients infected with a less pathogenic variant during the last wave of COVID-19 (Figure 3).

Kan Jang® increased physical activity and workout; however, it did not affect cognitive functions (attention and memory), quality of life score, time to normalization of body temperature, inflammatory marker D-dimer, or C-reactive protein compared to the placebo group.

Notably, the rate of patients with clinical deterioration and disease progression was significantly lower in the Kan Jang group than in the placebo group. The disease progression rate in the placebo group was about 2.5-fold higher than in the Kan Jang Group (Figure 5). The absolute risk reduction (ARR) via Kan Jang treatment was 14%, and the relative risk reduction (RRR) was 243.9%. The number of patients we needed to treat (NNT) with Kan Jang to prevent one additional bad outcome was found to be eight patients. As a rule, the higher the ARR and lower the NNT, the more effective the intervention. For comparison, Paxlovid™ effectively reduced the risk of progression (ARR of 6.2%) to severe COVID-19 in symptomatic adults at high risk of progression, and the number we needed to treat (NNT) was 17 patients [21]. Paxlovid™, unlike Kan Jang, induces dysgeusia, diarrhea, hypertension, and myalgia [21]. The safety of Kan Jang is an essential advantage as it is greater than the safety of other anti-COVID-19 drugs [21,41,42].

The treatment with Kan Jang® was well tolerated, and no adverse events were recorded in this study. That is consistent with the results of previous clinical studies where only four adverse events were recorded in one [17] of six studies [11–17] with 494 patients in the Kan Jang group. In that study, the only common adverse reaction was mild pruritus observed in four patients in the Kan Jang® group and six in the placebo group [17], and no severe adverse reactions were observed. There have been 0.706% adverse events in 562 patients in all seven clinical studies with Kan Jang® groups, including our recent study. That is significantly less than 131 mild to moderate adverse events observed in 690 (19%) patients from ten studies on *A. paniculata* herbal preparations that are essentially safe [43], and 383 (4.04%) adverse events in 9490 participants using andrographolide derivative injections, which can be life-threatening, mainly due gastrointestinal, skin and subcutaneous tissue disorders, as well as anaphylaxis [43].

Kan Jang has an excellent safety profile [16,17], presumably due to the contribution of the adaptogenic and antitoxic activity of Eleutherococcus [44]. Cytoprotective (neuroprotective, hepatoprotective, cardioprotective), stress-protective, antioxidant, antitoxic, and immunomodulating activity of Eleutherococcus preparations was demonstrated in many experimental models on isolated cells and (in vitro and ex vivo) animals [45–48]. This evidence of the safety of the Kan Jang combination shows that the pharmacological activity and toxicity of multi-component drugs are different from their ingredients, and the effects observed on purified compounds cannot be just extrapolated to their combinations with other substances, or vice versa, due to their multiple synergistic and antagonistic interactions in the organism [9,46–48] (Figure 2). In other words, the results obtained in this study are product-specific. They cannot be applied to mono herbal drugs, e.g., Andrographis extract, which in turn is the mixture of 39 compounds such as andrographolides, flavonoids, etc., and deregulates quite a different number of genes than expected from a simple calculation of many constituents of the plant extracts [9].

In this context, an essential difference in the modes and mechanisms of action of Kan-Jang and Andrographis compared with other drugs is their multitarget effects both directly on virus-receptor binding, viral membrane fusion into the host cell, viral replication, transcription, translocation, assembly, and release to infect other host cells [8,49], as well as indirect antiviral activity due to multiple effects on innate and adaptive immune systems, inflammatory response, and essentially on recovery of oxidative-stress-induced damage in compromised cells [8] (Figure 1).

The results of this study are consistent with the previous publications [11–18], where Kan Jang effectively relieved a sore throat, runny nose, cough, and muscle pain in patients with various upper-respiratory-tract viral infections. Kan Jang effectively ameliorated these URTI symptoms irrespective of the virulence and potential morbidity of different strains/variants of the SARS-CoV-2 virus, which were inducing mild to moderate COVID-19.

The limitations of our study were the short duration of symptoms of COVID-19 before treatment (three days) and the lack of concomitant chronic diseases, which can increase the risk of progression of the disease and intensive care therapy for patients. Further studies are required in patients at high risk of the progression of pneumonia, acute respiratory distress syndrome, and septic shock.

4. Materials and Methods

The methods of the study were described according to CONSORT recommendations [50] in a recent publication in detail, including study design, recruitment, and screening of patients, schedule of examinations, study population, patient inclusion and exclusion criteria, participant withdrawal, the intervention and comparator, doses and treatment regimens, methods of randomization and blinding, generation of sequence, allocation concealment, implementation, blinding of participants and personnel, completeness of outcome data, evaluation of compliance, statistical analysis, and sample size considerations [18] (Supplement S4).

4.1. Study Design, Recruitment, Screening of Patients, and Schedule of Examinations

This prospective, randomized, placebo-controlled, quadruple-blind, two-parallel-group (Figure 4 and Supplement S1), phase II interventional study was conducted at Tbilisi State Medical University, Tbilisi, Georgia, with the approval of the Biomedical Research Ethics Committee of Tbilisi State Medical University and the National Council on Bioethics (registration no. 3-2021/87, date of final protocol approval 25 March 2021). The ClinicalTrials.gov identifier is NCT04847518 (https://www.clinicaltrials.gov/ct2/show/NCT04847518; accessed on 6 June 2022).

Recruitment for the study was initiated on 26 May 2021, the 86th patient was recruited on 30 March 2022, and the study was completed on 30 October 2022.

Overall, 158 patients were assessed for eligibility; 140 patients met the inclusion criteria of mild COVID-19 symptoms [20] and were randomized and included in the ITT analysis of the study (Supplement S1, Full Analysis Dataset). In total, 100 patients (71.4% of randomized) completed their respective treatment cycles according to protocol, while 40 patients (38.6% of randomized) discontinued therapy after receiving at least one dose of study preparations at the patient's request. Nighty-nine patients, who completed the treatment, were evaluated for treatment efficacy for two weeks (visits 2 and 3) of treatment and one week after completing the treatment (the follow-up visit 4). The distribution between the study groups and the disposition of patients are shown in the flow chart in Figure 4.

4.2. Datasets Analyzed and Evaluation of Compliance

One hundred patients enrolled in the screening were assigned to randomized investigational products and included in the ITT analysis to ensure they met the inclusion criteria and to avoid potential bias due to the exclusion of patients at the baseline (visit 1). The compliance was assessed by a count of the returned capsules, regardless of when the dropout patients terminated the treatment. Overall compliance in each group was more than 80%, comprising 89.0 ± 2.8% in the Kan Jang group (n = 68) and 82.0 ± 3.4% in the placebo group (n = 72). There was an insignificant difference between groups (p = 0.2887) in the ITT analysis (Supplement S5).

An efficacy subset analysis per protocol (PP) was performed in patients when they completed the study therapy (visits 3 and 4). A per protocol (PP) analysis aims to identify a treatment effect on the symptoms. Therefore, some patients were excluded from the complete analysis set (ITT), and the PP population was used for the PP analysis, ensuring 100% compliance (Figure 4). The schedule of procedures and examinations is shown in Table 2.

Table 2. Schedule of examinations and procedures.

	Treatment							Follow-Up
	Day 1 Screening	Day 3	Day 5	Day 7	Day 9	Day 11	Day 14	Day 21
Doctor's visits	1 Baseline			2			3	4
Eligibility check/information	*							
Informed consent	*							
Clinical examination	*			*			*	*
Enrollment and allocation to intervention	*							
Treatment (Kan Jang or placebo)	*	*	*	*	*	*	*	
Biomarker assessments								
Body temperature (fever)	*	*	*	*	*	*	*	*

Table 2. *Cont.*

	Treatment							Follow-Up
	Day 1 Screening	Day 3	Day 5	Day 7	Day 9	Day 11	Day 14	Day 21
COVID-19 PCR test	*			*			*	*
Blood serum cytokine IL-6 (pg/mL)	*	*		*			*	
D-dimer (mg/L)	*			*			*	
C-reactive protein (mg/L)	*			*			*	
Blood cell count analysis	*			*			*	
ALT/AST	*			*				
Clinician- and observer-reported outcome assessments								
Cognitive performance (tests for attention and memory): d2 test	*			*			*	*
Wisconsin URS survey score	*	*		*			*	
Drug intake accountability							*	
Adverse events				*			*	*
Patient-reported outcome assessments								
Mild COVID-19 symptoms: • Fatigue • Headache • Loss of smell • Loss of taste • Rhinorrhea (nasal discharge) • Cough • Pain in muscles • Sore throat	*	*	*	*	*	*	*	*
Workout, min	*			*			*	*
Physical activity (questionnaire)	*			*			*	*
Paracetamol intake recording	*	*	*	*	*	*	*	
Rescue medication intake recording	*	*	*	*	*	*	*	

The medication compliance was calculated by counting the remaining capsules for each subject from the first to the last day of the study and verified by the study monitor at the end of the treatment. All the patients were provided with diary cards on which the daily usage of study medication was recorded. The duration of the treatment was recorded. Unused capsules were returned to the sponsor.

4.3. Study Population and Inclusion and Exclusion Criteria

The population for this study (aged 18 years and older of either sex; mean age: 47.98 ± 17.40 years) consisted of COVID-19 patients in a stable, moderate condition (i.e., not requiring intensive care unit (ICU) admission) with confirmed diagnoses based on positive SARS-CoV-2 tests and at least three of eight mild to moderate COVID-19 symptoms [20]: fatigue, headache, sore throat, nasal discharge, cough, pain in muscles, loss of taste and loss of smell (Table 1) in the last three days before recruitment for the study. Subjects were required to be under observation in the hospital, be able to take medication alone and give written informed consent.

Patients admitted with severe acute respiratory syndrome under invasive mechanical ventilation and diagnosed with an etiologic agent other than COVID-19 were excluded from the study. Other exclusion criteria were: acute and chronic pulmonary diseases, chronic rhinosinusitis, renal failure or creatinine 2.0 mg/dL, type 2 diabetes, autoimmune disease, patients taking antibiotics for a reason other than COVID-19 at enrollment, chronically suppressed immune system (AIDS, lymphoma, corticosteroid therapy, chemo- or radiotherapy for last six months), cancer patients taking immunosuppressive drugs, pregnant or lactating women, and subjects participating in other clinical studies or taking any medication influencing the outcome measures during the clinical trial.

Patients were free to withdraw from the clinical study at any time without giving a reason.

4.4. Intervention, Comparator, Doses, and Treatment Regimens

Pharmaceutical-grade standardized extracts of *A. paniculata* L. Nees. (herb) and *E. senticosus* (root) genuine extracts, as well as their fixed combination, Kan Jang®, were manufactured, tested, and then released for human use as per the ICH Q7A and EMEA guidelines for Good Agricultural and Collecting Practice and Good Manufacturing Practice (GMP) of active pharmaceutical ingredients at the Swedish Herbal Institute, which holds a valid EU-GMP license to produce pharmaceuticals.

One capsule of Kan Jang® (size 0, batch no. 40154, expiry date: February 2024) contains 260 mg of *A. paniculata* native extract SHA-10 (drug–native extract ratio of 4.5–8.1:1, extraction solvent 70% ethanol), including 15 mg of diterpene lactones (andrographolide and 14-deoxy-11,12-didehydroandrograholide), and 18.75 mg of *E. senticosus* native extract (drug–native extract ratio of 17–30:1, extraction solvent 70% ethanol, 0.25 mg Eleutherosides B and E); for details, see Supplementary File S5 in previous publication [18]. The matrix contains inactive excipients (microcrystalline cellulose and magnesium stearate). The placebo capsules (size 0, batch no. 40154) containing the inactive excipients were identical to the Kan Jang® capsules. Both preparations' appearance, smell, and color were similar and organoleptically undistinguishable.

The daily dose of the study intervention was two capsules three times per day for two consecutive weeks with a daily intake of 1560 mg of the dry extract *A. paniculata* SHA-10, corresponding to 90 mg of diterpene lactones andrographolides from *A. paniculata* herb and 112.5 mg of dry extract of E. senticosus, in the Kan Jang® group. All the patients were provided with diary cards on which the daily consumption of the study preparation was recorded. The number of capsules consumed during the treatment by each patient was verified for compliance according to the duration of the treatment/illness (days at the clinic). The Principal Investigator was responsible for maintaining drug accountability records.

4.5. Randomization, Blinding, and Allocation Concealment

Study preparations were randomly labeled by a qualified pharmacist (QP), and the random sequence of treatment codes was retained at the manufacturing site until the study was completed. Randomization, blinding, and allocation concealment were performed as previously reported [18], ensuring a quadruple-blind design.

4.5.1. Randomization and Blinding

Study preparations were labeled by a quality-assurance-responsible person (QP) at the manufacturing site. The QP used a random numbers sequence, which was generated by the PRISM GraphPad software (2017 Online version, GraphPad Software Inc., San Diego, CA, USA) "random number generator" (https://www.graphpad.com/quickcalcs/randomize1.cfm, accessed on 16 February 2022). The randomization sequence comprises a table of two columns (A and B) filled with randomly distributed unique numbers from 1 to 160.

It provides information on the content of each package—how placebo and verum packages/containers were labeled. The QP assigned the codes A and B to Kan Jang and placebo sets of packages.

4.5.2. Allocation Concealment

All the packages had allocation concealed random numbers printed on the label.

The PI blinded the randomization sequence list and code, so the PI assigned the packages sequentially as per numbers. The random sequence was disclosed to the statistician before the statistical evaluation of the results when 86 patients had completed the treatment.

4.5.3. Implementation and Blinding

At the first visit, participants received a consecutive number starting from 1 to 86. They were linked to a unique number according to the randomization sequence. Patients were sequentially enrolled by the PI, assigned a random number, and received the capsules in the package. The Investigator produced the participant list and gave a treatment code number (from 1 to 140) to each patient. They recorded the patients' names on case report forms (CRFs) and on the labels of the packages. The table shows the names of patients and corresponding study preparation numbers (treatment numbers mentioned on the labels of boxes).

Blinding for trial subjects was achieved using labeled packages containing capsules of the same appearance. The study product was delivered to the clinic prelabeled and coded according to the randomization list. The Principal Investigator kept the study participants' list, identifying the patients and the investigational product packages (numbers). The treatment code providing information about the actual assignment of groups A and B to Kan Jang® and placebo was broken by the QP. After that, a statistical analysis of the datasets was completed, and the results of the study were obtained. Thus, the study was quadruple-blind since study preparations were blinded to all investigators, care providers, participants, and outcome assessors.

4.6. Efficacy and Safety Outcomes and Endpoints

The efficacy endpoints were the differences in duration and the relief of inflammatory symptoms in Kang Jang and placebo group patients. The outcomes included changes in the severity of inflammatory symptoms, measured using various scales' scores, from the baseline to the end of therapy (Day 14) and follow-up period (21 days after randomization).

The primary efficacy outcome measures of the study were: (i)—the rate of patients (%) with clinical deterioration, (ii)—duration of hospitalization, (iii)—the time to virus clearance, (iv)—the duration of the acute phase of disease assessed as the time from the start of study medicine to complete symptom resolution, (v)—fever resolution and relief from the severity of fatigue, headache, sore throat, cough, rhinorrhea (nasal discharge/runny nose), myalgia (muscle pain), and loss of smell and taste.

Secondary endpoints comprised the measures of (i)—immune response marker IL-6 concentration in the serum, (ii)—blood hypercoagulation marker D-dimer, (iii)—inflammatory marker C-reactive protein, (iv)—physical activity, (v)—physical performance, (vi)—cognitive performance, and (vii)—severity of respiratory symptoms and quality of life according to the Wisconsin Upper Respiratory Symptoms score.

Safety and tolerability were assessed by monitoring the incidence and duration of adverse events. Laboratory samples for safety lever enzyme tests of aspartate aminotransferase (AST) and alanine aminotransferase (ALT) were taken on the day of inclusion in the study. All safety samples were drawn using standard syringes and analyzed according to the regular laboratory routine. Data were printed on separate sheets, one sheet per patient per test day. The data were published [18] in the CRF (see Supplementary File in previous publication [18]). Subjects were asked to report any adverse events they experienced to the Principal Investigator. For each patient, code envelopes were made. They contained information about the treatment the patient received during the study. The envelope was to be opened only in case of emergency, such as if severe side effects occurred. Adverse event reporting was conducted on visits 2, 3, and 4. The subjects were told to report any AE occurring during the study to the Investigator or the Investigator's personnel. Open, standardized AE questioning such as "Have you had any health problems since you were last questioned?" was performed by the Investigator (or designee) at each contact with the subject. That open, standardized questioning was carried out discreetly to prevent the patients from influencing each other.

5. Conclusions

This study provides new evidence on the clinical efficacy and safety of adaptogens, specifically Kan Jang®, in the acute phase of mild COVID-19. Kan Jang® reduces the risk of disease progression, the duration of illness, virus clearance, and days of hospitalization, and accelerates recovery of patients, relief of sore throat, muscle pain, and runny nose, and normalization of body temperature. Kan Jang® significantly relieves the severity of inflammatory URTI symptoms, including sore throat, cough, runny nose, and muscle pain, and increases patients' physical performance (workout) compared to the placebo.

Supplementary Materials: The following supporting information can be downloaded at: https://www.mdpi.com/article/10.3390/ph16091196/s1.

Author Contributions: Conceptualization, A.P.; methodology, A.P.; software, A.H., A.P. and N.M.; formal analysis, A.P.; investigation, L.R. and E.P.; resources, L.R.; data curation, E.P., N.M. and A.H.; writing—original draft preparation, A.P.; writing—review and editing, A.P. and E.P.; supervision, R.S. and A.H.; project administration, R.S. and L.R.; funding acquisition, R.S. and L.R. All authors have read and agreed to the published version of the manuscript.

Funding: This research was funded by the Swedish Herbal Institute AB in Vallberga, Sweden; grant no. 2021-1, dated 20 February 2021.

Institutional Review Board Statement: This study was conducted at the Tbilisi State Medical University, Tbilisi, Georgia, with the approval of the Biomedical Research Ethics Committee of Tbilisi State Medical University and the National Council on Bioethics (registration no. 3-2021/87, date of final protocol approval 25 March 2021). ClinicalTrials.gov Identifier: NCT04847518 (https://www.clinicaltrials.gov/ct2/show/NCT04847518; accessed on 6 June 2022).

Informed Consent Statement: All participants provided written informed consent to join the study before inclusion.

Data Availability Statement: Data are contained within the article and Supplementary Materials.

Acknowledgments: The authors acknowledge the support of the Swedish Herbal Institute for the supply of investigational products.

Conflicts of Interest: The authors declare no conflict of interest. A.P. is self-employed at the research and development company Phytomed AB and has no shares or financial interest in any pharmaceutical company. The funders had no role in the study's design; in the collection, analyses, or interpretation of data; in the writing of the manuscript; or in the decision to publish the results.

References

1. Azkur, A.K.; Akdis, M.; Azkur, D.; Sokolowska, M.; van de Veen, W.; Brüggen, M.C.; O'Mahony, L.; Gao, Y.; Nadeau, K.; Akdis, C.A. Immune response to SARS-CoV-2 and mechanisms of immunopathological changes in COVID-19. *Allergy* **2020**, *75*, 1564–1581. [CrossRef] [PubMed]
2. Tay, M.Z.; Poh, C.M.; Rénia, L.; MacAry, P.A.; Ng, L.F.P. The trinity of COVID-19: Immunity, inflammation and intervention. *Nat. Rev. Immunol.* **2020**, *20*, 363–374. [CrossRef]
3. Vardhana, S.A.; Wolchok, J.D. The many faces of the anti-COVID immune response. *J. Exp. Med.* **2020**, *217*, e20200678. [CrossRef] [PubMed]
4. Li, G.; Fan, Y.; Lai, Y.; Han, T.; Li, Z.; Zhou, P.; Pan, P.; Wang, W.; Hu, D.; Liu, X.; et al. Coronavirus infections and immune responses. *J. Med. Virol.* **2020**, *92*, 424–432. [CrossRef] [PubMed]
5. Schijns, V.; Lavelle, E.C. Prevention and treatment of COVID-19 disease by controlled modulation of innate immunity. *Eur. J. Immunol.* **2020**, *50*, 932–938. [CrossRef]
6. Lega, S.; Naviglio, S.; Volpi, S.; Tommasini, A. Recent Insight into SARS-CoV2 Immunopathology and Rationale for Potential Treatment and Preventive Strategies in COVID-19. *Vaccines* **2020**, *8*, 224. [CrossRef]
7. Yang, R.; Liu, H.; Bai, C.; Wang, Y.; Zhang, X.; Guo, R.; Wu, S.; Wang, J.; Leung, E.; Chang, H.; et al. Chemical composition and pharmacological mechanism of Qingfei Paidu Decoction and Ma Xing Shi Gan Decoction against Coronavirus Disease 2019 (COVID-19): In silico and experimental study. *Pharmacol. Res.* **2020**, *157*, 104820. [CrossRef]
8. Panossian, A.; Brendler, T. The Role of Adaptogens in Prophylaxis and Treatment of Viral Respiratory Infections. *Pharmaceuticals* **2020**, *13*, 236. [CrossRef]
9. Panossian, A.; Seo, E.J.; Wikman, G.; Efferth, T. Synergy assessment of fixed combinations of Herba Andrographidis and Radix Eleutherococci extracts by transcriptome-wide microarray profiling. *Phytomedicine* **2015**, *22*, 981–992. [CrossRef]

10. Panossian, A.; Efferth, T. Network Pharmacology of Adaptogens in the Assessment of Their Pleiotropic Therapeutic Activity. *Pharmaceuticals* **2022**, *15*, 1051. [CrossRef]
11. Melchior, J.; Palm, S.; Wikman, G. Controlled clinical study of standardized *Andrographis paniculata* extract in common cold—A pilot trial. *Phytomedicine* **1997**, *3*, 315–318. [CrossRef]
12. Melchior, J.; Spasov, A.A.; Ostrovskij, O.V.; Bulanov, A.E.; Wikman, G. Double-blind, placebo-controlled pilot and phase III study of activity of standardized *Andrographis paniculata* Herba Nees extract fixed combination (Kan Jang®) in the treatment of uncomplicated upper-respiratory tract infection. *Phytomedicine* **2000**, *7*, 341–350. [CrossRef]
13. Gabrielian, E.S.; Shukarian, A.K.; Goukasova, G.I.; Chandanian, G.L.; Panossian, A.G.; Wikman, G.; Wagner, H. A double blind, placebo-controlled study of *Andrographis paniculata* fixed combination Kan Jang in the treatment of acute upper respiratory tract infections including sinusitis. *Phytomedicine* **2002**, *9*, 589–597. [CrossRef]
14. Kulichenko, L.L.; Kireyeva, L.; Malyshkina, E.N.; Wikman, G. A randomized, controlled study of Kan Jang® versus amantadine in the treatment of influenza in Volgograd. *J. Herb. Pharmacother.* **2003**, *173*, 188–194. [CrossRef]
15. Spasov, A.A.; Ostrovskij, O.V.; Chernikov, M.V.; Wikman, G. Comparative controlled study of Andrographis paniculata fixedcombination, Kan Jang and an Echinacea preparation as adjuvant, in the treatment of uncomplicated respiratory disease in children. *Phytother. Res.* **2004**, *18*, 47–53. [CrossRef]
16. Panossian, A.; Wikman, G. Efficacy of *Andrographis paniculata* in upper respiratory tract (URT) infectious diseases and the mechanism of action. In *Evidence and Rational Based Research on Chinese Drugs*; Wagner, H., Ulrich Merzenich, G., Eds.; Springer: Vienna, Austria, 2012; pp. 137–180.
17. Narimanyan, M.; Jamalyan, K.; Balyan, A.; Barth, A.; Palm, S.; Wikman, G.; Panossian, A. Early intervention with Kan Jang® to treat upper-respiratory tract infections: A randomized, quadruple-blind study. *J. Tradit. Complement. Med.* **2021**, *11*, 552–562. [CrossRef] [PubMed]
18. Ratiani, L.; Pachkoria, E.; Mamageishvili, N.; Shengelia, R.; Hovhannisyan, A.; Panossian, A. Efficacy of Kan Jang® in Patients with Mild COVID-19: Interim Analysis of a Randomized, Quadruple-Blind, Placebo-Controlled Trial. *Pharmaceuticals* **2022**, *15*, 1013. [CrossRef] [PubMed]
19. Jain, V.K.; Iyengar, K.P.; Vaishya, R. Differences between first wave and second wave of COVID-19 in India. *Diabetol. Metab. Syndr.* **2021**, *15*, 1047–1048. [CrossRef]
20. World Health Organization. Clinical Management of Severe Acute Respiratory Infection When Novel Coronavirus (2019-nCoV) Infection Is Suspected: Interim Guidance. 2020. Available online: https://www.who.int/docs/default-source/coronaviruse/clinical-management-of-novel-cov.pdf (accessed on 28 July 2023).
21. Lamb, Y.N. Nirmatrelvir Plus Ritonavir: First Approval. *Drugs* **2022**, *82*, 585–591. [CrossRef]
22. National Institutes of Health. Antiviral Drugs That Are Approved, Authorized, or Under Evaluation for the Treatment of COVID-19. In *COVID-19 Treatment Guidelines Panel. Coronavirus Disease 2019 (COVID-19) Treatment Guidelines*; Last Updated: 29 April 2022. National Institutes of Health: Bethesda, MD, USA, 2022; Available online: https://files.covid19treatmentguidelines.nih.gov/guidelines/covid19treatmentguidelines.pdf; pp. 140–221. https://files.covid19treatmentguidelines.nih.gov/guidelines/section/section_36.pdf; (accessed on 29 April 2022).
23. Molnupiravir_COVID19-MSD—Art 5(3)—Conditions for Use (europa.eu). Available online: https://www.ema.europa.eu/en/documents/referral/lagevrio-also-known-molnupiravir-mk-4482-covid-19-article-53-procedure-conditions-use-conditions_en.pdf (accessed on 7 August 2022).
24. Yuvejwattana, S. Thailand Clears Use of Herbal Medicine for COVID-19 Treatment; Bloomberg; 30 December 2020, 7:02 AM GMT+1. Available online: https://www.bloomberg.com/news/articles/2020-12-30/thailand-clears-use-of-herbal-medicine-for-covid-19-treatment (accessed on 28 July 2023).
25. Yearsley, C. Thailand Approves Asian Herb Andrographis to Treat COVID-19. *HerbalGram.* **2021**, *129*, 35–37.
26. Karbwang, J.; Na-Bangchang, K. Repurposed drugs for COVID-19 treatment. *J. Thai. Trad. Alt. Med.* **2021**, *19*, 285–302.
27. World Health Organization. Therapeutics and COVID-19. Therapeutic Management of Hospitalized Adults with COVID-19; Last Updated: 24 February 2022. Available online: https://www.covid19treatmentguidelines.nih.gov/management/clinicalmanagement/hospitalized-adults{-}{-}therapeutic-management/ (accessed on 7 August 2022).
28. Tong, L.; Ma, Z.; Zhou, Y.; Yang, S.; Yang, Y.; Luo, J.; Huang, J.; Wang, F. Combination of Chinese herbal medicine and conventional western medicine for coronavirus disease 2019: A systematic review and meta-analysis. *Front. Med.* **2023**, *10*, 1175827. [CrossRef] [PubMed]
29. Ji, Z.; Hu, H.; Qiang, X.; Lin, S.; Pang, B.; Cao, L.; Zhang, L.; Liu, S.; Chen, Z.; Zheng, W.; et al. Traditional Chinese Medicine for COVID-19: A Network Meta-Analysis and Systematic Review. *Am. J. Chin. Med.* **2022**, *50*, 883–925. [CrossRef]
30. Wu, H.T.; Ji, C.H.; Dai, R.C.; Hei, P.J.; Liang, J.; Wu, X.Q.; Li, Q.S.; Yang, J.C.; Mao, W.; Guo, Q. Traditional Chinese medicine treatment for COVID-19: An overview of systematic reviews and meta-analyses. *J. Integr. Med.* **2022**, *20*, 416–426. [CrossRef] [PubMed]
31. Jiang, F.; Xu, N.; Zhou, Y.; Song, J.; Liu, J.; Zhu, H.; Jiang, J.; Xu, Y.; Li, R. Contribution of traditional Chinese medicine combined with conventional western medicine treatment for the novel coronavirus disease (COVID-19), current evidence with systematic review and meta-analysis. *Phytother. Res.* **2021**, *35*, 5992–6009. [CrossRef]

32. Liang, S.B.; Fang, M.; Liang, C.H.; Lan, H.D.; Shen, C.; Yan, L.J.; Hu, X.Y.; Han, M.; Robinson, N.; Liu, J.P. Therapeutic effects and safety of oral Chinese patent medicine for COVID-19: A rapid systematic review and meta-analysis of randomized controlled trials. *Complement. Ther. Med.* **2021**, *60*, 102744. [CrossRef]
33. Wanaratna, K.; Leethong, P.; Inchai, N.; Chueawiang, W.; Sriraksa, P.; Tabmee, A.; Sirinavin, S. Efficacy and Safety of *Andrographis paniculata* Extract in Patients with Mild COVID-19: A Randomized Controlled Trial. *medRxiv*. 2021. Available online: https://www.semanticscholar.org/paper/Efficacy-and-safety-of-Andrographis-paniculata-in-A-Wanaratna-Leethong/1f40a6181de4e0f15ab296673794210e82236455 (accessed on 7 August 2022).
34. Rattanaraksa, D.; Khempetch, R.; Poolwiwatchaikool, U.; Nimitvilai, S.; Loatrakul, O.; Srimanee, P. The efficacy and safety of *Andrographis paniculata* extract for the treatment of COVID-19 patients with mild symptoms, Nakhonpathom hospital. *Reg. Med. J.* **2021**, *40*, 269–281.
35. Tanwettiyanont, J.; Piriyachananusorn, N.; Sangsoi, L.; Boonsong, B.; Sunpapoa, C.; Tanamatayarat, P.; Kanchanasurakit, S.; Na-Ek, N. The efficacy of *Andrographis paniculata* (Burm.f.) Wall. ex Nees crude extract in hospitalized mild COVID-19 patients: A retrospective cohort study. *medRxiv* **2022**. [CrossRef]
36. Shanker, K.; Rangnekar, H.; Wele, A.; Soni, P.; Gaikwad, P.; Pal, A.; Bawankule, D.U.; Chanda, D. A randomized controlled pilot study of add-on therapy of CIM-MEG19 (standardized *Andrographis paniculata* formulation) in mild to moderate COVID-19. *Phytomed. Plus.* **2023**, *3*, 100398. [CrossRef]
37. Zhang, X.; Lv, L.; Zhou, Y.; Xie, L.; Xu, Q.; Zou, X.; Ding, Y.; Tian, J.; Fan, J.; Fan, H.; et al. Efficacy and safety of Xiyanping injection in the treatment of COVID-19: A multicenter, prospective, open-label and randomized controlled trial. *Phytother. Res.* **2021**, *35*, 4401–4410. [CrossRef]
38. Kaewdech, A.; Nawalerspanya, S.; Assawasuwannakit, S.; Chamroonkul, N.; Jandee, S.; Sripongpun, P. The use of *Andrographis paniculata* and its effects on liver biochemistry of patients with gastrointestinal problems in Thailand during the COVID-19 pandemic: A cross sectional study. *Sci. Rep.* **2022**, *12*, 18213. [CrossRef]
39. Tanwettiyanont, J.; Piriyachananusorn, N.; Sangsoi, L.; Boonsong, B.; Sunpapoa, C.; Tanamatayarat, P.; Na-Ek, N.; Kanchanasurakit, S. Use of *Andrographis paniculata* (Burm.f.) Wall. ex Nees and risk of pneumonia in hospitalised patients with mild coronavirus disease 2019: A retrospective cohort study. *Front. Med.* **2022**, *9*, 947373. [CrossRef]
40. Intharuksa, A.; Arunotayanun, W.; Yooin, W.; Sirisa-Ard, P. A Comprehensive Review of *Andrographis paniculata* (Burm. f.) Nees and Its Constituents as Potential Lead Compounds for COVID-19 Drug Discovery. *Molecules* **2022**, *27*, 4479. [CrossRef] [PubMed]
41. Izcovich, A.; Siemieniuk, R.A.; Bartoszko, J.J.; Ge, L.; Zeraatkar, D.; Kum, E.; Qasim, A.; Khamis, A.M.; Rochwerg, B.; Agoritsas, T.; et al. Adverse effects of remdesivir, hydroxychloroquine and lopinavir/ritonavir when used for COVID-19: Systematic review and meta-analysis of randomised trials. *BMJ Open* **2022**, *12*, e048502. [CrossRef] [PubMed]
42. Gupte, V.; Hegde, R.; Sawant, S.; Kalathingal, K.; Jadhav, S.; Malabade, R.; Gogtay, J. Safety and clinical outcomes of remdesivir in hospitalised COVID-19 patients: A retrospective analysis of active surveillance database. *BMC Infect. Dis.* **2022**, *22*, 1. [CrossRef] [PubMed]
43. Shang, Y.X.; Shen, C.; Stub, T.; Zhu, S.J.; Qiao, S.Y.; Li, Y.Q.; Wang, R.T.; Li, J.; Liu, J.P. Adverse Effects of Andrographolide Derivative Medications Compared to the Safe use of Herbal Preparations of *Andrographis paniculata*: Results of a Systematic Review and Meta-Analysis of Clinical Studies. *Front. Pharmacol.* **2022**, *13*, 773282. [CrossRef]
44. Brekhman, I.I. *On Antitoxic Action of Eleutherococcus*; Meditsina: Moscow, Russia, 1982; Volume 37.
45. HMPC—Committee on HerbalMedicinal Products. Assessment Report on *Eleutherococcus senticosus* (Rupr. etMaxim.) Maxim, Radix.EMA/HMPC/680615/2013. 25 March 2014, pp. 1–51. Available online: https://www.ema.europa.eu/en/documents/herbal-report/final-assessment-report-eleutherococcus-senticosus-rupr-et-maxim-maxim-radix_en.pdf (accessed on 30 May 2022).
46. Seo, E.J.; Klauck, S.M.; Efferth, T.; Panossian, A. Adaptogens in chemobrain (Part II): Effect of plant extracts on chemotherapy induced cytotoxicity in neuroglia cells. *Phytomedicine* **2019**, *58*, 152743. [CrossRef]
47. Seo, E.J.; Klauck, S.M.; Efferth, T.; Panossian, A. Adaptogens in chemobrain (Part III): Antitoxic effects of plant extracts towards cancer chemotherapy-induced toxicity-transcriptome-wide microarray analysis of neuroglia cells. *Phytomedicine* **2019**, *56*, 246–260. [CrossRef]
48. Panossian, A.; Seo, E.J.; Klauck, S.M.; Efferth, T. Adaptogens in chemobrain (part IV): Adaptogenic plants prevent the chemotherapeutics-induced imbalance of redox homeostasis by modulation of expression of genes encoding Nrf2-mediated signaling proteins and antioxidant, metabolizing, detoxifying enzymes in neuroglia cells. *Longhua Chin. Med.* **2020**, *3*, 4.
49. Lim, X.Y.; Chan, J.; Tan, T.Y.; Teh, B.P.; Mohd Abd Razak, M.R.; Mohamad, S.B.; Syed Mohamed, A.F. *Andrographis paniculata* (Burm. F.) Wall. Ex Nees, Andrographolide, and Andrographolide Analogues as SARS-CoV-2 Antivirals? A Rapid Review. *Nat. Prod. Commun.* **2021**, *16*, 1934578X211016610. [CrossRef]
50. Gagnier, J.J.; Boon, H.; Rochon, P.; Moher, D.; Barnes, J.; Bombardier, C.; CONSORT Group. Recommendations for reporting randomized controlled trials of herbal interventions: Explanation and elaboration. *J. Clin. Epidemiol.* **2006**, *59*, 1134–1149. [CrossRef] [PubMed]

Disclaimer/Publisher's Note: The statements, opinions and data contained in all publications are solely those of the individual author(s) and contributor(s) and not of MDPI and/or the editor(s). MDPI and/or the editor(s) disclaim responsibility for any injury to people or property resulting from any ideas, methods, instructions or products referred to in the content.

pharmaceuticals

MDPI

Article

Identification of Active Markers of Chinese Formula Yupingfeng San by Network Pharmacology and HPLC-Q-TOF–MS/MS Analysis in Experimental Allergic Rhinitis Models of Mice and Isolated Basophilic Leukemia Cell Line RBL-2H3

Xinqi Li, Caining Zhao and Jin Qi *

Research Center for Traceability and Standardization of TCMs, School of Traditional Chinese Pharmacy, China Pharmaceutical University, Nanjing 211198, China; 3120020134@stu.cpu.edu.cn (X.L.); 17312307045@163.com (C.Z.)
* Correspondence: qijin2006@163.com

Abstract: Background: Yupingfeng San (YPFS) is a classic formula for treating allergic rhinitis (AR), which is composed of *Astragalus mongholicus* Bunge (AST), *Atractylodes macrocephala* Koidz (AMR), and *Saposhni-kovia divaricata* (Turcz.) Schischk (SR) at a ratio of 3:1:1. However, the potential bioactive components of YPFS relevant to AR treatment are currently unknown. **Methods:** This study combined in vivo chemical profiling, network pharmacology, and experimental validation to identify the substances in YPFS that are active against AR. **Results:** Firstly, 98 compounds in YPFS were identified using high-performance liquid chromatography–quadrupole time-of-flight mass spectrometry (HPLC-Q-TOF-MS/MS) with the assistance of Global Natural Products Social (GNPS) molecular networking. Then, 42 prototype components and 57 metabolites were detected in the plasma, urine, and feces of mice with AR. A network pharmacological analysis based on 42 in vivo prototypical components was also conducted to screen 15 key components and 10 core targets, and 6 key components were further selected through molecular docking. Finally, the four key active components (cimifugin, wogonin, formononetin, and atractylenolide I) were revealed to be the main ingredients of YPFS through validation (in vitro and in vivo). **Conclusions:** This is the first systematic study of the components of YPFS in AR mice, laying the foundation for elucidating the overall material basis of this formulation. This study provides rich basic data for further pharmacological and mechanistic studies on YPFS.

Keywords: Yupingfeng San; allergic rhinitis; prototype component; network pharmacology; molecular docking

1. Introduction

Allergic rhinitis (AR), a common allergic condition, is triggered by immunoglobulin E (IgE) in response to allergens such as pollen [1,2]. As per the 2020 publication of the *White Book* by the World Allergy Organization, the worldwide occurrence of allergic disorders varies, ranging from 10 to 40% of the population, with 400 million cases of AR, urticaria, and anaphylactic shock [3,4]. In China, epidemiological studies showed a high prevalence of AR, affecting approximately 250 million people (17.6%) with a continuous upward trend [2]. Therefore, developing new medications for treating AR is of great practical importance.

Yupingfeng San (YPFS) has demonstrated remarkable effectiveness in treating AR, authored *Danxixinfa* by Zhu Zhenheng during the Yuan Dynasty, and has demonstrated remarkable effectiveness in treating AR. It is composed of three herbs—*Astragalus mongholicus* Bunge (AST), *Atractylodes macrocephala* Koidz (AMR), and *Saposhnikovia divaricata* (Turcz.)

Schischk (SR)—at a ratio of 3:1:1 [5]. The chemical components of YPFS are diverse, encompassing flavonoids, saponins, coumarins, chromones, and lactones [5]. However, the material basis on which YPFS treats AR remains unclear, and its complex metabolism in the body complicates the issue [6]. Therefore, it is imperative to employ modern scientific methods to elucidate the material basis and mechanism by which YPFS treats AR.

Most prior studies have primarily focused on validating the effectiveness of YPFS in mitigating inflammatory factors in AR [7,8]. Some reports have also identified the absorbed components of YPFS in the blood of rats [9]. However, the prototype components of YPFS in AR models have not been reported. Numerous studies have demonstrated that the absorption and metabolism of traditional Chinese medicine (TCM) differ significantly in pathological conditions compared to natural and healthy states [10,11]. Hence, it is important to identify the essential components of YPFS for managing AR.

This study introduced a systematic approach integrating high-performance liquid chromatography–quadrupole time-of-flight mass spectrometry (HPLC–ESI-Q-TOF-MS/MS) and Global Natural Product Social Molecular Networking (GNPS) to comprehensively identify the chemical components of YPFS. In vivo, we identified the prototype components and metabolites in the plasma, urine, and feces of AR mice after the oral intake of YPFS using mass spectrometry. Then, we constructed a network of the interactions between the prototype components of YPFS and AR-related targets to identify the key components and core targets and predict the potential pathways involved in the regulation of AR. The relationships between the core components and targets were further evaluated through molecular docking, and the components's anti-AR effects were validated in RBL-2H3 cells and AR mice. This study serves as a foundation for understanding the therapeutic potential of YPFS in AR and provides a scientific basis for screening active ingredients in other medications.

2. Results

2.1. Optimization of Chromatography Conditions

The parameters were optimized to increase the separation of YPFS and improve the peak shape and number of peaks. As depicted in Figure S1, the findings revealed that a sample size of 5 mg, flow rate of 1 mL/min, column temperature of 30 °C, and wavelength of 254 nm were the most conducive to the effective separation and identification of the compounds within YPFS. The RSD values of all components were less than 3%, indicating that the method exhibits good repeatability, precision, and stability (Table S1–S3).

2.2. Chemical Profile of YPFS

The chemical profile of YPFS was systematically analyzed using manual identification and GNPS methods. A total of 98 compounds (Figure 1A and Table S4) were identified, comprising 20 flavonoids, 8 saponins, 11 lactones, 4 sesquiterpenoids, 10 chromones, 13 coumarins, 12 amino acids, 3 organic acids, and 17 other compounds (Figure 1B).

The molecular network of YPFS was established using GNPS for the analysis of natural products [12]. As depicted in Figure S2, the generated molecular network comprised 39 clusters (each cluster containing $\geq$ 2 nodes), totaling 388 nodes, with 89 feature components predicted.

The data were subject to the rigorous deduplication and deletion of the endogenous components based on comparisons with literature, chemical compound databases, and MS data [5,13]. In total, 36 potential chemical components of YPFS were obtained. Their chemical structures are shown in Figure 2, including six flavonoids, seven amino acids, two saponins, three organic acids, one chromone, six coumarins, four lactones, and seven other compounds. Notably, 18 components were consistent with previous manual identification predictions, further validating the methodology. This analytical method expedited the identification of chemical constituents within intricate natural product matrices.

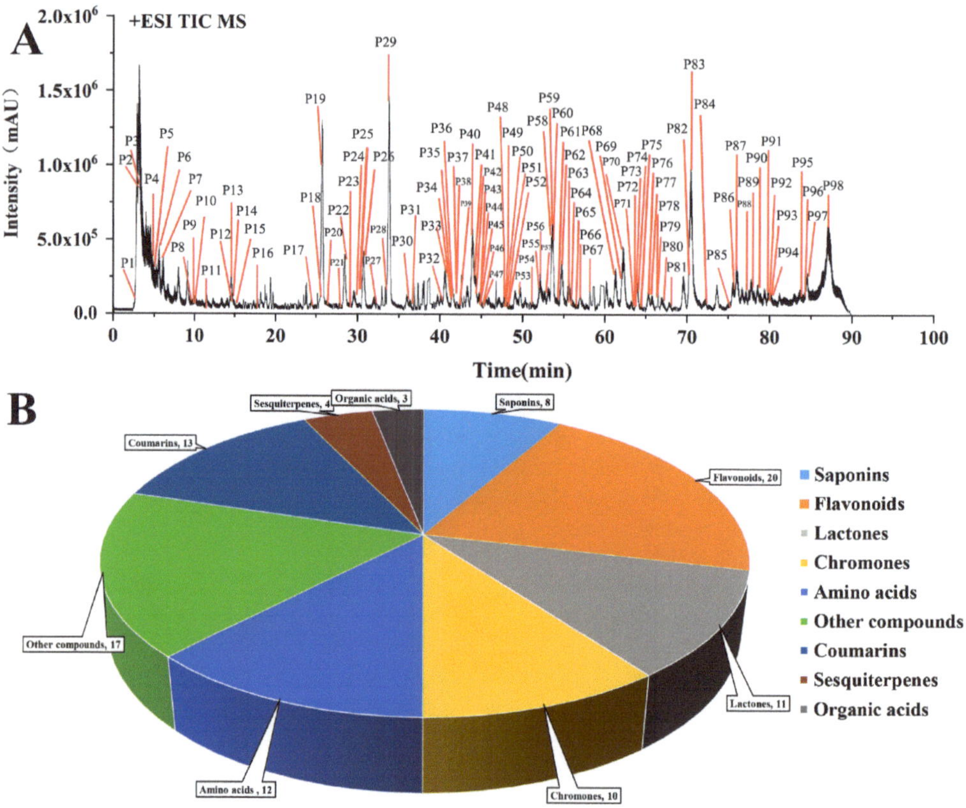

Figure 1. A total of 98 chemical components of YPFS identified through manual identification and GNPS in positive ion mode. (**A**) Base peak chromatograms of YPFS obtained using HPLC–Q-TOF-MS in positive modes. (**B**) Types of chemical compositions in YPFS.

Using the degree of similarity in the MS/MS spectra of individual compounds, which ranges from 0 to 1 in cosine similarity, compounds could be grouped into clusters within a molecular network. This clustering method enabled the rapid classification of components [14]. As illustrated in Figure 2 and Figure S2, the node associated with the ion at m/z 431.134 corresponded to the molecular formula $C_{22}H_{22}O_9$. Analysis of the MS/MS spectrum revealed fragment ions with mass-to-charge ratios of 269, 254, and 137, which are indicative of glycoside ligands, methyl groups, and compounds formed through the retro-Diels–Alder reaction (RDA). By cross-referencing the retention time and MS data with existing databases and scientific literature, the compound was conclusively identified as ononin [15]. The node of the ion at m/z 447.129 was adjacent to the node of ononin, suggesting a high degree of MS/MS spectral similarity with an excimer ion difference of 16 Da. Based on literature reports [16,17], this node had been identified as sissotrin. Similarly, other nodes were identified through this process.

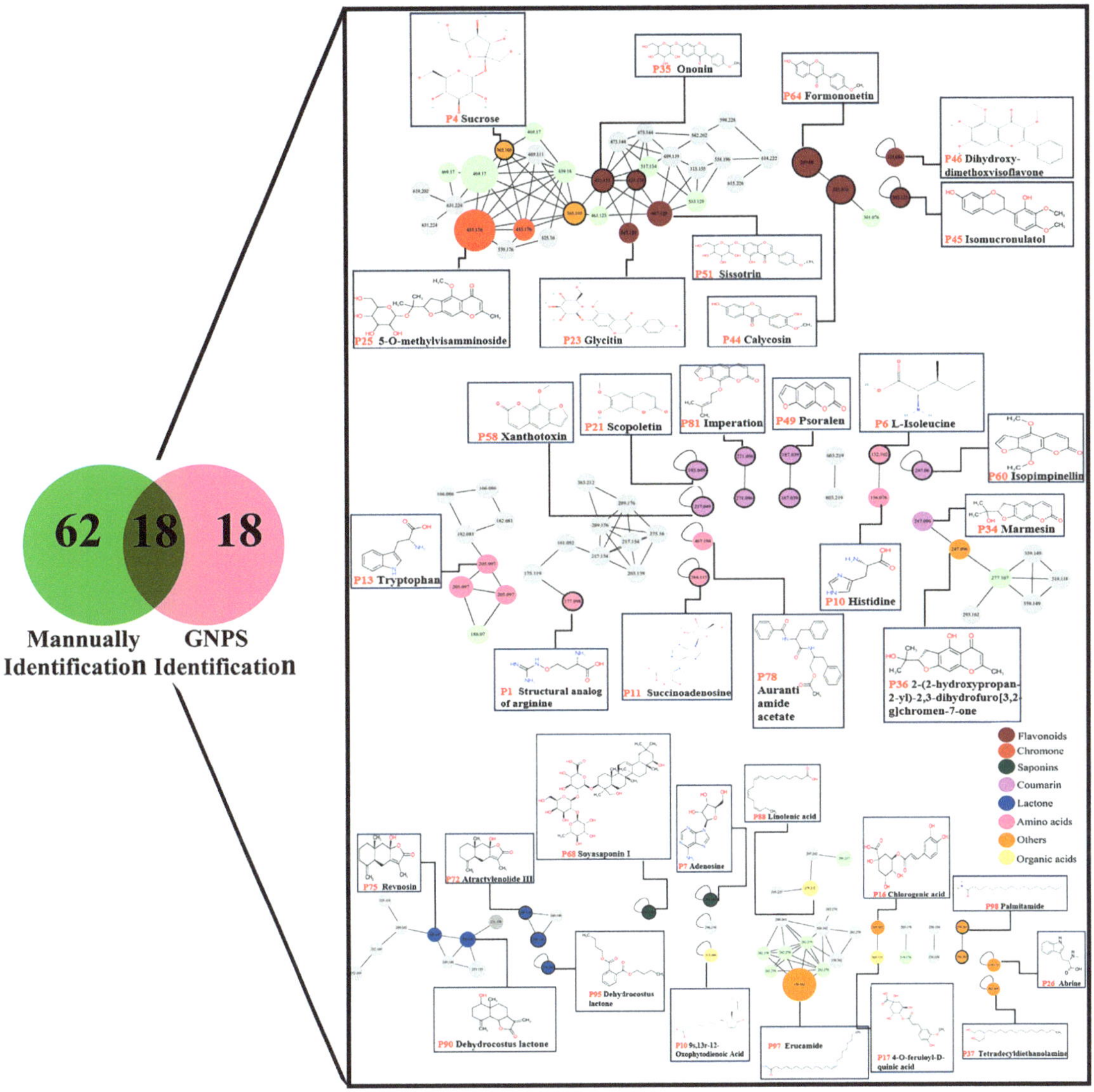

Figure 2. The structures of common components identified manually and using GNPS. The node color indicates the type of compound (red: flavonoids; green: saponins; purple: coumarin; blue: lactone; pink: amino acids; orange: others; pale yellow: organic acids).

2.3. *Characterization of Compounds*

Compounds with similar structures exhibited comparable cleavage patterns. Therefore, it is imperative to analyze the cleavage rules of the prototype components in order to accurately identify the metabolic components of YPFS in vivo. In this study, we analyzed several typical compounds present in YPFS.

2.3.1. Flavonoids

Initially, 19 flavonoids were identified in YPFS, primarily originating from AST. These flavonoids were identified based on characteristic neutral losses of CO (28 Da), CO_2 (44 Da), and CH_3 (15 Da). Calycosin (Figure 3A) and ononin (Figure S3A) exhibited the same cleavage pathway [18,19]. The cleavage rules of flavonoids were represented by the example of calycosin. Calycosin exhibited an $[M + H]^+$ ion at m/z 285.0753, which was formed by the loss of glucose ($C_6H_{10}O_5$, 162 Da) from calycosin-7-O-glucoside. The fragment ion at m/z 285.0753 $[M + H]^+$ further generated the fragment ions at m/z 270.0509, m/z 213.0528 and m/z 137.0221 through losing CH_3 (15 Da), $C_3H_4O_2$ (72 Da) and RDA. Additionally, fragment ions at m/z 225.1409 and 197.0717 were produced through the sequential loss of CO (28 Da) from the fragment ion at m/z 270.0509.

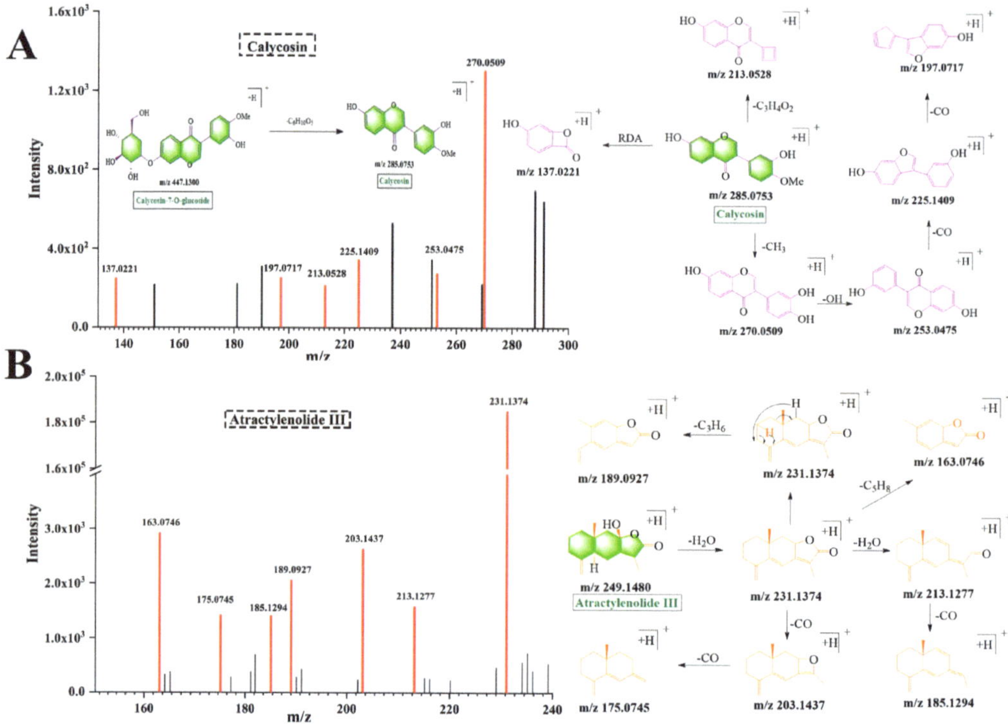

Figure 3. Chemical fragmentation pathways of chemical components in YPFS. (**A**) calycosin; (**B**) atractylenolide III.

2.3.2. Lactones

Analogously, 12 lactones were identified in YPFS, primarily originating from AMR. The cleavage rules of lactones were represented by the example of atractylenolide III (Figure 3B). Atractylenolide III exhibited an $[M + H]^+$ ion at m/z 249.1480. Fragment ions at m/z 231.1374 and m/z 213.1277 were generated through the sequential loss of H_2O (18 Da) from atractylenolide III. The fragment ion at m/z 213.1277 underwent further loss of CO (28 Da) to form the fragment ion at m/z 185.1294. Additionally, fragment ions at m/z 203.1437 and m/z 175.0745 were formed by the sequential removal of CO (28 Da) from the fragment ion at m/z 231.1374. Lastly, the fragment ions at m/z 163.0746 and m/z 189.0927 were generated by the loss of C_5H_8 (68 Da) and C_3H_6 (42 Da) groups from the fragment ion at m/z 231.1374.

2.3.3. Coumarins

A total of 12 coumarins were identified in YPFS, primarily originating from SR. These coumarins could be classified into two distinct categories, furanocoumarins and coumestrol, based on the structure of their parent nuclei. The characteristic fragment ions of coumarins, specifically those at m/z 131, 143, and 159, were detected in positive mode [20]. Psoralen (Figure S3B) and imperatorin (Figure S4A) shared the same cleavage pathway. Taking imperatorin as an example, the cleavage rule of coumarins is described below. Imperatorin exhibited an $[M + H]^+$ ion at m/z 271.0956, whose cleavage rule involved the continuous loss of CO (28 Da). The fragment ion at m/z 203.0330 was obtained through the loss of the C_5H_8 (68 Da) group. There were three possible fragmentation pathways: Firstly, fragment ions at m/z 175.0361, 147.0458, and 119.0856 were produced by the continuous loss of CO (28 Da) from the fragment ion at m/z 203.0330. Alternatively, the fragment ion at m/z 203.0330 could lose CO_2 (44 Da) to form the fragment ion at m/z 159.0965, which subsequently lost CO (28 Da) to form the fragment ion at m/z 131.1069. Additionally, the fragment ion at m/z 203.0330 could lose H_2O (18 Da) to obtain the fragment ion at m/z 185.0900. Fragment ions at m/z 157.0509 and 129.0552 were generated by the continuous lost of CO (28 Da) from the fragment ion at m/z 185.0900 [21].

2.3.4. Chromones

Currently, the chemical constituents characterized by SR primarily comprise two distinct classes: dihydrofuran chromogen ketones and dihydropyran chromogen ketones. Dihydrofuran chromogen ketones are more abundant, including cimifugin and 5-O-methylvisamminoside, among others. Taking the compound cimifugin (Figure S4B) as an example, the cleavage rule of coumarins was introduced below. Cimifugin exhibited $[M + H]^+$ at m/z 307.1177. The parent ion at m/z 307.1177 lost H_2O (18 Da) to form the fragment ion at m/z 289.0778. The parent ion at m/z 307.1177 further lost H_2O (18 Da) and 2 CH_3 (30 Da), resulting in the fragment ion at m/z 259.0596. The fragment ion at m/z 221.0412 was generated through the loss of the C_3H_2 (38 Da) group from the fragment ion at m/z 259.0596. The parent ion at m/z 307.1177 lost C_4H_8O (72 Da) to generate the fragment ion at m/z 235.1186. The fragment ion at m/z 205.0884 was formed through the loss of CH_2O (30 Da) from the fragment ion at m/z 235.1186. Lastly, the fragment ion at m/z 177.0554 was obtained through the loss of CO (28 Da) from the fragment ion at m/z 205.0884.

2.4. *Identification of YPFS Prototype Components In Vivo*

To differentiate prototype components from metabolites, we designated compounds detected in drug-associated biological fluids and YPFS extracts as prototype components. Conversely, compounds exclusively detected in the drug-associated biological fluid but absent in the YPFS extract were classified as metabolites. Under the same analytical conditions, mass spectrometry was utilized to analyze plasma, urine, and feces obtained from AR mice treated with YPFS. To screen for absorbed components, total ion chromatogram (TIC) and extraction ion chromatogram (EIC) curves were employed. Combining the results with those obtained from the in vitro analysis of YPFS, a total of 42 prototype components were identified (Table S5), with 11, 28, and 29 prototype components found in the plasma, urine, and feces, respectively. The base peak chromatograms of YPFS are presented in Figure 4A–F, and the structures of the compounds are shown in Figure S5. These compounds include 12 flavonoids, 11 chromones, 6 saponins, 5 lactones, 3 sesquiterpenoids, 2 coumarins, and 3 other compounds. Figure S6 shows the presence of 42 prototype components across the three biological fluids examined. Six prototype components were commonly detected in all three body fluids: blood, urine, and feces. These components were prim-O-glucosylcimifugin, atractylenolide I, formononetin, astramembrannin II, calycosin-7-O-glucoside, and cimifugin.

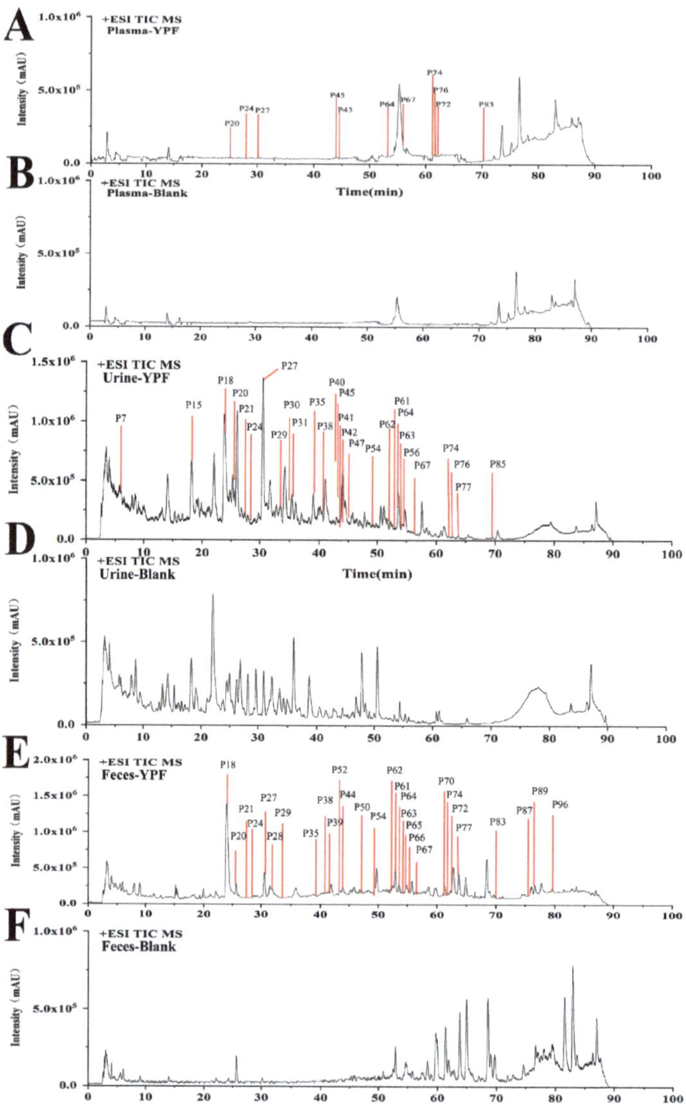

Figure 4. The total ion chromatography (TIC) of YPFS in biological and blank samples in positive ion mode and prototypical components of YPFS in vivo in AR mice. (**A**) YPFS plasma sample; (**B**) blank plasma sample; (**C**) YPFS urine sample; (**D**) blank urine sample; (**E**) YPFS feces sample; (**F**) blank feces sample.

2.5. Identification of YPFS Metabolites in Vivo

Through the in vitro chemical composition analysis of YPFS, the metabolic pathways of its components were elucidated by cross-referencing data from public databases and literature with the cleavage outcomes for YPFS component precursors. A total of 57 metabolites of YPFS were provisionally identified, including 39 metabolites in plasma, 46 in urine, and 28 in feces. All the identified metabolites are listed in Table S6.

2.5.1. Flavonoid Metabolites

As depicted in Figures 5 and S7A, 37 metabolites (M1–M37) were identified as flavonoid-related metabolites in plasma, urine, and feces. These metabolites were primarily associated with formononetin, calycosin, calycosin-7-O-glucoside, ononin, and methylnissolin. The pathway of calycosin's metabolism was used as an example for clarification.

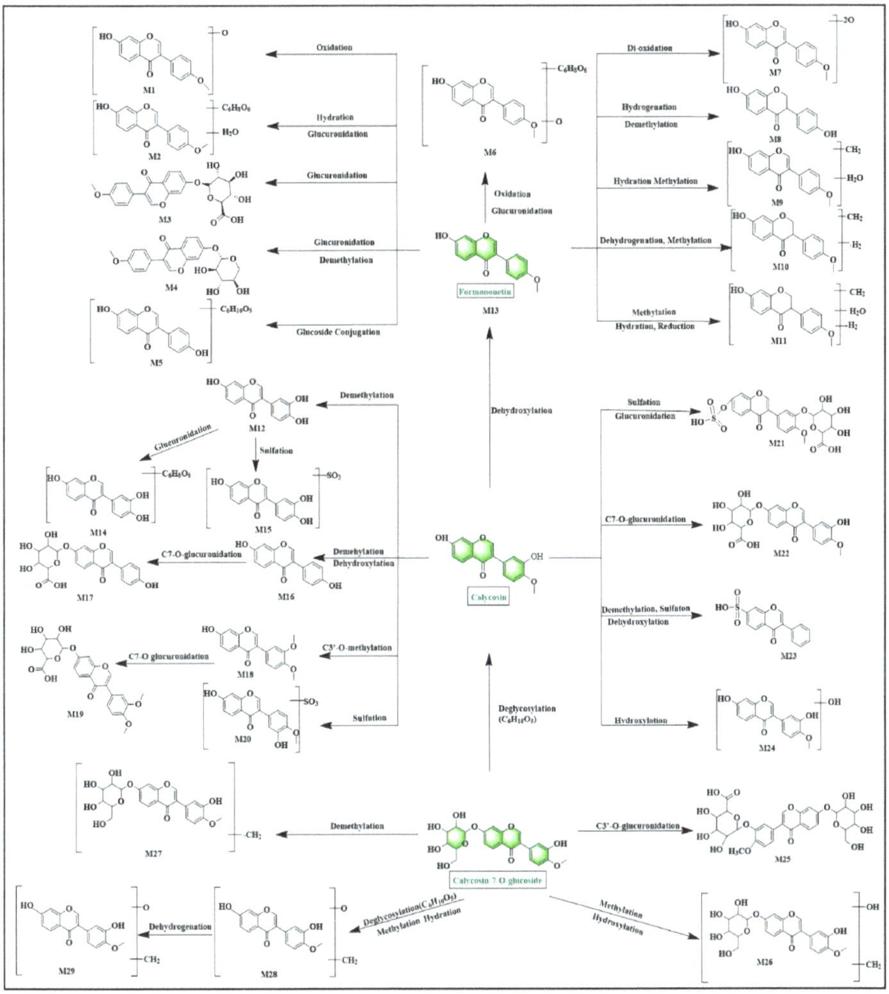

Figure 5. The possible metabolic pathways of formononetin, calycosin, and calycosin-7-O-glucoside from YPFS in AR mice.

As depicted in Figure 5, the metabolites of calycosin were identified as M12-M24. Initially, the phase I metabolite-reaction products included M12, M13, M16, and M24. Calycosin exhibited an $[M + H]^+$ ion at m/z 285.0753, and M12 exhibited an $[M + H]^+$ ion peak at m/z 271.0619, which was formed by losing 14 Da reduction compared to calycosin. The fragment ion with m/z 137.0244 corresponded to the peak of calycosin, suggesting that M12 could be a demethylated derivative of calycosin. M13 displayed an $[M + H]^+$ ion peak at m/z 269.0830, revealing a 16 Da decrease compared to calycosin. Its fragment ion at m/z 213.0933 aligned with calycosin, leading to the hypothesis that M13 (formononetin) might be a dehydroxylated product of calycosin. M16 presented an $[M + H]^+$ ion peak at *m/z* 255.0676, demonstrating a 30 Da decrement from calycosin. The fragment ions observed at *m/z* 227.0723 and 199.0764 matched those of calycosin, indicating the possibility that M16 was a demethylated and dehydroxylated variant of calycosin. M24 displayed an $[M + H]^+$ ion peak at *m/z* 301.0734, indicating a 16 Da increase over calycosin. Based on its molecular formula and fragmentation pattern, it was inferred that M24 could be a hydroxylated derivative of calycosin.

Phase II metabolite-reaction products were categorized into methylation, glucuronidation, and sulfation metabolites. M18 showed an $[M + H]^+$ ion peak at *m/z* 299.0936, indicating a 14 Da increase over calycosin, and the fragment ion of *m/z* 197.0655 corresponded to calycosin, suggesting that M18 was a calycosin methylation product.

M14 exhibited an $[M + H]^+$ ion peak at *m/z* 447.0955, indicating a mass increase of 176 Da over M12. The fragment ions observed at *m/z* 225.0702 closely resembled those of calycosin, suggesting that M14 may be the glucuronidation product of M12. Similarly, M17 displayed an $[M + H]^+$ ion peak at *m/z* 431.1015, demonstrating a mass increase of 176 Da over M16. It was proposed that M17 was the glucuronidation product of M16. M19 showed an $[M + H]^+$ ion peak at *m/z* 475.123, indicating a mass increase of 176 Da over M18. Combined with fragment ions that were identical to the methylation product of calycosin, M19 was determined to be the glucuronidation product of M18. M22 showed an $[M + H]^+$ ion peak at *m/z* 461.1082, indicating a mass increase of 176 Da over calycosin. The fragment ions at *m/z* 225.0580 and 285.0783 matched those of calycosin, suggesting that M22 could be the glucuronidation product of calycosin.

M15 displayed an $[M + H]^+$ ion peak at *m/z* 351.018, exhibiting a mass variance of 80 Da from M12. Consequently, it was conceivable that M15 could be a sulfated form of M12. Similarly, M20 displayed an $[M + H]^+$ ion peak at *m/z* 365.0333, showing a comparable mass increment of 80 Da compared to calycosin. The fragment ions with *m/z* values of 253.0498 and 225.0544 coincided with those of calycosin, indicating that M20 might also be a sulfation product of calycosin. M23, with the molecular formula $C_{15}H_{10}O_7S$, showed an $[M + H]^+$ ion peak at *m/z* 335.0248, displaying a mass increase of 50 Da compared to calycosin. Notably, the fragment ion at *m/z* 137.0239 was identical to calycosin, leading to the identification of M23 as a sulfated derivative resulting from the demethylation and dihydroxylation of calycosin. Meanwhile, M21 presented an $[M + H]^+$ ion peak at *m/z* 541.0647, reflecting a substantial mass augmentation of 256 Da over calycosin. Furthermore, the fragment ion at *m/z* 270.0515 aligned with calycosin. Taking into account its molecular formula $C_{22}H_{20}O_{14}S$, it was hypothesized that M21 could be a sulfated and glucuronidation product of calycosin. These results were consistent with literature indicating that phase II conjugation of calycosin in vivo was very common [22].

2.5.2. Saponins Metabolites

As depicted in Figure S7B, three metabolites (M38-M40) were identified as metabolites of astragaloside IV in plasma, urine, and feces. M38 exhibited an $[M + Na]^+$ ion at *m/z* 513.3775, indicating a mass 294 Da smaller than astragaloside IV. This observation suggested a molecular formula of $C_{30}H_{50}O_5$ for M38. M38 was determined to be the desugaring product of astragaloside IV, specifically resulting from the removal of glucose ($C_6H_{10}O_5$) and xylose ($C_5H_8O_4$), and was identified as cycloastragenol based on a combination of references and fragmentation patterns. M39 presented an $[M + Na]^+$ ion peak at *m/z* 511.3277, and its molecular formula $C_{30}H_{48}O_5$ indicated that it was the dehydrogenation product of M38. M40 exhibited an $[M + Na]^+$ ion peak at *m/z* 591.4888, indicating a mass 80 Da larger than M39 and was inferred to be the sulfation product of M39.

2.5.3. Lactones Metabolites

As depicted in Figure S8A, 13 metabolites (M41–M53) were identified as lactones-related metabolites present in plasma, urine, and feces. These metabolites were primarily associated with atractylenolide I and atractylenolide III. The metabolism pathway of atractylenolide III was used as an illustrative example. M41 exhibited an $[M + H]^+$ ion peak at m/z 265.1433, indicating a mass increase of 16 Da compared to atractylenolide III, suggesting M41 was a hydroxylated derivative of atractylenolide III. M42 displayed an $[M + H]^+$ ion peak at m/z 281.1411, demonstrating a further 16 Da increase from M41, leading to the inference that M42 was a hydroxylated product of M41. M43 presented an $[M + H]^+$ ion peak at 263.1291, indicating it was a dehydrogenated derivative of M41. M44 showed an $[M + H]^+$ ion peak at m/z 283.1564, representing a mass increase of 34 Da from atractylenolide III. Combined with its molecular formula $C_{15}H_{22}O_5$, this suggested M44 was a result of hydration and hydroxylation of atractylenolide III. M45 exhibited an $[M + H]^+$ ion peak at m/z 247.1349, demonstrating a 2 Da decrease from Atractonolide III. Its MS/MS ions at 229.1227 and 201.1316 were also 2 Da smaller than atractonolide III, implying M45 was a dehydrogenation product of atractonolide III. M46 displayed an $[M + H]^+$ ion peak at m/z 423.1691, indicating a mass increase of 176 Da from M45, suggesting M46 was a glucuronidation product of M45. M47 showed an $[M + H]^+$ ion peak at m/z 441.1794, demonstrating a further 18 Da increase from M46, inferring that M47 was a hydrated product of M46. These findings were consistent with the literature, which reported hydroxylation and glucuronidation as the primary metabolic reactions of atractonolactone III [22].

2.5.4. Chromone and Coumarin Metabolites

As depicted in Figure S8B, four metabolites (M54–M57) were identified as metabolites of cimifugin and scopoletin in plasma, urine, and feces. M54 exhibited an $[M + H]^+$ ion peak at m/z 323.1151, indicating a mass increase of 16 Da compared to cimifugin. The fragment ions at m/z 259.0609 and 221.0462 were consistent with cimifugin, suggesting that M54 was a hydroxylated product of cimifugin. M55 displayed an $[M + H]^+$ ion peak at m/z 483.1544, demonstrating a mass increase of 176 Da relative to cimifugin. The fragment ions at m/z 259.0625 and 235.0621 were consistent with cimifugin, indicating that M55 was the glucuronidation product of cimifugin.

2.6. Metabolic Types

As depicted in Figure S9, 39, 46, and 29 metabolites of YPF were detected in plasma, urine, and feces, respectively. In plasma (Figure S10), phase I reactions accounted for 33% (13 metabolites), primarily involving oxidation, reduction, hydroxylation, demethylation, and other reactions. Phase II reactions comprised 41% (16 metabolites), mainly glucuronidation and sulfation reactions. The combined phase I and phase II reactions accounted for 26% (10 metabolites), primarily oxidation–glucuronidation binding and oxidation/reduction/hydration–sulfation/methylation reactions. In urine (Figure S11), phase I reactions accounted for 39% (18 metabolites), mainly demethylation, hydroxylation, oxidation, hydration, and other reactions. Phase II reactions comprised 33% (15 metabolites), primarily glucuronidation and sulfation reactions. The combined phase I and phase II reactions accounted for 28% (13 metabolites), primarily demethylation/oxidation–sulfation/glucuronidation, hydration/reduction–methylation/glucuronidation, and other reactions. In feces (Figure S12), phase I reactions accounted for 50% (14 metabolites), mainly demethylation, hydration, and oxidation. Phase II reactions comprised 21% (6 metabolites), primarily methylation, glucuronidation, and sulfation reactions. The combined phase I and phase II reactions accounted for 29% (8 metabolites), primarily demethylation–methylation/sulfation/glucuronidation reactions.

2.7. Network Pharmacology

2.7.1. Construction of YPFS Composition-Targets-AR Networks and Screening of the Key Components

As depicted in Figure 6A, 1081 YPFS prototype component-related targets and 634 AR-correlated targets were obtained after removing duplicates. Subsequently, 166 overlapping targets between YPFS and AR were identified. To visualize the complex relationships between the active ingredients of YPFS and AR-related targets, a YPFS prototypical component–target–AR network was constructed using Cytoscape v3.9.1. This network comprised 213 nodes (including 42 ingredients, 3 herbs, YPFS, AR, and 166 targets) and 1016 edges (Figure 6B). Table S7 lists the serial numbers of the 42 prototype compounds. In the network, circles represent effective components, blue indicates common targets and yellow represents AR. The 42 components were sorted based on their degree of centrality using Network Analyzer (Figure 6C, Table S8). Ranking of the compounds based on the degree value showed that the top 15 compounds were cimifugin, wogonin, formononetin, skullcapflavone II, atractylenolide III, isoastragaloside I, melilotocarpan B, afrormosin, anomalin, methylnissolin, astramembrannin II, calycosin, apigenin, isomucronulatol, and atractylenolide I. It was predicted that these ingredients may be the key compounds for the treatment of AR with YPFS.

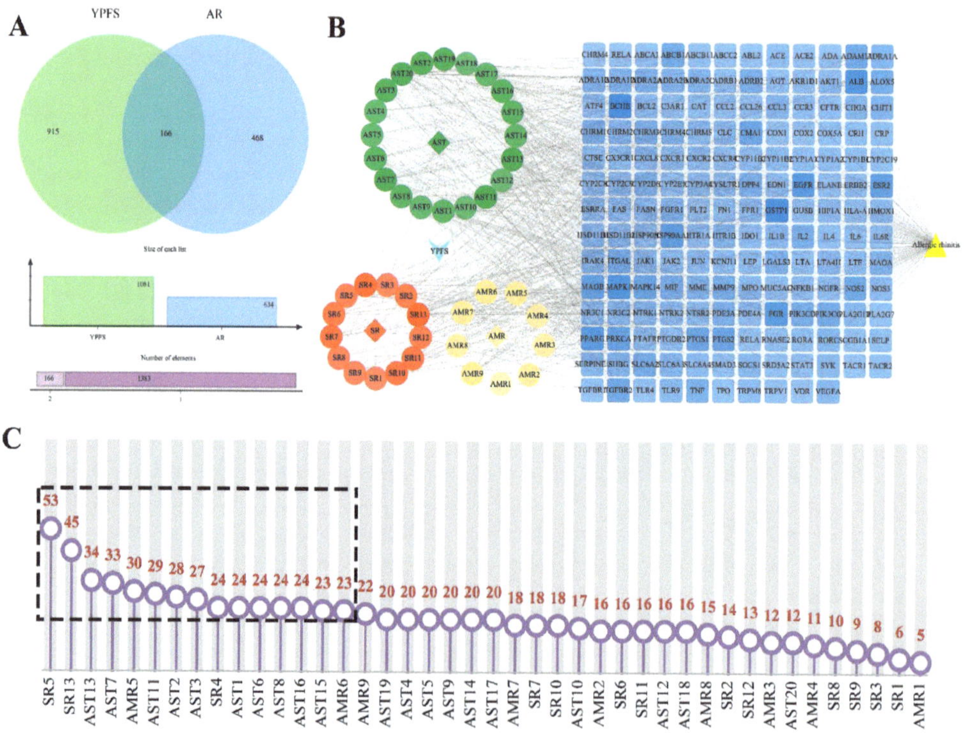

Figure 6. Screening of key components. (**A**) Venn diagram of YPFS and AR miscarriage targets; (**B**) YPFS–active component–target–disease network; (**C**) The degree information from network topology analysis on the 42 compounds.

2.7.2. Protein–Protein Interaction (PPI) Analysis and Screening of the Key Targets

The obtained 166 potential genes were imported into String to generate the PPI network, which consisted of 2184 edges (Figure 7A, Figure S13) and was visualized using

Cytoscape v3.9.1 (Figure S14). The top 20 genes were identified as key targets using the "Cytohubba" plugin (Figure 7B–D, Table S9). The results showed tumor necrosis factor (TNF), interleukin-6 (IL6), albumin (ALB), interleukin-1β (IL-1β), alpha kinase threonine-1 (AKT1), vascular endothelial growth factor A (VEGFA), c-x-c chemokine ligand 8 (CXCL8), signal transducer and activator of transcription 3 (STAT3), epidermal growth factor receptor (EGFR), toll-like receptor 4 (TLR4), interleukin-4 (IL4), c-c motif chemokine ligand 2 (CCL2), matrix metallopeptidase 9 (MMP9), jun proto-oncogene (JUN), prostaglandin-endoperoxide synthase 2 (PTGS2), fibronectin 1 (FN1), peroxisome proliferator-activated receptor gamma (PPARG), interleukin-2 (IL2), hypoxia-inducible factor 1 subunit alpha (HIF1A) and endothelin 1 (EDN1), suggesting that these targets may be key targets for YPFS in the treatment of AR.

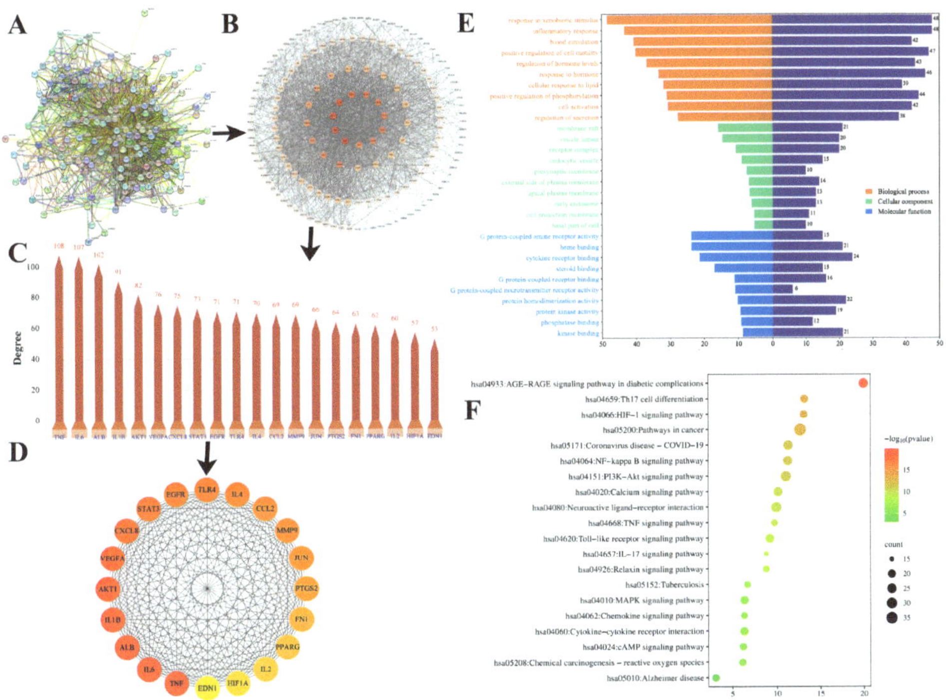

Figure 7. Screening of key targets and KEGG analysis. (**A**) Protein–protein interaction (PPI) network; (**B**) Circle plots for PPI network; (**C**) the degree information of 20 target; (**D**) Circle plots for the 20 targets with the highest degree; (**E**) columnar plots for GO analysis; (**F**) dot bubble plots for KEGG.

2.7.3. GO and KEGG Analysis

GO analysis identified 1810 biological processes (BP), 98 cell components (CC), and 202 molecular functions (MF). The top 10 enrichments for each of the three GO categories were presented in Figures 7E and S15 and Table S10. BP included inflammatory response, cellular response to lipids, cell activation, and regulation of secretion. CC primarily encompassed the receptor complex, the external side of the plasma membrane, the cell projection membrane, and the basal part of the cell. MF involved cytokine receptor binding, steroid binding, G protein-coupled receptor binding, and kinase binding. KEGG pathway enrichment analysis using DAVID revealed 160 enriched pathways. These genes were particularly enriched in immune and inflammatory-related pathways. The top 20 pathways (Figure 7F and Table S11) encompassed the toll-like receptor signaling pathway, PI3K-Akt signaling pathway, HIF-1 signaling pathway, MAPK signaling pathway, TNF signaling pathway, and NF-kappa B signaling pathway, among others. The greater the number of genes in these pathways, the higher the likelihood that YPFS treated AR by modulating these pathways or biological processes.

2.8. Molecular Docking

To elucidate the interactions between compounds and the primary targets of YPFS in the treatment of AR, we conducted molecular docking of the top 10 key targets with 15 core components to investigate the binding modes of protein–ligand complexes. Generally, a higher affinity between receptor and ligand corresponds to a lower score. It was widely accepted that a binding energy less than −5.0 kJ/mol indicates a favorable binding interaction between the compound and target. As depicted in the docking affinity heatmap (Figure 8A, Table S12), the core target proteins AKT1, EGFR, and ALB exhibited binding energies below −5 kJ/mol, indicating strong binding affinities for most of the key components. This implies that they were crucial targets for YPFS in the treatment of AR. Furthermore, Figure 8B–K illustrated that these compounds (methylnissolin, calycosin, cimifugin, atractylenolide I, formononetin, and wogonin) had the highest docking scores at each target, with the specific docking results parameters provided in Tables S12 and 1. The results revealed that cimifugin (SR5) achieved the lowest binding energy for VEGFA, ALB, TLR4, and CXCL8 among the 15 components (Tables S12 and 1). Cimifugin formed not only two hydrogen bonds with the amino acid residues ASP:34 and LEU:32 of VEGFA, but also four hydrogen bonds with the amino acid residues ILE:290, ARG:257, HIS:242 and GLN:196 of ALB, two hydrogen bonds with the amino acid residues GLU-286 and CYS-281 of TLR4, and four hydrogen bonds with the amino acid residues ARG:26, GLN:8, ALA:50 and LYS:11 of CXCL8.

Methylnissolin (AST1) showed the highest docking scores for AKT1 and STAT3 among the 15 components (Tables S12 and 1). This binding stability was closely associated with the efficacy of YPFS in the treatment of AR. Specifically, Methylnissolin exhibited the lowest energy when binding to AKT1, forming a hydrogen bond with the amino acid residue AER-205 of AKT1 and one pi–pi interaction with TRP-80. Atractylenolide I formed one hydrogen bond with amino acid residue ARG:179 of IL6. Calycosin formed three hydrogen bonds with the amino acid residues MET:769, PRO:770, and ASP:831 of EGFR. Formononetin formed two hydrogen bonds with amino acid residues ASN:108 and LYS:103 of IL-1β. One hydrogen bond was formed between wogonin and the amino acid residue TYR:151 of TNF-α, and two π–π bonds were formed with TYR:119. These results suggested that methylnissolin, calycosin, cimifugin, atractylenolide I, formononetin and wogonin may be key substances in improving AR.

Table 1. Molecular docking parameters for compounds with first binding energy for 10 core targets.

Compound	Protein-PDBID	Affinity KJ/Mol	Site (xyz)	Bonding
Methylnissolin (AST1)	AKT1-3O96	−8.52	(8.59, −7.1, 13.14)	SER, TRP
Cimifugin (SR5)	VEGFA-1MKK	−7.916	(5, 0, 16)	LEU, ASP
Calycosin (AST8)	EGFR-1M17	−7.461	(21.29, 0.54, 52.36)	MET, PRO, ASP
Cimifugin (SR5)	ALB-2BX8	−7.422	(5.46, −10.15, 7.28)	ILE, ARG, GLN, HIS
Wogonin (SR13)	TNF-α-2AZ5	−6.474	(−19.82, 74.94, 33.46)	TYR
Cimifugin (SR5)	TLR4-2Z63	−5.717	(−0.1, −53.56, 13.23)	GLU, CYS
Atractylenolide I (AMR6)	IL6-1ALU	−5.23	(0, 5, 0)	ARG
Formononetin (AST13)	IL-1β-6Y8M	−5.199	(6.95, 25.46, −9.28)	LYS, ASN
Methylnissolin (AST1)	STAT3-6NJS	−4.974	(13.24, 56.43, 0.24)	SER, GLN, TYR
Cimifugin (SR5)	CXCL8-1ICW	−4.302	(−6, 29, 30)	ARG, GLN, ALA, LYS

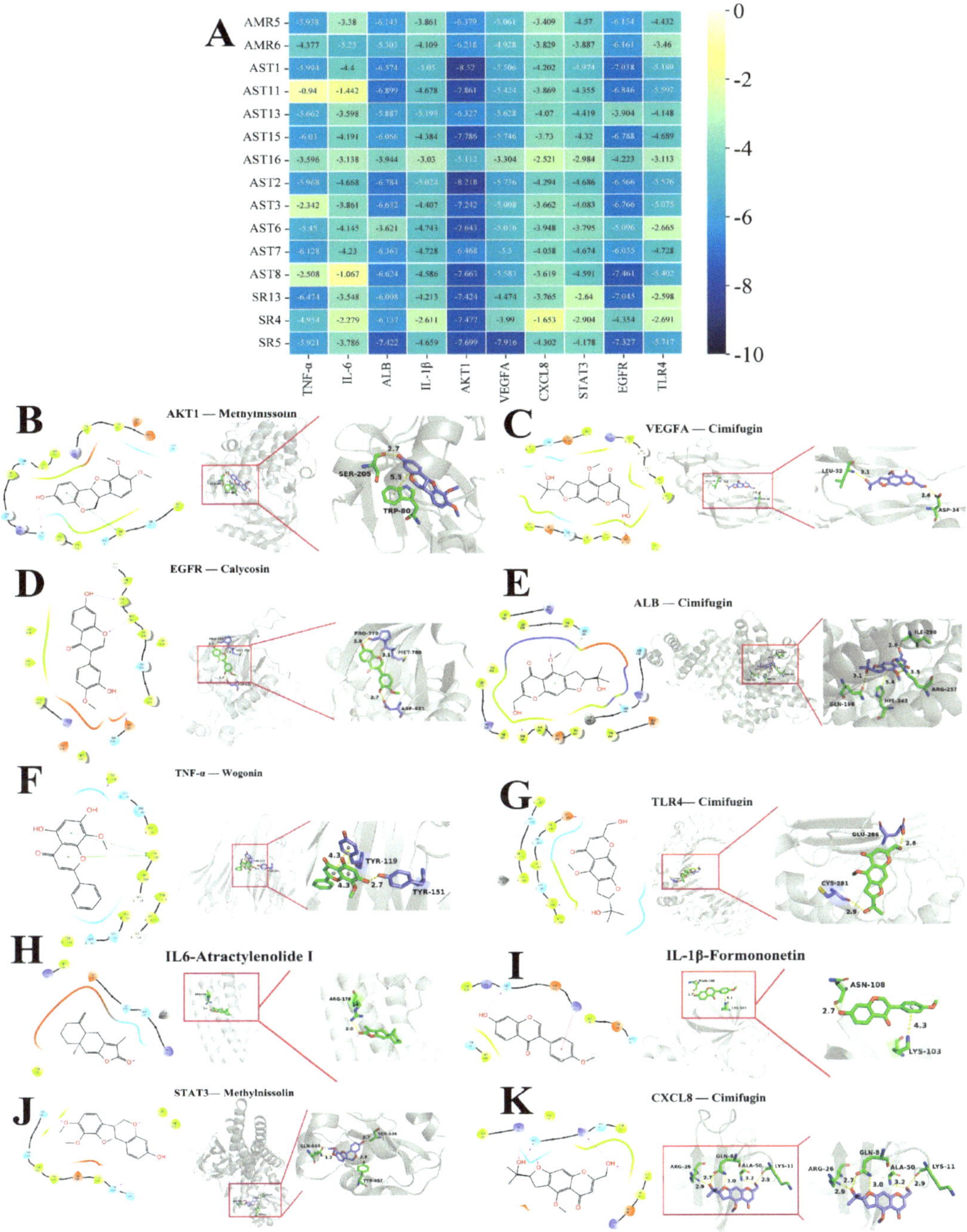

Figure 8. Molecular docking results. (**A**) Heatmap of 15 key components with 10 core targets; (**B**) methylnissolin (AR1)-AKT1; (**C**) cimifugin (SR5)-VEGFA; (**D**) calycosin (AR8)-EGFR; (**E**) cimifugin (SR5)-ALB; (**F**) wogonin (SR13)-TNF-α; (**G**) cimifugin (SR5)-TLR4; (**H**) atractylenolide I (AMR6)-IL6; (**I**) formononetin (AR13)-IL-1β; (**J**) methylnissolin (AR1)-STAT3; (**K**) cimifugin (SR5)-CXCL8.

2.9. Validating the Anti-AR Activity of Active Components (Vitro)

Comprehensive analysis of the vivo composition-based network pharmacological and molecular docking revealed that cimifugin, wogonin, formononetin, methylnissolin, calycosin, and atractylenolide I played key roles, with their structural formulae (red) shown in Figure 9A. The maximum administered doses were maintained at 20 μmol/L for cimifugin, atractylenolide I, and methylnissolin, and the maximum administered doses of formononetin, wogonin, and calycosin were maintained at 10 μmol/L (Figure 9B–G), which did not affect cell viability. As shown in Figure 9H–M, these six potential active ingredients improved the secretion of IgE, β-Hex, and inflammatory factors (IL-2, IL-5, IL-6, and TNF-α) in DNP-IgE/BSA-modeled RBL-2H3 cells to different degrees. Compared with the release rate of IL-6 and TNF-α in the model group, there was a significant reduction in the improvement in the 6 component groups ($p < 0.01$). Compared with the β-Hex, IL-5, IL-2, and IgE release rates in the model group, the rate of improvement was significant in all four component groups except for methylnissolin and calycosin ($p < 0.01$). These results further validated that these compounds may be the key substances for improving AR.

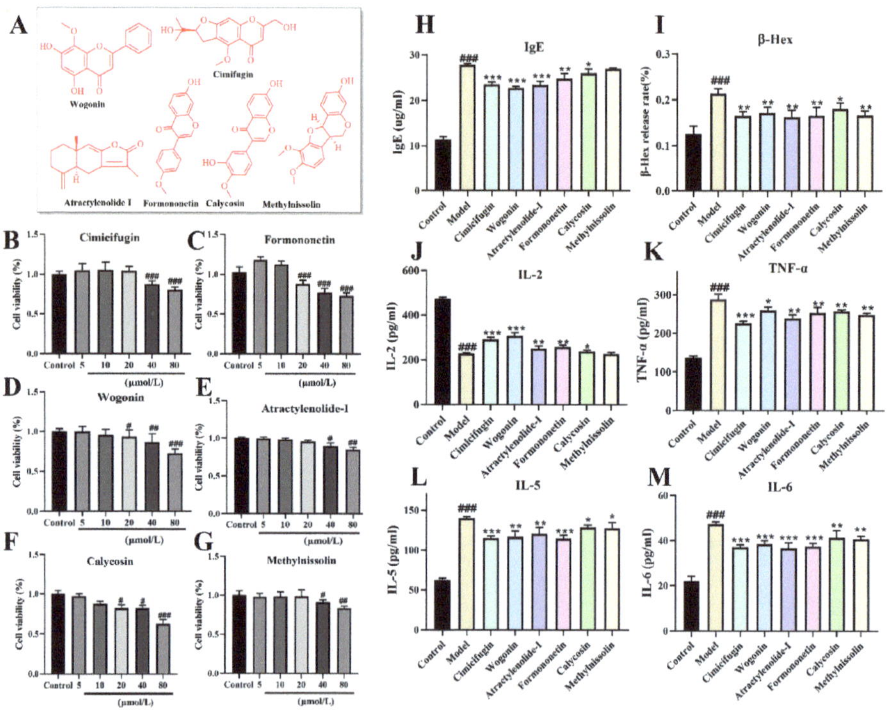

Figure 9. Verification of anti-AR effects of active ingredients on RBL-2H3. (**A**) The structural formulae of six active ingredients; MTT results for cimifugin (**B**), formononetin (**C**), atractylenolide-I (**D**), wogonin (**E**), calycosin (**F**), and methylnissolin (**G**); anti-AR effects of the active ingredients in terms of (**H**) Ig-E, (**I**) β-Hex, (**J**) IL-2, (**K**) TNF-α, (**L**) IL-5, and (**M**) IL-6 ($^{\#}$ $p < 0.05$, $^{\#\#}$ $p < 0.01$, $^{\#\#\#}$ $p < 0.001$ compared to the control and model group; * $p < 0.05$, ** $p < 0.01$ and *** $p < 0.001$ compared to the model group).

2.10. Validating the Anti-AR Activity of Active Components (Vivo)

2.10.1. Active Components Improve Nasal Symptom Score and Serum Indicators

After a comprehensive analysis of the cellular validation results, four compounds (cimifugin, wogonin, formononetin, and atractylenolide I) were selected to continue the validation at the animal level (Figure 10). At the conclusion of the 30 d experimental procedure (Figure 10A), the body weight of AR mice was reduced. Compared to the control group, the model group exhibited a significant decrease in body weight ($p < 0.001$). As shown in Figure 10B–D, cimifugin and wogonin significantly enhanced the body weight of AR mice ($p < 0.05$). Compared to the model group, four compounds (cimifugin, wogonin, formononetin, and atractylenolide I) could effectively reduce the number of rubbing and sneezing in AR mice (Figure 10E,F). As shown in Figure 10G–L, compared with the release rate of histamine, IgE, IL-5, IL-6, and TNF-α in the serum of the model group, there was a significant reduction in the improvement of four active components groups ($p < 0.05$). Compared with the release rates of IL-2 in the model group, four compounds (cimifugin, wogonin, formononetin, and atractylenolide I) could significantly increase ($p < 0.001$, Figure 10I). These results further validated that these compounds may be the key substances for improving AR.

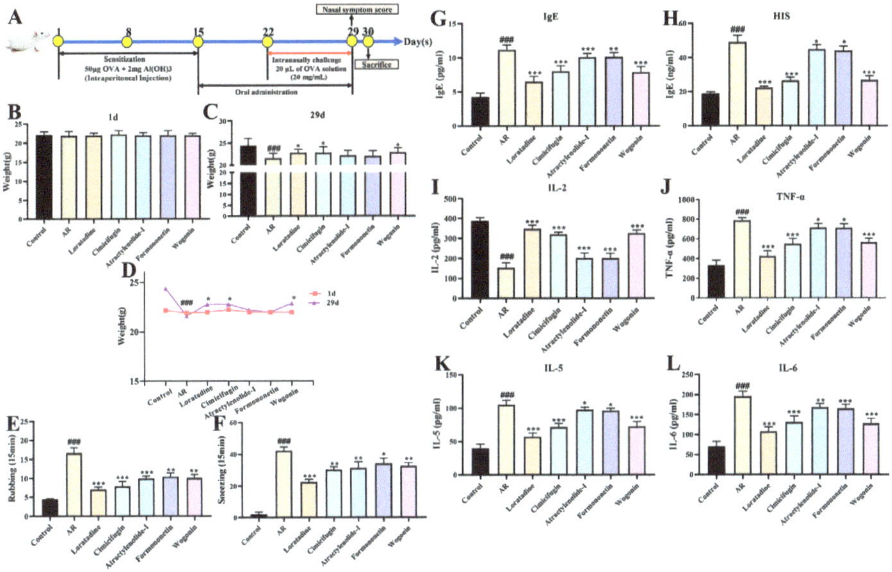

Figure 10. Effects of cimifugin, atractylenolide I, formononetin, and wogonin on body weight, behavior, and serum cytokine levels of AR mice. (**A**) The flow chart of the experiment; (**B**) initial weight; (**C**) last weight; (**D**) weight curve, (**E**) rubbing; (**F**) sneezing, (**G**) IgE; (**H**) histamine; (**I**) IL-2; (**J**) TNF-α; (**K**) IL-5; (**L**) IL-6 ($^{###}$ $p < 0.001$ compared to the control group; * $p < 0.05$, ** $p < 0.01$ and *** $p < 0.001$ compared to the model group).

2.10.2. Active Components Improve Pathological State of Nasal Mucosa and Lung

The staining results of HE and pas indicated that the ciliated structure of the nasal mucosa in the control group was intact, and the epithelial cells were arranged neatly. AR mice were massive cilium shedding in the nasal mucosa, a marked increase in goblet cells, accompanied by inflammatory cell infiltration. Additionally, the epithelial cells showed obvious congestion and edema, with detachment of the basement membrane. Cimifugin, atractylenolide I, formononetin, and wogonin showed significant improvements in nasal mucosa congestion, inflammatory cell infiltration, goblet cell hyperplasia, and cilium shedding (Figure 11A,B).

HE staining revealed normal lung tissue morphology in the control group, with clear alveolar, bronchial, and interstitial structures. In contrast, AR mice showed significant lung tissue enlargement, severe hemorrhage, and extensive inflammatory cell infiltration around the bronchi and alveoli. Cimifugin, atractylenolide I, formononetin, and wogonin exhibited no obvious alveolar hemorrhage or edema, with marked improvement in inflammatory cell infiltration, indicating that active components had a strong improvement effect on the nasal mucosa and lung (Figure 11C).

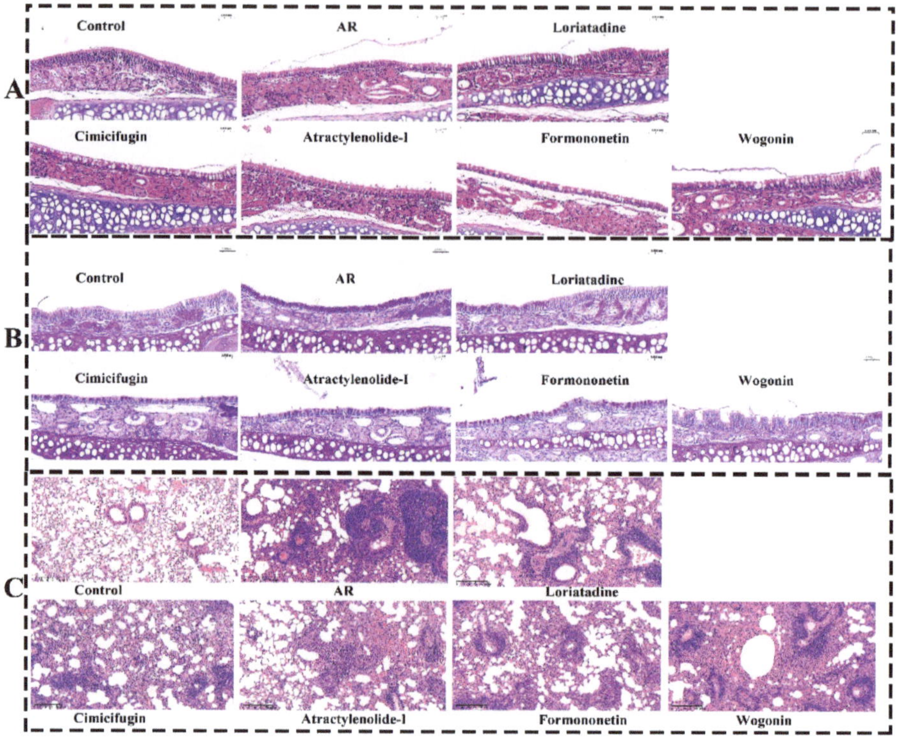

Figure 11. Effect of cimifugin, atractylenolide I, formononetin, and wogonin on histological changes in the nasal mucosa and lung of AR mice. (**A**) HE staining (nasal mucosa, 200×); (**B**) PAS staining (nasal mucosa, 200×); (**C**) HE (lung, 400×).

3. Discussion

In this study, we combined the GNPS molecular network with traditional manual identification to systematically identify 98 chemical constituents of YPFS. In total, 42 prototypical components and 57 metabolites were identified in plasma, urine, and feces of AR mice. The relevant metabolic pathways mainly involved reduction, demethylation, hydroxylation, methylation, and glucuronidation. After oral administration, as TCM traverses the digestive tract, numerous metabolic reactions may ensue, triggered by gastric acid, a range of digestive enzymes, and intestinal flora [23]. Many researchers have explored the potential medicinal ingredients of TCM through the dynamic fermentation of human intestinal bacteria [24,25]. The primary metabolic reactions undergone by flavonoids and their glycosides within the body involve hydrolysis, conjugation, cleavage, and oxidation. Our research findings align with these processes. Compared with existing studies on YPFS, this investigation has identified a more comprehensive array of chemical constituents and successfully performed in vivo characterization of both prototype components and their metabolites in AR mice, which substantially contributes to elucidating the holistic material basis underlying YPFS's therapeutic effects on AR [9]. Furthermore, the established methodology provided a validated framework for systematic compositional analysis and metabolic investigation of other TCM compound formulations.

Subsequently, a YPFS-active component-target-AR network was constructed based on 42 prototype components identified in vivo, and 15 key constituents were prioritized through degree centrality analysis. This was followed by molecular docking between these prioritized components and the top 10 key targets screened from the PPI network. The results indicated that cimifugin, methylnissolin, calycosin, atractylenolide I, formononetin, and wogonin exhibited the lowest energy of binding with each target, suggesting they were potentially the key components of YPFS for treating AR. Concurrently, the results of KEGG and GO analyses suggested that the NF-κB pathway may be a key pathway in the improvement of AR by YPFS. Wang et al. et al. found that YPFS could reduce the levels of NF-κB and TSLP in nasal lavage fluid of patients, along with decreased serum concentrations of IL-6 and TNF-α in AR patients through a multicentre clinical trial [26]. This suggested that YPFS may improve AR by modulating the NF-κB/TSLP signaling pathway, in keeping with our study. Through PPI analysis, TLR4, TNF-α, and IL1β were selected among the top 10 core targets with the highest docking scores, and they were also identified as key targets of the NF-κB pathway (Figure S16), indicating their significant role in the treatment of AR by YPFS. TLR4, a pivotal target in the toll-like receptor pathway, has been established in numerous studies as a crucial target in AR [27–29]. It played the key role of modulating the NF-κB pathway, resolving the Th1/Th2 imbalance, and suppressing allergic symptoms [30,31]. Hu et al. et al. found that YPFS could suppress the release of inflammatory factors and the apoptosis of cells as well as relieve the symptoms of AR rats probably by inhibiting the TLR4/NF-κB pathway [32]. Nie et al. also explained that YPFS could ameliorate atopic dermatitis by inhibiting the TLR4/MyD88/NF-κB pathway through animal experiments, and molecular docking showed that compounds such as calycosin, formononetin, and divaricatol had specific binding properties with TNF-α, IL-6, and TLR4 [33]. These findings align with our results, which increased the credibility of the key components selected. In a word, the six key potential active components of YPFS for treating AR were screened through network pharmacology and molecular docking.

Conclusively, the pharmacological efficacy of the prioritized compounds was rigorously validated through in vitro bioassays and in vivo functional studies in AR models. Firstly, cellular validation was carried out on six ingredients, and the results showed that compared with the β-Hex, IL-5, IL-2, and IgE release rates in the model group, the rate of improvement was significant in all four component groups except for methylnissolin and calycosin ($p < 0.01$). Then, the efficacy of the four compounds (cimifugin, atractylenolide I, formononetin, and wogonin) was further verified at the animal level, and it was found that could improve nasal symptom score, serum indicators, pathological state of the nasal mucosa and lung of AR mice. Previous studies have demonstrated that cimifugin has the capacity to suppress allergic inflammation by downregulating epithelial-derived factors crucial to the inflammatory process [34]. Research has indicated that wogonin was a potential ingredient active against AR and has identified EGFR and AKT1 as key targets for AR treatment with YPFS [35,36]. Wogonin could reduce the level of IgE, IL-4, and IL-5 in serum, effectively inhibit AR-like symptoms, notably inhibiting the production of TSLP and IL-33, and play an anti-allergic role in AR mice [37]. Cheng et al. also explained that wogonin was the key active component in improving AR integrating analysis of network pharmacology and proteomics [38]. Formononetin could improve IL-13-induced inflammation and mucus formation in human nasal epithelial cells by activating the SIRT1/Nrf2 signaling pathway [39]. Qu et al. also found that formononetin was a potential component for treating allergic rhinitis through network pharmacology and molecular docking [40]. Atractylenolide I, a novel antagonist of TLR4, inhibited TLR4 signaling by disrupting the binding of lipopolysaccharide (LPS) or paclitaxel to TLR4 in human leukocyte membranes [41,42]. This compound subsequently reduced the production of

TNF-α, IL-6, and NO and the expression of monocyte chemotactic protein-1 (MCP-1) in vascular smooth muscle cells. Furthermore, atractylenolide I inhibited the activation of p38 mitogen-activated protein kinase (p38-MAPK) and nuclear factor kappa-light-chain-enhancer of activated B cells (NF-κB) [43]. In summary, the vitro (RBL-2H3 cells) and in vivo (AR mice) results further validated that these four compounds (cimifugin, atractylenolide I, formononetin, and wogonin) may be the key substances for improving AR.

4. Materials and Methods

4.1. Reagents and Instrument

Saline for injection was purchased from Shuanghe Pharmaceutical Co., Ltd. (Huai'an, Jiangsu, China). Ovalbumin, aluminum hydroxide, DNP-BSA (A2831), DNP-IgE (D8406), and 4-nitrophenyl-N-acetyl-β-D-glucosaminide were obtained from Sigma-Aldrich (Saint Louis, MO, USA). The freeze dryer was acquired from Sihuan Scientific Instrument Factory Co. Ltd. (Beijing, China). An electronic analytical balance from Shanghai Huaiping Chemical Co. Ltd. (Shanghai, China) was used. A cryogenic centrifuge from Beckman Coulter (Darmstadt, Germany) was utilized. The laparotomy syringe was procured from Jiangsu Great Wall Medical Co., Ltd. (Nanjing, Jiangsu, China). The Milli-Q ultrapure water system was supplied by Millipore (Bedford, MA, USA). IL-2, IL-5, IL-6, and TNF-α ELISA kits were purchased from Aifang Co., Ltd. (Changsha, Hunan, China). Immunoglobulin E (IgE) and histamine were measured by ELISA using kits from Wuhan Fain Biologicals (Wuhan, China). RBL-2H3 cells were obtained from the Shanghai Chinese Academy of Sciences (Shanghai, China). Dimethyl sulfoxide (DMSO) and MTT were procured from Biyuntian (Shanghai, China). Cimifugin (purity $\geq$ 98%), formononetin (purity $\geq$ 98%), atractylenolide-I (purity $\geq$ 98%), atractylenolide III (purity $\geq$ 98%), calycosin (purity $\geq$ 98%), methylnissolin (purity $\geq$ 98%), and wogonin (purity $\geq$ 98%) were purchased from Chengdu Efa Biotechnology Co., Ltd. (Chengdu, China).

4.2. Extraction of YPFS

Astragalus mongholicus Bunge *(AST)*, *Atractylodes macrocephala* Koidz *(AMR)*, and *Saposhnikovia divaricata* (Turcz.) Schischk *(SR)* were obtained from the Clinic of Nanjing Traditional Chinese Medicine (Nanjing, Jiangsu, China) and authenticated by Prof. Qi Jin of China Pharmaceutical University, who specializes in Traditional Chinese Medicine. AST, AMR, and SR were obtained from Hunyuan (Shanxi, China, 39°41′35.7″ N, 113°41′16.3″ E), Pan'an (Zhejiang, China, 29°3′26.3″ N, 120°26′44.1″ E), Chifeng (Neimenggu, China, 42°15′23.50″ N, 118°52′58.14″ E). AST, AMR, and SR were powdered using a 20-mesh sieve and accurately weighed 180, 90, and 90 g at a ratio of 3:1:1. Then extracted twice with 10 volumes of 95% ethanol for 2 h each. The extract was concentrated using a rotary evaporator and then freeze-dried, resulting in a weight of 89.53 g. The yield was calculated to be 29.84% and stored in a −20 °C freezer for using.

4.3. Animal Experiments

Ten female BALB/c mice (18–22 g) were purchased from Yangzhou University (Yangzhou, Jiangsu Province, China) under license number SCXK (Su) 2020-0009. The animals were housed in a room maintained at a temperature of 23 ± 1 °C and subjected to a 12-h light/dark cycle. All the procedures were approved by the Animal Ethics Committee of China Pharmaceutical University.

The AR model was constructed according to the methods described in references [44,45] and consisted of two steps: the basal sensitization step, the ovalbumin (OVA) was used as the sensitizing antigen, and the 200 μL saline containing 50 μg of OVA and 2 mg of aluminum hydroxide with the intraperitoneal injection. Three sensitizations were performed on days 1, 8, and 15 d. The challenge phase began 7 d after the end of the basal sensitization phase and consisted of instilling 20 μL of OVA solution (20 mg/mL) into the bilateral nasal cavity of the mice once a day for 8 consecutive days. The mice were randomly divided into the model group and YPFS group. Saline and YPFS (25 g/kg body weight) were administered to the AR and YPFS groups, respectively, continuously for 3 d from 27 d to 29 d, once in the morning and once in the evening.

4.4. Sample Pretreatment

Plasma samples: After the final administration, blood samples were collected from the aforementioned groups at designated time intervals of 0.5, 1, 2, 4, and 8 h, specifically from the posterior venous location behind the eye, in volumes of 500 μL. These samples were transferred into centrifuge tubes containing sodium heparin and centrifuged at 3500 rpm for 10 min. The supernatant plasma was stored in a freezer at −20 °C. Subsequently, the plasma samples were combined with acetonitrile in a 1:3 ratio, vortexed for 1 min, and centrifuged at 4000 rpm for 15 min. The resulting supernatant was evaporated using a nitrogen blower, and the residual material was dissolved in a methanol–water mixture (7:3, *v*/*v*). This solution was then centrifuged at 12,000 rpm for 15 min and filtered through a 0.22 μm microporous filter membrane prior to analysis. The urine and feces sample preparation procedures followed the same protocol as that of the plasma samples.

4.5. HPLC and MS Condition

4.5.1. Optimized Detection Conditions

The wavelength (220, 254, and 258 nm), injection volume (2.5, 5, and 10 mg), flow rate (0.6, 0.8, and 1 mL/min), and column temperature (25, 30, and 35 °C) were investigated and optimized.

4.5.2. Conditions of Chromatography and Methodological Study

The Agilent 1260 Infinity HPLC system from Agilent Technologies (Santa Clara, CA, USA) was used for chromatographic separation. A Diamonsil C18 column (4.6 × 250 mm, 5 μm) was employed. The mobile phase consisted of acetonitrile as phase A and water with 0.01% formic acid as phase B. The elution gradient proceeded as follows: 97–85% B for 0–20 min; 85–82% B for 20–25 min; 82–80% B for 25–30 min; 80–72% B for 30–40 min; 72–60% B for 40–50 min; 60–40% B for 50–70 min; 40–5% B for 70–85 min; 5–0% B for 85–90 min. 20 μL sample was injected for analysis.

Precision experiment: An appropriate amount of YPFS sample powder was weighed and prepared for the sample solutions. Under the specified chromatographic conditions, the sample was continuously injected six times. Repeatability experiment: Six test sample solutions were prepared in parallel from the same YPFS sample powder according to the preparation method, followed by HPLC analysis. Stability experiment: HPLC analysis was performed at 0, 2, 4, 8, 12, and 24 h after sample preparation. The relative peak areas of the four compounds (calycosin, atractylenolide III, cimifugin, formononetin) were recorded, and the relative standard deviation (RSD) values were calculated.

4.5.3. Conditions of Mass Spectrometry

Mass spectrometry data were collected using the Agilent 1260 Q-TOF MS/MS system (Agilent Technologies, Santa Clara, CA, USA), which was equipped with an electrospray ionization (ESI) interface. The operational parameters were set as follows: The ESI source was operated in positive mode with a nitrogen (N_2) flow rate of 10.0 L/min for the drying gas. The temperature of the drying gas was maintained at 350 °C, and the nebulizer pressure was set to 40 psi. The capillary voltage was adjusted to 4000 V, while the skimmer voltage was maintained at 65 V. Additionally, the fragment voltage was set to 135 V, and the MS/MS collision voltage was configured at 30 V. The scan range covered a mass range from 50 to 1500 Da. Furthermore, the detection wavelength was 254 nm. The collected data were subsequently analyzed using the Agilent Mass Hunter Workstation B.06.00 (Agilent Technologies, Santa Clara, CA, USA).

4.6. Molecular Networks

The molecular network (MN) of the components of YPFS was constructed following the online process of the GNPS database. The original MS/MS data files were converted into mzML format using the MSConvert software (Institute for Systems Biology, Washington, DC, USA) from the ProteoWizard package, and these mzML format files were subsequently checked using MZmine 2 v.40.1 [46]. The mzML format file was uploaded to the GNPS database using WinSCP software (Prague University of Economics and Business, Prague, Czech Republic,), and the MN of YPFS was established according to the online workflow at GNPS. The parameters for the MN were set with the following specifications: a precursor ion mass error of 0.02 Da, a fragment ion mass error of 0.02 Da, a minimum cosine value of 0.7, a minimum requirement of six paired fragment ions, a maximum limit of ten connections at a single point, a maximum number of compounds in the network of 100, a filter precursor ion window of +/−17 Da, and retention of only the first six peaks in the range of +/−50 Da. The MN was visualized using the online GNPS and Cytoscape 3.9.0 software.

4.7. Network Pharmacology

4.7.1. Screening of Prototype Compounds Targets and Disease Targets

TCMSP (https://old.tcmsp-e.com/tcmsp.php (accessed on 21 March 2023)), BATMAN-TCM (Score $\geq$ 10, http://www.badd-cao.net:2345/ (accessed on 21 March 2023)), the Swiss Target Prediction database (http://swisstargetprediction.ch/ (accessed on 21 March 2023)) and PharmMapper (Norm Fit $\geq$ 0.8, http://www.lilab-ecust.cn/pharmmapper/ (accessed on 21 March 2023)) were used to predict the targets of active ingredients. AR-associated targets were searched for using the keyword "allergic rhinitis" at GeneCards (https://www.genecards.org/ (accessed on 23 March 2023)), TTD (http://db.idrblab.net/ttd/ (accessed on 23 March 2023)) and Drugbank (https://go.drugbank.com/ (accessed on 23 March 2023)). After deduplication, all the targets were combined, and Uniprot (https://www.uniprot.org/ (accessed on 23 March 2023)) was used to convert and correct the ID of the target name for further study. A Venn diagram was used to depict the overlapping genes regarded as potential targets of YPFS therapy for AR.

4.7.2. Compound–Target–Disease Network Construction and Screening of Key Components

The information from the column "network" and "type" from the file in XLXS format was imported to Cytoscape 3.9.1 and the "YPFS-Targets-AR" network. The key ingredients were screened by analyzing the network topology parameters with the "Network Analyzer" tool.

4.7.3. Protein–Protein Interaction (PPI) Network and Screening of Key Targets

The YPF/AR-related targets were imported to String (https://cn.string-db.org/ (accessed on 28 March 2023)), with the species set as "Homo sapiens"; the obtained data were screened with a minimum required interaction score of 0.4. The results were exported in TSV format and imported to Cytoscape3.9.1 to construct the PPI network, and the key target genes were identified using the "Cytohubba" plugin.

4.7.4. GO and KEGG

Gene ontology (GO) annotation analysis involving biological process (BP), cellular component (CC), and molecular function (MF) was carried out using Metascape. David (https://david.ncifcrf.gov/ accessed on 28 March 2023) was used for KEGG pathway enrichment analysis, improving the obtained related genes with the species set as "Homo sapiens".

4.8. Dockings

Molecular docking was performed on key components and their respective targets. The compounds were downloaded in SDF format from PubChem. The 3D structures of AKT1 (PDB ID: 3O96), ALB (PDB ID: 2BX8), CXCL8 (PDB ID: 1ICW), EGFA (PDB ID: 1M17), IL-1β (PDB ID: 6Y8M), IL6 (PDB ID: 1ALU), STAT3 (PDB ID: 6NJS), TLR4 (PDB ID: 2Z63), TNF-α (PDB ID: 2AZ5), and VEGFA (PDB ID: 1MKK) were obtained from the RCSB Protein Data Bank (https://www.rcsb.org/ accessed on 14 April 2023). AutoDock Vina was utilized for molecular docking. Solvent molecules and other non-essential components were removed from the targets using Pymol 2.5 (Schrödinger, LLC, New York, NY, USA), and the structures were visualized.

4.9. Validating the Anti-AR Activity of Active Components In Vitro

RBL-2H3 cells were cultured in MEM medium supplemented with 15% FBS at 37 °C in a 5% CO_2 incubator. The cells were seeded into 96-well plates at a density of 1×10^4 cells/well, and a control group and groups treated with various concentrations of the drug were established. The plates were incubated for 24 h, 100 μL of MEM medium containing different concentrations of the active components was added to each well. Following a 12 h incubation at a constant temperature, the medium was discarded, and 100 μL of MTT solution was added to each well. After 4 h, 150 μL of DMSO was added to each well, and the plates were shaken for 10 min. The OD value at 570 nm was measured with a reference wavelength of 650 nm, and the cell viability was calculated.

RBL-2H3 cells were diluted to 0.5×10^4 cells/well and seeded into 96-well plates by adding 100 μL per well. The plates were incubated for 24 h; the cells were incubated with 0.5 μg/mL DNP-IgE working solution added to each well. Following another 24 h incubation, the cells were rinsed three times with PBS. Six wells were established for each of the following groups: control, model, and drug administration (cimifugin, formononetin, atractylenolide-I, methylnissolin, calycosin, and wogonin). Each well in the control group was supplemented with 100 μL of serum-free MEM medium, while each group received 100 μL of different concentrations of the test solution. After a 12 h incubation, 100 μL of 400 ng/mL DNP-BSA working solution was added to each well, except for the control group, which received an equal volume of basal medium. Then, the plates were incubated for 1 h. The reaction was terminated by incubation at 4 °C for 10 min, and the supernatant from each well was collected for analysis of various indices. RBL-2H3 expresses high-affinity IgE receptors (FcεRI) on the cell surface. When DNP-IgE binds to the receptor, it triggers the aggregation of FcεRI through DNP-BSA cross-linking, activating downstream signaling pathways (e.g., MAPK, Jak-STAT, and TNF pathways), leading to degranulation and inflammatory factor release [47,48]. This process is highly compatible with the pathological features of allergic rhinitis.

4.10. Validating the Anti-AR Activity of Active Components In Vivo

Animal models were constructed in the same way as 4.3 Animal experiments. Seventy mice were randomly divided into seven groups: control group, model group (AR), cimifugin group (30 mg/kg), formononetin group (30 mg/kg), atractylenolide-I group (30 mg/kg), wogonin group (30 mg/kg), and loratadine group (1.67 mg/kg). The dose group was administered continuously from 15 d to 29 d, once daily for 14 d. After 2 weeks, mice were executed to collect nasal tissue, serum, and lungs for analysis. The blood was centrifuged at 3500 rpm for 15 min, and the serum was stored in the refrigerator at -80 °C. Body weight was recorded on days 1 d and 29 d. After the last nasal OVA drop, the number of rubbing and sneezing within 15 min was recorded. Morphological and inflammatory infiltration of the nasal mucosa and left lung were observed under the microscope by the HE and PAS staining.

4.11. Data Analysis

The chemical compound database for the three herbs in YPFS was established using the TCMSP (https://old.tcmsp-e.com/tcmsp.php accessed on 1 February 2023), PubChem (https://pubchem.ncbi.nlm.nih.gov/ accessed on 1 February 2023), ChemSpider (https://www.chemspider.com/Default.aspx accessed on 1 February 2023) and Web of Science (https://webofscience.clarivate.cn/wos/alldb/basic-search accessed on 1 February 2023) databases, among others, to retrieve compound information for the relevant medicinal materials. The chemical composition of the metabolites was determined through the identification of compounds, pertinent biotransformation information, relevant scholarly works, databases, and the MS/MS spectra of the metabolites. All data were statistically analyzed using GraphPad Prism 8.0.1 software (GraphPad Software Company, San Diego, CA, USA), and experimental data were expressed as mean $\pm$ SD and evaluated for variance normality and uniformity. Analysis of variance was used to compare data of each group, the student test was used to compare data of two groups, and the Dunnett test was used to compare data of three groups or more. $p < 0.05$ was considered statistically significant.

5. Conclusions

In this study, the chemical constituents of YPFS were firstly comprehensively analyzed using GNPS combined with HPLC–Q-TOF-MS/MS. A total of 98 compounds were identified, and the main cleavage modes of each type of compound were summarized. Next, the active ingredients of YPFS were further identified from the perspective of in vivo processes. A total of 42 prototypical components and 57 metabolites were identified through HPLC–Q-TOF-MS/MS and the metabolites were all derived from the detected prototypical components. The relevant metabolic pathways mainly involved reduction, demethylation, hydroxylation, methylation, and glucuronidation. Based on 42 prototypical components in YPFS, potential ingredients were screened with network pharmacology and molecular docking. A total of 166 potential targets were found. A total of 15 key components were selected based on the topological analysis of TCM–constituent–target network diagrams, and the core targets were screened through the PPI network. Molecular docking of the 15 core components with the first 10 key targets showed that cimifugin, methylnissolin, calycosin, atractylenolide I, formononetin, and wogonin may be the key components of YPFS for the treatment of AR. Finally, cimifugin, atractylenolide I, formononetin, and wogonin were identified as the key compounds for improving AR via the vitro (RBL-2H3 cells) and in vivo (AR mice) validated. In summary, this study provided a reference for exploring the mechanism of YPFS's anti-AR effects and helped to clarify the overall material basis of YPFS prescription. Furthermore, the dynamics of the four components identified in this study require further in vivo investigation.

Supplementary Materials: The following supporting information can be downloaded at: https://www.mdpi.com/article/10.3390/ph18040540/s1, Table S1. Precision investigation results of 4 components of YPFS. Table S2. Repeatability investigation results of 4 components of YPFS. Table S3. Stability investigation results of 4 components of YPFS. Table S4. Mass spectrometric information of compounds from YPFS by HPLC-Q-TOF-MS/MS. Table S5. Mass spectrometric information of prototype components in vivo of YPFS by HPLC-Q-TOF-MS/MS. Table S6. Mass spectrometric information of metabolites components in vivo of YPFS by HPLC-Q-TOF-MS/MS. Table S7. The serial number of the 42 prototype compounds. Table S8. The information from network topology analysis on the 42 compounds. Table S9. The degree information from PPI on the top 20 core targets. Table S10. The detailed information on the top 30 GO terms. Table S11. The details on the top 20 KEGG pathways. Table S12. Molecular docking binding energy (KJ/mol). Figure S1. Optimization of chromatographic conditions. (A. Wavelength; B. Injection volume; C. Flow rate; D. Column temperature). Figure S2. Molecular network of the constituents in YPFS. The node color indicated the chemical type of the compound (red: flavonoids; green: saponins; purple: coumarin; blue: lactone; pink: amino acids; orange: others; pale yellow: organic acids). Figure S3. Chemical fragmentation pathways of chemical components in YPFS. (A) ononin; (B) psoralen. Figure S4. Chemical fragmentation pathways of chemical components in YPFS. (A) imperatorin; (B) cimifugin. Figure S5. The structural formula of prototypical components from YPFS in AR mice. Figure S6. Venn diagram detailing the proportions of plasma, urine and feces. Figure S7. The possible metabolic pathways of (A) ononin and methylnissolin, and (B) astragaloside IV from YPFS in AR mice. Figure S8. The possible metabolic pathways of (A) atractylenolide I and atractylenolideIII, and (B) cimfugin and scopoletin from YPFS in AR mice. Figure S9. The proportion of metabolic types of YPFS in the plasma, urine, and feces of AR mice. Figure S10. Metabolic types of YPFS in the plasma of AR mice. Figure S11. Metabolic types of YPFS in the urine of AR mice. Figure S12. Metabolic types of YPFS in the feces of AR mice. Figure S13. Protein-protein interaction (PPI) network. Figure S14. Circle plots for PPI network. Figure S15. Circle plots for GO analysis. Figure S16. NF-κB pathway diagram with genes predicted to be YPF-AR targets (red).

Author Contributions: X.L.: methodology, resources, software, writing–original draft, investigation, detection, and experimental work. C.Z.: writing—review and editing, visualization. J.Q.: conceptualization, funding acquisition, writing—review and editing, supervision, and conceptualization. All authors have read and agreed to the published version of the manuscript.

Funding: This work was supported by the National Natural Sciences Foundation of China (No. 82274067).

Institutional Review Board Statement: The study was conducted in accordance with the Declaration of Helsinki, and approved by the Ethics Committee of China Pharmaceutical University (protocol code 2024-12-007, 29 November 2024). The study was conducted in accordance with the local legislation and institutional requirements.

Informed Consent Statement: Not applicable.

Data Availability Statement: All data are contained in the article and the Supplementary Materials.

Conflicts of Interest: The authors declare no conflicts of interest.

References

1. Bousquet, J.; Anto, J.M.; Bachert, C.; Baiardini, I.; Bosnic-Anticevich, S.; Walter Canonica, G.; Melén, E.; Palomares, O.; Scadding, G.K.; Togias, A.; et al. Allergic rhinitis. *Nat. Rev. Dis. Primers* **2020**, *6*, 95. [CrossRef] [PubMed]
2. Cheng, L.; Chen, J.; Fu, Q.; He, S.; Li, H.; Liu, Z.; Tan, G.; Tao, Z.; Wang, D.; Wen, W.; et al. Chinese Society of Allergy Guidelines for Diagnosis and Treatment of Allergic Rhinitis. *Allergy Asthma Immunol. Res.* **2018**, *10*, 300–353.
3. Cardona, V.; Ansotegui, I.J.; Ebisawa, M.; El-Gamal, Y.; Fernandez Rivas, M.; Fineman, S.; Geller, M.; Gonzalez-Estrada, A.; Greenberger, P.A.; Sanchez Borges, M.; et al. World Allergy Organization Anaphylaxis Guidance 2020. *World Allergy Organ. J.* **2020**, *13*, 100472. [CrossRef] [PubMed]
4. Zhang, Y.; Lan, F.; Zhang, L. Advances and highlights in allergic rhinitis. *Allergy* **2021**, *76*, 3383–3389. [PubMed]
5. Bai, Y.; Wei, W.; Yao, C.; Wu, S.; Wang, W.; Guo, D.-A. Advances in the chemical constituents, pharmacological properties and clinical applications of TCM formula Yupingfeng San. *Fitoterapia* **2023**, *164*, 105385. [CrossRef]
6. Chang, S.; Wang, M.; Tian, Y.; Qi, J.; Qiu, Z. Systematic analysis and identification of the absorption and metabolic components of Zengye decoction in type 2 diabetic rats by HPLC-ESI-Q-TOF–MS/MS. *Chin. Med.* **2020**, *15*, 50.
7. Chen, H.; Feng, W.; Lu, Y.; Yang, Y.; Xin, Z.-H.; Li, M.; Xin, L.; Gong, Y.-D. Effects and mechanism of Chinese medicine Jiawei Yupingfeng in a mouse model of allergic rhinitis. *J. Integr. Med.* **2021**, *19*, 354–361.
8. Nikles, S.; Monschein, M.; Zou, H.; Liu, Y.; He, X.; Fan, D.; Lu, A.; Yu, K.; Isaac, G.; Bauer, R. Metabolic profiling of the traditional Chinese medicine formulation Yu Ping Feng San for the identification of constituents relevant for effects on expression of TNF-α, IFN-γ, IL-1β and IL-4 in U937 cells. *J. Pharm. Biomed. Anal.* **2017**, *145*, 219–229.
9. Li, M.; Yue, X.; Gao, Y.; Zhang, B.; Yuan, C.; Wu, T. Method for rapidly discovering active components in Yupingfeng granules by UPLC-ESI-Q-TOF-MS. *J. Mass Spectrom.* **2020**, *55*, e4627.

10. Yang, L.; Liang, J.; Zheng, Q.; Zhou, L.; Xiong, Y.; Wang, H.; Yuan, J. A Comparative Study of Serum Pharmacochemistry of Kai-Xin-San in Normal and AD Rats Using UPLC-LTQ-Orbitrap-MS. *Pharmaceuticals* **2022**, *16*, 30. [CrossRef]
11. Cui, L.; Ma, C.; Shi, W.; Yang, C.; Wu, J.; Wu, Z.; Lou, Y.; Fan, G. A Systematic Study of Yiqi Qubai Standard Decoction for Treating Vitiligo Based on UPLC-Q-TOF/MS Combined with Chemometrics, Molecular Docking, and Cellular and Zebrafish Assays. *Pharmaceuticals* **2023**, *16*, 1716. [CrossRef] [PubMed]
12. Wang, M.; Carver, J.J.; Phelan, V.V.; Sanchez, L.M.; Garg, N.; Peng, Y.; Nguyen, D.D.; Watrous, J.; Kapono, C.A.; Luzzatto-Knaan, T.; et al. Sharing and community curation of mass spectrometry data with Global Natural Products Social Molecular Networking. *Nat. Biotechnol.* **2016**, *34*, 828–837. [PubMed]
13. Chen, X.; Lin, J.; Liang, Q.; Chen, X.; Wu, Z. Pseudoephedrine alleviates atopic dermatitis-like inflammatory responses in vivo and in vitro. *Life Sci.* **2020**, *258*, 118139.
14. Wang, Y.; Ju, Z.; Li, L.; Zhang, T.; Zhang, S.; Ding, L.; Zhan, C.; Wang, Z.; Yang, L. A complementary chromatographic strategy for integrated components characterization of Imperatae Rhizoma based on convergence and liquid chromatography combined with mass spectrometry and molecular network. *J. Chromatogr.* **2022**, *1678*, 463342. [CrossRef]
15. Li, H.; Zhu, Y. The chemical composition analysis of Yixin Tongmai Granules using UHPLC-MS/MS and exploration of its potential mechanism in treatment of coronary artery disease based on network pharmacology and molecular docking. *Medicine* **2025**, *104*, e41620.
16. Malca-Garcia, G.R.; Zagal, D.; Graham, J.; Nikolić, D.; Friesen, J.B.; Lankin, D.C.; Chen, S.-N.; Pauli, G.F. Dynamics of the isoflavone metabolome of traditional preparations of *Trifolium pratense* L. *J. Ethnopharmacol.* **2019**, *238*, 111865. [CrossRef]
17. Singh, C.; Anand, S.K.; Tiwari, K.N.; Mishra, S.K.; Kakkar, P. Phytochemical profiling and cytotoxic evaluation of Premna serratifolia L. against human liver cancer cell line. *3 Biotech* **2021**, *11*, 115. [CrossRef]
18. Wang, X.; Zhou, W.; Wang, Q.; Zhang, Y.; Ling, Y.; Zhao, T.; Zhang, H.; Li, P. A novel and comprehensive strategy for quality control in complex Chinese medicine formula using UHPLC-Q-Orbitrap HRMS and UHPLC-MS/MS combined with network pharmacology analysis: Take Tangshen formula as an example. *J. Chromatogr. B* **2021**, *1183*, 122889.
19. Zhou, K.; Xiao, S.; Cao, S.; Zhao, C.; Zhang, M.; Fu, Y. Improvement of glucolipid metabolism and oxidative stress via modulating PI3K/Akt pathway in insulin resistance HepG2 cells by chickpea flavonoids. *Food Chem. X* **2024**, *23*, 101630.
20. Chen, Y.; Tang, L.; Wu, M.; Shu, L.; Xu, Y.; Yao, Y.; Li, Y. A practical method for rapid discrimination of constituents in Psoraleae Fructus by UPLC-Q-Orbitrap MS. *J. Mass Spectrom.* **2023**, *58*, e4966.
21. Xu, F.; Hu, J.; Li, Y.; Cheng, C.; Au, R.; Tong, Y.; Wu, Y.; Cui, Y.; Fang, Y.; Chen, H.; et al. Qin-Yu-Qing-Chang decoction reshapes colonic metabolism by activating PPAR-γ signaling to inhibit facultative anaerobes against DSS-induced colitis. *Chin. Med.* **2024**, *19*, 130. [PubMed]
22. Hu, L.; Yao, Z.; Qin, Z.; Liu, L.; Song, X.; Dai, Y.; Kiyohara, H.; Yamada, H.; Yao, X. In vivo metabolic profiles of Bu-Zhong-Yi-Qi-Tang, a famous traditional Chinese medicine prescription, in rats by ultra-high-performance liquid chromatography coupled with quadrupole time-of-flight tandem mass spectrometry. *J. Pharm. Biomed. Anal.* **2019**, *171*, 81–98. [PubMed]
23. Jia, L.; Qin, X.; Liu, Y. Metabolite Identification of Huangqi Jianzhong Tang in Rat Urine and Feces after Oral Administration Based on UHPLC-Q-Exactive-MS. *J. Chromatogr. Sci.* **2023**, *61*, 211–224. [PubMed]
24. Yan, Y.; Fu, C.; Cui, X.; Pei, X.; Li, A.; Qin, X.; Du, C.; Du, H. Metabolic profile and underlying antioxidant improvement of Ziziphi Spinosae Folium by human intestinal bacteria. *Food Chem.* **2020**, *320*, 126651.
25. Su, Y.; Hu, K.; Li, D.; Guo, H.; Sun, L.; Xie, Z. Microbial-Transferred Metabolites and Improvement of Biological Activities of Green Tea Catechins by Human Gut Microbiota. *Foods* **2024**, *13*, 792. [CrossRef]
26. Wang, H.; Liu, T.; Liao, C.; Liang, F.; Tian, L. Safety and onset time of modified Yupingfeng nasal spray versus mometasone furoate nasal spray on house dust mites-induced moderate to severe allergic rhinitis: A prospective, multicenter, randomized, open-label, parallel-group clinical trial. *J. Ethnopharmacol.* **2025**, *344*, 119574.
27. Kang, C.; Li, X.; Liu, P.; Liu, Y.; Niu, Y.; Zeng, X.; Zhao, H.; Liu, J.; Qiu, S. Tolerogenic dendritic cells and TLR4/IRAK4/NF-κB signaling pathway in allergic rhinitis. *Front. Immunol.* **2023**, *14*, 1276512.
28. Zhang, J.; Gao, L.; Yu, D.; Song, Y.; Zhao, Y.; Feng, Y. Three Artemisia pollens trigger the onset of allergic rhinitis via TLR4/MyD88 signaling pathway. *Mol. Biol. Rep.* **2024**, *51*, 319. [CrossRef]
29. Ebrahim Soltani, Z.; Badripour, A.; Haddadi, N.S.; Elahi, M.; Kazemi, K.; Afshari, K.; Dehpour, A.R. Allergic rhinitis in BALB/c mice is associated with behavioral and hippocampus changes and neuroinflammation via the TLR4/NF-κB signaling pathway. *Int. Immunopharmacol.* **2022**, *108*, 108725.
30. Dong, J.; Xu, O.; Wang, J.; Shan, C.; Ren, X. Luteolin ameliorates inflammation and Th1/Th2 imbalance via regulating the TLR4/NF-κB pathway in allergic rhinitis rats. *Immunopharmacol. Immunotoxicol.* **2021**, *43*, 319–327.
31. Tian, B.; Ma, X.; Jiang, R. Daphnetin Mitigates Ovalbumin-Induced Allergic Rhinitis in Mice by Regulating Nrf2/HO-1 and TLR4/NF-kB Signaling. *Am. J. Rhinol. Allergy* **2022**, *37*, 19–25. [CrossRef] [PubMed]
32. Hu, X.; Zhang, Z.; Li, W.; Deng, Y. Action Mechanism of Yupingfeng Capsule for Combined Allergic Rhinitis and Asthma Syndrome via the Toll-Like Receptor 4/Nuclear Factor-Kappa B Pathway. *Indian J. Pharm. Sci.* **2022**, *85*, 456–462. [CrossRef]

33. Nie, J.; Jiang, X.; Wang, G.; Xu, Y.; Pan, R.; Yu, W.; Li, Y.; Wang, J. Yu-Ping-Feng-San alleviates inflammation in atopic dermatitis mice by TLR4/MyD88/NF-κB pathway. *J. Ethnopharmacol.* **2024**, *329*, 118092. [CrossRef]
34. Wang, X.; Jiang, X.; Yu, X.; Liu, H.; Tao, Y.; Jiang, G.; Hong, M. Cimifugin suppresses allergic inflammation by reducing epithelial derived initiative key factors via regulating tight junctions. *J. Cell. Mol. Med.* **2017**, *21*, 2926–2936. [CrossRef]
35. Liu, Z.; Sun, Q.; Liu, X.; Song, Z.; Song, F.; Lu, C.; Zhang, Y.; Song, X.; Yang, Y.; Li, Y. Network pharmacology analysis and experimental verification reveal the mechanism of the traditional Chinese medicine YU-Pingfeng San alleviating allergic rhinitis inflammatory responses. *Front. Plant Sci.* **2022**, *13*, 934130. [CrossRef]
36. Kim, K.A.; Jung, J.H.; Choi, Y.S.; Kang, G.; Kim, S.T. Anti-inflammatory effect of wogonin on allergic responses in ovalbumin-induced allergic rhinitis in the mouse. *Allergy Rhinol.* **2018**, *9*, 2152656718764145. [CrossRef]
37. Lai, M.; Sun, S.; Zuo, T.; Li, L.; Zhao, Q.; Li, W.; Zheng, J.; Hong, M. Sanfeng Tongqiao Dripping Pills alleviate House Dust Mite-induced allergic rhinitis in mice by inhibiting Th2 differentiation and repairing the nasal epithelial barrier. *Phytomedicine* **2024**, *132*, 155899. [CrossRef]
38. Cheng, J.; Zhang, M.; Zheng, Y.; Wang, J.; Wang, Q. Integrative analysis of network pharmacology and proteomics to identify key targets of Tuomin-Zhiti-Decoction for allergic rhinitis. *J. Ethnopharmacol.* **2022**, *296*, 115448. [CrossRef]
39. Huang, J.; Chen, X.; Xie, A. Formononetin ameliorates IL-13-induced inflammation and mucus formation in human nasal epithelial cells by activating the SIRT1/Nrf2 signaling pathway. *Mol. Med. Rep.* **2021**, *24*, 832. [CrossRef]
40. Qu, M.; Tao, W.; Ma, J.; Lin, S.-C. Network Pharmacology and Molecular Docking Study of Yupingfeng Powder in the Treatment of Allergic Diseases. *Evid. Based Complement. Altern. Med.* **2022**, *2022*, 1–18. [CrossRef]
41. Deng, M.; Chen, H.; Long, J.; Song, J.; Xie, L.; Li, X. Atractylenolides (I, II, and III): A review of their pharmacology and pharmacokinetics. *Arch. Pharm. Res.* **2021**, *44*, 633–654. [PubMed]
42. Chen, X.; Lu, D.; Liu, W.; Xie, J.; Lu, Z.; Yang, H.; Yan, H.; Wang, L.; Che, C. Therapeutic effect of Atractylenolide I on Aspergillus fumigatus keratitis by affecting MyD88/NF-κB pathway and IL-1β, IL-10 expression. *Cytokine* **2023**, *162*, 156112. [PubMed]
43. Li, W.; Zhi, W.; Liu, F.; He, Z.; Wang, X.; Niu, X. Atractylenolide I restores HO-1 expression and inhibits Ox-LDL-induced VSMCs proliferation, migration and inflammatory responses in vitro. *Exp. Cell Res.* **2017**, *353*, 26–34. [PubMed]
44. Wei, X.; Zhang, B.; Liang, X.; Liu, C.; Xia, T.; Xie, Y.; Deng, X.; Tan, X. Higenamine alleviates allergic rhinitis by activating AKT1 and suppressing the EGFR/JAK2/c-JUN signaling. *Phytomedicine* **2021**, *86*, 153565.
45. Van Nguyen, T.; Piao, C.H.; Fan, Y.J.; Shin, D.-U.; Kim, S.Y.; Song, H.-J.; Song, C.H.; Shin, H.S.; Chai, O.H. Anti-allergic rhinitis activity of α-lipoic acid via balancing Th17/Treg expression and enhancing Nrf2/HO-1 pathway signaling. *Sci. Rep.* **2020**, *10*, 12528. [CrossRef]
46. Xi, Z.H.; Du, N.N.; Qiao, Y.J.; Zhang, G.X.; Yao, G.D.; Huang, X.X.; Song, S.J. Rapid isolation of cytotoxic daphnane diterpenoids from Daphne altaica Pall. using MS-DIAL. *Fitoterapia* **2025**, *181*, 106385.
47. Chen, M.; Su, Q.; Shi, Y. Molecular mechanism of IgE-mediated FcεRI activation. *Nature* **2025**, *637*, 453–460.
48. Bai, Q.; Feng, X.; Wang, Y.; Wu, C.; Liu, Y.; Sun, J.; Yue, T.; Long, F. Adenosine cyclic phosphate with ultrasonic-assisted pectinase extraction alleviated allergic reactions in RBL-2H3 through inhibiting the influx of intracellular Ca2+. *Food Sci. Hum. Wellness* **2023**, *12*, 832–841.

Disclaimer/Publisher's Note: The statements, opinions and data contained in all publications are solely those of the individual author(s) and contributor(s) and not of MDPI and/or the editor(s). MDPI and/or the editor(s) disclaim responsibility for any injury to people or property resulting from any ideas, methods, instructions or products referred to in the content.

pharmaceuticals

MDPI

Article

Aronia Berry Extract Modulates MYD88/NF-kB/P-Glycoprotein Axis to Overcome Gemcitabine Resistance in Pancreatic Cancer

Yuan Li [1,2], Caiming Xu [1,3], Haiyong Han [4], Silvia Pascual-Sabater [5], Cristina Fillat [5] and Ajay Goel [1,6,*]

[1] Department of Molecular Diagnostics and Experimental Therapeutics, Beckman Research Institute of City of Hope, Biomedical Research Center, Monrovia, CA 91016, USA; yuali@coh.org (Y.L.); caixu@coh.org (C.X.)
[2] Department of Clinical Laboratory, Yangpu Hospital, Tongji University School of Medicine, Shanghai 200090, China
[3] Department of General Surgery, The First Affiliated Hospital of Dalian Medical University, Dalian 116004, China
[4] Division of Molecular Medicine, The Translational Genomics Research Institute, Phoenix, AZ 85004, USA; hhan@tgen.org
[5] Institut d'Investigacions Biomèdiques August Pi i Sunyer (IDIBAPS), 08036 Barcelona, Spain; sipascual@clinic.cat (S.P.-S.); cfillat@clinic.cat (C.F.)
[6] City of Hope Comprehensive Cancer Center, Duarte, CA 91010, USA
* Correspondence: ajgoel@coh.org; Tel.: +1-626-218-3452

Abstract: Pancreatic ductal adenocarcinoma (PDAC) is a highly lethal disease with poor survival rates, primarily due to the limited effectiveness of gemcitabine (Gem)-based chemotherapy, as well as the acquisition of chemotherapeutic resistance. Aronia berry extracts (ABEs), abundant in phenolic constituents, have been recently recognized for their anticancer properties as well as their encouraging potential to help overcome chemoresistance in various cancers. In the present study, we explored ABE's potential to overcome Gem resistance in PDAC and identify specific growth regulatory pathways responsible for its anticancer activity. Through a series of in vitro experiments in gemcitabine-resistant (Gem-R) cells, we elucidated the synergistic interactions between Gem and ABE treatments. Using advanced transcriptomic analysis and network pharmacology, we revealed key molecular pathways linked to chemoresistance and potential therapeutic targets of ABE in Gem-R PDAC cells. Subsequently, the findings from cell culture studies were validated in patient-derived 3D tumor organoids (PDOs). The combination treatment of ABE and Gem demonstrated significant synergism and anticancer effects on cell viability, proliferation, migration, and invasion in Gem-R cells. Transcriptomic analysis revealed a correlation between the NF-Kb signaling pathway and Gem-R ($p < 0.05$), exhibiting a marked upregulation of MYD88. Additionally, MYD88 exhibited a significant correlation with the overall survival rates in patients with PDAC patients in the TCGA cohort (HR = 1.58, $p < 0.05$). The MYD88/NF-Kb pathway contributes to chemoresistance by potentially upregulating efflux transporters like P-glycoprotein (P-gp). Our findings revealed that the combined treatment with ABE suppressed the NF-Kb pathway by targeting MYD88 and reducing P-gp expression to overcome Gem resistance. Lastly, the combination therapy proved highly effective in PDOs in reducing both their number and size ($p < 0.05$). Our study offers previously unrecognized insights into the ability of ABE to overcome Gem resistance in PDAC cells through its targeting of the MYD88/NF-κb/P-gp axis, hence providing a safe and cost-effective adjunctive therapeutic strategy to improve treatment outcomes in PDAC.

Keywords: pancreatic ductal adenocarcinoma; aronia berry extracts; gemcitabine resistance; MYD88; anticancer effect; chemoprevention

1. Introduction

Pancreatic ductal adenocarcinoma (PDAC) is the third leading cause of cancer-related deaths. Despite the 5-year overall survival (OS) rates in the United States having improved from less than 5% in 1990 to as high as ~10% in 2023, the prognosis for patients with PDAC remains distressingly low. This high rate of mortality is primarily attributed to the difficulty in early diagnosis of pancreatic cancer, as well as the inadequate efficacy of the currently available treatment options [1,2]. Gemcitabine (Gem)-based chemotherapy has played a crucial role in advancing the treatment of patients with PDAC since its initial endorsement as a primary therapeutic option in 1997. Nab-paclitaxel combined with Gem hydrochloride is a commonly used therapeutic regimen in PDAC [3–7]. However, its efficacy is notably limited by the emergence of drug resistance in most patients, highlighting a primary challenge for the effective management of this lethal malignancy [8,9]. The newer combination therapeutic regimens, such as FOLFIRINOX, while resulting in slightly improved treatment outcomes, exhibit significant toxicity and are expensive. Collectively, these findings highlight the critical importance of gaining deeper insights into the molecular mechanisms underlying Gem resistance and investigating novel therapeutic approaches to overcome chemoresistance in PDAC and improve patient outcomes.

Previous studies have demonstrated the pivotal role of ABC drug transporters in conferring therapeutic resistance to Gem in PDAC [10,11]. Among these, P-glycoprotein (P-gp) is the most recognized and established regulator contributing to chemoresistance [12]. Also known as ABCB1 or MDR1, P-gp plays a vital role in the emergence of multidrug resistance (MDR) and was the first human ABC transporter to be characterized [13]. P-gp has been shown to affect the uptake, distribution, and elimination of compounds through modulation of cellular efflux function [14]. Furthermore, the efficacy of several chemotherapeutic agents, including paclitaxel (PTX), cisplatin (CP), and doxorubicin (DOX), can be affected by P-gp overexpression. However, P-gp inhibitors such as quinidine and verapamil have been plagued by their toxicity and poor selectivity, in turn limiting their clinical use [14–16]. Thus, exploring safer and more effective P-gp inhibitors remains highly valuable in various cancers.

More recently, there is growing enthusiasm to explore the possibility of using various natural products and their derivatives as complementary and alternative therapeutic approaches, given their multi-targeted efficacy and safety in various cancers [17–29]. These natural compounds include curcumin [22,30–36], Andrographis [29,37–39], resveratrol [40,41], and ginseng [42], all of which have been studied for their adjunctive and synergistic efficacy in various cancers, including PDAC. Additionally, several of these natural remedies have been identified as inhibitors of P-gp, MRP1, MRP2, and BCRP and can thus sensitize cancer cells to various chemotherapeutic drugs. A range of natural compounds such as flavonoids, coumarins, resins, saponins, and terpenoids have been explored for their ability to combat drug resistance in cancer by inhibiting P-gp [43]. In particular, aronia berries (black chokeberries) are thought to have anticancer potential due to their richness in phenolic compounds that possess potent antioxidant and anti-inflammatory properties, such as procyanidins, anthocyanins, phenolic acids, and their analogs [44]. Previous studies also indicated that derivatives of aronia berry extract (ABE) can impede the formation of breast cancer stem cells, making it a promising candidate for cancer chemoprevention [45]. Furthermore, *Aronia Melanocarpa* extract has been reported to impede the growth of PDAC through induction of apoptosis in pancreatic cancer cell lines, such as AsPC-1 [46]. While recent research has suggested the anticancer effects of ABE in PDAC, its specific impact on chemosensitivity, especially when combined with chemotherapeutic agents, is largely unexplored. Given that most natural medicines are used as adjuncts to conventional chemotherapy, we hypothesized that ABE might have the potential to inhibit tumor growth and overcome chemoresistance in PDAC.

In our current study, we conducted a comprehensive set of experiments in PDAC cell lines and patient-derived 3D organoids (PDOs) to gain additional insights into the molecular mechanisms of ABE's anticancer effects and its potential in overcoming Gem resistance in PDAC.

2. Results

2.1. The Treatment of ABE and Gemcitabine Shows Synergistic Inhibition in Gem-R Pancreatic Cancer Cells

In our present investigation, we initially utilized the two Gem-resistant (Gem-R) cell lines as described in our prior study [34]. Notably, the half-maximal inhibitory concentration (IC_{50}) of Gem was substantially higher in Gem-R BxPC-3 and Gem-R MIA-PaCa-2 cells compared to their respective parental cell lines (Figure 1A,E). Specifically, the IC_{50} of Gem in the parental cells was determined to be 680.35 nM and 728.81 nM, respectively. These values were significantly lower than those observed in the Gem-R counterparts, Gem-R BxPC-3 and Gem-R MIA-PaCa-2, which exhibited IC_{50} values exceeding 1200 nM. Next, we determined that the ABE could reduce Gem resistance in Gem-R PDAC cells. Subsequently, we administered concurrent treatments with varying concentrations of ABE (0, 30, 60, 90, and 120 μg/mL) and Gem (0, 400, 800, and 1200 nM), demonstrating a synergistic reduction in cell viability. Next, we investigated the anticancer effects of ABE on Gem-R BxPC-3 and Gem-R MIA-PaCa-2 cell lines. These resistant cell lines were treated with ABE for 48 h, and cell viability was assessed using the CCK-8 assay. The results demonstrated a dose-dependent inhibition of cell viability by ABE, with IC_{50} values of 110.97 μg/mL for Gem-R BxPC-3 and 89.17 μg/mL for Gem-R MIA-PaCa-2 (Supplementary Figure S1). Ultimately, the criteria for the optimal dosage ratio of ABE and Gem in the combination treatment was established by both dose–response curves needing to achieve over 50% inhibition and the Bliss synergy score exceeding 10 points. In these treatments, the ABE/Gem ratio of 90:800 exhibited a synergistic anticancer effect in both cell lines. In the Gem-R BxPC-3 cell line, under this ratio, the inhibitory rate was 64.36%, in contrast to −9.23% with only Gem treatment and 45.92% with only ABE treatment, resulting in a synergy score of 23.43 (Figure 1B–D). Similarly, in the Gem-R MIA-PaCa-2 cell line, under the same ratio, the inhibitory rate was 67.02%, compared to 9.53% with only Gem treatment and 49.6% with only ABE treatment, yielding a synergy score of 12.61 (Figure 1F–H). Hence, all subsequent experiments were conducted using a concentration of 90 μg/mL ABE and 800 nM Gem.

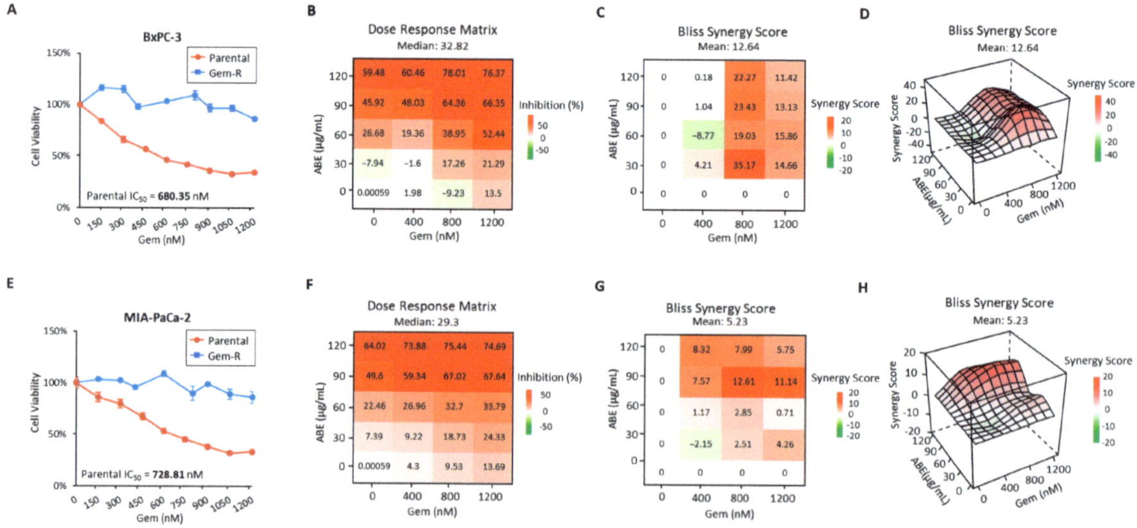

Figure 1. Identification of drug resistance in Gem-R PDAC cells and the treatment ABE with Gem synergistically inhibits the growth of Gem-R PDAC cells. The IC_{50} of parental BxPC-3/Gem-R BxPC-3 (**A**) and parental MIA-PaCa-2/Gem-R MIA-PaCa-2 (**E**) to Gem were calculated using the CCK8 assay. Error bars are the mean ± SD. Percentage of growth inhibition of Gem-R BxPC-3 cell

lines (**B**) and Gem-R MIA-PaCa-2 (**F**) cell lines treated in multiple combinations of ABE (0, 30, 60, 90, and 120 μg/mL) and Gem (0, 400, 800, and 1200 nM). The dose–response matrix was measured using the CCK8 assay. Bliss synergy values were calculated from data in B and F, accessed as a 2D-contour (**C**,**G**) and 3D-contour (**D**,**H**) drug dose–response model evaluating cell viability after treatment with ABE in Gem-R PDAC cells.

2.2. The Combination of ABE and Gemcitabine Inhibits Cell Proliferation, Colony Formation, Migration, and Invasion in Gem-R PDAC Cell Lines

To evaluate the potential effects of ABE in enhancing chemosensitivity to Gem in Gem-R PDAC cells, we conducted a series of functional experiments, including colony formation, wound healing (scratch), and invasion assays. To evaluate the impact of the combination of ABE and Gem on PDAC cell proliferation, we treated Gem-R BxPC-3 and Gem-R MIA-PaCa-2 cells with Gem (800 nM) and ABE (90 μg/mL), either separately or in combination for 48 h. We performed CCK-8 assays, which revealed that the combination of Gem and ABE significantly outperformed the individual treatments ($p < 0.01$ vs. Gem; $p < 0.01$ vs. ABE in Gem-R MIA-PaCa-2 cells; $p < 0.01$ vs. Gem; $p < 0.01$ vs. ABE in Gem-R BxPC-3 cells; Figure 2A). Furthermore, the colony formation assay demonstrated a significant reduction in clonogenicity with the combination of Gem and ABE compared to individual treatments in BxPC-3 cells (combination vs. Gem: fold change [FC] = 0.49, $p < 0.01$; combination vs. ABE: FC = 0.66, $p < 0.01$, Figure 2B) and MIA-PaCa-2 cell lines (combination vs. Gem: FC = 0.22, $p < 0.01$; combination vs. ABE: FC = 0.37, $p < 0.01$, Figure 2B). To determine whether the combination of Gem and ABE had a more significant impact on the motility and invasive potential of PDAC cells, we conducted scratch–wound and transwell assays. The scratch–wound assay revealed that the combination of Gem and ABE significantly inhibited cell migration to a greater extent than individual treatments in both BxPC-3 (combination vs. Gem: FC = 0.22, $p < 0.01$; combination vs. ABE: FC = 0.37, $p < 0.01$, Figure 2C) and MIA-PaCa-2 cell lines (combination vs. Gem: FC = 0.14, $p < 0.01$; combination vs. ABE: FC = 0.37, $p < 0.01$, Figure 2C). The transwell assay showed that a combination of Gem and ABE shows a significantly more significant reduction in invasion than individual treatment in both BxPC-3 (combination vs. Gem: FC = 0.53, $p < 0.01$; combination vs. ABE: FC = 0.68, $p < 0.01$, Figure 2D) and MIA-PaCa-2 cell lines (combination vs. Gem: FC = 0.20, $p < 0.01$; combination vs. ABE: FC = 0.27, $p < 0.01$, Figure 2D). Furthermore, epithelial-to-mesenchymal transition (EMT) and matrix metalloproteinase-related proteins are closely linked to tumor metastasis and invasion. We examined the expression of these proteins following treatments with gemcitabine, ABE, and their combination. Our findings revealed that the combination treatment significantly reduced the expression of proteins associated with migration and metastasis (e.g., MMP9, vimentin) while significantly increasing the expression of those that promote cell adhesion (e.g., E-cadherin; Supplementary Figure S2).

Consistent with the previous results, our findings demonstrated that the combination of Gem and ABE significantly reduced the migration and invasion abilities of PDAC cells.

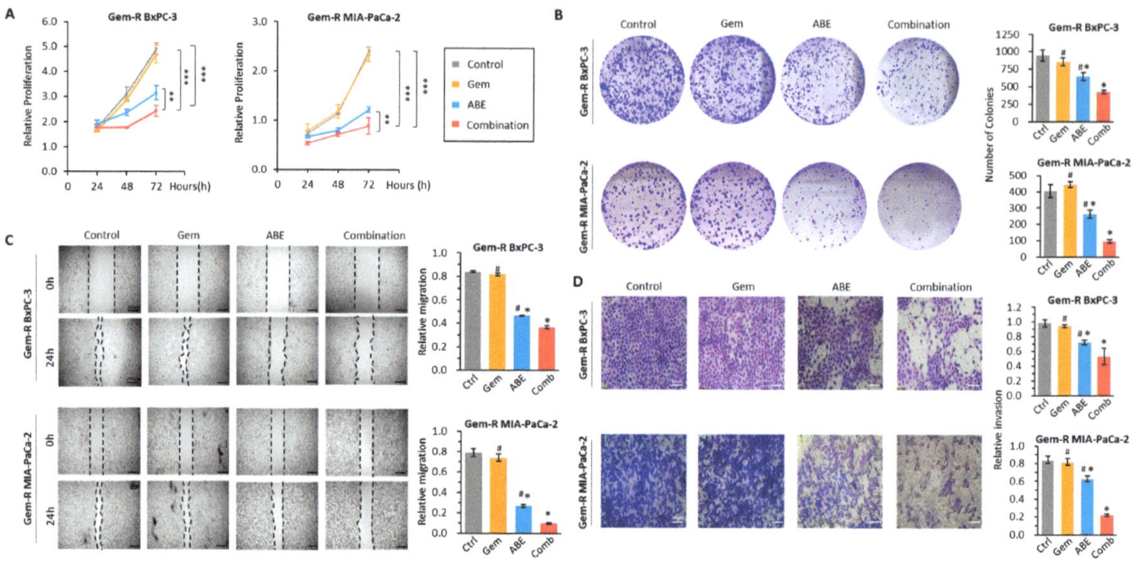

Figure 2. Aronia berry extract exhibits anti-proliferation, migration, and invasion effects in Gem-R PDAC cells. (**A**) Cell proliferation was compared in Gem-R cell lines treated with Gem, ABE, and their combination. Cell viability was assessed using a CCK-8 assay at 24, 48, and 72 h time points. (**B**) A colony formation assay was conducted to evaluate the clonogenicity of each group after a 48 h exposure to the treatments. Representative images of colonies were taken, and the number of colonies was counted. (**C**) A wound healing assay was performed on Gem-R PDAC cells treated with Gem, ABE, and their combination for 48 h. Representative images of the wound and the recovering areas (marked by black lines) were taken, and the percentage of wound closure was measured. (**D**) A transwell assay was carried out on Gem-R PDAC cells treated with Gem, ABE, and their combination for 48 h. The number of invading cells was counted in three randomly selected fields on the membrane. * p indicates $p < 0.05$ vs. control group; # p indicates $p < 0.05$ vs. combination group. ** p indicates $p < 0.01$; *** p indicates $p < 0.001$.

2.3. ABE, in Combination with Gemcitabine, Promotes Cell Apoptosis

To evaluate the apoptotic effects of the combination of Gem and ABE on PDAC cells, we initially conducted an annexin V binding assay. The analysis indicated that the combination of Gem and ABE increased the rate of apoptosis in both Gem-R cell lines (Figure 3A) compared to the individual administrations of Gem and ABE (BxPC-3 cell line: combination vs. Gem: 26.55% vs. 8.095%, $p < 0.01$; combination vs. ABE: 26.55% vs. 17.33%, $p < 0.01$, Figure 3A; and MIA-PaCa-2 cell line: combination vs. ABE: 25.60% vs. 11.29%, $p < 0.01$; combination vs. ABE: 25.60% vs. 14.8%, $p < 0.01$, Figure 3A). Next, we conducted Western blotting (WB) to analyze changes in apoptotic proteins following treatment. The WB analysis demonstrated that the combination of Gem and ABE increased the expression of cleaved caspase-9 and Bax, pro-apoptotic proteins, while simultaneously reducing the expression of total PARP, an anti-apoptotic protein, in comparison to the control and individual treatments (Figure 3B) [47–51]. Collectively, these data confirm that the combination of Gem and ABE induced more Gem-R PDAC cell apoptosis compared to individual treatments in vitro.

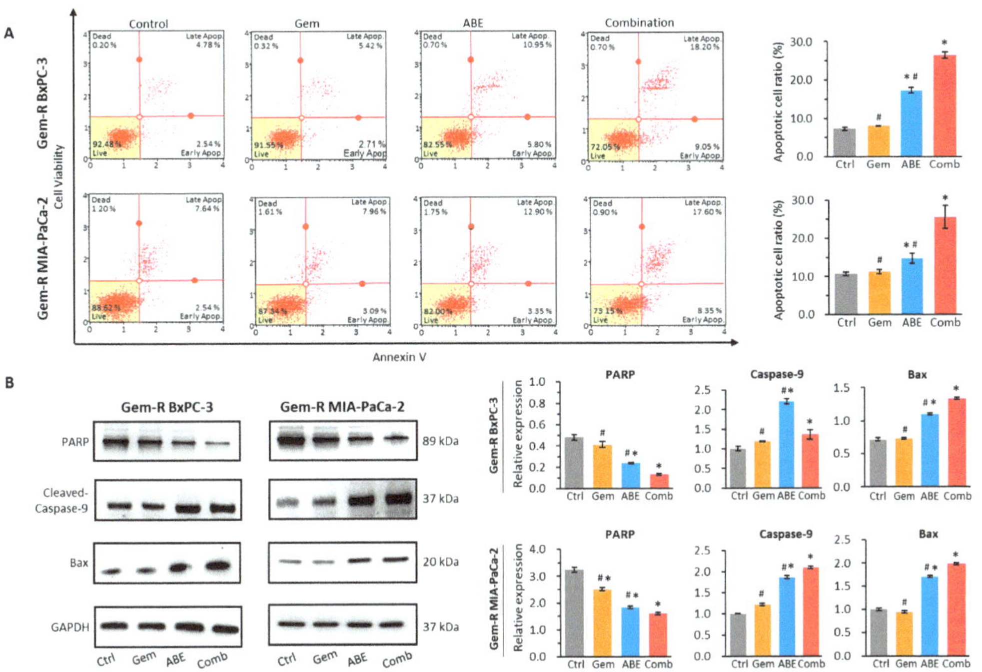

Figure 3. Aronia berry extract induces cell apoptosis in Gem-R PDAC cells. (**A**) Annexin V assay was conducted to measure the percentage of apoptotic cells on Gem-R PDAC cells treated with Gem, ABE, and their combination for 48 h. (**B**) Western blot analysis was performed on Gem-R PDAC cells treated with Gem, ABE, and their combination for 48 h. The relative levels of PARP, caspase-9, and Bax were quantitatively analyzed by comparing them to the control GAPDH expression. * p indicates $p < 0.05$ vs. control group; # p indicates $p < 0.05$ vs. combination group.

2.4. The MYD88/NF-κB Signaling Pathway Is Aviated in Gem-R PDAC Cells

To investigate the mechanism of Gem resistance in PDAC cells, we conducted gene expression profiling analysis by comparing parental and Gem-R PDAC cells using publicly available Gene Expression Omnibus (GEO) datasets (GSE148200 and GSE140077). We identified 2520 upregulated and 1292 downregulated genes in MIA-PaCa-2 cells, and 1958 upregulated and 1984 downregulated genes in the BxPC-3 cell line. These genes were selected based on specific criteria: a log2FC > ± 0.5 and a p-value < 0.05. (Figure 4A). Subsequently, we focused on the genes that were consistently regulated in the same direction in both cell lines, identifying 276 upregulated and 111 downregulated genes. We performed KEGG pathway analysis to gain further insights into the biological pathways associated with these significantly dysregulated genes. This analysis was carried out utilizing the DAVID database (https://david.ncifcrf.gov/, accessed on 28 August 2022) [52,53]. The top 15 enrichment pathways were prioritized (Figure 4B). Among these, of particular interest was the MYD88/NF-κB signaling pathway, which ranked as the top pathway and has been the subject of numerous previous studies highlighting its potential involvement in PDAC and Gem resistance [54–57]. Additionally, previous research has shown that MYD88 expression is increased in cancer tissues and correlated with paclitaxel resistance in other cancers such as breast, ovarian, and lung [58–60]. MYD88 is a significant target for innovative therapies in pancreatic cancer due to its crucial molecular role in linking various upstream ligand–receptor complexes [61]. We evaluated the expression level of MYD88 in 179 primary tumor tissues and 171 normal tissues from the TCGA database. Our findings revealed a significantly higher expression of MYD88 in cancer tissues compared to normal tissues ($p < 0.001$, Figure 4C). Furthermore, Kaplan–Meier survival analyses identified a significant association between MYD88 expression and overall survival in PDAC patients from the TCGA cohort (HR = 1.58, $p = 0.045$, Supplementary Figure S3). Finally, we validated the higher mRNA expression of MYD88 in Gem-R PDAC cell lines compared to parental cell lines through qRT-PCR assays (BxPC-3 cell line: $p < 0.05$, Figure 4D and $p < 0.05$, Figure 4D). Consequently, our findings suggest that the combination of ABE and Gem could potentially target the MYD88/NF-κB signaling pathway.

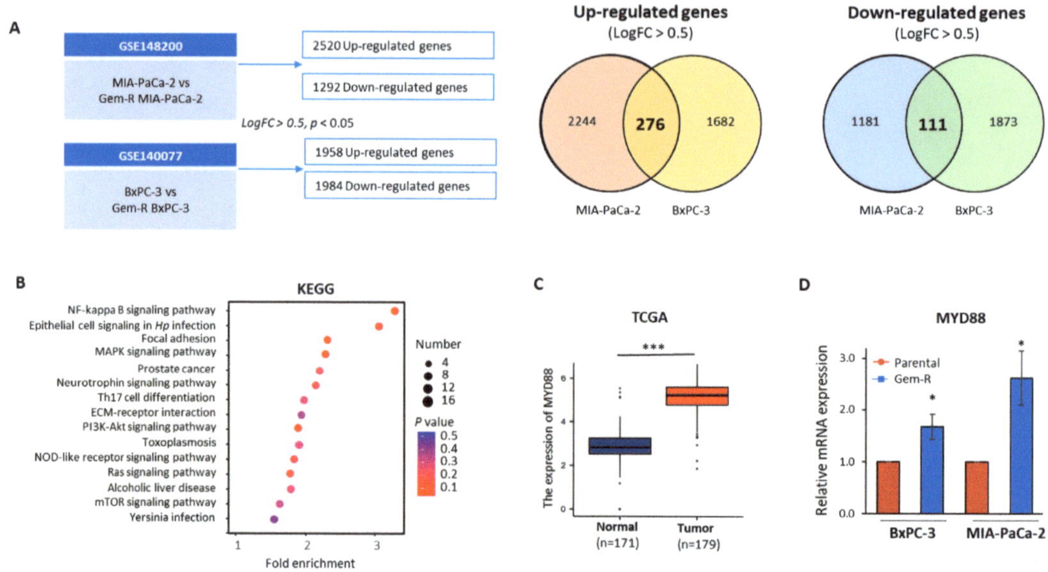

Figure 4. The MYD88/NF-κB signaling pathway is associated with Gem resistance in PDAC cells. (**A**) A schematic illustrates the differentially expressed genes in the BxPC-3-R vs. BxPC-3-Parental cell line (GSE 140077) and MIA-PaCa-2-R vs. MIA-PaCa-2-Parental cell line (GSE 148200). The Venn diagram represents the upregulated and downregulated expression of genes selected using Log2FC > 0.5 and $p < 0.05$. (**B**) Scatter plots of KEGG pathway enrichment analysis of up- and downregulated genes in Gem-R PDAC cell lines. The top 15 significantly enriched KEGG pathways were prioritized. The circle represents the number of differentially expressed genes, while the circle color represents the range of p-values. (**C**) The expression level of the selected genes, MYD88, was analyzed using PDAC cases from the TCGA dataset. (**D**) qRT-PCR assays were undertaken to measure the mRNA expression levels of selected genes in Gem-R and parental PDAC cell lines, with *β-actin* expression as the internal control. * p indicates $p < 0.05$ vs. control group; *** p indicates $p < 0.001$.

2.5. Combined Treatment with Gemcitabine and ABE Downregulates P-gp through the MYD88/TLR3/NF-κB Signaling Pathway

The Toll-like receptor (TLR) family, which includes *TLR1-10*, comprises transmembrane glycoproteins primarily expressed in immune cells. In particular, *TLR3* and *TLR4* can activate transcription factor NF-κB through both MYD88-dependent pathways and MYD88-independent pathways [62]. To investigate whether the MYD88/TLR3/NF-κB signaling pathway is a key target pathway of the combined treatment of Gem and ABE, WB was performed to assess the expression levels of key genes involved in the MYD88/TLR3/NF-κB axis when treated with Gem, ABE, and their combination. Our results revealed that TLR3, MYD88, and p65 were significantly downregulated by the combined treatment in both Gem-R BxPC-3 and Gem-R MIA-PaCa-2 cell lines (Figure 5A). In addition, the MYD88/NF-κB signaling pathway is linked to chemoresistance, as it has the ability to upregulate efflux transporters such as P-gp [56,63]. To evaluate the protein expression level of P-gp, we conducted immunofluorescence analysis (excitation: 488 nm; emission: 564 nm). Compared to the parental PDAC cell lines, the expression of P-gp in the Gem-R cell lines was significantly higher, and the treatment with ABE reduced the expression of P-gp in both PDAC cell lines after treatment (Figure 5B).

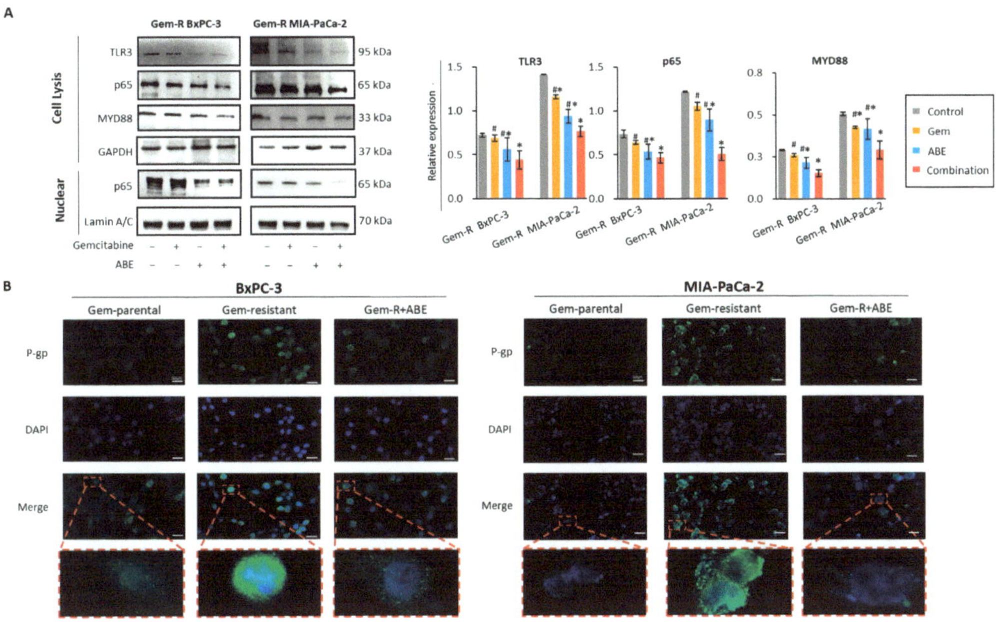

Figure 5. The combination treatment of Gem and ABE has been shown to regulate the MYD88/TLR3/NF-κB/P-gp complex in Gem-R PDAC cell lines. (**A**) Western blot analysis was performed on Gem-R PDAC cells treated with Gem, ABE, and their combination for 48 h. The relative levels of TLR3, MYD88, and p65 were quantitatively analyzed by comparing them to the control GAPDH expression. (**B**) Representative images for an immunofluorescence assay were obtained to evaluate P-gp expression in parental PDAC cell lines, Gem-R PDAC cell lines, and Gem-R PDAC cell lines treated with ABE for 48 h. * p indicates $p < 0.05$ vs. control group; # p indicates $p < 0.05$ vs. combination group.

2.6. The Combination of Gemcitabine and ABE Suppressed the Growth of PDOs

Three-dimensional organoid culture systems have emerged as valuable tools for modeling various aspects of cancer biology, particularly in investigating drug responses and evaluating treatment efficacy. In this study, we used tumor organoid models derived from two patients to examine the anti-tumor effects of Gem and ABE. As anticipated, our results demonstrated that the combination of Gem and ABE significantly suppressed both the size and number of organoids (Figure 6A). In terms of the number of organoids, when compared to individual treatments (Patient1: combination vs. Gem: FC = 0.48, $p < 0.05$; combination vs. ABE: FC = 0.56, $p < 0.05$, Figure 6B; Patient2: combination vs. Gem: FC = 0.65, $p < 0.05$; combination vs. ABE, FC = 0.62, $p < 0.05$; Figure 6B), the combination was more effective. Similarly, the combination of Gem and ABE significantly reduced organoid sizes compared to individual treatments (Patient1: combination vs. Gem: FC = 0.52, $p < 0.05$; combination vs. ABE: FC = 0.64, $p < 0.05$, Figure 6C; Patient2: combination vs. Gem: FC = 0.65, $p < 0.05$; combination vs. ABE, FC = 0.73, $p < 0.05$; Figure 6C). These findings support the notion that the combination of Gem and ABE effectively inhibit the growth of PDOs, corroborating the results of our cell culture experiments.

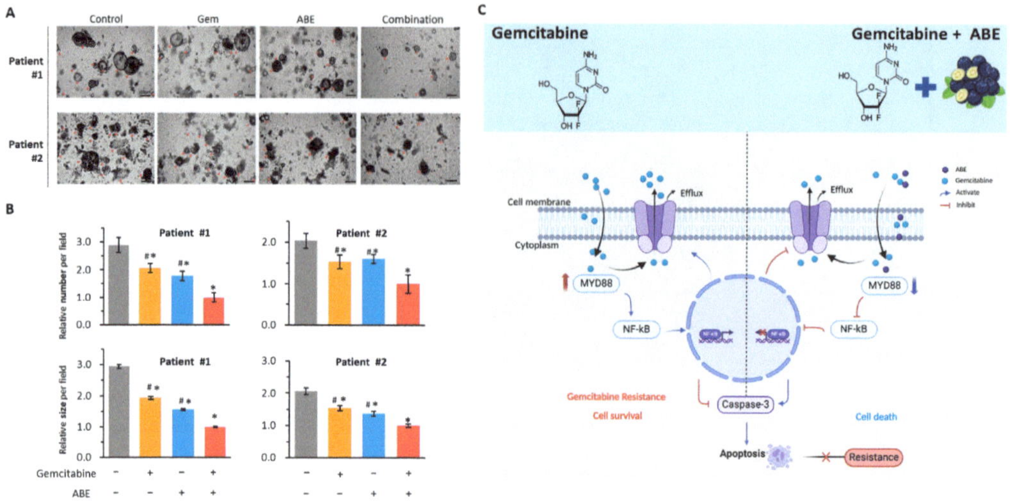

Figure 6. The combination of Gem and ABE demonstrates effective anticancer activity in PDAC patient-derived 3D organoid (PDO) models. (**A**) Representative images of PDOs treated with Gem, ABE, and their combination for 48 h. (**B**) The average number and size of PDOs treated with Gem, ABE, and their combination for 48 h. (**C**) A schematic illustration of ABE-induced reversal of Gem resistance in PDAC. This illustration reveals that Gem can activate the MYD88/NF-Kb axis, resulting in chemoresistance by upregulating the efflux transporters (P-gp) and hindering cellular apoptosis in PDAC cells (**left**). In contrast, ABE helps counteract chemoresistance by downregulating MYD88 and downstream signaling pathways in PDAC cells (**right**). * p indicates $p < 0.05$ vs. control group; # p indicates $p < 0.05$ vs. combination group.

3. Discussion

Since the FDA's approval in 1996, Gem-based chemotherapy has been used extensively as a standard treatment regimen in multiple kinds of solid cancers, including ovarian, breast, and non-small cell lung cancer [64]. Combination chemotherapeutic strategies incorporating Gem remain the preferred treatment for advanced PDAC [65]. However, chemotherapeutic resistance often occurs in PDAC patients. In particular, the development of resistance to Gem presents an important clinical challenge in the effective therapeutic management of patients with PDAC [66]. Several studies have demonstrated the significant role of ABC transporters like P-gp, BCRP, and MRP1 in mediating resistance to Gem chemotherapy [12], and the ability of ABC transporter inhibitors to sensitize tumor cells to chemotherapeutic agents has been researched extensively. While some positive outcomes have been observed in preclinical studies and early-phase clinical trials, no effective MDR-reversing agents targeting ABC transporters have been approved for clinical use thus far [67].

As such, natural products have been explored for their potential anticancer activity and low toxicity in healthy tissues and have recently been reported to overcome drug resistance [68,69]. In our study, we aimed to combine Gem with ABE to investigate whether ABE could enhance the sensitivity of Gem-R cells to Gem. Following the administration of a combination of Gem and ABE to Gem-R cells, our results revealed that ABE synergistically increased the sensitivity of Gem-R cells to Gem. Moreover, the combination of ABE and Gem effectively inhibited cell viability, clonogenicity, migration, and invasion while inducing a higher apoptosis rate in PDAC cells. We further found that the combined ABE and Gem treatment could significantly strengthen the anticancer activity in inhibiting PDAC PDOs, supporting our cell culture-based findings.

To investigate the mechanism by which ABE enhances sensitivity to Gem, we analyzed gene expression profiles from two publicly available GEO datasets and compared parental and Gem-R PDAC cells. Our analysis revealed that MYD88, an adaptor molecule for Toll-like receptors [70–73], was upregulated in Gem-R PDAC cells. Furthermore, we found that TLR3 can activate the transcription factor NF-κB through both the MYD88-dependent pathway and the MYD88-independent pathway [62]. The MYD88/NF-κB signaling pathway is linked to chemoresistance as it can elevate the expression of efflux transporters such as P-gp [56,74,75]. In the current study, we revealed that MYD88 is upregulated in tumor tissues and that high expression levels of MYD88 correlate with poor prognosis in PDAC. Second, we discovered that MYD88 is also upregulated in Gem-R cell lines. Therefore, we hypothesized that ABE could target MYD88, potentially restoring sensitivity to Gem in Gem-R cell lines. Our WB results showed that co-treatment of Gem and ABE inhibited the expression of MYD88, TRL3, and P65 expression. These findings provide further evidence for MYD88/TRL3/NF-κB as a critical pathway in the ability of ABE to overcome Gem resistance. However, the present study lacks definitive evidence to establish whether alterations in TLR3 expression are MYD88-dependent. Additionally, the co-treatment of Gem and ABE suppressed the expression of P-gp, as shown by immunofluorescence staining. These findings provide compelling evidence for the involvement of the MYD88/TRL3/NF-κB/P-gp axis as the target mechanism through which ABE counteracts Gem resistance in PDAC cells (Figure 6C).

Finally, we confirmed the anti-tumor efficacy of Gem and ABE by utilizing tumor organoid models derived from two PDAC patients. However, we did not generate the Gem-R organoids, which may limit our finding validation in vivo. Based on these, we propose that we can validate these findings in animal models in the future.

In summary, our study provides valuable insights into the potential of ABE to overcome Gem resistance in PDAC. Our data reveal a crucial mechanism underlying the synergistic interaction between ABE and Gem. These findings contribute to the growing body of evidence supporting the potential of ABE as a novel therapy for PDAC, particularly for patients confronting Gem resistance.

4. Materials and Methods

4.1. Cell Culture

The PDAC cell lines BxPC-3 and MIA-PaCa-2 were sourced from the American Type Culture Collection (ATCC, Manassas, VA, USA). The cell lines were cultured in RPMI medium (Gibco, Carlsbad, CA, USA) with 1% penicillin/streptomycin (Sigma-Aldrich, St. Louis, MO, USA) and 10% fetal bovine serum (Gibco, Waltham, MA, USA). The cells were cultured at 37 °C with 5% CO2 in a humidified environment. The adherent cells were harvested using 0.05% trypsin (Invitrogen, Carlsbad, CA, USA). Gem-R MIA-PaCa-2 and Gem-R BxPC-3 were established by continuously culturing the cells with increasing doses of Gem, as described previously [3].

4.2. Herbal Preparations

The aronia berry extract utilized in this study (Aronia Berry Complex, EuroPharma USA, Green Bay, WI, USA) was a powdered product characterized by colors ranging from purple to dark red. It was derived from the fruit of the *Aronia Melanocarpa* plant, extracted in 70% ethanol, and standardized to a 40% polyphenol content. The extract was initially dissolved in dimethyl sulfoxide to create stock concentrations and then further diluted to suitable levels in the culture media.

4.3. Reagents

Gemcitabine (Sigma-Aldrich, St. Louis, MO, USA) was dissolved in dimethyl sulfoxide (DMSO, Sigma-Aldrich). The stock solutions of Gem (10 μM) and ABE (100 mg/mL) were carefully stored at −20 °C in the dark, ensuring their stability and reliability. These stock solutions were diluted with a complete culture medium to the necessary experimental concentrations before each application.

4.4. Cell Counting Kit-8 Assays

For the Cell Counting Kit-8 (CCK-8) assay, a CCK-8 kit (Dojindo, Kumamoto, Japan) was used to measure cell viability according to the instructions as follows: Initially, to confirm the Gem-resistant properties, the proliferation rates were compared between Gem-R and their parental counterparts by subjecting them to escalating doses of Gem. Cells were seeded in 96-well flat plates at a density of 5×10^3 cells per well and incubated for 24 h. Thereafter, cells were treated with increasing concentrations of Gem (0–1200 nM). After 48 h of treatment, 10 μL of CCK-8 solution was added to each well and incubated for 2 h. The absorbance of the product was measured at a wavelength of 450 nm (OD450) using a microplate reader from Tecan Trading AG (Tecan Trading AG, Männedorf, Switzerland).

Subsequently, to investigate the influence of ABE and Gem combination on PDAC cell proliferation, cells were seeded in 96-well plates at a density of 2×10^3 cells/well in 100 μL of complete culture medium and incubated for 24 h. Subsequently, the cells were exposed to appropriate concentrations of Gem (800 nM), ABE (90 μg/mL), and their combination. The cell proliferation rates were evaluated at different time points.

4.5. Drug Response Testing

The Gem-R PDAC cell lines were plated at a density of 5×10^3 cells per well in 96-well plates and incubated for 24 h before drug administration. Subsequently, the cells were exposed to varying concentrations of Gem (0, 400, 800, and 1200 nM), ABE (0, 30, 60, 90, and 120 μg/mL), and twenty different combinations of these drugs for a duration of 48 h, to identify the synergistic concentrations. Cell viability was assessed using the CCK8 assay. Synergy scores were subsequently calculated using SynergyFinder 3.0, a freely accessible tool specifically designed for interactive analysis and visualization of combination response outcomes [76].

4.6. Colony Formation Assay

For the cell colony formation assay, 5×10^2 cells were seeded per well in 6-well flat plates and then treated with Gem, ABE, and their combination for 48 h. Following this treatment, colony formation was allowed to proceed for a duration of 7 to 10 days, with the culture medium refreshed every three days. At the end of this incubation period, the cell colonies were fixed with 4% paraformaldehyde in PBS (Thermo Fisher Scientific, Fair Lawn, NJ, USA) for 30 min and stained with 1% crystal violet (Thermo Fisher Scientific). The number of colonies was quantified using ImageJ 1.53q software.

4.7. Wound Healing Assay

In the cell wound healing assay, we seeded 5×10^5 cells per well in 6-well flat plates following a 48 h treatment with Gem, ABE, or their combination. After the cells reached 80% to 90% confluence, we utilized a sterile 20 μL micropipette tip to create a controlled scratch in the monolayer. This was followed by a wash with a serum-free medium to eliminate detached cells. Subsequently, the cells were cultured in a complete medium. Photographs of the cells were captured 24 h after the formation of the wound, and the percentage of wound closures was determined using ImageJ software.

4.8. Invasion Assays

To conduct cell invasion assays, BioCoat Matrigel Invasion Chambers with 8.0 μm Pore Polyester Membrane (BD Biosciences, Franklin Lakes, NJ, USA) were utilized. For the invasion assay, 5×10^4 cells per well were seeded in 24-well flat plates after treatment with Gem, ABE, and their combination for 48 h. The cells were then transferred onto inserts in serum-free medium and subsequently moved to wells containing culture medium supplemented with 10% FBS. After a 48 h incubation period, the cells that had invaded the bottom surface of the membrane were fixed and stained using a Diff-Quick staining kit (Thermo Fisher Scientific). The stained cells were later quantified under a microscope.

4.9. Apoptosis Assay

For apoptosis assays, Muse™ Annexin V and Dead Cell kits (Luminex Corp, Austin, TX, USA) were utilized following the manufacturer's guidelines. In these experiments, 5×10^5 cells were seeded per well in 6-well flat plates. Following treatment with Gem, ABE, and their combination for 48 h, the cells were harvested. Subsequently, 100 µL of cell suspension was mixed with 100 µL of Muse Annexin V & Dead Cell Reagent. The percentage of apoptotic cells was assessed using a Muse™ Cell Analyzer (Millipore Corp, Billerica, MA, USA) in accordance with the manufacturer's instructions.

4.10. Gene Enrichment and Pathway Analysis

To identify differentially expressed genes in Gem-R PDAC cells, gene expression profiles from two publicly available datasets (GSE148200 and GSE140077) sourced from the Gene Expression Omnibus (GEO) database (https://www.ncbi.nlm.nih.gov/geo/, accessed on 28 August 2022) were analyzed. Differential gene expression across various groups was examined using the "DEseq2" package in R, with a significance threshold of $p < 0.05$ and a Log_2FC (Logarithm of Fold Change) exceeding ±0.5.

The Kyoto Encyclopedia of Genes and Genomes (KEGG) pathway analysis was performed using the DAVID bioinformatics database (https://david.ncifcrf.gov/, accessed on 28 August 2022). The KEGG pathway enrichment analysis was visualized using the "ggplot2" package in R.

4.11. Isolation of Cytosolic and Nuclear Extracts

After treatment with Gem, ABE, and their combination for 48 h, cytosolic and nuclear extracts were isolated from cells using a nuclear extraction kit (Active Motif, Carlsbad, CA, USA) according to the manufacturer's instructions. These extracts were collected and stored in aliquots at −80 °C.

4.12. Protein Isolation and Western Blot

Total protein was extracted from PDAC cell lines treated for 48 h with Gem, ABE, or their combination. The cells were harvested using a plastic scraper. Subsequently, the cells were lysed with ice-cold protein extraction solution RIPA containing a protease inhibitor cocktail (Thermo Fisher Scientific). Protein concentration was determined using the BCA procedure (Thermo Fisher Scientific). Equal amounts of protein samples were separated via SDS-PAGE using 6% or 10% Mini-PROTEAN TGXTM Precast Gels (BIO-RAD, Hercules, CA, USA) and then transferred onto nitrocellulose membranes, followed by an additional transfer onto a 0.45 µm PVDF membrane (Cytiva, Marlborough, MA, USA). The membranes were blocked with 5% bovine serum albumin in Tris-buffered saline with 0.1% Tween-20 for 1 h at room temperature. Subsequently, the membranes were incubated with primary antibodies, including anti-Bax (1:1000, 5023S; Cell Signaling Technology [CST], Danvers, MA, USA), anti-Cleaved Caspase 9 (1:1000, 9505; CST), anti-PARP (1:1000, 9532S; CST), anti-TLR3(1:1000, PA5-20183; Thermo Fisher Scientific), anti-MYD88 (1:1000, 23230-1-AP; Proteintech, Rosemont, IL, USA), anti-P65 (1:1000, 3034; CST), anti-Lamin A/C (1:2000, 10298-1-AP, Proteintech), anti-Vimentin (1:2000, 10366-1-AP, Proteintech), anti-MMP-9 (1:1000, 10375-2-AP, Proteintech), anti-E-cadherin (1:2000, 10375-2-AP, Proteintech), and anti-GAPDH (1:2000, 5174T; CST) at 4 °C overnight. After three washes with TBST, the membranes were incubated with corresponding anti-rabbit (1:2000, 7074; CST) or anti-mouse (1:2000, 7076; CST) secondary antibodies for 1 h at room temperature. The blot was visualized using an HRP-based chemiluminescence kit (Thermo Fisher Scientific) with Gel Imaging Systems (BIO-RAD). GAPDH protein served as an internal control, and the intensities of the protein bands were quantified using ImageJ software.

4.13. Quantitative Reverse Transcription PCR (qRT-PCR)

Total RNA extraction was performed using the Qiagen miRNeasy Kit (Qiagen, Hilden, Germany). Subsequently, cDNA was synthesized by reverse transcription of total RNA using the High-Capacity cDNA Reverse Transcription Kit from Thermo Fisher Scientific. RT-PCR assays were conducted with the QuantStudio 6 Flex RT-PCR System (Applied Biosystems, Foster City, CA, USA), following the manufacturer's instructions and utilizing the SensiFAST SYBR Lo-ROX Kit from Bioline (London, UK). The relative mRNA expression levels of MYD88 were determined using the $2^{-\Delta\Delta Ct}$ method and normalized to the expression of *β-actin*, acting as the internal control. The primer sequences can be found in Supplementary Table S1.

4.14. Immunofluorescence Assay

For the immunofluorescence assay, 5×10^5 cells per well were seeded in 6 cm flat plates with slides. Following an 18-h incubation period, the cells were treated with Gem, ABE, or their combination for 48 h. The slides were subsequently fixed with 4% paraformaldehyde for 10 min at room temperature, then permeabilized with 0.5% Triton-X-100 (Thermo Fisher Scientific) for 10 min at room temperature. To remove endogenous peroxides, the slides were treated and subsequently blocked in TBST (Tris-buffered saline with Tween-20) containing 3% BSA for 1 h at room temperature. The slides were then incubated with an anti-P-glycoprotein Polyclonal antibody (1:500, 22336-1-AP, Proteintech) overnight at 4 °C, followed by incubation with a FITC-labeled secondary antibody (1:1000, A-21202; Thermo Fischer Scientific) for 1 h at 37 °C. Following incubation, the slides were subjected to three additional washes with phosphate-buffered saline, each lasting five minutes. Finally, the cell nucleus was stained with DAPI (Thermo Fischer Scientific), and all images were captured using a Carl Zeiss fluorescent microscope (Zeiss, Oberkochen, Germany).

4.15. Patient-Derived 3-Dimensional Tumor Organoids (PDOs)

PDOs from PDAC patients were generated as described in a previous study [77]. With approval from the institution's ethics committees, written informed consent was obtained from all patients. To ensure confidentiality and anonymity, patients were coded following ethical guidelines outlined in the Declaration of Helsinki. For the human organoids, the Human Complete Feeding Medium (hCPLT) consisted of PancreaCultTM Organoid Growth Medium (STEMCELL Technologies, Cambridge, MA, USA), supplemented with EGF (STEMCELL Technologies) and prostaglandin E2 (STEMCELL Technologies), following the manufacturer's instructions. PDAC organoids were seeded in a 24-well plate to form a dome in 40 μL Matrigel (Corning, Tehama County, CA, USA) with 500 μL of hCPLT. The domes were then divided into 4 groups and appropriate concentrations of Gem (800 nM), ABE (90 μg/mL), and their combination. After 7 days of treatment, the number and size of the organoids were analyzed through microscopy (magnification ×100) and measured using ImageJ software.

4.16. Statistical Analysis

Statistical analysis was conducted utilizing SPSS software version 21.0 and GraphPad Prism version 6.0. Student's *t*-test was employed to assess the significance of observed differences between the two groups, while one-way analysis of variance (ANOVA) was utilized to evaluate differences among multiple comparisons. All experiments were performed in triplicate on independent biological replicates, and the data were presented as mean ± standard deviation. A significance level of $p < 0.05$ was considered statistically significant.

5. Conclusions

We firstly provided a novel insight into the potential of ABE to overcome Gem-R in PDAC cells using a systematic series of Gem-resistant cell culture and patient-derived tumor organoids experiments. Additionally, we observed that the MYD88/NF-κB/P-gp axis plays a crucial role in Gem resistance in PDAC, and ABE could overcome this resistance by downregulating MYD88 and its downstream signaling pathways in PDAC cells (Figure 6C), suggesting ABE as a promising and cost-effective approach to enhance therapeutic outcomes in PDAC. These results warrant further exploration of ABE in clinical settings to validate its potential as an adjunct therapy for improving treatment efficacy in pancreatic cancer patients. In the next phase of our research, we should further investigate the optimal dose of ABE to effectively counter Gem resistance at the minimum required dosage.

Supplementary Materials: The following supporting information can be downloaded at: https://www.mdpi.com/article/10.3390/ph17070911/s1, Supplementary Figure S1: The cell viability of Gem-R BxPC-3 (A) and Mia PaCa-2/Gem-R (B) cells treated with different concentrations of ABE for 48 h. Supplementary Figure S2: Western blot analysis was performed on Gem-R PDAC cells treated with Gem, ABE, and their combination for 48 h. The relative levels of E-cadherin, Vimentin, and MMP9 were quantitatively analyzed by comparing them to the control GAPDH expression. Supplementary Figure S3: Kaplan–Meier survival curves of the selected genes, MYD88, were analyzed by PDAC cases from the TCGA public database. Supplementary Figure S4: Original Western blots. Supplementary Table S1: Primer sequences and their PCR conditions in this study.

Author Contributions: Y.L. and A.G. conceived and designed the experiments. Y.L. and C.X. performed the experiments and analyzed the data. H.H. generated Gem-resistant cell lines. C.F. and S.P.-S. generated PDAC patient-derived tumor organoids. Y.L. and A.G. wrote and edited the manuscript. All authors have read and agreed to the published version of the manuscript.

Funding: This work was supported by CA72851, CA181572, CA184792, CA187956, CA202797, CA214254, and CA271443 grants from the National Institutes of Health.

Institutional Review Board Statement: This study was conducted according to the guidelines of the Declaration of Helsinki and was approved by the Institutional Review Board of City of Hope (protocol code 187811 and date of approval: 20 April 2020).

Informed Consent Statement: In accordance with the Declaration of Helsinki, all patients have provided written informed consent, and all procedures involving patient-derived tumor organoids have been approved by the institutional review board.

Data Availability Statement: Data are contained within the article.

Acknowledgments: The authors thank Silei Sui, Yoh Asahi, Katsuki Miyazaki, Keisuke Okuno, Courtney Quan, Katsutoshi Shoda, and Alessandro Mannucci for their insightful discussions and valuable advice throughout this project.

Conflicts of Interest: The authors disclosed no potential conflicts of interest.

References

1. McGuigan, A.; Kelly, P.; Turkington, R.C.; Jones, C.; Coleman, H.G.; McCain, R.S. Pancreatic cancer: A review of clinical diagnosis, epidemiology, treatment and outcomes. *World J. Gastroenterol.* **2018**, *24*, 4846–4861. [CrossRef] [PubMed]
2. Wood, L.D.; Canto, M.I.; Jaffee, E.M.; Simeone, D.M. Pancreatic Cancer: Pathogenesis, Screening, Diagnosis, and Treatment. *Gastroenterology* **2022**, *163*, 386–402.e381. [CrossRef] [PubMed]
3. Adamska, A.; Domenichini, A.; Falasca, M. Pancreatic Ductal Adenocarcinoma: Current and Evolving Therapies. *Int. J. Mol. Sci.* **2017**, *18*, 1338. [CrossRef] [PubMed]
4. Gourgou-Bourgade, S.; Bascoul-Mollevi, C.; Desseigne, F.; Ychou, M.; Bouche, O.; Guimbaud, R.; Becouarn, Y.; Adenis, A.; Raoul, J.L.; Boige, V.; et al. Impact of FOLFIRINOX compared with gemcitabine on quality of life in patients with metastatic pancreatic cancer: Results from the PRODIGE 4/ACCORD 11 randomized trial. *J. Clin. Oncol.* **2013**, *31*, 23–29. [CrossRef] [PubMed]
5. Moore, M.J.; Goldstein, D.; Hamm, J.; Figer, A.; Hecht, J.R.; Gallinger, S.; Au, H.J.; Murawa, P.; Walde, D.; Wolff, R.A.; et al. Erlotinib plus gemcitabine compared with gemcitabine alone in patients with advanced pancreatic cancer: A phase III trial of the National Cancer Institute of Canada Clinical Trials Group. *J. Clin. Oncol.* **2007**, *25*, 1960–1966. [CrossRef] [PubMed]
6. Sarvepalli, D.; Rashid, M.U.; Rahman, A.U.; Ullah, W.; Hussain, I.; Hasan, B.; Jehanzeb, S.; Khan, A.K.; Jain, A.G.; Khetpal, N.; et al. Gemcitabine: A Review of Chemoresistance in Pancreatic Cancer. *Crit. Rev. Oncog.* **2019**, *24*, 199–212. [CrossRef] [PubMed]
7. Wang, S.; Zheng, Y.; Yang, F.; Zhu, L.; Zhu, X.Q.; Wang, Z.F.; Wu, X.L.; Zhou, C.H.; Yan, J.Y.; Hu, B.Y.; et al. The molecular biology of pancreatic adenocarcinoma: Translational challenges and clinical perspectives. *Signal Transduct. Target. Ther.* **2021**, *6*, 249. [CrossRef] [PubMed]
8. Walker, E.J.; Ko, A.H. Beyond first-line chemotherapy for advanced pancreatic cancer: An expanding array of therapeutic options? *World J. Gastroenterol.* **2014**, *20*, 2224–2236. [CrossRef]

9. Wang-Gillam, A.; Hubner, R.A.; Siveke, J.T.; Von Hoff, D.D.; Belanger, B.; de Jong, F.A.; Mirakhur, B.; Chen, L.T. NAPOLI-1 phase 3 study of liposomal irinotecan in metastatic pancreatic cancer: Final overall survival analysis and characteristics of long-term survivors. *Eur. J. Cancer* **2019**, *108*, 78–87. [CrossRef]
10. Gu, J.; Huang, W.; Wang, X.; Zhang, J.; Tao, T.; Zheng, Y.; Liu, S.; Yang, J.; Chen, Z.S.; Cai, C.Y.; et al. Hsa-miR-3178/RhoB/PI3K/Akt, a novel signaling pathway regulates ABC transporters to reverse gemcitabine resistance in pancreatic cancer. *Mol. Cancer* **2022**, *21*, 112. [CrossRef]
11. Liu, J.; Hu, G.; Gong, Y.; Yu, Q.; He, B.; Li, W.; He, Z.; Hao, W.; He, Z.; Liu, Y. Silencing of *TRPM8* inhibits aggressive tumor phenotypes and enhances gemcitabine sensitivity in pancreatic cancer. *Pancreatology* **2018**, *18*, 935–944. [CrossRef]
12. Kohan, H.G.; Boroujerdi, M. Time and concentration dependency of P-gp, MRP1 and MRP5 induction in response to gemcitabine uptake in Capan-2 pancreatic cancer cells. *Xenobiotica* **2015**, *45*, 642–652. [CrossRef]
13. Robey, R.W.; Pluchino, K.M.; Hall, M.D.; Fojo, A.T.; Bates, S.E.; Gottesman, M.M. Revisiting the role of ABC transporters in multidrug-resistant cancer. *Nat. Rev. Cancer* **2018**, *18*, 452–464. [CrossRef] [PubMed]
14. Elmeliegy, M.; Vourvahis, M.; Guo, C.; Wang, D.D. Effect of P-glycoprotein (P-gp) Inducers on Exposure of P-gp Substrates: Review of Clinical Drug-Drug Interaction Studies. *Clin. Pharmacokinet.* **2020**, *59*, 699–714. [CrossRef] [PubMed]
15. Joshi, P.; Vishwakarma, R.A.; Bharate, S.B. Natural alkaloids as P-gp inhibitors for multidrug resistance reversal in cancer. *Eur. J. Med. Chem.* **2017**, *138*, 273–292. [CrossRef] [PubMed]
16. Zheng, W.; Li, M.; Lin, Y.; Zhan, X. Encapsulation of verapamil and doxorubicin by MPEG-PLA to reverse drug resistance in ovarian cancer. *Biomed. Pharmacother.* **2018**, *108*, 565–573. [CrossRef] [PubMed]
17. Goel, A.; Boland, C.R.; Chauhan, D.P. Specific inhibition of cyclooxygenase-2 (COX-2) expression by dietary curcumin in HT-29 human colon cancer cells. *Cancer Lett.* **2001**, *172*, 111–118. [CrossRef]
18. Goel, A.; Jhurani, S.; Aggarwal, B.B. Multi-targeted therapy by curcumin: How spicy is it? *Mol. Nutr. Food Res.* **2008**, *52*, 1010–1030. [CrossRef] [PubMed]
19. Goel, A.; Kunnumakkara, A.B.; Aggarwal, B.B. Curcumin as "Curecumin": From kitchen to clinic. *Biochem. Pharmacol.* **2008**, *75*, 787–809. [CrossRef]
20. Goel, A.; Aggarwal, B.B. Curcumin, the golden spice from Indian saffron, is a chemosensitizer and radiosensitizer for tumors and chemoprotector and radioprotector for normal organs. *Nutr. Cancer* **2010**, *62*, 919–930. [CrossRef] [PubMed]
21. Link, A.; Balaguer, F.; Goel, A. Cancer chemoprevention by dietary polyphenols: Promising role for epigenetics. *Biochem. Pharmacol.* **2010**, *80*, 1771–1792. [CrossRef]
22. Toden, S.; Goel, A. The Holy Grail of Curcumin and its Efficacy in Various Diseases: Is Bioavailability Truly a Big Concern? *J. Restor. Med.* **2017**, *6*, 27–36. [CrossRef]
23. Toden, S.; Theiss, A.L.; Wang, X.; Goel, A. Essential turmeric oils enhance anti-inflammatory efficacy of curcumin in dextran sulfate sodium-induced colitis. *Sci. Rep.* **2017**, *7*, 814. [CrossRef]
24. Weng, W.; Goel, A. Curcumin and colorectal cancer: An update and current perspective on this natural medicine. *Semin. Cancer Biol.* **2022**, *80*, 73–86. [CrossRef] [PubMed]
25. Alam, M.; Ali, S.; Ashraf, G.M.; Bilgrami, A.L.; Yadav, D.K.; Hassan, M.I. Epigallocatechin 3-gallate: From green tea to cancer therapeutics. *Food Chem.* **2022**, *379*, 132135. [CrossRef]
26. Efferth, T.; Oesch, F. Anti-inflammatory and anti-cancer activities of frankincense: Targets, treatments and toxicities. *Semin. Cancer Biol.* **2022**, *80*, 39–57. [CrossRef]
27. Naeem, M.; Iqbal, M.O.; Khan, H.; Ahmed, M.M.; Farooq, M.; Aadil, M.M.; Jamaludin, M.I.; Hazafa, A.; Tsai, W.C. A Review of Twenty Years of Research on the Regulation of Signaling Pathways by Natural Products in Breast Cancer. *Molecules* **2022**, *27*, 3412. [CrossRef] [PubMed]
28. Pan, L.; Feng, F.; Wu, J.; Fan, S.; Han, J.; Wang, S.; Yang, L.; Liu, W.; Wang, C.; Xu, K. Demethylzeylasteral targets lactate by inhibiting histone lactylation to suppress the tumorigenicity of liver cancer stem cells. *Pharmacol. Res.* **2022**, *181*, 106270. [CrossRef]
29. Zhao, Y.; Wang, C.; Goel, A. A combined treatment with melatonin and andrographis promotes autophagy and anticancer activity in colorectal cancer. *Carcinogenesis* **2022**, *43*, 217–230. [CrossRef]
30. Reuter, S.; Gupta, S.C.; Park, B.; Goel, A.; Aggarwal, B.B. Epigenetic changes induced by curcumin and other natural compounds. *Genes Nutr.* **2011**, *6*, 93–108. [CrossRef]
31. Chandran, B.; Goel, A. A randomized, pilot study to assess the efficacy and safety of curcumin in patients with active rheumatoid arthritis. *Phytother. Res.* **2012**, *26*, 1719–1725. [CrossRef]
32. Shakibaei, M.; Mobasheri, A.; Lueders, C.; Busch, F.; Shayan, P.; Goel, A. Curcumin enhances the effect of chemotherapy against colorectal cancer cells by inhibition of NF-κB and Src protein kinase signaling pathways. *PLoS ONE* **2013**, *8*, e57218. [CrossRef]
33. Toden, S.; Okugawa, Y.; Buhrmann, C.; Nattamai, D.; Anguiano, E.; Baldwin, N.; Shakibaei, M.; Boland, C.R.; Goel, A. Novel Evidence for Curcumin and Boswellic Acid-Induced Chemoprevention through Regulation of miR-34a and miR-27a in Colorectal Cancer. *Cancer Prev. Res.* **2015**, *8*, 431–443. [CrossRef]
34. Yoshida, K.; Toden, S.; Ravindranathan, P.; Han, H.; Goel, A. Curcumin sensitizes pancreatic cancer cells to gemcitabine by attenuating PRC2 subunit *EZH2*, and the lncRNA PVT1 expression. *Carcinogenesis* **2017**, *38*, 1036–1046. [CrossRef]

35. Miyazaki, K.; Xu, C.; Shimada, M.; Goel, A. Curcumin and Andrographis Exhibit Anti-Tumor Effects in Colorectal Cancer via Activation of Ferroptosis and Dual Suppression of Glutathione Peroxidase-4 and Ferroptosis Suppressor Protein-1. *Pharmaceuticals* **2023**, *16*, 383. [CrossRef]
36. Miyazaki, K.; Morine, Y.; Xu, C.; Nakasu, C.; Wada, Y.; Teraoku, H.; Yamada, S.; Saito, Y.; Ikemoto, T.; Shimada, M.; et al. Curcumin-Mediated Resistance to Lenvatinib via *EGFR* Signaling Pathway in Hepatocellular Carcinoma. *Cells* **2023**, *12*, 612. [CrossRef]
37. Sharma, P.; Shimura, T.; Banwait, J.K.; Goel, A. Andrographis-mediated chemosensitization through activation of ferroptosis and suppression of β-catenin/Wnt-signaling pathways in colorectal cancer. *Carcinogenesis* **2020**, *41*, 1385–1394. [CrossRef] [PubMed]
38. Zhao, Y.; Wang, C.; Goel, A. Andrographis overcomes 5-fluorouracil-associated chemoresistance through inhibition of *DKK1* in colorectal cancer. *Carcinogenesis* **2021**, *42*, 814–825. [CrossRef]
39. Okuno, K.; Xu, C.; Pascual-Sabater, S.; Tokunaga, M.; Takayama, T.; Han, H.; Fillat, C.; Kinugasa, Y.; Goel, A. Andrographis Reverses Gemcitabine Resistance through Regulation of *ERBB3* and Calcium Signaling Pathway in Pancreatic Ductal Adenocarcinoma. *Biomedicines* **2023**, *11*, 119. [CrossRef]
40. Buhrmann, C.; Shayan, P.; Kraehe, P.; Popper, B.; Goel, A.; Shakibaei, M. Resveratrol induces chemosensitization to 5-fluorouracil through up-regulation of intercellular junctions, Epithelial-to-mesenchymal transition and apoptosis in colorectal cancer. *Biochem. Pharmacol.* **2015**, *98*, 51–68. [CrossRef] [PubMed]
41. Buhrmann, C.; Yazdi, M.; Popper, B.; Shayan, P.; Goel, A.; Aggarwal, B.B.; Shakibaei, M. Evidence that TNF-β induces proliferation in colorectal cancer cells and resveratrol can down-modulate it. *Exp. Biol. Med.* **2019**, *244*, 1–12. [CrossRef] [PubMed]
42. Okuno, K.; Pratama, M.Y.; Li, J.; Tokunaga, M.; Wang, X.; Kinugasa, Y.; Goel, A. Ginseng mediates its anticancer activity by inhibiting the expression of DNMTs and reactivating methylation-silenced genes in colorectal cancer. *Carcinogenesis* **2023**, *44*, 394–403. [CrossRef] [PubMed]
43. Dewanjee, S.; Dua, T.K.; Bhattacharjee, N.; Das, A.; Gangopadhyay, M.; Khanra, R.; Joardar, S.; Riaz, M.; Feo, V.; Zia-Ul-Haq, M. Natural Products as Alternative Choices for P-Glycoprotein (P-gp) Inhibition. *Molecules* **2017**, *22*, 871. [CrossRef]
44. Ren, Y.; Frank, T.; Meyer, G.; Lei, J.; Grebenc, J.R.; Slaughter, R.; Gao, Y.G.; Kinghorn, A.D. Potential Benefits of Black Chokeberry (*Aronia melanocarpa*) Fruits and Their Constituents in Improving Human Health. *Molecules* **2022**, *27*, 7823. [CrossRef] [PubMed]
45. Choi, H.S.; Kim, J.H.; Kim, S.L.; Deng, H.Y.; Lee, D.; Kim, C.S.; Yun, B.S.; Lee, D.S. Catechol derived from aronia juice through lactic acid bacteria fermentation inhibits breast cancer stem cell formation via modulation Stat3/IL-6 signaling pathway. *Mol. Carcinog.* **2018**, *57*, 1467–1479. [CrossRef] [PubMed]
46. Thani, N.A.; Keshavarz, S.; Lwaleed, B.A.; Cooper, A.J.; Rooprai, H.K. Cytotoxicity of gemcitabine enhanced by polyphenolics from Aronia melanocarpa in pancreatic cancer cell line AsPC-1. *J. Clin. Pathol.* **2014**, *67*, 949–954. [CrossRef] [PubMed]
47. Gourley, C.; Balmana, J.; Ledermann, J.A.; Serra, V.; Dent, R.; Loibl, S.; Pujade-Lauraine, E.; Boulton, S.J. Moving From Poly (ADP-Ribose) Polymerase Inhibition to Targeting DNA Repair and DNA Damage Response in Cancer Therapy. *J. Clin. Oncol.* **2019**, *37*, 2257–2269. [CrossRef] [PubMed]
48. Jenner, A.; Pena-Blanco, A.; Salvador-Gallego, R.; Ugarte-Uribe, B.; Zollo, C.; Ganief, T.; Bierlmeier, J.; Mund, M.; Lee, J.E.; Ries, J.; et al. *DRP1* interacts directly with BAX to induce its activation and apoptosis. *EMBO J.* **2022**, *41*, e108587. [CrossRef] [PubMed]
49. Rose, M.; Burgess, J.T.; O'Byrne, K.; Richard, D.J.; Bolderson, E. PARP Inhibitors: Clinical Relevance, Mechanisms of Action and Tumor Resistance. *Front. Cell Dev. Biol.* **2020**, *8*, 564601. [CrossRef]
50. Spitz, A.Z.; Gavathiotis, E. Physiological and pharmacological modulation of BAX. *Trends Pharmacol. Sci.* **2022**, *43*, 206–220. [CrossRef]
51. Wei, S.; Peng, L.; Yang, J.; Sang, H.; Jin, D.; Li, X.; Chen, M.; Zhang, W.; Dang, Y.; Zhang, G. Exosomal transfer of miR-15b-3p enhances tumorigenesis and malignant transformation through the DYNLT1/Caspase-3/Caspase-9 signaling pathway in gastric cancer. *J. Exp. Clin. Cancer Res.* **2020**, *39*, 32. [CrossRef] [PubMed]
52. Huang, D.W.; Sherman, B.T.; Lempicki, R.A. Systematic and integrative analysis of large gene lists using DAVID bioinformatics resources. *Nat. Protoc.* **2009**, *4*, 44–57. [CrossRef] [PubMed]
53. Sherman, B.T.; Hao, M.; Qiu, J.; Jiao, X.; Baseler, M.W.; Lane, H.C.; Imamichi, T.; Chang, W. DAVID: A web server for functional enrichment analysis and functional annotation of gene lists (2021 update). *Nucleic Acids Res.* **2022**, *50*, W216–W221. [CrossRef]
54. Chaudhary, P.; Guragain, D.; Chang, J.H.; Kim, J.A. *TPH1* and *5-HT$_7$* Receptor Overexpression Leading to Gemcitabine-Resistance Requires Non-Canonical Permissive Action of EZH2 in Pancreatic Ductal Adenocarcinoma. *Cancers* **2021**, *13*, 5305. [CrossRef]
55. Lou, C.; Lu, H.; Ma, Z.; Liu, C.; Zhang, Y. Ginkgolide B enhances gemcitabine sensitivity in pancreatic cancer cell lines via inhibiting *PAFR*/NF-κB pathway. *Biomed. Pharmacother.* **2019**, *109*, 563–572. [CrossRef] [PubMed]
56. Meng, Q.; Liang, C.; Hua, J.; Zhang, B.; Liu, J.; Zhang, Y.; Wei, M.; Yu, X.; Xu, J.; Shi, S. A miR-146a-5p/TRAF6/NF-kB p65 axis regulates pancreatic cancer chemoresistance: Functional validation and clinical significance. *Theranostics* **2020**, *10*, 3967–3979. [CrossRef]
57. Yao, N.; Chen, Q.; Shi, W.; Tang, L.; Fu, Y. *PARP14* promotes the proliferation and gemcitabine chemoresistance of pancreatic cancer cells through activation of NF-κB pathway. *Mol. Carcinog.* **2019**, *58*, 1291–1302. [CrossRef]
58. Bates, M.; Spillane, C.D.; Gallagher, M.F.; McCann, A.; Martin, C.; Blackshields, G.; Keegan, H.; Gubbins, L.; Brooks, R.; Brooks, D.; et al. The role of the *MAD2-TLR4-MyD88* axis in paclitaxel resistance in ovarian cancer. *PLoS ONE* **2020**, *15*, e0243715. [CrossRef] [PubMed]

59. Chen, Q.; Luo, J.; Wu, C.; Lu, H.; Cai, S.; Bao, C.; Liu, D.; Kong, J. The miRNA-149-5p/MyD88 axis is responsible for ursolic acid-mediated attenuation of the stemness and chemoresistance of non-small cell lung cancer cells. *Environ. Toxicol.* **2020**, *35*, 561–569. [CrossRef]
60. Chen, X.; Zhao, F.; Zhang, H.; Zhu, Y.; Wu, K.; Tan, G. Significance of *TLR4/MyD88* expression in breast cancer. *Int. J. Clin. Exp. Pathol.* **2015**, *8*, 7034–7039.
61. Zhu, X.; Burfeind, K.G.; Michaelis, K.A.; Braun, T.P.; Olson, B.; Pelz, K.R.; Morgan, T.K.; Marks, D.L. *MyD88* signalling is critical in the development of pancreatic cancer cachexia. *J. Cachexia Sarcopenia Muscle* **2019**, *10*, 378–390. [CrossRef] [PubMed]
62. Fitzpatrick, J.M.; Minogue, E.; Curham, L.; Tyrrell, H.; Gavigan, P.; Hind, W.; Downer, E.J. *MyD88*-dependent and -independent signalling via TLR3 and TLR4 are differentially modulated by Δ9-tetrahydrocannabinol and cannabidiol in human macrophages. *J. Neuroimmunol.* **2020**, *343*, 577217. [CrossRef]
63. Haddad, J.J.; Abdel-Karim, N.E. NF-κB cellular and molecular regulatory mechanisms and pathways: Therapeutic pattern or pseudoregulation? *Cell Immunol.* **2011**, *271*, 5–14. [CrossRef] [PubMed]
64. Scagliotti, G.V.; Parikh, P.; von Pawel, J.; Biesma, B.; Vansteenkiste, J.; Manegold, C.; Serwatowski, P.; Gatzemeier, U.; Digumarti, R.; Zukin, M.; et al. Phase III Study Comparing Cisplatin Plus Gemcitabine with Cisplatin Plus Pemetrexed in Chemotherapy-Naive Patients with Advanced-Stage Non-Small-Cell Lung Cancer. *J. Clin. Oncol.* **2023**, *41*, 2458–2466. [CrossRef] [PubMed]
65. Von Hoff, D.D.; Ervin, T.; Arena, F.P.; Chiorean, E.G.; Infante, J.; Moore, M.; Seay, T.; Tjulandin, S.A.; Ma, W.W.; Saleh, M.N.; et al. Increased survival in pancreatic cancer with nab-paclitaxel plus gemcitabine. *N. Engl. J. Med.* **2013**, *369*, 1691–1703. [CrossRef] [PubMed]
66. Liu, K.; Geng, Y.; Wang, L.; Xu, H.; Zou, M.; Li, Y.; Zhao, Z.; Chen, T.; Xu, F.; Sun, L.; et al. Systematic exploration of the underlying mechanism of gemcitabine resistance in pancreatic adenocarcinoma. *Mol. Oncol.* **2022**, *16*, 3034–3051. [CrossRef] [PubMed]
67. Choi, Y.H.; Yu, A.M. ABC transporters in multidrug resistance and pharmacokinetics, and strategies for drug development. *Curr. Pharm. Des.* **2014**, *20*, 793–807. [CrossRef] [PubMed]
68. Hashem, S.; Ali, T.A.; Akhtar, S.; Nisar, S.; Sageena, G.; Ali, S.; Al-Mannai, S.; Therachiyil, L.; Mir, R.; Elfaki, I.; et al. Targeting cancer signaling pathways by natural products: Exploring promising anti-cancer agents. *Biomed. Pharmacother.* **2022**, *150*, 113054. [CrossRef]
69. Newman, D.J.; Cragg, G.M. Natural Products as Sources of New Drugs over the Nearly Four Decades from 01/1981 to 09/2019. *J. Nat. Prod.* **2020**, *83*, 770–803. [CrossRef]
70. Lingel, A.; Ehlers, E.; Wang, Q.; Cao, M.; Wood, C.; Lin, R.; Zhang, L. Kaposi's Sarcoma-Associated Herpesvirus Reduces Cellular Myeloid Differentiation Primary-Response Gene 88 (*MyD88*) Expression via Modulation of Its RNA. *J. Virol.* **2016**, *90*, 180–188. [CrossRef]
71. Yuan, Q.; Gu, J.; Zhang, J.; Liu, S.; Wang, Q.; Tian, T.; Chen, Z.; Zhang, J. *MyD88* in myofibroblasts enhances colitis-associated tumorigenesis via promoting macrophage M2 polarization. *Cell Rep.* **2021**, *34*, 108724. [CrossRef] [PubMed]
72. Yuan, Q.; Zhang, J.; Liu, Y.; Chen, H.; Liu, H.; Wang, J.; Niu, M.; Hou, L.; Wu, Z.; Chen, Z.; et al. MyD88 in myofibroblasts regulates aerobic glycolysis-driven hepatocarcinogenesis via ERK-dependent PKM2 nuclear relocalization and activation. *J. Pathol.* **2022**, *256*, 414–426. [CrossRef] [PubMed]
73. Zhu, G.; Cheng, Z.; Huang, Y.; Zheng, W.; Yang, S.; Lin, C.; Ye, J. MyD88 mediates colorectal cancer cell proliferation, migration and invasion via NF-κB/AP-1 signaling pathway. *Int. J. Mol. Med.* **2020**, *45*, 131–140. [CrossRef] [PubMed]
74. Jung, H.; Kim, J.S.; Kim, W.K.; Oh, K.J.; Kim, J.M.; Lee, H.J.; Han, B.S.; Kim, D.S.; Seo, Y.S.; Lee, S.C.; et al. Intracellular annexin A2 regulates NF-κB signaling by binding to the p50 subunit: Implications for gemcitabine resistance in pancreatic cancer. *Cell Death Dis.* **2015**, *6*, e1606. [CrossRef] [PubMed]
75. Yu, C.; Chen, S.; Guo, Y.; Sun, C. Oncogenic *TRIM31* confers gemcitabine resistance in pancreatic cancer via activating the NF-κB signaling pathway. *Theranostics* **2018**, *8*, 3224–3236. [CrossRef]
76. Ianevski, A.; Giri, A.K.; Aittokallio, T. SynergyFinder 3.0: An interactive analysis and consensus interpretation of multi-drug synergies across multiple samples. *Nucleic Acids Res.* **2022**, *50*, W739–W743. [CrossRef]
77. Raimondi, G.; Mato-Berciano, A.; Pascual-Sabater, S.; Rovira-Rigau, M.; Cuatrecasas, M.; Fondevila, C.; Sanchez-Cabus, S.; Begthel, H.; Boj, S.F.; Clevers, H.; et al. Patient-derived pancreatic tumour organoids identify therapeutic responses to oncolytic adenoviruses. *EBioMedicine* **2020**, *56*, 102786. [CrossRef]

Disclaimer/Publisher's Note: The statements, opinions and data contained in all publications are solely those of the individual author(s) and contributor(s) and not of MDPI and/or the editor(s). MDPI and/or the editor(s) disclaim responsibility for any injury to people or property resulting from any ideas, methods, instructions or products referred to in the content.

pharmaceuticals

MDPI

Review

Unlocking the Green Gold: Exploring the Cancer Treatment and the Other Therapeutic Potential of Fucoxanthin Derivatives from Microalgae

Fatouma Mohamed Abdoul-Latif [1,*], Ayoub Ainane [2], Ibrahim Houmed Aboubaker [1,3], Ali Merito Ali [1], Houda Mohamed [1,3], Pannaga Pavan Jutur [4] and Tarik Ainane [2,*]

1 Center for Research and Study of Djibouti, Medicinal Research Institute, Djibouti City P.O. Box 486, Djibouti
2 Superior School of Technology, University of Sultan Moulay Slimane, P.O. Box 170, Khenifra 54000, Morocco
3 Peltier Hospital of Djibouti, Djibouti City P.O. Box 2123, Djibouti
4 Omics of Algae Group, Industrial Biotechnology, International Centre for Genetic Engineering and Biotechnology, Aruna Asaf Ali Marg, New Delhi 110067, India; pavan.jutur@icgeb.org
* Correspondence: fatoumaabdoulatif@gmail.com (F.M.A.-L.); t.ainane@usms.ma (T.A.)

Abstract: Fucoxanthin, a carotenoid widely studied in marine microalgae, is at the heart of scientific research because of its promising bioactive properties for human health. Its unique chemical structure and specific biosynthesis, characterized by complex enzymatic conversion in marine organisms, have been examined in depth in this review. The antioxidant, anti-inflammatory, and anti-cancer activities of fucoxanthin have been rigorously supported by data from in vitro and in vivo experiments and early clinical trials. Additionally, this review explores emerging strategies to optimize the stability and efficacy of fucoxanthin, aiming to increase its solubility and bioavailability to enhance its therapeutic applications. However, despite these potential benefits, challenges persist, such as limited bioavailability and technological obstacles hindering its large-scale production. The medical exploitation of fucoxanthin thus requires an innovative approach and continuous optimization to overcome these barriers. Although further research is needed to refine its clinical use, fucoxanthin offers promising potential in the development of natural therapies aimed at improving human health. By integrating knowledge about its biosynthesis, mechanisms of action, and potential beneficial effects, future studies could open new perspectives in the treatment of cancer and other chronic diseases.

Keywords: anti-cancer; biochemistry; fucoxanthin; prospects; therapeutic; microalgae

1. Introduction

Tiny microorganisms known as microalgae, often unnoticed by the naked eye, have garnered significant interest from researchers worldwide in recent decades [1]. Their appeal stems from their remarkable potential in medicine and sustainable biotechnology, aligning perfectly with the global pursuit of healthy lifestyles and sustainable development [2]. These microscopic organisms in seawater or freshwater play an important role in marine ecosystems and offer numerous opportunities to satisfy the increasing demand for natural resource-based products [3]. One of the most fascinating aspects of microalgae is their biological diversity, as they are rich in bioactive compounds of various secondary metabolites, including polysaccharides, lipids, proteins, enzymes, vitamins, and carotenoids [4]. These substances have diverse applications across multiple sectors, such as medicine, pharmaceuticals, nutraceuticals, cosmetics, and energy. Extensive research has validated the potential of microalgae as a reliable and environmentally friendly raw material for producing valuable resources, paving the way for new advancements in biotechnology [4].

Yellow to orange-red terpenoid pigments, called carotenoids, are synthesized by specific organisms, particularly microalgae. These substances can act as antioxidants by neutralizing free radicals [5]. Carotenoids are divided into two categories: xanthophylls, which contain oxygen, and carotenes, which are pure hydrocarbons devoid of oxygen. Some xanthophylls produced by higher plants, such as violaxanthin, antheraxanthin, zeaxanthin, neoxanthin, and lutein, can also be synthesized by green microalgae [6]. However, these have additional xanthophylls such as loroxanthin, astaxanthin, and canthaxanthin [7]. Diatom microalgae can produce diatoxanthin, diadinoxanthin, and fucoxanthin [8]. Microalgae such as *Haematococcus pluvialis, Dunaliella salina, Blakesleatrispora* sp., *Desmodesmus* sp., *Euglena gracilis, Tisochrysis lutea, Isochrysis galbana, Phaeodactylum tricornutum*, and *Cylindrotheca closterium* are among the most studied due to their ability to produce carotenoids in significant quantities [9,10]. These microorganisms synthesize a range of rare carotenoids, opening the way to innovative industrial and scientific applications. Carotenoids, especially fucoxanthin, stand out for their powerful antioxidant and anti-inflammatory properties, which benefit human health by helping to combat various chronic diseases [11].

Fucoxanthin, a marine carotenoid found in the photosynthetic cells of diatoms, is gaining significant interest due to its numerous health benefits [12]. Research has highlighted its effectiveness in weight management, cardiovascular health, and blood sugar regulation, and its promising potential in cancer prevention [13]. From a biotechnology perspective, fucoxanthin is extensively used in producing dietary supplements, nutraceuticals, and cosmetics, underscoring its vital role in the health and wellness industry [14]. Beyond medical applications, fucoxanthin shows promise in sustainable production and the development of value-added by-products [15]. The cultivation of microalgae for the extraction of fucoxanthin offers a promising prospect for eco-compatible production, circumventing the environmental challenges frequently linked to the scale-up of this molecule.

This comprehensive review also examines fucoxanthin derivatives from microalgae and their anticancer and other medical implications. It delves into fucoxanthin's complex chemistry and biochemistry and its essential role in enhancing human health. Additionally, it addresses future challenges and potential prospects for using and further developing fucoxanthin and its derivatives in various applications.

2. Chemistry of Fucoxanthin and Its Derived Compounds

Fucoxanthin, a distinctive xanthophyll with the chemical formula $C_{42}H_{58}O_6$, is an essential secondary metabolite in the chloroplasts of various aquatic species, including microalgae, giving them brown or olive green colors [16]. Its molecular structure, close to dinoxanthin, peridinin, and neoxanthin (Figure 1), is distinguished by an unusual allenic bond, nine conjugated double bonds, and an epoxide at position 5,6, as well as several oxygenated functional groups, including alcohols, epoxides, ketones, and esters (Figure 2) [17].

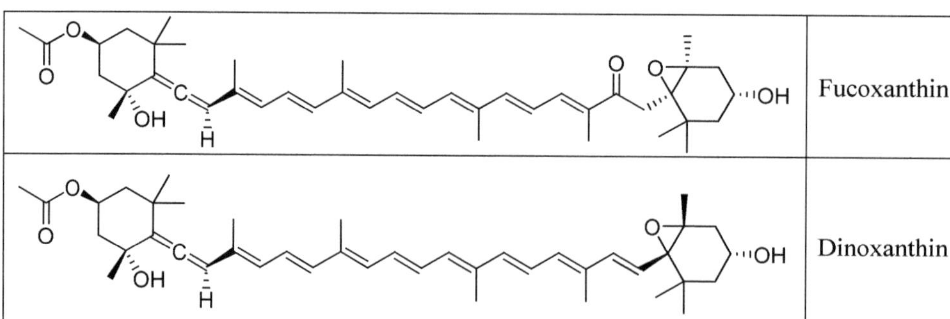

Figure 1. *Cont.*

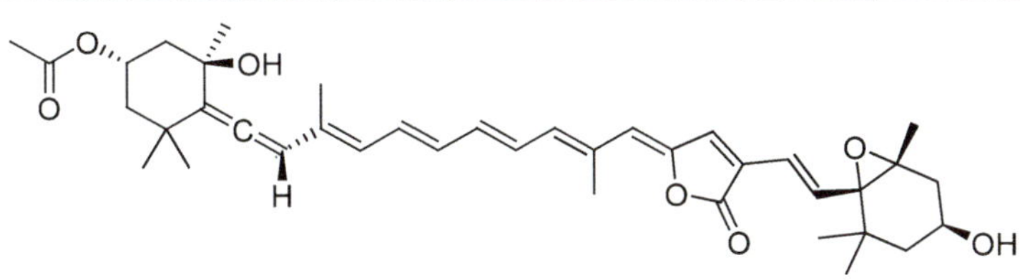	Peridinin
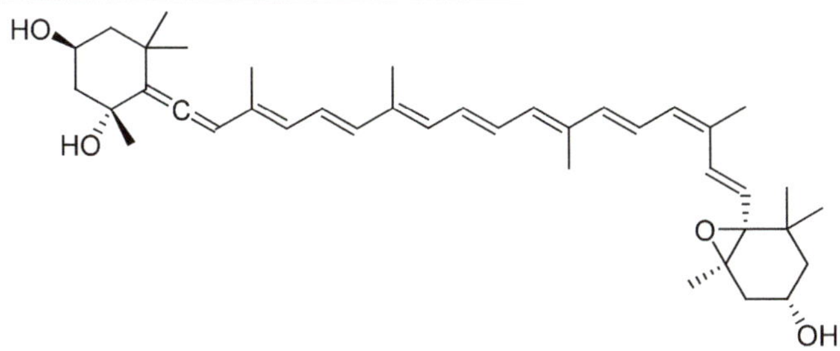	Neoxanthin

Figure 1. Fucoxanthin, Dinoxanthin, Peridinin, and Neoxanthin structures.

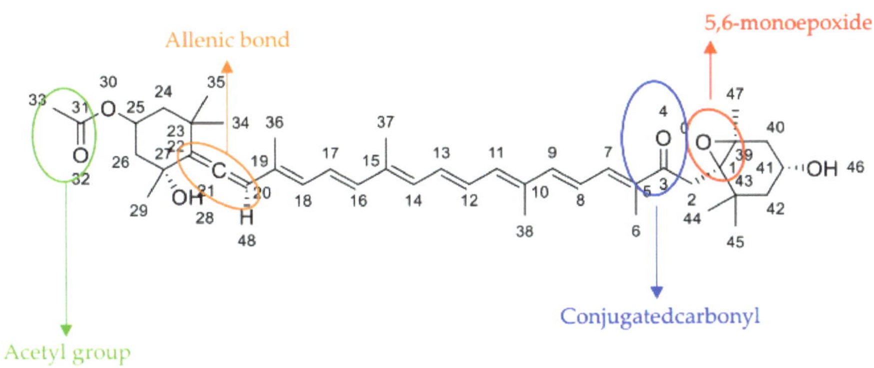

Figure 2. Key characterized groups and bands of Fucoxanthin.

The intricate chemical composition and molecular conformation of fucoxanthin significantly influence its physicochemical and biological properties. Its effectiveness in various biological processes, including its biochemical mechanisms in living organisms, is critically dependent on storage conditions and the formation of other isomers, which is essential for its potential medical and pharmaceutical applications [18].

Like other carotenoids, fucoxanthin is prone to degradation during storage due to temperature, light, oxygen, enzymes, unsaturated lipids, and other pro-oxidant agents. Isomerization, a common occurrence, can lead to the formation of cis isomers influenced by processing conditions, environment, and the specific type of carotenoid [19]. During the purification process of fucoxanthin, three prominent peaks are typically observed, corresponding to the transform and two cis isomers, with an increased proportion of cis isomers at higher extraction temperatures. The sensitivity of fucoxanthin to heat results in the degradation of its total form and the all-trans form when heated between 25 and 100 °C. High temperatures also promote the formation of the cis-13 and cis-15 isomers and degrade the cis-9 isomer. The extraction of fucoxanthin from Undaria pinnatifida algae by an ultrasonic process using rapeseed oil as a solvent made it possible to obtain an oil rich in fucoxanthin. High internal phase emulsions (HIPEs), stabilized by 2% gelatin (GE)-based microgels, exhibit a small, uniform particle size and a stable microstructure. These HIPEs show excellent viscoelasticity, high recovery rate, and thixotropy, thus optimizing their performance for 3D food printing. HIPEs protect fucoxanthin from chemical degradation and improve its bioaccessibility by promoting the release of free fatty acids (FFA) during simulated in vitro digestion. The use of rapeseed oil as an extraction solvent offers a green and stable alternative to traditional organic solvents, showing strong potential for the future design of oil-soluble bioactives and flavors [20].

The metabolic transformation of fucoxanthin produces two major derivatives: fucoxanthinol and amarouciaxanthin A (Figure 3) [21]. Fucoxanthinol, the major metabolite generated by hydrolysis in the gastrointestinal tract shortly after ingestion, plays an essential role in the biological effects of fucoxanthin. Conversely, amarouciaxanthin A comes mainly from fucoxanthinol present in the liver. Studies in mice have shown that the conversion of fucoxanthinol to amarouciaxanthin A occurs primarily in liver microsomes and requires $NAD(P)^+$ as a cofactor. Additionally, fucoxanthinol added to culture medium via HepG2 cells has also been shown to convert to amarouciaxanthin A, indicating that this metabolic pathway is active in cultured human cells [22].

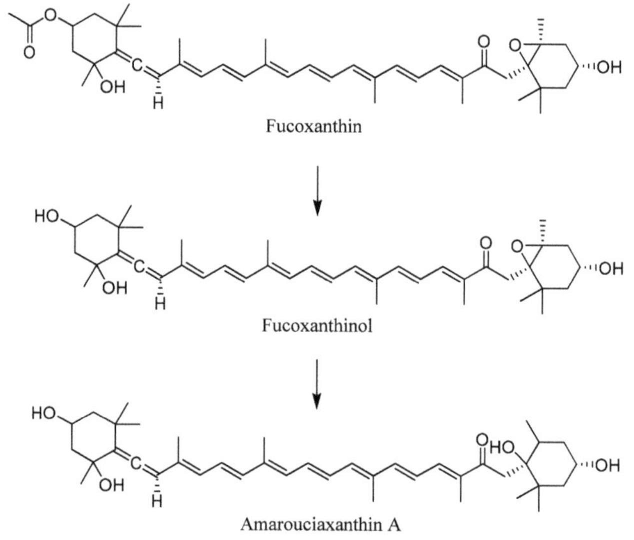

Figure 3. The chemical structures of the two fucoxanthin derivatives: fucoxanthinol and amarouciaxanthin A.

3. Microalgae as a Source of Fucoxanthin

Microalgae, although often unknown to the general public, captivate researchers with their distinctive properties and exceptional diversity. These microscopic organisms, usually smaller than 50 µm, play a vital role in the stability of aquatic ecosystems as photosynthetic organisms. Their ability to transform sunlight, carbon dioxide, and water into organic compounds makes them essential primary producers, contributing to biomass and nutrient cycling [23].

Microalgae are notable for their ability to synthesize diverse bioactive compounds, including carotenoids, which are essential for photosynthesis and beneficial to human health due to their antioxidant properties [24]. Among these carotenoids, fucoxanthin is particularly valuable in the medical and pharmaceutical fields. Found in the chloroplasts of various microalgae species such as *Isochrysis* sp. and *Tisochrysis* sp., fucoxanthin is a xanthophyll that accumulates in the cytosolic lipid droplets of marine diatoms like *Phaeodactylum tricornutum* [25,26]. As an antenna pigment in light-harvesting complexes, fucoxanthin efficiently transfers energy to the photosynthetic electron transport chain, absorbing light with high efficiency and transferring it to chlorophyll a (>80%). This energy transfer is primarily due to a spectral overlap between the donor and acceptor components of the complex [27]. Additionally, the short excited-state lifetime of fucoxanthin facilitates rapid energy transfer, as indicated by low fluorescence. Fucoxanthin also plays a role in protecting against excessive light and has potent antioxidant activity. Fucoxanthin chlorophyll proteins, complexes that combine fucoxanthin, chlorophyll a, and chlorophyll c with proteins, are the main light-harvesting complexes in microalgae. While similar complexes are present in brown algae, they adapt differently to high light intensities through various non-photochemical quenching mechanisms. In macroalgae, fucoxanthin chlorophyll proteins include violaxanthin and zeaxanthin via the violaxanthin cycle, whereas in diatom microalgae, they bind to diadinoxanthin and diatoxanthin through the diadinoxanthin-diatoxanthin cycle for photoprotection (Figure 4) [28].

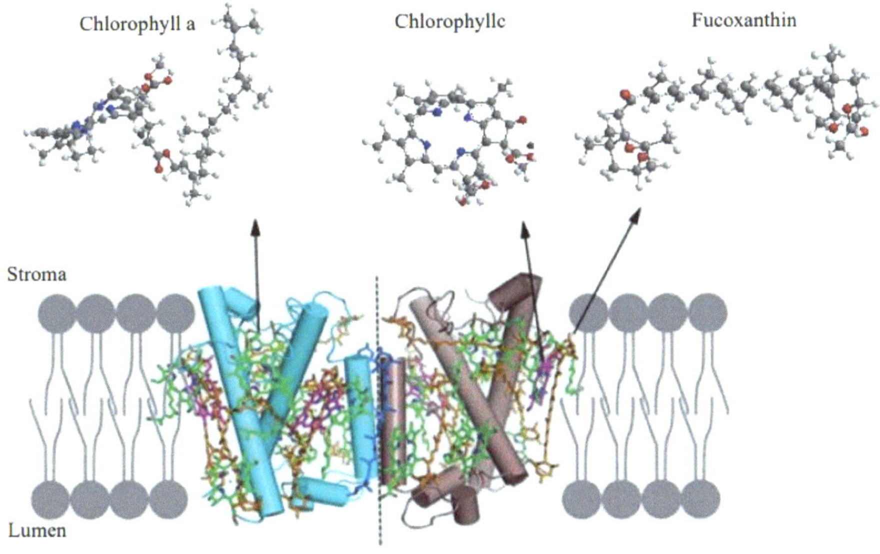

Figure 4. Model of the structure of fucoxanthin-chlorophyll proteins (FCP) in thylakoid membranes.

Recent studies have documented varying concentrations of fucoxanthin in different microalgae strains, highlighting these organisms' diversity and biotechnological potential (Table 1). Ishika et al. (2017) [29] reported 1.2 mg/g fucoxanthin in *Amphora* sp., while *Chaetoceros muelleri* presented 2.92 mg/g. Tokushima et al. (2016) [30] recorded a 12.0 mg/g concentration in *Cyclotella cryptica*, highlighting its capacity to store this pigment. Similarly, Ryabushko et al. (2017) [31] observed a concentration of 15.2 mg/g in *Cylindrotheca closterium*, while Wang et al. (2018) [32] noted levels varying from 15.2 to 25.5 mg/g among different strains of this species.

Popovich et al. (2020) [33] reported an exceptionally high concentration of 38 mg/g of fucoxanthin in *Halamphora coffeaeformis*, underscoring its potential as a valuable source of this carotenoid. This intra-species diversity emphasizes the importance of carefully selecting microalgae strains to optimize fucoxanthin production for biotechnological applications. Research by Wu et al. (2016) [34] on *Phaeodactylum tricornutum* also demonstrated significant variability in fucoxanthin concentrations, ranging from 5.5 mg/g to a broad range of 9.9 to 59.2 mg/g, as noted by McClure et al. (2018) [35], again highlighting the intra-species diversity of this carotenoid. Studies on other species, such as *Isochrysis zhangjiangensis* by Li et al. (2019) [36] and *Odontella aurita* by Xia et al. (2013) [37], confirmed their potential as rich sources of fucoxanthin, with concentrations of 23.3 mg/g and 18.47 mg/g, respectively.

Mohamadnia et al. (2021) [38] optimized the culture medium for *Tisochrysis lutea*, achieving a remarkable fucoxanthin concentration of 79.40 mg/g. This significant advancement demonstrates the effectiveness of optimizing culture conditions to enhance the production of this carotenoid.

Table 1. Variability of fucoxanthin concentrations in the main strains of microalgae.

Microalgae Species	Fucoxanthin (mg/g)	Reference
Amphora sp.	1.2	[29]
Chaetoceros muelleri	2.9	[29]
Phaeodactylum tricornutum	5.5	[34]
Cyclotella cryptica	12.0	[30]
Odontella aurita	18.5	[37]
Isochrysis zhangjiangensis	23.3	[36]
Cylindrotheca closterium	25.5	[32]
Halamphora coffeaeformis	38.0	[33]
Phaeodactylum tricornutum	59.2	[35]
Tisochrysis lutea	79.4	[38]

4. Biosynthesis of Fucoxanthin

Fucoxanthin biosynthesis in microalgae is a complex and multifactorial mechanism involving a series of enzymatic reactions [39]. This process begins with forming carotenoid precursors such as β-carotene and producing fucoxanthin. Each step in this metabolic pathway is catalyzed by specific enzymes that add, modify, or rearrange functional groups to form the distinct structure of fucoxanthin [40–45]. The main stages of the fucoxanthin formation mechanism are described below (Figure 5):

1. The initial step in fucoxanthin biosynthesis involves the production of the C5 building blocks, isopentenyl pyrophosphate (IPP) or its isomer, dimethylallyl diphosphate (DMAPP). These molecules can be synthesized from acetyl-CoA through the cytosolic mevalonic acid (MVA) pathway or from pyruvate and glyceraldehyde-3-phosphate (G3P) via the plastid methylerythritol 4-phosphate pathway (MEP). Although both pathways yield the same end product, carotenoid synthesis predominantly utilizes IPP or DMAPP derived from the MEP pathway.
2. The formation of phytoene marks the initial specific step in the carotenoid pathway. This process is catalyzed by phytoene synthase (PSY), which facilitates the head-to-head condensation of two geranylgeranyl diphosphate (GGPP) molecules to produce phytoene. Phytoene is pivotal as it serves as the first distinct intermediate in this biosynthetic pathway.
3. The conversion of phytoene into lycopene follows next and involves several enzymes. Phytoene desaturase (PDS), ζ-carotene desaturase (ZDS), and carotene isomerase (CRTISO) sequentially desaturate phytoene to generate ζ-carotene, subsequently leading to the formation of lycopene. Lycopene represents the initial colored carotenoid produced in this biosynthesis pathway.
4. Subsequently, lycopene undergoes cyclization to form β-carotene, a process catalyzed by lycopene β-cyclase (LCYB). β-carotene is an important precursor in fucoxanthin biosynthesis, marking a significant step in converting linear carotenoids into cyclic compounds.
5. β-carotene is then hydroxylated to produce zeaxanthin. This biosynthetic step involves β-carotene hydroxylase (CHYB), which introduces hydroxyl groups to β-carotene, forming β-cryptoxanthin and subsequently zeaxanthin.

6. Zeaxanthin undergoes epoxidation to form antheraxanthin and then violaxanthin. Zeaxanthin epoxidase (ZEP) catalyzes this reaction by adding epoxy groups to zeaxanthin, thereby transforming the carotenoids into more complex and functional forms.
7. Violaxanthin is further converted into neoxanthin by neoxanthin synthase (NSY). This transformation represents another critical step in the biosynthetic pathway, laying the groundwork for subsequent modifications that lead to the formation of fucoxanthin.
8. Finally, neoxanthin undergoes a series of intricate enzymatic modifications to synthesize fucoxanthin. These reactions involve various enzymes, including hydroxylases, epoxidases, and specific isomerases, which introduce hydroxyl and epoxy groups and facilitate structural rearrangements necessary for forming fucoxanthin's unique allene bond.

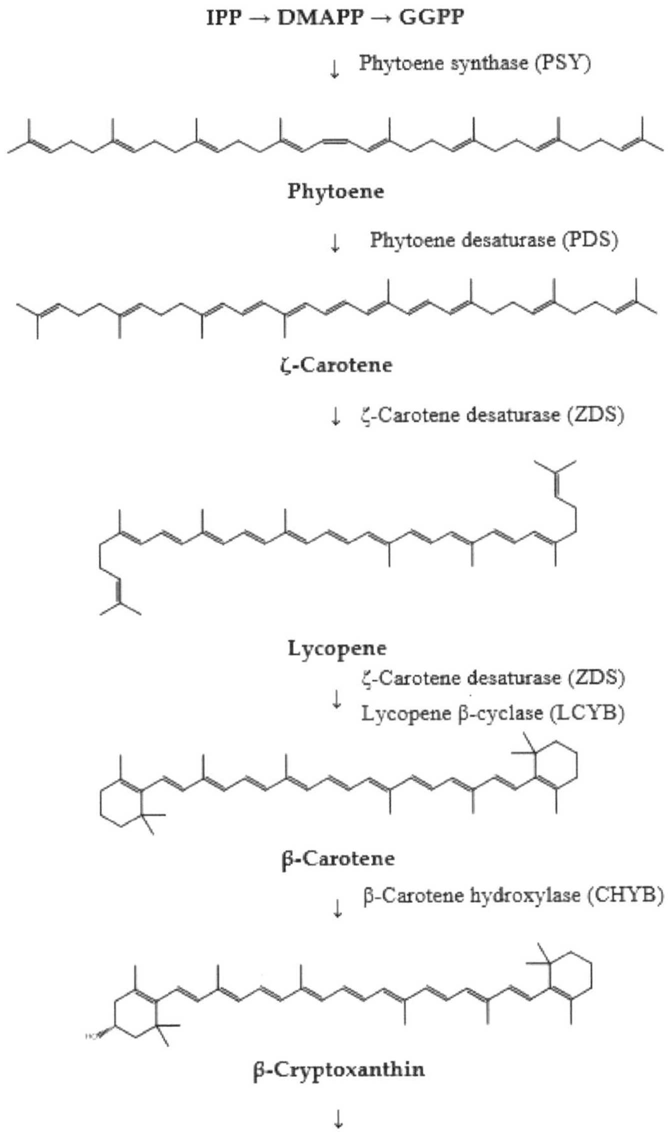

Figure 5. *Cont.*

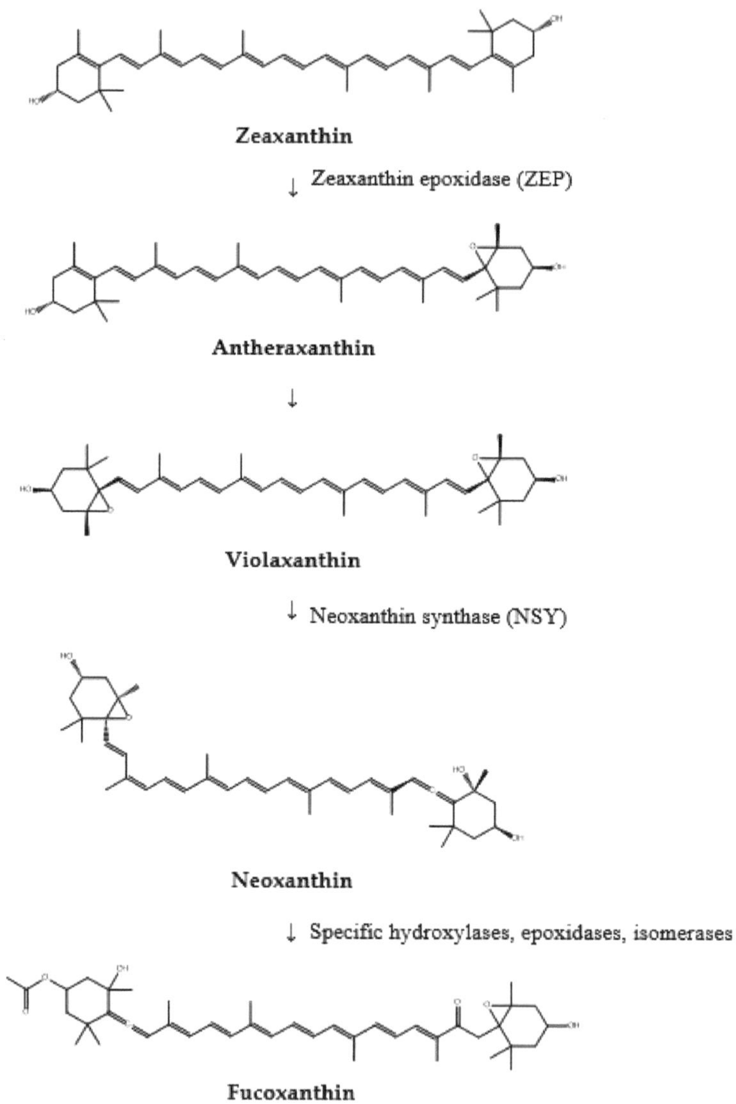

Figure 5. Main steps of the fucoxanthin biosynthesis pathway in the majority of microalgae species.

5. Anticancer Activities of Fucoxanthin

The anticancer properties of fucoxanthin are multiple and diverse (Table 2), impacting various types of neoplasms with specific mechanisms of apoptosis induction and cell cycle regulation [46]. Its selective properties, sparing healthy cells while targeting neoplastic cells, make it particularly promising for developing new oncological therapies.

Glioma is the most common primary tumor of the central nervous system. Fucoxanthin is cytotoxic to human glioma cell lines U87 and U251, with no inhibitory effect on normal neurons. It induces apoptosis of glioblastoma multiforme cells by increasing the levels of pro-apoptotic proteins, including cleaved PARP, caspase-3, and caspase-9, and reducing the accumulation of phosphorylated Akt and mTOR proteins. Control experiments with

PI3K inhibitors confirmed that fucoxanthin-induced apoptosis is linked to inhibition of the PI3K/Akt/mTOR pathway [47].

Table 2. Main anticancer activities of fucoxanthin.

Organs	Mechanisms of Action
Brain	• Induction of apoptosis of glioma cells. • Inhibition of the PI3K/Akt/mTOR pathway.
Breasts	• Inhibition of tumor angiogenesis. • Reduced viability of breast cancer cells. • Protection against doxorubicin cardiotoxicity.
Lungs	• Prevention of the development of non-small cell lung cancer. • Induction of apoptosis by regulation of the expressions of p53, p21, Fas, PUMA, Bcl-2, and caspase-3/8. • Cell cycle inhibition and reduction of mortalin–p53 interaction. • Reduction of proteins linked to proliferation, survival, and metastasis of cancer cells.
Cervix	• Synergy with TRAIL to increase apoptosis. • Targeting the PI3K/Akt/NF-κB pathway.
Colon	• Induction of DNA fragmentation. • Inhibition of tumor growth.
Liver	• Protection of hepatocytes. • Reduction of lipid peroxide metabolites. • Reduction of stress markers.
Blood	• Cytotoxicity for leukemia cells. • Induction of apoptosis by mitochondrial and caspase-3 pathways.
Stomach	• Reduction in the expression of Mcl-1 and STAT3 proteins. • Induction of apoptosis and autophagy.

Breast carcinoma is a common malignancy that seriously threatens the physical and mental health of women. Fucoxanthin has anti-angiogenic effects, inhibiting the proliferation, migration, and formation of tubular structures by human lymphatic endothelial cells (HLEC). It also reduces lymphatic microvessel density in breast carcinoma models, highlighting its potential as an anti-lymphangiogenic agent to prevent metastasis. Fucoxanthin and its metabolite fucoxanthinol lessen the viability of MCF-7 and MDA-MB-231 cells in a dose- and time-dependent manner. The combination of fucoxanthin and adriamycin selectively activates oxidative stress-mediated apoptosis of breast carcinoma cells while protecting cardiomyocytes from the toxic effects of doxorubicin [48].

Fucoxanthin inhibits the proliferation of nasopharyngeal carcinoma cells by inducing autophagy. It also prevents the development of non-small cell lung carcinoma at a relatively safe dose for normal cells by inducing apoptosis of neoplastic cells by regulating the expression of p53, p21, Fas, PUMA, Bcl-2, and caspase-3/8. Fucoxanthin induces the nuclear translocation of p53, stimulating its function as a transcriptional activator in cancer cells and reducing the levels of proteins associated with cancer cell proliferation, survival and metastasis [49].

Cervical carcinoma is currently the most common gynecological malignancy. Fucoxanthin significantly increases apoptosis in cervical cancer cell lines HeLa, SiHa, and CaSki in synergy with the TRAIL protein. TRAIL, a member of the TNF family of cytokines, selectively induces apoptosis of tumor cells without affecting normal cells. TRAIL-mediated apoptosis is enhanced by fucoxanthin via the PI3K/Akt/NF-κB pathway, which increases apoptosis of cervical cancer cells [50].

Colorectal carcinoma is a common cancer of the digestive tract. Fucoxanthin significantly reduced the proliferation of human colorectal cancer cell lines Caco-2, HT-29, and DLD-1 by inducing DNA fragmentation. More recently, it has been reported that fucoxanthin can effectively prevent colorectal cancer in a carcinogenic AOM/DSS mouse model. It inhibits colon damage and promotes resistance to colorectal cancer by activating the anoikis pathway and inhibiting tumor growth [51].

Hepatocellular carcinoma is one of the most common and deadly cancers worldwide. Studies have shown that fucoxanthin has a significant therapeutic effect against diethylnitrosamine-induced liver cancer in rats. It restores normal body weight, serum albumin levels, antioxidant enzymes, liver enzymes, serum bilirubin, and stress markers. Fucoxanthin protects hepatocytes by increasing antioxidant enzymes such as superoxide dismutase and catalase and eliminating lipid peroxide metabolites. Another study demonstrated that fucoxanthin-enriched extracts were cytotoxic to HepG2 liver cancer cells in a dose- and time-dependent manner. Fucoxanthin also reduces the viability of HepG2 cells by inhibiting cyclin D expression and inducing cell cycle arrest in the G_0/G_1 phase [52].

Chronic myelogenous leukemia (CML) is a malignancy resulting from the clonal proliferation of hematopoietic stem cells. Imatinib, although commonly used, has many adverse effects and cannot eradicate the pathogenic gene. Combined with imatinib and doxorubicin, Fucoxanthin showed cytotoxic effects on human leukemia cell lines K562 and TK6. Fucoxanthin also induces apoptosis of human acute promyelocytic leukemia cells, which is related to the early loss of mitochondrial membrane potential and the activation of caspase-3. It influences HEL human erythroid leukemia cells by reducing their viability, increasing cell apoptosis and causing cell cycle arrest in the G_0/G_1 phase. Fucoxanthin opens the mitochondrial membrane permeability transition channel, reducing mitochondrial transmembrane potential and regulating Bcl-2 and Bax gene expression, leading to increased expression of pro-apoptotic caspase-3 [53].

Gastric cancer, with its high morbidity and mortality, represents a significant challenge in oncology. Traditional chemotherapy for this type of cancer lacks specificity and causes substantial damage to normal cells. Fucoxanthin shows promising anticancer activity by reducing Mcl-1 and STAT3 protein expression in SGC-7901 and BGC-823 gastric cancer cells, causing cell cycle arrest in the S phase and apoptosis in the G_2/M phase. It is not toxic to normal cells, which is essential for therapeutic application. Studies have also revealed that fucoxanthin can induce autophagy and apoptosis by increasing the expression of beclin-1, LC3, and cleaved caspase-3 while decreasing the expression of Bcl-2 [54].

In silico studies on the anticancer properties of fucoxanthin have significantly enriched our bioinformatics understanding of its therapeutic potential against cancer. This cutting-edge research used advanced models such as molecular dynamics simulation, molecular docking, ADME (Absorption-Distribution-Metabolism-Excretion) and QSAR (Quantitative Structure–Activity Relationship) analysis of computational chemistry to explore the mechanisms of action of this promising molecule.

Seminal studies by Januar et al. (2012) [55] extensively investigated fucoxanthin's cytotoxic properties, employing an in silico approach that revealed its stable complex formation with tubulin at the colchicine binding site. This discovery opened new avenues for targeted cancer therapies centered on microtubule depolymerization and cell cycle arrest. Furthermore, Jung et al. (2017) [56] examined fucoxanthin's impact on monoamine oxidase-A (MAO-A) and monoamine oxidase-B (MAO-B) through molecular docking studies, demonstrating significant enzyme inhibition. Their findings suggested that fucoxanthin could influence critical metabolic pathways in cancer treatment. Building on these studies, Garg et al. (2019) [57] delved deeper into fucoxanthin's interaction with the p53-mortalin complex using molecular docking and dynamics simulations. Their research confirmed its ability to disrupt essential interactions between p53 and mortalin, reinforcing its potential as a competitive inhibitor in anticancer therapeutic strategies.

Moreover, Dibha et al. (2022) [58] extended the investigation into fucoxanthin's potential in breast cancer by focusing on its interactions with nuclear factor kappa B (NF-kB) via molecular docking simulations. Their analysis underscored essential physicochemical and pharmacokinetic characteristics of fucoxanthin essential for its therapeutic development, including challenges such as limited blood–brain barrier penetration and restricted gastrointestinal absorption. Additionally, Padmi et al. (2023) [59] explored fucoxanthin's potential against melanoma cells by targeting the MC1R receptor, adding a novel dimension to studying its anticancer properties. These collective studies highlight fucoxanthin's diverse mechanisms and potential applications in cancer therapy, underscored by its interactions with key molecular targets and pathways in cancer progression and metabolism.

6. Medical Applications of Fucoxanthin

6.1. Antioxidative Activities

Fucoxanthin possesses a distinctive chemical structure characterized by an allenic bond, an epoxide, and a hydroxyl group. These features endow fucoxanthin with potent antioxidant properties by enabling it to interact effectively with various free radicals. The allenic bond's high reactivity facilitates the neutralization of free radicals, while the epoxide group enhances fucoxanthin's ability to react with reactive oxygen species (ROS). Moreover, the hydroxyl group enhances fucoxanthin's solubility in biological fluids, thereby improving its antioxidant efficacy [60].

Oxidative stress arises from an imbalance between ROS production and the body's antioxidant defense mechanisms. Fucoxanthin mitigates oxidative stress through several mechanisms. It activates antioxidant enzymes such as catalase, superoxide dismutase (SOD), and glutathione peroxidase (GSH-Px), important for ROS breakdown. Additionally, fucoxanthin modulates antioxidant signaling pathways by stimulating Nrf2 (nuclear factor erythroid 2–related factor 2), which enhances the expression of genes encoding antioxidant enzymes like NAD(P)H quinone oxidoreductase 1 (NQO1) [61].

In vitro studies have demonstrated fucoxanthin's robust antioxidant activity, notably its ability to scavenge DPPH radicals, a standard measure of antioxidant capacity. Research using various cell lines such as RAW 264.7, HepG2, Caco-2, and HeLa cells has shown that fucoxanthin reduces cellular metabolic activity and chemiluminescence, indicating a reduction in oxidative reactions, while also elevating intracellular glutathione levels, a critical antioxidant that protects cells from oxidative damage. Fucoxanthin has been shown to inhibit lipid peroxidation, safeguarding cell membranes from oxidative harm [62–64].

In vivo studies corroborate these findings, demonstrating that fucoxanthin administration reduces intracellular ROS levels and enhances antioxidant enzyme activity in animal models. For instance, in mice with asthma induced by ovalbumin or liver injury induced by alcohol, fucoxanthin supplementation increased total antioxidant capacity (T-AOC) and levels of SOD, GSH-Px, and catalase, essential for neutralizing ROS and improving organ antioxidant defenses. Activation of the Nrf2 pathway in these studies underscores fucoxanthin's role in enhancing overall antioxidant defenses [65,66].

Research on human retinal epithelial cells has highlighted fucoxanthin's protective effects against oxidative stress associated with conditions like diabetic retinopathy, demonstrating increased antioxidant enzyme activity and reduced ROS levels [67,68]. In summary, fucoxanthin and its derivatives exhibit potent antioxidant properties through diverse mechanisms (Figure 6), making them promising candidates for therapeutic applications to mitigate oxidative stress-related pathologies.

Figure 6. Recent antioxidant activities of fucoxanthin: in vitro and in vivo research. (created with www.map-this.com accessed on 21 May 2024).

6.2. Antimicrobial Activities

Fucoxanthin is a promising agent in combating bacterial infections due to its distinct chemical and physical properties and therapeutic potential. Fucoxanthin primarily disrupts several critical processes within bacterial cells, reducing their ability to survive and proliferate. One of the main mechanisms of action of fucoxanthin is its effect of increasing the permeability of the bacterial cell membrane [69]. This change in permeability leads to leakage of cellular contents, similar to the effect observed with specific terpenoids such as carvacrol, thereby enhancing the antimicrobial effects of fucoxanthin. Additionally, research indicates that fucoxanthin disrupts oxidative phosphorylation, a process fundamental to cellular respiration in aerobic bacteria. This disruption reduces intracellular oxygen concentration, making the environment inhospitable for bacteria and promoting their death [70]. At the same time, fucoxanthin promotes the accumulation of intracellular free radicals, causing severe oxidative damage to essential cellular components such as membranes, DNA, proteins, and lipids, thus contributing to its effectiveness against bacteria [71]. Another important mechanism by which fucoxanthin exerts its antibacterial activity is the inhibition of biofilm formation [70]. Bacterial biofilms are aggregates of bacteria surrounded by an extracellular matrix, which makes them highly resistant to antibiotic treatments and the host's immune defenses. Fucoxanthin disrupts the formation of this biofilm matrix, decreases bacterial cell adhesion, and inhibits toxin production, thereby weakening bacterial virulence. Additionally, fucoxanthin disrupts cellular quorum sensing communication networks, a key mechanism for coordinating bacterial activities within biofilms, further reducing their resilience.

Particularly notable, fucoxanthin demonstrated exceptional activity against *Mycobacterium tuberculosis*, with MICs as low as 2.8–4.1 μM (equivalent to 1.85–2.7 μg/mL) [72]. This bacterium causes tuberculosis, a severe infectious disease affecting millions of people around the world. Additionally, fucoxanthin showed effectiveness against *Listeria monocytogenes* at a concentration of 1000 μg/mL, indicating its potential to combat this potentially deadly food bacteria [73]. Generally, in vitro studies have demonstrated that fucoxanthin has significant activity against various aerobic bacteria. In a study including more than 30 bacterial species, fucoxanthin showed its effectiveness against 13 aerobic bacteria, with

MICs generally varying between 62.5 and 250 μg/mL for Gram-positive bacteria and between 125 and 500 μg/mL for Gram-negative bacteria [69]. Table 3 summarizes the main results of studies on the antibacterial activity of fucoxanthin against various bacterial strains, highlighting its potential as an effective antimicrobial agent, particularly against the predominant Gram-positive and Gram-negative pathogens. These results highlight the ability of fucoxanthin to effectively target common bacterial pathogens while mitigating the risk of antibiotic resistance [69].

Table 3. Main results of the antibacterial activities of fucoxanthin.

Bacterial Strain	ZOI * (mm)	MIC ** (μg/mL)
Streptococcus agalactiae	12.2 ± 0.7	62.5
Staphylococcus epidermidis	11.2 ± 0.7	125
Staphylococcus aureus	11.0 ± 0.6	125
Escherichia coli	10.2 ± 0.7	125
Streptococcus pyogenes	10.0 ± 0.6	125
Klebsiella oxytoca	9.2 ± 0.7	125–250
Enterococcus faecalis	9.0 ± 0.8	125–250
Streptococcus pneumoniae	9.7 ± 0.5	125
Staphylococcus aureus	11.0 ± 0.6	125
Klebsiella pneumoniae	8.8 ± 0.7	250
Acinetobacter lwoffii	8.2 ± 0.4	250
Pseudomonas aeruginosa	7.5 ± 0.5	250–500
Serratia marcescens	7.3 ± 0.5	500
Proteus mirabilis	7.2 ± 0.4	500
Cutibacterium acnes	6.0 ± 0.1	>1000
Veillonella parvula	6.0 ± 0.1	>1000
Porphyromonas gingivalis	6.0 ± 0.1	>1000

* ZOI: zone of growth inhibition. ** MIC: minimal inhibitory concentration.

Additionally, although fucoxanthin has considerable antibacterial activities, its antifungal properties are limited, according to some studies. More research needs to focus on this latter activity. A survey conducted by Pérez et al. (2016) [74] in vitro revealed that different oil extracts of *Saccharina japonica* and *Sargassum horneri*, rich in fucoxanthin, exhibited excellent antifungal properties against *Candida albicans* and *Aspergillus brasiliensis*. According to the results, the mixed acetone–methanol extract of *Sargassum horneri* showed the most potent antimicrobial activity against these two fungi. Maadane et al. (2017) [75] explored some antifungal activities of marine microalgae from the Moroccan coast. Of these, the strain *Phaeodactylum tricornutum*, known for its significant amounts of fucoxanthin, showed moderate results against *Candida albicans*, while *Aspergillus niger* proved resistant. Furthermore, another study by Peraman and Nachimuthu (2019) [76] examined in vitro seven fucoxanthin-containing microalgae (*Pavlova lutheri*, *Isochrysis galbana*, *Navicula* sp., *Chaetoceros calcitrans*, *Dunaliella salina*, *Thalassiosira* sp., and *Chaetoceros gracilis*) for their antifungal activities against *Aspergillus brasiliensis*, *Aspergillus fumigatus*, and *Candida albicans*. At the minimum inhibitory concentration (MIC) of 40 mg/mL, *Dunaliella salina* showed effective antifungal activity against all three fungi, with growth inhibition percentages of 89.26%, 87.67%, and 81.02% for *Aspergillus brasiliensis*, *Aspergillus fumigatus*, and *Candida albicans*, respectively. Notably, *Chaetoceros gracilis* extracts also demonstrated excellent inhibitory activity.

While research on the antiviral properties of fucoxanthin is limited, several studies have introduced new perspectives in pharmacology. For instance, Tsushima et al. (1995) [77]

conducted an in vitro study using Raji cells, revealing that fucoxanthin and its metabolites can inhibit Epstein–Barr virus activation induced by 12-O-tetradecanoylphorbol-13-acetate (TPA) at low concentrations. Notably, halocynthiaxanthin, a metabolite derived from fucoxanthin found in marine animals, exhibited significant inhibitory activity, albeit with observed cytotoxicity at higher concentrations [78]. Tamama et al. (2020) [79] explored various fucoxanthin-rich algae capable of modulating the ACE/AngII/ATR1 axis, which could be beneficial in treating patients with COVID-19. However, the precise role of fucoxanthin in this context requires further elucidation. Recent studies by Kang et al. have delved into the antiviral effects of fucoxanthin extracted from *Sargassum siliquastrum*. Kang et al. (2023) [80] demonstrated its ability to inhibit severe acute respiratory syndrome coronavirus 2 (SARS-CoV-2) infection in Vero cells without notable toxicity. Computer simulations suggested that fucoxanthin interferes with the binding between angiotensin-converting enzyme 2 and the virus's spike protein in a concentration-dependent manner. Moreover, Kang et al. (2024) [81] investigated fucoxanthin's antiviral activity against the Zika virus in vitro and in silico, highlighting its potential as a promising antiviral agent for emerging viral infections.

6.3. Anti-Inflammatory Activities

Inflammation is the immune system's response to various stimuli, characterized by symptoms such as erythema, edema, pain, and functional changes in tissues (Figure 7). This response can be triggered by different pathogens, including lipopolysaccharide (LPS), a component of the cell walls of Gram-negative bacteria. Fucoxanthin, known for its significant anti-inflammatory properties, is attracting increasing interest and has been extensively studied for its effects on inflammatory processes and potential therapeutic applications [82].

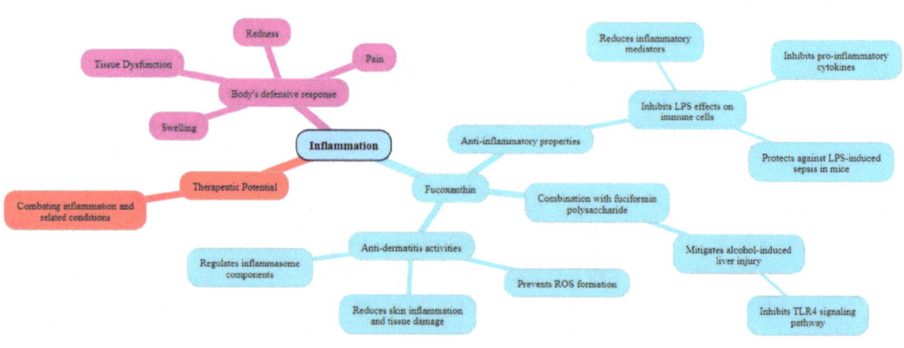

Figure 7. Anti-inflammatory activities of fucoxanthin. (created with www.map-this.com accessed on 22 February 2024).

LPS binds to the CD14 receptor, forming immune complexes that interact with Toll-like receptors (TLRs), triggering an intracellular signaling cascade and the release of various inflammatory mediators [83]. Fucoxanthin has demonstrated significant anti-inflammatory and antioxidant effects on LPS-activated murine RAW 264.7 cells. Research shows that fucoxanthin inhibits iNOS expression and COX-2 transcription, reducing NO and PGE2 production. Additionally, it dose-dependently inhibits the stimulation of pro-inflammatory cytokines such as TNF-α and IL-6 and prevents LPS-induced loss of cell viability and mitochondrial membrane potential [84].

Another experimental study demonstrated that fucoxanthin significantly reduced the increased expression of IL-1β, TNF-α, iNOS, and COX-2, attenuating inflammation in LPS-stimulated macrophages [85]. This mechanism involves suppression of IκB-α phosphorylation, MAPKs, and Akt, thereby leading to inactivation of NF-κB and inhibiting LPS-induced secretion of NO and PGE2 by macrophages. These signaling pathways, essential for regulating inflammation and immune responses, play a good role in the anti-inflammatory effects of fucoxanthin.

In vivo studies using mouse models of LPS-induced sepsis confirmed that fucoxanthin decreases the levels of inflammatory cytokines like IL-6 and IL-1β by regulating the NF-κB signaling pathway, thus suggesting its therapeutic potential against sepsis and acute inflammation [86]. Additionally, after excessive alcohol consumption, LPS can activate the TLR4 receptor, thereby amplifying inflammation and leading to liver damage. Administration of fucoxanthin combined with fuciformin polysaccharide significantly inhibited the TLR4 signaling pathway, thereby reducing alcohol-induced liver damage [87].

In addition to its effectiveness against LPS-induced inflammation, fucoxanthin shows promising activities against dermatitis. Combined with rosmarinic acid (RA), it reduces UVB-induced apoptosis of HaCaT keratinocytes by regulating inflammasome components and preventing ROS formation while increasing the expression of antioxidant genes such as Nrf2 and HO-1 [88]. In mouse models, fucoxanthin creams showed significant efficacy against tissue plasminogen activator (TPA)-induced hyperplasia and UVB-induced acute erythema, thereby reducing skin edema, epidermal thickness, MPO activity, and COX-2 expression. Furthermore, fucoxanthin pretreatment improved the symptoms of UVB-induced corneal denervation and trigeminal neuralgia by increasing Nrf2 expression [89].

6.4. Anti-Obesity Activity

Obesity has become a major public health issue on a global scale, leading to significant consequences on the metabolic health of individuals (Table 4) [90]. This chronic condition, characterized by excessive accumulation of adipose tissue, is associated with an increased risk of developing several serious metabolic disorders, such as type 2 diabetes, cardiovascular disease, and fatty liver disease. Faced with this growing concern, fucoxanthin and its derivatives are emerging as a promising molecule in the fight against obesity. Several recent studies have highlighted the beneficial effects of fucoxanthin on reducing body fat in obese individuals, sparking growing interest in its therapeutic potential [91].

Table 4. Mechanisms of action of fucoxanthin against obesity.

Mechanism of Action Enzymatic	Description	Reactions Involved
Regulation of UCP1	Stimulates the expression of Uncoupling Protein 1 (UCP1) in brown adipose tissue (BAT).	Unspecified
Browning of white adipose tissue (WAT)	Transforms some white adipose tissue depots into brown adipose tissue via induction of UCP1.	Unspecified
Inhibition of adipogenesis	Inhibits adipocyte differentiation by negatively regulating adipogenic genes like PPARγ.	PPARγ inhibition Pancreatic lipase inhibition
Reduced fatty acid synthesis	Increases phosphorylation of AMP-activated protein kinase (AMPK).	Reduced acetyl-CoA carboxylase activity
Inhibition of digestive enzymes	Inhibits the enzymes α-amylase and α-glucosidase, reducing glucose absorption.	Inhibition of α-amylase Inhibition of α-glucosidase

The mechanisms of action of fucoxanthin against obesity are diverse and complementary. First, fucoxanthin stimulates the expression of Uncoupling Protein 1 (UCP1) in brown adipose tissue (BAT). This protein is essential because it promotes metabolic heat production by dissipating energy, which helps reduce body fat accumulation. Another key mechanism is the browning of white adipose tissue (WAT). Fucoxanthin can transform white adipose tissue deposits into brown adipose tissue, thereby increasing the ability of these cells to oxidize fatty acids and generate heat. This process, known as browning, represents an innovative approach to treating obesity by improving lipid metabolism [92,93].

Furthermore, fucoxanthin and its metabolite, fucoxanthinol, inhibit adipocyte differentiation by negatively regulating gene expression in adipogenesis, such as PPARγ. These compounds also show a significant ability to inhibit pancreatic lipase activity, which reduces the absorption of lipids from the diet and thus limits the accumulation of body fat [94]. Regulation of fatty acid synthesis is another important aspect of fucoxanthin action. By increasing the phosphorylation of AMP-activated protein kinase (AMPK) and inhibiting the activity of acetyl-CoA carboxylase, fucoxanthin helps reduce fatty acid synthesis, thereby improving overall lipid metabolism. Finally, fucoxanthin can inhibit digestive enzymes such as α-amylase and α-glucosidase, which limits the absorption of carbohydrates and contributes to better body weight management in obese individuals. This enzyme inhibition could play an important role in the prevention of type 2 diabetes and other obesity-related metabolic complications [95].

6.5. Antidiabetic Activity

Fucoxanthin is gaining attention for its potential therapeutic benefits in managing diabetes, a disease whose global prevalence is projected to increase significantly by 2045 [96]. While diabetes cannot currently be cured, efforts are concentrated on early detection and lifestyle changes to mitigate its complications and mortality rates. Type 2 diabetes pathophysiology is closely linked to obesity, marked by chronic low-grade inflammation and insulin resistance in adipose tissues. Fucoxanthin addresses these issues on multiple fronts. In obesity, hypertrophied adipocytes release proinflammatory cytokines like TNF-α and IL-6, exacerbating systemic inflammation and insulin resistance [97]. Preclinical studies indicate that fucoxanthin can suppress the production of these cytokines, thereby reducing oxidative stress and enhancing insulin sensitivity. By inhibiting MCP-1 production, fucoxanthin also limits macrophage infiltration into adipose tissue, further improving insulin resistance associated with obesity and type 2 diabetes [98].

Fucoxanthin also directly affects carbohydrate metabolism. In animal models fed high-fat diets and diabetic mice (KK-Ay, db/db), fucoxanthin supplementation has shown significant reductions in blood glucose, plasma insulin, and HbA1C compared to controls [99]. These benefits stem from fucoxanthin's ability to promote GLUT4 translocation in skeletal muscle and adipose tissues, facilitating glucose uptake and utilization. Moreover, fucoxanthin alters adipokine levels in the body. It increases the secretion of adiponectin, an anti-inflammatory adipokine that regulates lipid and glucose metabolism, while decreasing levels of resistin, a pro-inflammatory adipokine linked to insulin resistance and obesity. This modulation helps enhance insulin sensitivity and reduce diabetes-related complications [100].

Fucoxanthin also influences various biological and metabolic parameters critical in diabetes management. It enhances hepatic glucokinase activity, promoting glucose breakdown and glycogen synthesis in the liver to stabilize blood sugar levels and prevent postprandial glucose spikes [101]. Compared to traditional pharmacological treatments like oral hypoglycemic agents and insulin, fucoxanthin offers the advantage of being a natural compound derived from marine algae, potentially offering a safer therapeutic alternative with fewer side effects. Nevertheless, further clinical studies are necessary to assess its efficacy and safety in humans fully.

6.6. Other Medical Applications

Fucoxanthin, known for its diverse therapeutic properties, reveals various pharmacological activities beyond its remarkable applications as an anticancer agent, antioxidant, antimicrobial, anti-inflammatory, and other vital effects. These characteristics have prompted extensive research to explore its specific applications in different healthcare fields.

The neuroprotective effects of fucoxanthin have been demonstrated by its ability to safeguard brain cells against oxidative damage induced by agents such as hydrogen peroxide (H_2O_2) and β-amyloid oligomers, biomarkers associated with Alzheimer's disease. Activation of the PI3K/Akt signaling pathway and inhibition of the ERK pathway have reduced inflammation and oxidative stress [102].

Antidepressant properties of fucoxanthin have also been demonstrated, reversing depressive behaviours in animal models exposed to lipopolysaccharide (LPS). This process was mediated by regulation of the AMPK-NF-κB signaling pathway, which decreased the production of pro-inflammatory cytokines and reduced the expression of pro-inflammatory enzymes in the brain [103].

Kidney disease, a major public health issue, has been addressed with the significant renoprotective effects of fucoxanthin, attenuating diabetic nephropathy by reducing oxidative stress and fibrosis via the Akt/Sirt1/FoxO3α signaling pathway. In ethylene glycol-treated albino rats, fucoxanthin normalized biochemical markers and decreased oxidative stress, lipid peroxidation, and tubular damage. In streptozotocin (STZ)-induced diabetic rat models, renal function improvements and mitochondrial integrity protection were observed [104].

Fucoxanthin also favorably influences the lipid profile by increasing the proportion of essential fatty acids like docosahexaenoic acid (DHA) in the liver, thereby reducing the risk of cardiovascular diseases. Protective efficacy against stroke and ischemic neuronal injury was noted, mitigating damage caused by hypoxia and reperfusion [105].

A promising potential of fucoxanthin as an anti-osteoporotic agent has been demonstrated, particularly in postmenopausal women and older people. In ovariectomized (OVX) rat models, bone resorption and inflammatory markers were reduced. Osteoclast differentiation and activity were inhibited via modulation of MAP kinase and Nrf2 signaling pathways [106].

Significant antifibrotic effects of fucoxanthin have been noted in conditions such as pulmonary fibrosis and nasal polyps. A reduction in the expression of proteins such as smooth muscle actin (α-SMA), collagen type I (Col-1), fibronectin, and interleukin-6 (IL-6), induced by TGF-β1, was noted. These effects suppressed phosphorylated MAPK signaling pathways, PI3K/Akt, Akt/SP-1, and Smad2/Smad3 pathways [107].

The anti-hyperuricemic activity of fucoxanthin regulated uric acid transport in the kidneys by modulating the levels of proteins involved in this process, such as GLUT9 and URAT1. This regulation helped lower serum uric acid levels, thereby preventing hyperuricemia, a risk factor for gout and other metabolic complications [108].

The skin, the first barrier against external aggressions, is particularly vulnerable to UV radiation. Fucoxanthin reduced the levels of inflammatory cytokines such as TNF-α and IL-6 via the Nrf2 signaling pathway. In atopic dermatitis models, topical application of fucoxanthin inhibited eosinophil infiltration and IL-33 expression while increasing the expression of regulatory cytokines. ROS production and photodegradation reduction have demonstrated a photoprotective potential against UVA and UVB radiation [109].

7. Bioavailability, Stability and Toxicity

The bioavailability of fucoxanthin is essential because it directly influences its effectiveness in the body, depending on factors such as stability, bioaccessibility, and toxicity. Fucoxanthin's resistance to photolysis, thermodegradation, and oxidation poses a major challenge for its application in medical products and pharmaceutical treatments, potentially compromising its stability and efficacy under adverse environmental conditions. Although fucoxanthin is stable as a free molecule, in vitro digestibility studies show progressive degradation in the gastric simulator (10%) to the ileum (20%), where it is transformed into fucoxanthinol. [89,110] Various strategies are being explored to improve its solubility and stability in the gastrointestinal tract, including encapsulation in lipid nanoparticles, cyclodextrin complexes, and specialized emulsions, such as those based on whey proteins, which are promising in preserving its therapeutic potential.

Concerning toxicity, initial studies indicate that fucoxanthin has low toxicity and good tolerability at therapeutic doses. Research by Beppu et al. (2009) [111] in mice and rats revealed no mortality or histological abnormalities after oral exposure up to 2000 mg/kg/day for 30 days. However, an increase in plasma cholesterol was observed. Subsequent evaluations, including the Ames test and oral toxicity studies, confirmed the absence of mutagenicity, genotoxicity, and toxicity at lethal doses (LD50 > 2000 mg/kg), as well as the absence of abnormalities of internal organs at doses up to 200 mg/kg. Available data suggest that fucoxanthin has a promising safety profile, although continued monitoring is essential for a full assessment of its clinical utility. Further studies will be needed to assess its long-term safety and potential effects at high doses.

To date, although several literature reviews have been published on the various pharmacological properties of fucoxanthin, the distinct pharmacokinetic profile of fucoxanthin at the clinical and preclinical levels is not well documented [112–114]. The pharmacokinetics of fucoxanthin aims to show how a substance changes after administration, passing through the phases of absorption, distribution, metabolism, and excretion [109]. In humans, the absorption of fucoxanthin isolated from Laminaria japonica after oral administration has been little studied [115]. Available studies mainly focus on animal models, such as mice, where fucoxanthin has been shown to be rapidly converted to fucoxanthinol and amarouciaxanthin A in blood plasma [116]. One study reported that the metabolites fucoxanthinol and amarouciaxanthin A reach peak plasma concentrations after 4 h, with a peak concentration of fucoxanthinol twice that of amarouciaxanthin A [22]. However, it is essential to note that the mechanisms observed in rodents are not necessarily identical in humans due to significant interspecies differences in carotenoid metabolic pathways. Thus, additional clinical studies are needed to accurately determine the pharmacokinetics of fucoxanthin in humans, particularly with regard to its intestinal absorption and metabolism.

8. Challenges and Prospects of Microalgal Fucoxanthin

Fucoxanthin, a prominent carotenoid in certain microalgae, is increasingly recognized for its diverse pharmacological properties. This bioactive compound has garnered attention for its potential as a natural and renewable source, particularly its antioxidant and anticancer effects. However, despite its promising therapeutic potential, fucoxanthin faces several complex challenges that hinder its full utilization in the medical field:

- Efficient Extraction: Extracting fucoxanthin from microalgae efficiently remains a significant hurdle. The carotenoid's relatively low intracellular concentration necessitates sophisticated and economically viable extraction methods to achieve high yields. Moreover, purifying fucoxanthin to a degree suitable for medical applications requires advanced chromatography and filtration techniques.
- Stability Issues: Fucoxanthin is prone to physicochemical changes under light, heat, and oxygen, compromising its bioavailability and effectiveness in pharmaceutical and nutraceutical formulations. Therefore, developing effective stabilization methods is important to maintain its therapeutic potential.
- High Production Costs: The cost of producing fucoxanthin commercially is prohibitively high. Microalgae cultivation, harvesting, extraction, and purification processes are economically intensive. Reducing these costs while ensuring product quality is essential for broader adoption in medical and nutritional industries.
- Regulatory Standards: Establishing rigorous regulatory standards is essential to ensure fucoxanthin products' safety, quality, and effectiveness. Harmonized regulations are needed to instill confidence among consumers and healthcare professionals in this promising natural substance.
- Clinical Evaluation: While preclinical studies have demonstrated diverse therapeutic potential, rigorous clinical trials are essential to evaluate fucoxanthin's efficacy in

humans. These studies will determine optimal doses, administration routes, and potential adverse effects associated with medical use.

- Despite these challenges, promising prospects are emerging for fucoxanthin:
- Nutritional and Metabolic Health: Fucoxanthin may contribute to managing body weight and metabolic disorders such as hyperuricemia and hyperlipidemia, suggesting broader applications beyond traditional medicine.
- Formulation Innovations: Innovations in formulation technologies, such as encapsulation in nanoparticles and liposomes, aim to enhance the stability and bioavailability of fucoxanthin, thereby improving its absorption and effectiveness.
- Sustainability: Microalgae, as primary sources of fucoxanthin, offer a sustainable alternative to conventional sources of active compounds. Their controlled cultivation reduces environmental impact compared to agricultural or synthetic methods, supporting sustainable production practices.
- Education and Awareness: Educating consumers and healthcare professionals about the potential benefits and proper use of fucoxanthin is essential for widespread adoption. This includes raising awareness about quality challenges and regulatory considerations associated with products containing this carotenoid.

In summary, while fucoxanthin faces significant hurdles, ongoing research and innovation hold promise for unlocking its full potential in both therapeutic and nutritional applications.

9. Conclusions

Fucoxanthin, a bioactive carotenoid found abundantly in microalgae, has garnered significant attention for its diverse therapeutic potential across various medical fields. Its promising anticancer properties are particularly intriguing to researchers striving to advance innovative therapies for human health improvement. Beyond its anticancer effects, fucoxanthin exhibits many beneficial properties, including antioxidant, antimicrobial, anti-inflammatory, anti-obesity, antidiabetic, neuroprotective, and antidepressant activities. Current medical research is focused on elucidating the mechanisms through which fucoxanthin operates and exploring its synergistic interactions with other bioactive compounds. These efforts aim to optimize its beneficial effects in stable pharmaceutical formulations, facilitating precise and practical clinical applications. The functional versatility of fucoxanthin, combined with its abundant presence in microalgae biomass, positions it as a renewable and potentially sustainable source of this valuable compound. This aspect holds promise for developing innovative and enduring therapeutic solutions within modern medical technology.

Author Contributions: Conceptualization: F.M.A.-L. and T.A.; Data curation: A.A. and F.M.A.-L., Formal analysis: I.H.A., H.M. and T.A., Funding acquisition: F.M.A.-L.; Investigation: F.M.A.-L.; Methodology: A.A. and F.M.A.-L.; Project administration: F.M.A.-L.; Resources: P.P.J. and A.M.A.; Software: I.H.A., H.M. and A.M.A.; Supervision: F.M.A.-L.; Validation: F.M.A.-L. and T.A.; Visualization: A.A., F.M.A.-L. and T.A.; Roles/Writing—original draft: F.M.A.-L. and T.A.; Writing—review & editing: F.M.A.-L., P.P.J. and T.A. All authors have read and agreed to the published version of the manuscript.

Funding: This research received no external funding.

Conflicts of Interest: The authors declare no conflict of interest.

References

1. Yadav, K.; Vasistha, S.; Srivastava, A.; Rai, M.P. Microalgae as sustainable feedstock for biofuel production and value-added co-products. In *Microalgal Biomass for Bioenergy Applications*; Woodhead Publishing: Cambridge, UK, 2024; pp. 253–286.
2. Occhipinti, P.S.; Russo, N.; Foti, P.; Zingale, I.M.; Pino, A.; Romeo, F.V.; Randazzo, C.L.; Caggia, C. Current challenges of microalgae applications: Exploiting the potential of non-conventional microalgae species. *J. Sci. Food Agric.* **2024**, *104*, 3823–3833. [CrossRef] [PubMed]

3. Omoregie, I.P.; Ihotu, K.-E.I.; Osagie, A.D. Marine Green Microalgae Biomass Production and Application. In *Marine Greens*; CRC Press: Boca Raton, FL, USA, 2024; pp. 93–101.
4. Bürck, M.; Ramos, S.D.P.; Braga, A.R.C. Enhancing the Biological Effects of Bioactive Compounds from Microalgae through Advanced Processing Techniques: Pioneering Ingredients for Next-Generation Food Production. *Foods* **2024**, *13*, 1811. [CrossRef] [PubMed]
5. Duan, X.; Xie, C.; Hill, D.R.A.; Barrow, C.J.; Dunshea, F.R.; Martin, G.J.O. Bioaccessibility, bioavailability and bioactivities of carotenoids in microalgae: A review. *Food Rev. Int.* **2024**, *40*, 230–259. [CrossRef]
6. Yin, R.; Zhuang, G.; Lei, Y.; Han, J.; Li, Y.; Zhang, J.; Yan, X. Valorization of Nannochloropsis oceanica for integrated co-production of violaxanthin cycle carotenoids. *Bioresour. Technol.* **2024**, *399*, 130597. [CrossRef] [PubMed]
7. Kneip, J.S.; Kniepkamp, N.; Jang, J.; Mortaro, M.G.; Jin, E.; Kruse, O.; Baier, T. CRISPR/Cas9-Mediated Knockout of the Lycopene ε-Cyclase for Efficient Astaxanthin Production in the Green Microalga Chlamydomonas reinhardtii. *Plants* **2024**, *13*, 1393. [CrossRef] [PubMed]
8. Lin, C.-H.; Chang, Y.-F.; Prasetya, S.J.; Yu, F.-Y.; Lai, S.-Y.; Wang, M.-Y. An integrated process for enhanced production and purification of fucoxanthin and sulfated polysaccharides in diatom Hyalosynedra toxoneides cultures. *J. Taiwan Inst. Chem. Eng.* **2024**, *155*, 105308. [CrossRef]
9. Razz, S.A. Comprehensive overview of microalgae-derived carotenoids and their applications in diverse industries. *Algal Res.* **2024**, *78*, 103422. [CrossRef]
10. Huang, H.; Lang, Y.; Zhou, M. A comprehensive review on medical applications of microalgae. *Algal Res.* **2024**, *80*, 103504. [CrossRef]
11. Gong, B.; Ma, S.; Yan, Y.; Wang, Z. Progress on the biological characteristics and physiological activities of fucoxanthin produced by marine microalgae. *Front. Mar. Sci.* **2024**, *11*, 1357425. [CrossRef]
12. Zhang, Z.; Wei, Z.; Xue, C. Delivery systems for fucoxanthin: Research progress, applications and future prospects. *Crit. Rev. Food Sci. Nutr.* **2024**, *64*, 4643–4659. [CrossRef]
13. Kumarasinghe, H.S.; Gunathilaka, M.D.T.L. A systematic review of fucoxanthin as a promising bioactive compound in drug development. *Phytochem. Lett.* **2024**, *61*, 52–65. [CrossRef]
14. Pang, Y.; Song, W.; Fan, L. Progress in the Preparation and Application of Fucoxanthin from Algae. *Int. J. Biol. Life Sci.* **2024**, *5*, 46–51. [CrossRef]
15. García-García, P.; Ospina, M.; Señoráns, F.J. Tisochrysis lutea as a source of omega-3 polar lipids and fucoxanthin: Extraction and characterization using green solvents and advanced techniques. *J. Appl. Phycol.* **2024**, *2024*, 1–12. [CrossRef]
16. Phong, W.N.; Regmi, R.; Adhikari, K.C.; Bhattarai, R.R.; Show, P.L. Bioactive Compounds from Diatoms. In *Diatoms Biotechnology*; CRC Press: Boca Raton, FL, USA, 2023; pp. 20–37.
17. Mc Gee, D.; Gillespie, E. The bioactivity and chemotaxonomy of microalgal carotenoids. *Biodivers. Chemotaxon.* **2019**, *2019*, 215–237.
18. Gomez-Zavaglia, A.; Barros, L.; Prieto, M.A.; Cassani, L. Recent Progress in Understanding the Impact of Food Processing and Storage on the Structure–Activity Relationship of Fucoxanthin. *Foods* **2023**, *12*, 3167. [CrossRef]
19. Bardakci, H.; Yücel, Ç.; Karatoprak, G.Ş.; Küpeli Akkol, E.; Hakan Barak, T.; Sobarzo-Sánchez, E. Oxidative stress and marine carotenoids: Application by using nanoformulations. *Mar. Drugs* **2020**, *18*, 423. [CrossRef]
20. Wang, C.; Jin, M.; Sun, C.; Bai, Y.; Dong, X.; Qi, H. Rapeseed oil as the extraction solvent motivates fucoxanthin-loaded high internal phase emulsion preparation for food 3D printing. *LWT* **2023**, *187*, 115349. [CrossRef]
21. Sun, H.; Yang, S.; Zhao, W.; Kong, Q.; Zhu, C.; Fu, X.; Zhang, F.; Liu, Z.; Zhan, Y.; Mou, H.; et al. Fucoxanthin from marine microalgae: A promising bioactive compound for industrial production and food application. *Crit. Rev. Food Sci. Nutr.* **2023**, *63*, 7996–8012. [CrossRef]
22. Asai, A.; Sugawara, T.; Ono, H.; Nagao, A. Biotransformation of fucoxanthinol into amarouciaxanthin A in mice and HepG2 cells: Formation and cytotoxicity of fucoxanthin metabolites. *Drug Metab. Dispos.* **2004**, *32*, 205–211. [CrossRef]
23. Doolaanea, A.A.; Alfatama, M.; Alkhatib, H.; Mawazi, S.M. Fucoxanthin: Structure, Sources, Properties and Benefits. In *Handbook of Food Bioactive Ingredients: Properties and Applications*; Springer International Publishing: Cham, Switzerland, 2023; pp. 1–27.
24. Mapelli-Brahm, P.; Gómez-Villegas, P.; Gonda, M.L.; León-Vaz, A.; León, R.; Mildenberger, J.; Rebours, C.; Saravia, V.; Vero, S.; Vila, E.; et al. Microalgae, seaweeds and aquatic bacteria, archaea, and yeasts: Sources of carotenoids with potential antioxidant and anti-inflammatory health-promoting actions in the sustainability era. *Mar. Drugs* **2023**, *21*, 340. [CrossRef]
25. Li, D.; Chen, N.; Zhou, Z.; Zhong, C.; Fang, J.; Huang, L.; Sun, H.; Chi, C.; Zhou, Y.; He, Y. Isochrysis sp. cultivation in pilot-scale to concurrently produce sustainable triacylglycerols for human milk fat substitutes and fucoxanthin. *Algal Res.* **2023**, *69*, 102937. [CrossRef]
26. Truong, T.Q.; Park, Y.J.; Koo, S.Y.; Choi, J.-H.; Enkhbayar, A.; Song, D.-G. Interdependence of fucoxanthin biosynthesis and fucoxanthin-chlorophyll a/c binding proteins in Phaeodactylum tricornutum under different light intensities. *J. Appl. Phycol.* **2023**, *35*, 25–42. [CrossRef]
27. Zheng, M.; Zhou, C.; Wang, W.; Kuang, T.; Shen, J.; Tian, L. Origin of energy dissipation in the oligomeric fucoxanthin-chlorophyll a/c binding proteins. *J. Phys. Chem. Lett.* **2023**, *14*, 7967–7974. [CrossRef]

28. Agostini, A.; Bína, D.; Carbonera, D.; Litvín, R. Conservation of triplet-triplet energy transfer photoprotective pathways in fucoxanthin chlorophyll-binding proteins across algal lineages. *Biochim. Et Biophys. Acta (BBA)-Bioenerg.* **2023**, *1864*, 148935. [CrossRef]
29. Ishika, T.; Moheimani, N.R.; Bahri, P.A.; Laird, D.W.; Blair, S.; Parlevliet, D. Halo-adapted microalgae for fucoxanthin production: Effect of incremental increase in salinity. *Algal Res.* **2017**, *28*, 66–73. [CrossRef]
30. Tokushima, H.; Inoue-Kashino, N.; Nakazato, Y.; Masuda, A.; Ifuku, K.; Kashino, Y. Advantageous characteristics of the diatom Chaetoceros gracilis as a sustainable biofuel producer. *Biotechnol. Biofuels* **2016**, *9*, 235. [CrossRef]
31. Ryabushko, V.I.; Zheleznova, S.N.; Nekhoroshev, M.V. Effect of nitrogen on fucoxanthin accumulation in the diatom Cylindrotheca closterium (Ehrenb.) Reimann et Lewin. *Int. J. Algae* **2017**, *19*, 79–84. [CrossRef]
32. Wang, S.; Verma, S.K.; Hakeem Said, I.; Thomsen, L.; Ullrich, M.S.; Kuhnert, N. Changes in the fucoxanthin production and protein profiles in Cylindrotheca closterium in response to blue light-emitting diode light. *Microb. Cell Factories* **2018**, *17*, 110. [CrossRef]
33. Popovich, C.A.; Faraoni, M.B.; Sequeira, A.; Daglio, Y.; Martín, L.A.; Martínez, A.M.; Damiani, M.C.; Matulewicz, M.C.; Leonardi, P.I. Potential of the marine diatom Halamphora coffeaeformis to simultaneously produce omega-3 fatty acids, chrysolaminarin and fucoxanthin in a raceway pond. *Algal Res.* **2020**, *51*, 102030. [CrossRef]
34. Wu, H.; Li, T.; Wang, G.; Dai, S.; He, H.; Xiang, W. A comparative analysis of fatty acid composition and fucoxanthin content in six Phaeodactylum tricornutum strains from diff erent origins. *Chin. J. Oceanol. Limnol.* **2016**, *34*, 391–398. [CrossRef]
35. McClure, D.D.; Luiz, A.; Gerber, B.; Barton, G.W.; Kavanagh, J.M. An investigation into the effect of culture conditions on fucoxanthin production using the marine microalgae Phaeodactylum tricornutum. *Algal Res.* **2018**, *29*, 41–48. [CrossRef]
36. Li, Y.; Sun, H.; Wu, T.; Fu, Y.; He, Y.; Mao, X.; Chen, F. Storage carbon metabolism of Isochrysis zhangjiangensis under different light intensities and its application for co-production of fucoxanthin and stearidonic acid. *Bioresour. Technol.* **2019**, *282*, 94–102. [CrossRef]
37. Xia, S.; Wang, K.; Wan, L.; Li, A.; Hu, Q.; Zhang, C. Production, characterization, and antioxidant activity of fucoxanthin from the marine diatom Odontella aurita. *Mar. Drugs* **2013**, *11*, 2667–2681. [CrossRef]
38. Mohamadnia, S.; Tavakoli, O.; Faramarzi, M.A.; Shamsollahi, Z. Production of fucoxanthin by the microalga Tisochrysis lutea: A review of recent developments. *Aquaculture* **2020**, *516*, 734637. [CrossRef]
39. Seth, K.; Kumar, A.; Rastogi, R.P.; Meena, M.; Vinayak, V. Bioprospecting of fucoxanthin from diatoms—Challenges and perspectives. *Algal Res.* **2021**, *60*, 102475. [CrossRef]
40. Kwon, D.Y.; Vuong, T.T.; Choi, J.; Lee, T.S.; Um, J.-I.; Koo, S.Y.; Hwang, K.T. Fucoxanthin biosynthesis has a positive correlation with the specific growth rate in the culture of microalga *Phaeodactylum tricornutum*. *J. Appl. Phycol.* **2021**, *33*, 1473–1485. [CrossRef]
41. Bai, Y.; Cao, T.; Dautermann, O.; Buschbeck, P.; Cantrell, M.B.; Chen, Y.; Lein, C.D.; Shi, X.; Ware, M.A.; Yan, F. Green diatom mutants reveal an intricate biosynthetic pathway of fucoxanthin. *Proc. Natl. Acad. Sci. USA* **2022**, *119*, e2203708119. [CrossRef]
42. Shen, X.; Pan, K.; Zhang, L.; Zhu, B.; Li, Y.; Han, J. A DUF4281 domain-containing protein (homologue of ABA4) of Phaeodactylum tricornutum regulates the biosynthesis of fucoxanthin. *Algal Res.* **2022**, *65*, 102728. [CrossRef]
43. Cao, T.; Bai, Y.; Buschbeck, P.; Tan, Q.; Cantrell, M.B.; Chen, Y.; Jiang, Y.; Liu, R.-Z.; Ries, N.K.; Shi, X.; et al. An unexpected hydratase synthesizes the green light-absorbing pigment fucoxanthin. *Plant Cell* **2023**, *35*, 3053–3072. [CrossRef] [PubMed]
44. Sun, H.; Wang, J.; Li, Y.; Yang, S.; Di Chen, D.; Tu, Y.; Liu, J.; Sun, Z. Synthetic biology in microalgae towards fucoxanthin production for pharmacy and nutraceuticals. *Biochem. Pharmacol.* **2023**, 115958. [CrossRef] [PubMed]
45. Song, J.; Zhao, H.; Zhang, L.; Li, Z.; Han, J.; Zhou, C.; Xu, J.; Li, X.; Yan, X. The heat shock transcription factor PtHSF1 mediates triacylglycerol and fucoxanthin synthesis by regulating the expression of GPAT3 and DXS in *Phaeodactylum tricornutum*. *Plant Cell Physiol.* **2023**, *64*, 622–636. [CrossRef] [PubMed]
46. Méresse, S.; Fodil, M.; Fleury, F.; Chénais, B. Fucoxanthin, a marine-derived carotenoid from brown seaweeds and microalgae: A promising bioactive compound for cancer therapy. *Int. J. Mol. Sci.* **2020**, *21*, 9273. [CrossRef] [PubMed]
47. Salami, M.; Salami, R.; Aarabi, M.H.; Mafi, A.; Ghorbanhosseini, S.S.; Shafabakhsh, R.; Asemi, Z. Targeting glioma cells with nutraceuticals: Therapeutic effects based on molecular mechanisms, new evidence and perspectives. *Mini Rev. Med. Chem.* **2023**, *23*, 1167–1192. [CrossRef] [PubMed]
48. Lau, T.Y.; Kwan, H.Y. Fucoxanthin is a potential therapeutic agent for the treatment of breast cancer. *Mar. Drugs* **2022**, *20*, 370. [CrossRef] [PubMed]
49. Ahmed, S.A.; Mendonca, P.; Elhag, R.; Soliman, K.F. Anticancer effects of fucoxanthin through cell cycle arrest, apoptosis induction, angiogenesis inhibition, and autophagy modulation. *Int. J. Mol. Sci.* **2022**, *23*, 16091. [CrossRef] [PubMed]
50. Moga, M.A.; Dima, L.; Balan, A.; Blidaru, A.; Dimienescu, O.G.; Podasca, C.; Toma, S. Are bioactive molecules from seaweeds a novel and challenging option for the prevention of HPV infection and cervical cancer therapy?—A review. *Int. J. Mol. Sci.* **2021**, *22*, 629. [CrossRef] [PubMed]
51. Manochkumar, J.; Doss CG, P.; Efferth, T.; Ramamoorthy, S. Tumor preventive properties of selected marine pigments against colon and breast cancer. *Algal Res.* **2022**, *61*, 102594. [CrossRef]
52. Sangavi, P.; Langeswaran, K.; KumaR, S.G.O. Anticarcinogenic Efficacy of Fucoxanthin on HepG2 Cell Lines. *J. Clin. Diagn. Res.* **2022**, *16*, XC05–XC09. [CrossRef]
53. Almeida, T.P.; Ramos, A.A.; Ferreira, J.; Azqueta, A.; Rocha, E. Bioactive compounds from seaweed with anti-leukemic activity: A mini-review on carotenoids and phlorotannins. *Mini Rev. Med. Chem.* **2020**, *20*, 39–53. [CrossRef]

54. Wang, J.; Dong, X.; Li, D.; Fang, Z.; Wan, X.; Liu, J. Fucoxanthin inhibits gastric cancer lymphangiogenesis and metastasis by regulating Ran expression. *Phytomedicine* **2023**, *118*, 154926. [CrossRef]
55. Januar, H.I.; Dewi, A.S.; Marraskuranto, E.; Wikanta, T. In silico study of fucoxanthin as a tumor cytotoxic agent. *J. Pharm. Bioallied Sci.* **2012**, *4*, 56–59. [CrossRef] [PubMed]
56. Jung, H.A.; Roy, A.; Choi, J.S. In vitro monoamine oxidase A and B inhibitory activity and molecular docking simulations of fucoxanthin. *Fish. Sci.* **2017**, *83*, 123–132. [CrossRef]
57. Garg, S.; Afzal, S.; Elwakeel, A.; Sharma, D.; Radhakrishnan, N.; Dhanjal, J.K.; Sundar, D.; Kaul, S.C.; Wadhwa, R. Marine carotenoid fucoxanthin possesses anti-metastasis activity: Molecular evidence. *Mar. Drugs* **2019**, *17*, 338. [CrossRef] [PubMed]
58. Dibha, A.F.; Wahyuningsih, S.; Ansori AN, M.; Kharisma, V.D.; Widyananda, M.H.; Parikesit, A.A.; Sibero, M.T.; Probojati, R.T.; Murtadlo, A.A.; Trinugroho, J.P.; et al. Utilization of secondary metabolites in algae Kappaphycus alvarezii as a breast cancer drug with a computational method. *Pharmacogn. J.* **2022**, *14*, 536–543. [CrossRef]
59. Padmi, H.; Fitriani, F.; Saraswati PB, A.; Nurhidayati, S.Z.; Ilhami BT, K.; Sunarwidhi, A.L.; Widyastuti, S.; Prasedya, E.S. Anti-melanoma potential of fucoxanthin by inhibition MC1R receptor: In Silico study. In *AIP Conference Proceedings*; AIP Publishing: New York, NY, USA, 2023; Volume 2956.
60. Miyashita, K.; Beppu, F.; Hosokawa, M.; Liu, X.; Wang, S. Bioactive significance of fucoxanthin and its effective extraction. *Biocatal. Agric. Biotechnol.* **2020**, *26*, 101639. [CrossRef]
61. Mumu, M.; Das, A.; Emran, T.B.; Mitra, S.; Islam, F.; Roy, A.; Dasc, M.K.R.; Park, M.N.; Chandran, D.; Sharma,, R. Fucoxanthin: A promising phytochemical on diverse pharmacological targets. *Front. Pharmacol.* **2022**, *13*, 929442. [CrossRef]
62. Zeng, J.; Zhang, Y.; Ruan, J.; Yang, Z.; Wang, C.; Hong, Z.; Zuo, Z. Protective effects of fucoxanthin and fucoxanthinol against tributyltin-induced oxidative stress in HepG2 cells. *Environ. Sci. Pollut. Res.* **2018**, *25*, 5582–5589. [CrossRef] [PubMed]
63. Neumann, U.; Derwenskus, F.; Flaiz Flister, V.; Schmid-Staiger, U.; Hirth, T.; Bischoff, S.C. Fucoxanthin, a carotenoid derived from Phaeodactylum tricornutum exerts antiproliferative and antioxidant activities in vitro. *Antioxidants* **2019**, *8*, 183. [CrossRef]
64. Younas, U.; Tehseen, S.; Khan, F.; Niaz, K. Brown Algae (Fucoxanthin) Against Cancer. In *Nutraceuticals and Cancer Signaling: Clinical Aspects and Mode of Action*; Springer: Cham, Switzerland, 2021; pp. 99–127.
65. Shen, Y.; Zhang, H.; Cheng, L.; Wang, L.; Qian, H.; Qi, X. In vitro and in vivo antioxidant activity of polyphenols extracted from black highland barley. *Food Chem.* **2016**, *194*, 1003–1012. [CrossRef]
66. Zhang, X.; Xiong, Z.; Zhang, S.; Li, K.; Bu, Y.; Zheng, N.; Zhao, S.; Wang, J. Enrichment of milk antioxidant activity by dietary supplementation of red clover isoflavone in cows and its improvement on mice intestinal health. *Food Chem.* **2024**, *446*, 138764. [CrossRef]
67. Chiang, Y.F.; Chen, H.Y.; Chang, Y.J.; Shih, Y.H.; Shieh, T.M.; Wang, K.L.; Hsia, S.M. Protective effects of fucoxanthin on high glucose-and 4-hydroxynonenal (4-HNE)-induced injury in human retinal pigment epithelial cells. *Antioxidants* **2020**, *9*, 1176. [CrossRef] [PubMed]
68. Liu, Y.; Liu, M.; Zhang, X.; Chen, Q.; Chen, H.; Sun, L.; Liu, G. Protective effect of fucoxanthin isolated from Laminaria japonica against visible light-induced retinal damage both in vitro and in vivo. *J. Agric. Food Chem.* **2016**, *64*, 416–424. [CrossRef] [PubMed]
69. Karpiński, T.M.; Adamczak, A. Fucoxanthin—An antibacterial carotenoid. *Antioxidants* **2019**, *8*, 239. [CrossRef] [PubMed]
70. Karpiński, T.M.; Ożarowski, M.; Alam, R.; Łochyńska, M.; Stasiewicz, M. What do we know about antimicrobial activity of astaxanthin and fucoxanthin? *Mar. Drugs* **2021**, *20*, 36. [CrossRef] [PubMed]
71. Abdoul-Latif, F.M.; Ainane, A.; Houmed Aboubaker, I.; Mohamed, J.; Ainane, T. An Overview of Cancer in Djibouti: Current Status, Therapeutic Approaches, and Promising Endeavors in Local Essential Oil Treatment. *Pharmaceuticals* **2023**, *16*, 1617. [CrossRef] [PubMed]
72. Šudomová, M.; Shariati, M.A.; Echeverría, J.; Berindan-Neagoe, I.; Nabavi, S.M.; Hassan, S.T. A microbiological, toxicological, and biochemical study of the effects of fucoxanthin, a marine carotenoid, on Mycobacterium tuberculosis and the enzymes implicated in its cell wall: A link between mycobacterial infection and autoimmune diseases. *Mar. Drugs* **2019**, *17*, 641. [CrossRef] [PubMed]
73. Rubiño, S.; Aymerich, T.; Peteiro, C.; Bover-Cid, S.; Hortós, M. Antimicrobial potential of Ericaria selaginoides extracts against Listeria monocytogenes in "mató", a Catalan fresh cheese. *J. Appl. Phycol.* **2023**, *35*, 949–959. [CrossRef]
74. Pérez, M.J.; Falqué, E.; Domínguez, H. Antimicrobial action of compounds from marine seaweed. *Mar. Drugs* **2016**, *14*, 52. [CrossRef] [PubMed]
75. Maadane, A.; Merghoub, N.; El Mernissi, N.; Ainane, T.; Amzazi, S.; Wahby, I.; Bakri, Y. Antimicrobial activity of marine microalgae isolated from Moroccan coastlines. *J. Microbiol. Biotechnol. Food Sci.* **2017**, *6*, 1257. [CrossRef]
76. Peraman, M.; Nachimuthu, S. Identification and quantification of fucoxanthin in selected carotenoid-producing marine microalgae and evaluation for their chemotherapeutic potential. *Pharmacogn. Mag.* **2019**, *15*, 243–249. [CrossRef]
77. Tsushima, Y.; Moka, T.; Katsuyama, M.; Kozuka, M.; Matsuno, T.; Tokuda, H.; Nishino, S.; Iwashima, A. Inhibitory effect of natural carotenoids on Epstein-Barr virus activation activity of a tumor promoter in Raji cells. A screening study for anti-tumor promoters. *Biol. Pharm. Bull.* **1995**, *18*, 227–233. [CrossRef] [PubMed]
78. Ikeda, C.; Manabe, Y.; Tomonaga, N.; Wada, T.; Maoka, T.; Sugawara, T. Evaluation of intestinal absorption of dietary halocynthiaxanthin, a carotenoid from the sea squirt *Halocynthia roretzi*. *Mar. Drugs* **2020**, *18*, 588. [CrossRef] [PubMed]
79. Tamama, K. Potential benefits of dietary seaweeds as protection against COVID-19. *Nutr. Rev.* **2021**, *79*, 814–823. [CrossRef] [PubMed]

80. Kang, N.; Heo, S.Y.; Kim, E.A.; Cha, S.H.; Ryu, B.; Heo, S.J. Antiviral effect of fucoxanthin obtained from Sargassum siliquastrum (Fucales, Phaeophyceae) against severe acute respiratory syndrome coronavirus 2. *Algae* **2023**, *38*, 295–306. [CrossRef]
81. Kang, N.; Kim, E.A.; Park, A.; Heo, S.Y.; Heo, J.H.; Heo, S.J. Antiviral Potential of Fucoxanthin, an Edible Carotenoid Purified from Sargassum siliquastrum, against Zika Virus. *Mar. Drugs* **2024**, *22*, 247. [CrossRef] [PubMed]
82. Khanna, S.; Bishnoi, M.; Kondepudi, K.K.; Shukla, G. Isolation, characterization and anti-inflammatory mechanism of probiotics in lipopolysaccharide-stimulated RAW 264.7 macrophages. *World J. Microbiol. Biotechnol.* **2020**, *36*, 74. [CrossRef] [PubMed]
83. Ciesielska, A.; Matyjek, M.; Kwiatkowska, K. TLR4 and CD14 trafficking and its influence on LPS-induced pro-inflammatory signaling. *Cell. Mol. Life Sci.* **2021**, *78*, 1233–1261. [CrossRef]
84. Jeong, S.; Kim, M.-B.; Baek, S.; Lee, J.; Lee, H.; Cao, B.; Kim, Y.; Cao, L.; Lee, S. Suppression of Pro-Inflammatory M1 Polarization of LPS-Stimulated RAW 264.7 Macrophage Cells by Fucoxanthin-Rich Sargassum hemiphyllum. *Mar. Drugs* **2023**, *21*, 533. [CrossRef] [PubMed]
85. Jayasinghe AM, K.; Kirindage KG, I.S.; Fernando IP, S.; Kim, K.N.; Oh, J.Y.; Ahn, G. The anti-inflammatory effect of low molecular weight fucoidan from sargassum siliquastrum in lipopolysaccharide-stimulated RAW 264.7 macrophages via inhibiting NF-κB/MAPK signaling pathways. *Mar. Drugs* **2023**, *21*, 347. [CrossRef]
86. Kim, K.N.; Heo, S.J.; Yoon, W.J.; Kang, S.M.; Ahn, G.; Yi, T.H.; Jeon, Y.J. Fucoxanthin inhibits the inflammatory response by suppressing the activation of NF-κB and MAPKs in lipopolysaccharide-induced RAW 264.7 macrophages. *Eur. J. Pharmacol.* **2010**, *649*, 369–375.
87. Lee, A.H.; Shin, H.Y.; Park, J.H.; Koo, S.Y.; Kim, S.M.; Yang, S.H. Fucoxanthin from microalgae Phaeodactylum tricornutum inhibits pro-inflammatory cytokines by regulating both NF-κB and NLRP3 inflammasome activation. *Sci. Rep.* **2021**, *11*, 543. [CrossRef]
88. Vostálová, J.; Zdařilová, A.; Svobodová, A. Prunella vulgaris extract and rosmarinic acid prevent UVB-induced DNA damage and oxidative stress in HaCaT keratinocytes. *Arch. Dermatol. Res.* **2010**, *302*, 171–181. [CrossRef] [PubMed]
89. Mohibbullah, M.; Haque, M.N.; Sohag, A.A.M.; Hossain, M.T.; Zahan, M.S.; Uddin, M.J.; Hannan, M.A.; Moon, I.S.; Choi, J.-S. A systematic review on marine algae-derived fucoxanthin: An update of pharmacological insights. *Mar. Drugs* **2022**, *20*, 279. [CrossRef] [PubMed]
90. Mohamed Abdoul-Latif, F.; Ainane, A.; Houmed Aboubaker, I.; Mohamed Ahmed, N.; Ainane, T. Effectiveness of a diet for type 2 diabetics based on vegetables and fruits of the Cucurbitaceae family. *J. Anal. Sci. Appl. Biotechnol.* **2021**, *3*, 107–113.
91. Hitoe, S.; Shimoda, H. Seaweed fucoxanthin supplementation improves obesity parameters in mild obese Japanese subjects. *Funct. Foods Health Dis.* **2017**, *7*, 246–262. [CrossRef]
92. Woo, M.N.; Jeon, S.M.; Shin, Y.C.; Lee, M.K.; Kang, M.A.; Choi, M.S. Anti-obese property of fucoxanthin is partly mediated by altering lipid-regulating enzymes and uncoupling proteins of visceral adipose tissue in mice. *Mol. Nutr. Food Res.* **2009**, *53*, 1603–1611. [CrossRef]
93. Sharma, P.P.; Baskaran, V. Polysaccharide (laminaran and fucoidan), fucoxanthin and lipids as functional components from brown algae (Padina tetrastromatica) modulates adipogenesis and thermogenesis in diet-induced obesity in C57BL6 mice. *Algal Res.* **2021**, *54*, 102187. [CrossRef]
94. Gopal, S.S.; Maradgi, T.; Ponesakki, G. Antiobese properties of carotenoids: An overview of underlying molecular mechanisms. *Carotenoids: Prop. Process. Appl.* **2020**, *2020*, 75–105.
95. Ferreira, Y.A.M.; Jamar, G.; Estadella, D.; Pisani, L.P. Role of carotenoids in adipose tissue through the AMPK-mediated pathway. *Food Funct.* **2023**, *14*, 3454–3462. [CrossRef]
96. Vidyashree, J.S.; Shetti, P.P.; Ghagane, S.C. Seaweeds as a potential resource in diabetes management: A review. *Future J. Pharm. Sci.* **2024**, *10*, 12. [CrossRef]
97. Oliyaei, N.; Moosavi-Nasab, M.; Tamaddon, A.M.; Tanideh, N. Antidiabetic effect of fucoxanthin extracted from Sargassum angustifolium on streptozotocin-nicotinamide-induced type 2 diabetic mice. *Food Sci. Nutr.* **2021**, *9*, 3521–3529. [CrossRef]
98. Bae, M.; Kim, M.B.; Park, Y.K.; Lee, J.Y. Health benefits of fucoxanthin in the prevention of chronic diseases. *Biochim. Et Biophys. Acta (BBA)-Mol. Cell Biol. Lipids* **2020**, *1865*, 158618. [CrossRef] [PubMed]
99. Agarwal, S.; Singh, V.; Chauhan, K. Antidiabetic potential of seaweed and their bioactive compounds: A review of developments in last decade. *Crit. Rev. Food Sci. Nutr.* **2023**, *63*, 5739–5770. [CrossRef] [PubMed]
100. Imai, M.; Kawakami, F.; Chiba, M.; Kanzaki, M.; Maruyama, H. Undaria pinnatifida (Wakame) Intake Ameliorates High-Fat Diet-Induced Glucose Intolerance via Promoting GLUT4 Expression and Membrane Translocation in Muscle. *J. Nutr. Metab.* **2023**, *2023*, 9774157. [CrossRef] [PubMed]
101. Zhang, Y.; Xu, W.; Huang, X.; Zhao, Y.; Ren, Q.; Hong, Z.; Huang, M.; Xing, X. Fucoxanthin ameliorates hyperglycemia, hyperlipidemia and insulin resistance in diabetic mice partially through IRS-1/PI3K/Akt and AMPK pathways. *J. Funct. Foods* **2018**, *48*, 515–524. [CrossRef]
102. Pruccoli, L.; Balducci, M.; Pagliarani, B.; Tarozzi, A. Antioxidant and Neuroprotective Effects of Fucoxanthin and Its Metabolite Fucoxanthinol: A Comparative In Vitro Study. *Curr. Issues Mol. Biol.* **2024**, *46*, 5984–5998. [CrossRef] [PubMed]
103. Paudel, P.; Seong, S.H.; Jung, H.A.; Choi, J.S. Characterizing fucoxanthin as a selective dopamine D3/D4 receptor agonist: Relevance to Parkinson's disease. *Chem.-Biol. Interact.* **2019**, *310*, 108757. [CrossRef] [PubMed]

104. Chen, Y.C.; Cheng, C.Y.; Liu, C.T.; Sue, Y.M.; Chen, T.H.; Hsu, Y.H.; Huang, N.-J.; Chen, C.H. Combined protective effects of oligo-fucoidan, fucoxanthin, and L-carnitine on the kidneys of chronic kidney disease mice. *Eur. J. Pharmacol.* **2021**, *892*, 173708. [CrossRef]
105. Tsukui, T.; Baba, N.; Hosokawa, M.; Sashima, T.; Miyashita, K. Enhancement of hepatic docosahexaenoic acid and arachidonic acid contents in C57BL/6J mice by dietary fucoxanthin. *Fish. Sci.* **2009**, *75*, 261–263. [CrossRef]
106. Guo, L.; Dang, M.; Song, Q.; Zhang, W.; Li, B. Protective effect of fucoxanthin on ovariectomy-induced osteoporosis in rats. *Pharmacogn. Mag.* **2020**, *16*, 242–249.
107. Ma, S.Y.; Park, W.S.; Lee, D.S.; Choi, G.; Yim, M.J.; Lee, J.M.; Jung, W.-K.; Park, S.G.; Seo, S.-K.; Park, S.J. Fucoxanthin inhibits profibrotic protein expression in vitro and attenuates bleomycin-induced lung fibrosis in vivo. *Eur. J. Pharmacol.* **2017**, *811*, 199–207. [CrossRef] [PubMed]
108. Chau, Y.T.; Chen, H.Y.; Lin, P.H.; Hsia, S.M. Preventive effects of fucoidan and fucoxanthin on hyperuricemic rats induced by potassium oxonate. *Mar. Drugs* **2019**, *17*, 343. [CrossRef]
109. Hashimoto, T.; Ozaki, Y.; Mizuno, M.; Yoshida, M.; Nishitani, Y.; Azuma, T.; Komoto, A.; Maoka, T.; Tanino, Y.; Kanazawa, K. Pharmacokinetics of fucoxanthinol in human plasma after the oral administration of kombu extract. *Br. J. Nutr.* **2012**, *107*, 1566–1569. [CrossRef] [PubMed]
110. Beppu, F.; Niwano, Y.; Tsukui, T.; Hosokawa, M.; Miyashita, K. Single and repeated oral dose toxicity study of fucoxanthin (FX), a marine carotenoid, in mice. *J. Toxicol. Sci.* **2009**, *34*, 501–510. [CrossRef] [PubMed]
111. Mok, I.K.; Lee, J.K.; Kim, J.H.; Pan, C.H.; Kim, S.M. Fucoxanthin bioavailability from fucoxanthin-fortified milk: In vivo and in vitro study. *Food Chem.* **2018**, *258*, 79–86. [CrossRef] [PubMed]
112. Bharathiraja, K.; Babu, L.H.; Vijayaprakash, S.; Tamilselvan, P.; Balasubramanian, M.P. Fucoxanthin, a marine carotenoid protects cadmium-induced oxidative renal dysfunction in rats. *Biomed. Prev. Nutr.* **2013**, *3*, 201–207. [CrossRef]
113. Lourenço-Lopes, C.; Garcia-Oliveira, P.; Carpena, M.; Fraga-Corral, M.; Jimenez-Lopez, C.; Pereira, A.G.; Prieto, M.A.; Simal-Gandara, J. Scientific approaches on extraction, purification and stability for the commercialization of fucoxanthin recovered from brown algae. *Foods* **2020**, *9*, 1113. [CrossRef] [PubMed]
114. Lykov, A.; Rachkovsky, E.; Gevorgiz, R.; Zheleznova, S.; Kotlyarova, A. Toxicity of fucoxanthin on Balb/c mice splenocytes and thymocytes. In Proceedings of the 2020 Cognitive Sciences, Genomics and Bioinformatics (CSGB), Novosibirsk, Russia, 6–10 July 2020; IEEE: Piscataway, NJ, USA, 2020; pp. 277–280.
115. Ming, J.X.; Wang, Z.C.; Huang, Y.; Ohishi, H.; Wu, R.J.; Shao, Y.; Wang, H.; Qin, M.Y.; Wu, Z.L.; Li, Y.Y.; et al. Fucoxanthin extracted from Laminaria Japonica inhibits metastasis and enhances the sensitivity of lung cancer to Gefitinib. *J. Ethnopharmacol.* **2021**, *265*, 113302. [CrossRef]
116. Hashimoto, T.; Ozaki, Y.; Taminato, M.; Das, S.K.; Mizuno, M.; Yoshimura, K.; Kanazawa, K. The distribution and accumulation of fucoxanthin and its metabolites after oral administration in mice. *Br. J. Nutr.* **2009**, *102*, 242–248. [CrossRef]

Disclaimer/Publisher's Note: The statements, opinions and data contained in all publications are solely those of the individual author(s) and contributor(s) and not of MDPI and/or the editor(s). MDPI and/or the editor(s) disclaim responsibility for any injury to people or property resulting from any ideas, methods, instructions or products referred to in the content.

pharmaceuticals

MDPI

Article

Zanthoxylum piperitum Benn. Attenuates Monosodium Urate-Induced Gouty Arthritis: A Network Pharmacology Investigation of Its Anti-Inflammatory Mechanisms

Sung Wook Kim [1,†], Soo Hyun Jeong [1,†], Jong Uk Kim [1], Mi Hye Kim [1], Wonwoong Lee [2], Cheol-Jung Lee [3], Tae Han Yook [1,*] and Gabsik Yang [1,*]

1 College of Korea Medicine, Woosuk University, Jeonju-si 54986, Republic of Korea; sheep.sw91@gmail.com (S.W.K.); sammytega1@naver.com (S.H.J.); ju1110@hanmail.net (J.U.K.); kimmh526@woosuk.ac.kr (M.H.K.)

2 Department of Pharmacy, College of Pharmacy, Woosuk University, Wanju 55338, Republic of Korea; wwlee@woosuk.ac.kr

3 Research Center for Materials Analysis, Korea Basic Science Institute, Daejeon 34133, Republic of Korea; veritas0613@kbsi.re.kr

* Correspondence: nasis@naver.com (T.H.Y.); yanggs@woosuk.ac.kr (G.Y.)

† These authors contributed equally to this work.

Abstract: Background: Monosodium urate crystal accumulation in the joints is the cause of gout, an inflammatory arthritis that is initiated by elevated serum uric acid levels. It is the most prevalent form of inflammatory arthritis, affecting millions worldwide, and requires effective treatments. The necessity for alternatives with fewer side effects is underscored by the frequent adverse effects of conventional therapies, such as urate-lowering drugs. IL-1β is a potential therapeutic target due to its significant role in the inflammatory response induced by MSU. *Zanthoxylum piperitum* Benn. (ZP), a shrub that possesses antibacterial, antioxidant, and anti-inflammatory properties, has demonstrated potential in the treatment of inflammatory conditions. **Methods:** For anti-inflammatory properties of ZP, Raw264.7 cell stimulated LPS were treated ZP and using RNA-seq with Bone marrow derived macrophage, we observed to change inflammatory gene. Pharmacological networks were conducted to select target gene associated with ZP. For in vivo, mice were injected MSU in footpad for induce gouty arthritis model. The components of ZP were analyzed using GC-MS, and distilled extracts of ZP (deZP) were prepared. **Results:** In vitro, deZP decreased inflammatory cytokines. However, in vivo, it also decreased paw thickness and IL-1β levels. The anti-inflammatory effects of deZP are believed to be mediated through the NLRP3 inflammasome pathway, as indicated by RNA sequencing and network pharmacology analyses. **Conclusions:** ZP has an anti-inflammatory effect and regulation of the NLRP3 inflammasome in vitro and in vivo. Further research, including clinical trials, is required to confirm the safety of deZP, determine the optimal dosing, and evaluate its long-term effects.

Keywords: *Zanthoxylum piperitum*; NLRP3 inflammasome; monosodium urate crystals; gouty arthritis; network pharmacology

1. Introduction

Gouty arthritis is a prevalent and chronic form of inflammatory arthritis that primarily affects adults, resulting from the deposition of monosodium urate (MSU) crystals. According to recent studies, its prevalence ranges from <1–6.8%, and its incidence ranges from 0.58 to 2.89 per 1000 person-years [1]. However, drugs prescribed to patients in the current clinical environment have limited effectiveness. For instance, previous studies have shown that more than 80% of patients who take colchicine experience abdominal pain before experiencing complete clinical improvement. Moreover, the side effects of non-steroidal anti-inflammatory drugs (NSAIDs) are more pronounced in the elderly [2]. Importantly, these drugs only offer temporary pain relief and do not provide a radical cure for gouty arthritis.

Gouty arthritis is characterized by the deposition of uric acid crystals in joints and periarticular tissues. Recent research has identified a receptor, NOD-like receptor family, pyrin domain containing 3 (NLRP3), that responds to uric acid crystals and triggers an inflammatory signal. NLRP3 is a member of the NOD-like receptor family (NLR), which detects endogenous danger signals, including microbial invasion and uric acid crystals. When these signals are present, NLRP3 combines with apoptosis-related speckle-like proteins, such as adapter proteins, CARD (ASC), and pro-caspase-1, to form an inflammasome. Pro-caspase-1 is then cleaved into an active form of caspase-1, which in turn cleaves the pro-IL-1β precursor to generate active IL-1β, which is secreted into the extra-cellular environment [3–5]. While current anti-IL-1β therapies are expected to be effective against acute gouty arthritis attacks, they are associated with high costs, inconvenient treatment routes and regimens, and side effects. Therefore, it is crucial to identify inflammatory mechanisms underlying the pathogenesis of gouty arthritis and to develop more effective therapeutic agents [6–8].

Zanthoxylum piperitum Benn. (ZP) is a summergreen shrub that is commonly found in East Asia, such as Korea, China, and Japan. ZP includes a lot of spices and oil ingredients in root, stem, leaf, and fruit [9]. Hence, ZP was reported to have been used for various purposes, including as an anti-inflammatory [10], antibacterial [11], and antioxidant agent [12], as well as oriental medicine used to treat disease such as osteoarthritis [12], obesity [13], and cancer [14]. Terpinen-4-ol, one of the active compounds of ZP, was previously studied for anti-inflammatory effects. It decreased pro-inflammatory cytokine production via nuclear factor kappa B (NF-kappa B) and NLRP3 inflammasome activation [15]. The mechanism against gouty arthritis, one of the inflammatory diseases, is not yet clear. We selected the distilled extracts method to use ZP as a pharmacopuncture.

Given that IL-1β is a major cytokine produced in response to MSU deposition and NLRP3 inflammasome activation is strongly linked to the development of gouty arthritis [12,16,17], targeting the NLRP3 inflammasome may be an effective therapeutic strategy for this condition. In this study, we aimed to investigate the effect of distilled extracts of ZP, which is known to have anti-inflammatory properties [9,18], on IL-1β production and inflammasome signal transduction in gouty arthritis.

2. Results

2.1. Anti-Inflammatory Effects of Destilled Extracts ZP (deZP) in Macrophages

To determine the potential cytotoxicity of deZP, we treated RAW264.7 cells with varying concentrations of deZP ranging from 0.32 to 5%. We observed that the highest concentration of 5% showed cytotoxicity in RAW264.7 cells. Consequently, we excluded this concentration from further investigations. Next, we examined whether deZP could exhibit anti-inflammatory effects in macrophages. We induced inflammation in RAW264.7 cells using lipopolysaccharide (LPS) and treated them with different concentrations of deZP. Our results showed that as the concentration of deZP increased, the levels of nitrite, IL-6, TNF-α, and MPO activity decreased in a concentration-dependent manner (Figure 1). This finding indicates that deZP has potential anti-inflammatory effects in macrophages.

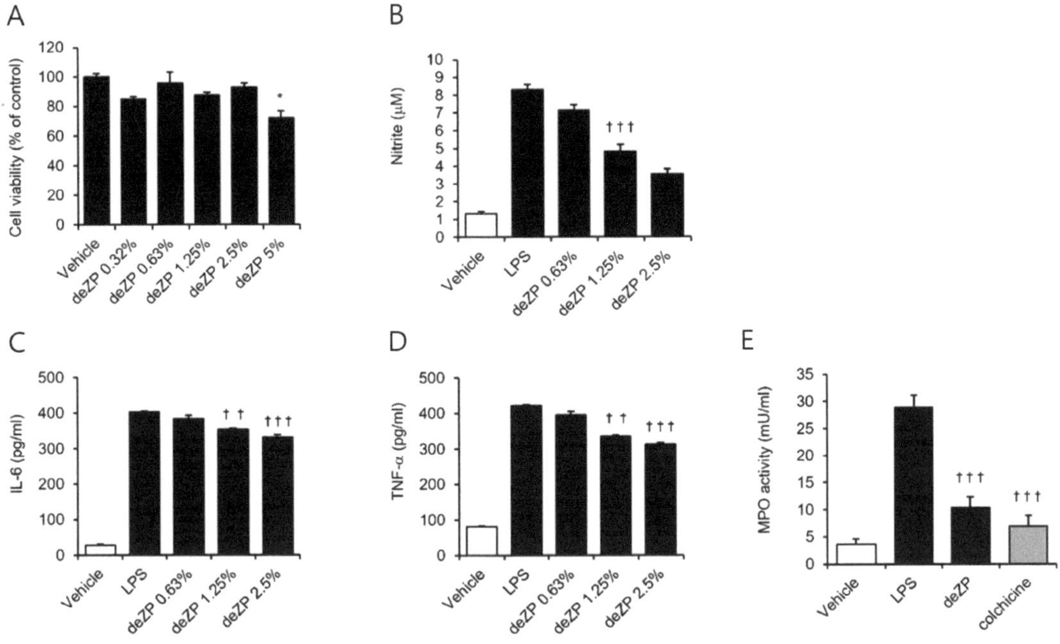

Figure 1. Anti-inflammatory effects of deZP. (**A**) RAW264.7 cells were treated with deZP at concentrations ranging from 0.32% to 5% for 24 h. (**B–D**) RAW264.7 cells were treated with deZP at concentrations of 0.63%, 1.25%, and 2.5% for 1 h, followed by stimulation with LPS (100 ng/mL) with deZP for 24 h. (**C,D**) Cell supernatants were collected for analysis of IL-6 and TNF-α using ELISA. (**E**) RAW264.7 cells were stimulated with LPS (100 ng/mL) and treated with deZP or colchicine (10 μM). The data are presented as mean ± SD (n = 3). * $p < 0.05$ vs. vehicle. †† $p < 0.01$, ††† $p < 0.001$ vs. LPS.

2.2. Regulation of NLRP3 Inflammasome by deZP in Transcriptional Profiling and Gene Ontology

To investigate whether the anti-inflammatory efficacy of deZP is related to the NLRP3 inflammasome, we conducted RNA sequencing with bone marrow-derived macrophage (BMDM) cells. We induced the NLRP3 inflammasome by treating BMDM cells with MSU crystals after LPS priming. We then compared cells treated with or without deZP. Our results showed that genes that increased after MSU treatment in various categories were decreased after treatment with deZP (Figure 2A). A scatter plot showed that the genes were regulated by deZP in the inflammasome categories. It shows that increased NLRP3 gene expression by MSU regulates the decrease after treatment with deZP (Figure 2B). Unfortunately, IL-1β was not regulated by deZP. Additionally, gene ontology analysis revealed that deZP down-regulated genes involved in cytokine–cytokine receptor interaction and the NF-kappa B signaling pathway. These findings demonstrate that deZP can regulate NF-kappa B-related inflammatory responses, as well as the NLRP3 inflammasome (Figure 2C).

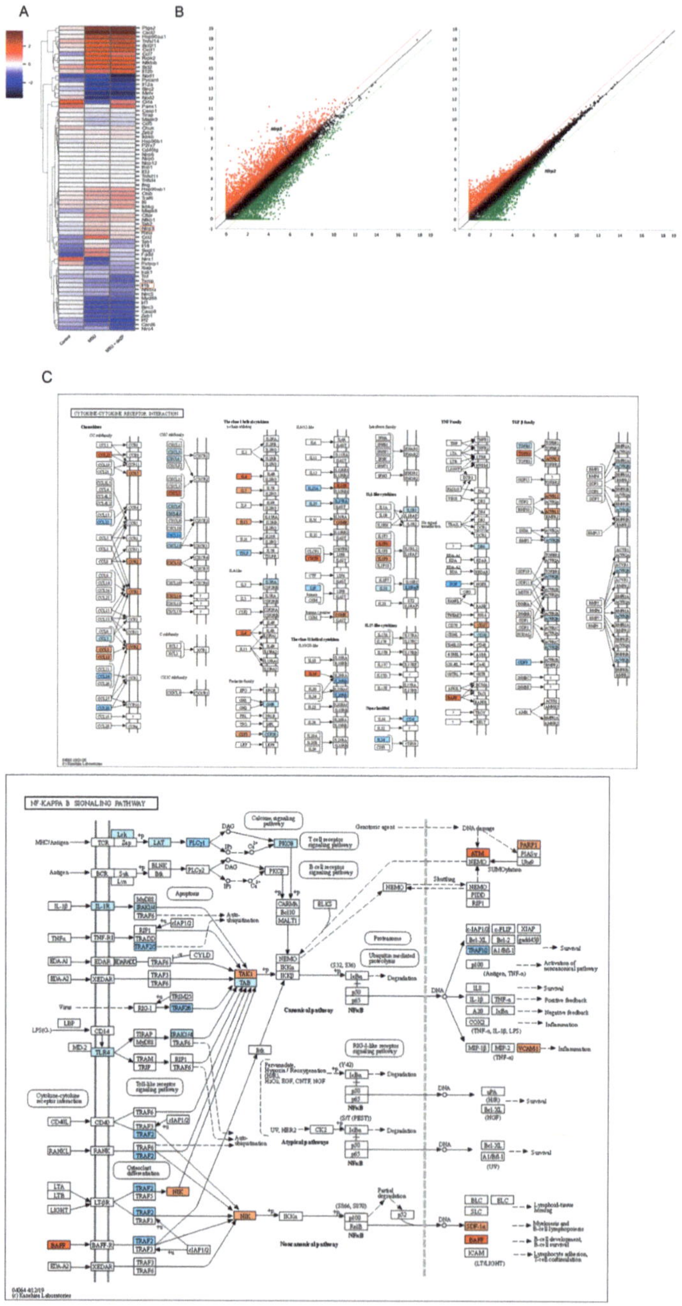

Figure 2. Analysis of significantly differentially expressed genes (DEGs) and gene ontology (GO). (**A**,**B**) Heatmap showing genes with differential expression and exhibiting a scatter plot of DEG. They show inflammasome categories. (**A**) red boxes are presented as target genes. (**B**) red spots mean up regulated genes, and green are down regulated genes. (**C**) Gene ontology shows regulated genes by deZP in specific signaling (cytokine–cytokine receptor interactions and NF-kappa B signaling pathway). Red boxes are up-regulated genes, and blue are down-regulated genes.

2.3. Association with the ZP's Active Compound and Gouty Arthritis Through Network Construction

Because natural products such as ZP have various active compounds, we need to select one of ZP's compounds to construct the network. A previous study reported that Terpinen-4-ol, as one of ZP's compounds, has anti-inflammatory effects and relates with the NLRP3 inflammasome [15]. Hence, we construct the network with Terpinen-4-ol as the active compound. Moreover, 44 genes of the Terpinen-4-ol network with 104 nodes and 1434 edges (Figure 3A). To investigate the effects of the Terpinen-4-ol on gouty arthritis, the target genes of Terpinen-4-ol network and the 'gout' gene set were compared, resulting in the deduction of eight intersecting genes, including *NLRP3*, *PTGS2*, *CXCL8*, *IL6*, *TNF*, *IL1B*, *TLR4*, and *ALB* (Figure 3C). Although the matching percentage between Terpinen-4-ol-targeted genes and the gouty arthritis gene set was only 9.5% (Figure 3B), the common genes were highly linked with the 8 nodes and 28 edges around *NLRP3* (Figure 3D).

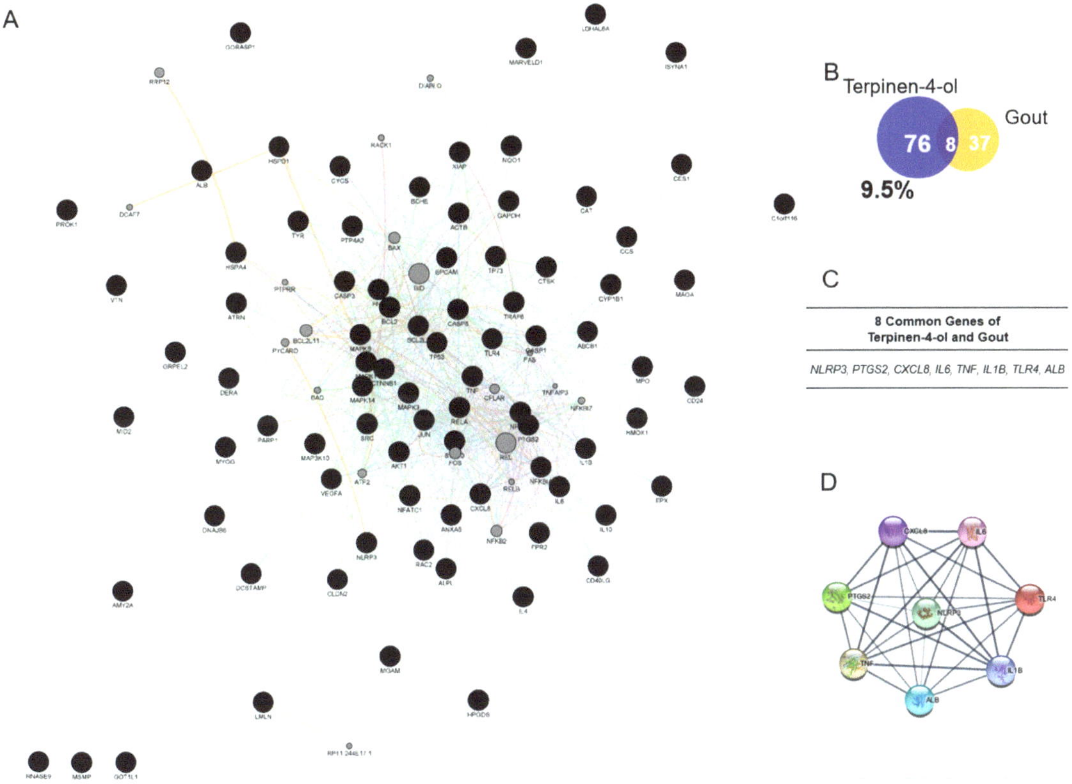

Figure 3. Network of Terpinen-4-ol as ZP's active compound. (**A**) Whole network Terpine-4-ol as an active compound of ZP with 104 nodes and 1434 edges. (**B**) Venn diagram of intersection target genes between the Terpinen-4-ol network and 'gout' gene set. (**C**) Eight common genes of Terpinen-4-ol and gout. (**D**) Network of common genes of Terpinen-4-ol and gout with 8 nodes and 28 edges.

2.4. Prediction of the Potential Underlying Mechanism of ZP on Gouty Arthritis

Based on significant FDR values, the potential functional mechanisms of ZP on gouty arthritis were mainly associated with the 'NOD-like receptor signaling pathway (hsa04621)', 'NF-kappa B signaling pathway (hsa04064)', 'Toll-like receptor signaling pathway (hsa04620)', 'TNF signaling pathway (hsa04668)', and 'Cytokine–cytokine receptor interaction (hsa04060)' on the KEGG Pathways database (Figure 4A). Additionally, the GO

Biological Process database showed the association with the 'Regulation of acute inflammatory response (GO:0002673)', 'Response to lipopolysaccharide (GO:0032496)', 'Cellular response to lipopolysaccharide (GO:0071222)', 'Inflammatory response (GO:0006954)', 'Regulation of cytokine production involved in immune response (GO:0002718)', 'Regulation of inflammatory response (GO:0050727)', and 'Cytokine production (GO:0001816)' with the underlying mechanisms of ZP (Figure 4C). Those predicted biological terms shared NLRP3, TLR4, IL6, IL1B, TNF, CXCL8, and PTGS2 in KEGG Pathways and GO Biological Process (Figure 4B,D).

A

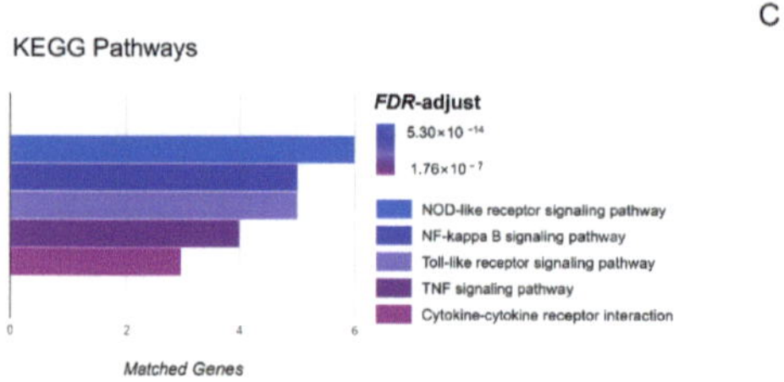

Category	FDR value	Description	Matched Genes	Background Genes
KEGG Pathways	9.44×10^{-10}	NOD-like receptor signaling pathway (hsa04621)	6	174
	6.49×10^{-9}	NF-kappa B signaling pathway (hsa04064)	5	101
	6.49×10^{-9}	Toll-like receptor signaling pathway (hsa04620)	5	101
	1.29×10^{-6}	TNF signaling pathway (hsa04668)	4	112
	3.46×10^{-5}	Cytokine-cytokine receptor interaction (hsa04060)	4	282

C

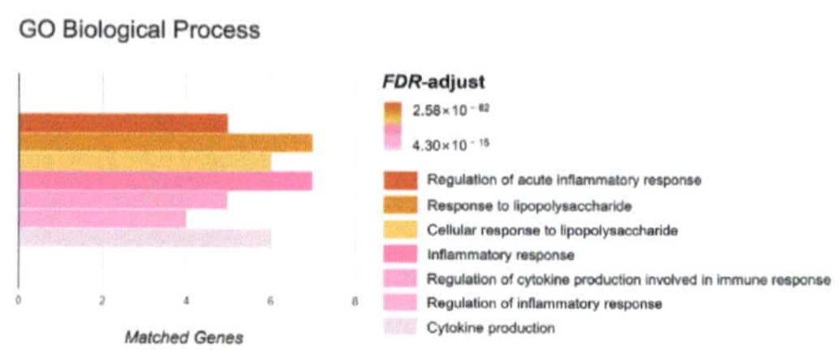

Category	FDR value	Description	Matched Genes	Background Genes
GO biological process	2.96×10^{-8}	Regulation of acute inflammatory response (GO:0002673)	5	47
	2.96×10^{-8}	Response to lipopolysaccharide (GO:0032496)	7	313
	7.06×10^{-8}	Cellular response to lipopolysaccharide (GO:0071222)	6	185
	1.32×10^{-7}	Inflammatory response (GO:0006954)	7	515
	2.04×10^{-7}	Regulation of cytokine production involved in immune response (GO:0002718)	5	89
	1.12×10^{-6}	Regulation of inflammatory response (GO:0050727)	6	351
	1.90×10^{-6}	Cytokine production (GO:0001816)	5	158

B

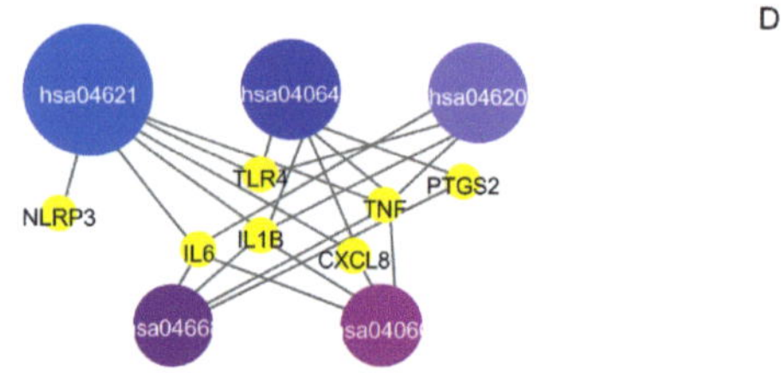

D

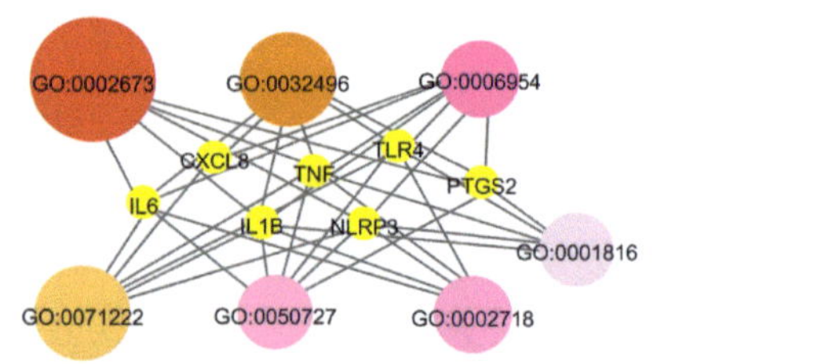

Figure 4. Functional enrichment analysis of the ZP network and 'gout' gene set. (**A**) Biological terms of the intersecting genes of ZP and gout using a KEGG Pathways database in high FDR value order. (**B**) Significantly enriched pathway terms of ZP in gout on a KEGG Pathways database. (**C**) Biological terms of the intersecting genes of ZP and gout using a GO Biological Process database in high FDR value order. (**D**) Significantly enriched pathway terms of ZP in gout on a GO Biological Process database.

2.5. Specific Mechanisms of ZP in NOD-like Receptor Signaling Pathway in Gouty Arthritis

With the purpose of finding a more specific mechanism of ZP, genes were compared with the ZP network and the 'NOD-like receptor signaling pathway', which is primarily listed in the prediction results of the KEGG Pathway above. On a map of the KEGG Pathway (hsa04621), the common genes were highlighted in a red box, showing that ZP-targeted genes particularly belong to 'Pro-inflammatory cytokines', 'Chemokines', 'Toll-like receptor signaling pathway', and 'Pro-inflammatory effects' (Figure 5).

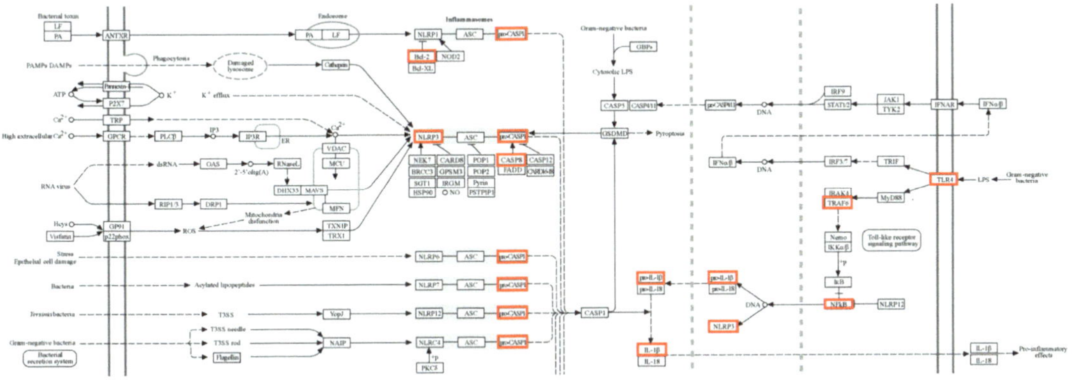

Figure 5. A diagram of the NOD-like receptor signaling pathway in the KEGG Pathway database (hsa04621). All matched genes are marked in a red box.

2.6. Anti-Inflammation of deZP in Acute Gouty Arthritis Mouse Model

We investigated whether deZP could inhibit the NLRP3 inflammasome in a gouty arthritis mouse model. We orally administered deZP to mice and induced gouty arthritis by injecting MSU crystals into their footpads. We observed a statistically significant decrease in footpad thickness in mice treated with deZP from 6 h after MSU administration (Figure 6A). Moreover, IL-1β levels in the footpads of mice administered with deZP also showed a statistically significant decrease (Figure 6B). deZP decreased also caspase-1 and IL-1β protein levels in Western blot (Figure 6C). In real-time PCR, deZP also regulated IL-1β mRNA expression level as vehicle. However, NLRP3 and caspase-1 mRNA expression were not regulated by deZP (Figure 6D). These results demonstrate that deZP attenuated IL-1β mRNA expression and protein levels produced by the NLRP3 inflammasome in an acute gouty arthritis mouse model induced by MSU.

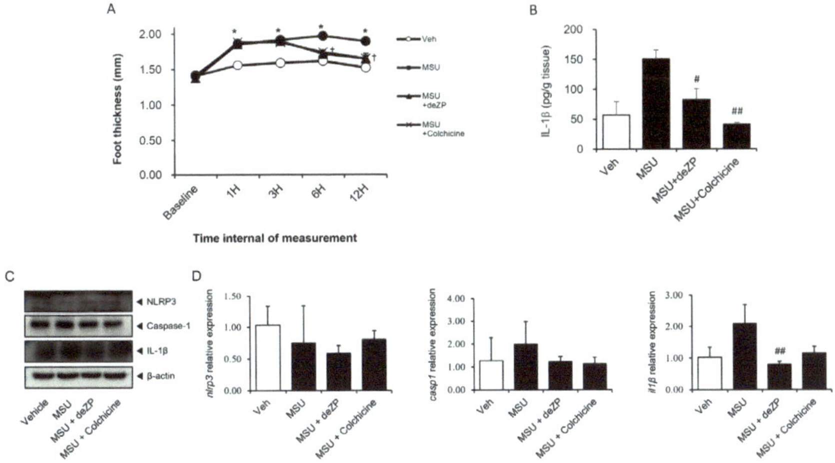

Figure 6. Treatment with deZP attenuates in vivo inflammatory symptoms. (**A**) Time course of footpad thickness after oral administration of deZP and injection of MSU crystals. (**B**) IL-1β levels in footpad tissue measured by ELISA. (**C**) Footpad tissues were analyzed by Western blot for NLRP3, caspase-1, and IL-1β. (**D**) Real-time PCR shows that mRNA expression levels of NLRP3, caspase-1, and IL-1β in footpad tissue. Data are presented as mean ± SD (n = 5 mice per group). * $p < 0.05$ vs. vehicle. † $p < 0.05$ vs. MSU. # $p < 0.05$, ## $p < 0.01$ vs. MSU.

2.7. Ingredients of deZP

To investigate the ingredients contained in deZP, we performed electron ionization mass spectra. These ingredients include Linalool, Terpinen-4-ol, α-terpineol, Camphene, and Piperitone [19]. When we investigated the relative peak area on the total ion chromatogram for deZP, we found that Terpinen-4-ol, which appeared at 14.5 min on the chromatogram, was the most abundant component (Figure 7A). These ingredients were reported to have various functions. Ingredients of deZP are used antimicrobial, insecticide, and food additive [20–23]. Exceptionally, Delta 3-carene is useful to improve sleep and Piperitone is one of the main materials for synthetic Menthol and Thymol. Furthermore, we measured retention times and relative area abundances of peaks of each ingredient (Figure 7B) [21,24]. Terpinen-4-ol, the most frequently identified ingredient in deZP, was calculated for its quantification. Terpinen-4-ol was contained in about 487.36 μg/mL of deZP (Figure 7C).

A

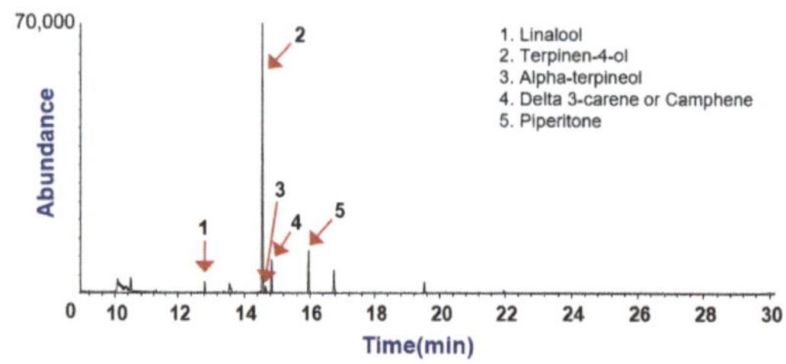

B

Ingredients	Function	Retention time	Relative abundance (%)
1. Linalool	antimicrobial, food addictive, insecticide	12.81	2.48
2. Terpinen-4-ol	antimicrobial	14.54	75.20
3. Alpha Terpineol	flavor, insecticide	14.64	1.73
4. Delta 3-carene	Improving sleep	14.82	8.58
4. Camphene	food addictive for spices	-	-
5. Piperitone	Main raw materials to produce synthetic menthol and thymol	15.95	12.01

C

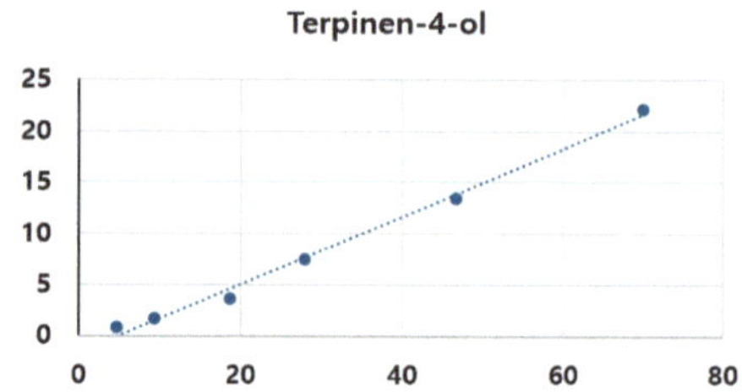

Sample	RT	Calibration	Dynamic range	R^2	LOD	LOQ
Terpinen-4-ol	14.54	y = 0.3318x−1.6165	4.665–69.975	0.9936	2.106	6.319

Concentration of Terpinen-4-ol in ZP (μg/mL)	
ZP-1	474.1
ZP-2	484.5
ZP-3	503.5

Figure 7. Identification of deZP and their functions. (**A**) Total ion chromatogram of deZP with labeled peaks: 1. Linalool; 2. Terpinen-4-ol; 3. Alpha-terpineol; 4. Delta 3-carene or Camphene; 5. Piperitone. (**B**) Function of the identified ingredients of deZP. (**C**) terpinene-4-ol, the main compound of ZP, was measured for concentrations.

3. Materials and Methods

3.1. Animal Care and Cell Culture

The Institutional Animal Care and Use Committee (IACUC) of Woosuk University of Korea approved the experimental procedures and animal care (Permission no. WS2020-05). C57BL/6 mice obtained Daehan bio (Seoul, Republic of Korea) were housed in a controlled animal facility with a 12:12-h light/dark cycle, a temperature of 23 ± 3 °C, and a relative humidity of 40–60%. We cultured BMDMs according to the previously described protocol [14]. Macrophages and RAW264.7 cells were cultured in Dulbecco's modified Eagle medium supplemented with 10% (*v*/*v*) fetal bovine serum obtained Invitrogen (Carlsbad, CA, USA), 10,000 units/mL penicillin, and 10,000 μg/mL streptomycin.

3.2. Reagents

We obtained purified lipopolysaccharides (LPS) from Escherichia coli from List Biological Laboratory (Campbell, CA, USA), which we dissolved in endotoxin-free water. We obtained MSU from Invivogen (Carlsbad, CA, USA), and we crystallized it in phosphate-buffered saline (PBS) for 5 min of sonication. We obtained the nitrite assay kit, protease inhibitor cocktail, and phosphatase inhibitor cocktail from Sigma-Aldrich (St. Louis, MO, USA). We obtained tissue protein extraction reagent (T-PER) from Thermo Scientific (Waltham, MA, USA). We purchased distilled extracts of ZP from AJ Pharmacopuncture (Seoul, Republic of Korea). deZP was mixed with 10 L of distilled water in the extractor and boiled at 105 °C for 2 h. ZP extracts were distilled with about 5 L of water at 107 °C. deZP was stored at 4 °C. For Western blot detection, the IL-1β antibody was obtained from R&D Systems (Minneapolis, MN, USA), the NLRP3 and Caspase-1 antibodies were sourced from AdipoGen (San Diego, CA, USA), and the β-actin antibody was procured from Santa Cruz Biotechnology (Dallas, TX, USA). The secondary antibodies were purchased from Cell Signaling Technology (Danvers, MA, USA).

3.3. Cell Viability

We measured cell viability using the Cell Proliferation Kit I (MTT) obtained from Roche (Basel, Switzerland). After treating RAW264.7 cells with deZP at concentrations of from 0.32% to 5% for 24 h, we measured cell viability using a microplate reader, the SpectraMax® ABS, purchased from Molecular Devices (San Jose, CA, USA).

3.4. Myeloperoxidase (MPO) Activity Assay

We measured MPO activity in supernatants of raw264.7 cells treated with LPS with deZP 0.63%, 1.25%, and 2.5% using the MPO Colorimetric Activity Assay Kit obtained from Bio-Vision (Milpitas, CA, USA), following the manufacturer's instructions.

3.5. Enzyme-Linked Immunosorbent Assays

We determined the levels of mouse IL-1β, IL-6, and TNF-α in supernatants of mice foot homogenate and raw264.7 cells treated with LPS 100 ng/mL with deZP at concentrations ranging from 0.63%, 1.25%, and 2.5% for 24 h. It was analyzed using the Duoset ELISA Kit obtained from R&D Systems (Minneapolis, MN, USA) and measured 450 nm observance with Molecular Devices SpectraMax® ABS (San Jose, CA, USA).

3.6. RNA Sequencing

We activated BMDMs' NLRP3 inflammasome using LPS priming and treated them with MSU with or without deZP. We extracted total RNAs from BMDMs using Trizol rea-gent obtained from Invitrogen (Carlsbad, CA, USA), following the manufacturer's instructions. E-Biogen Inc. (Seoul, Republic of Korea) performed RNA quality and RNA sequencing assays.

3.7. Network Construction and Comparison with the Gene Set of Gouty Arthritis

The network of ZP was constructed by collecting the target genes of Terpine-4-ol, also known as 4-Carvomenthenol, since the most component of ZP in this study consisted of Terpine-4-ol. The target genes of Terpine-4-ol were obtained from the PubChem database "https://pubchem.ncbi.nlm.nih.gov/" (accessed on 7 July 2023). A total of 44 genes as chemical–gene co-occurrences in the literature created a network of ZP. Additionally, the gene set of gouty arthritis was collected from the MalaCards database "https://www.malacards.org/" (accessed on 7 July 2023) by searching a keyword 'gout' (Table 1). The gouty arthritis gene set was a total of 45 genes and compared with the ZP network. We constructed a new network with a total of 8 intersecting genes between ZP and gouty arthritis.

Table 1. List of the related target genes of 'gout' from the MalaCards database.

Description	Symbol	Relevance Score
Aspartyl-TRNA synthetase 2, mitochondrial	*DARS2*	400
Hypoxanthine phosphoribosyltransferase 1	*HPRT1*	61
Uromodulin	*UMOD*	41.1
Solute carrier family 22 member 12	*SLC22A12*	36.5
Xanthine dehydrogenase	*XDH*	34.63
Phosphoribosyl pyrophosphate synthetase 1	*PRPS1*	32.88
ATP binding cassette subfamily g member 2 (junior blood group)	*ABCG2*	26.2
Adenine phosphoribosyltransferase	*APRT*	25.02
MEFV innate immunity regulator, pyrin	*MEFV*	23.79
Interleukin 6	*IL6*	22.83
Interleukin 18	*IL18*	22.82
C-X-C Motif chemokine ligand 8	*CXCL8*	22.3
Major histocompatibility complex, class I, B	*HLA-B*	22.13
Solute carrier family 22 member 8	*SLC22A8*	22.05
Albumin	*ALB*	21.92
Interleukin 1 beta	*IL1B*	21.69
Purinergic receptor P2X 7	*P2RX7*	21.66
Solute carrier family 22 member 7	*SLC22A7*	20.29
Cytochrome P450 family 2 subfamily c member 8	*CYP2C8*	19.07
Cytochrome P450 family 2 subfamily E member 1	*CYP2E1*	18.79
Interleukin 1 receptor accessory protein like 2	*IL1RAPL2*	18.58
Polycystin 2, transient receptor potential cation channel	*PKD2*	18.41
Solute carrier family 2 member 9	*SLC2A9*	17.69
Purine nucleoside phosphorylase	*PNP*	16.83
NLR pyrin domain containing 3	*NLRP3*	16.64
Solute carrier family 17 member 1	*SLC17A1*	16.24
Adenylosuccinate lyase	*ADSL*	16.11
Solute carrier family 22 member 11	*SLC22A11*	16
Lactate dehydrogenase D	*LDHD*	15.72
Solute carrier family 17 member 3	*SLC17A3*	15.5
Tumor necrosis factor	*TNF*	15.3
PDZ domain containing 1	*PDZK1*	14.97
Solute carrier family 16 member 9	*SLC16A9*	14.87
Insulin	*INS*	14.78
Aldehyde dehydrogenase 2 family member	*ALDH2*	14.78
Solute carrier family 22 member 6	*SLC22A6*	14.55
S100 Calcium binding protein A9	*S100A9*	14.55
C-Reactive protein	*CRP*	14.31
Interleukin 1 receptor type 1	*IL1R1*	14.22
Alpha kinase 1	*ALPK1*	14.2
Capping protein regulator and myosin 1 linker 1	*CARMIL1*	14.18
Toll-like receptor 4	*TLR4*	14.17
Solute carrier family 22 member 13	*SLC22A13*	14.12
Prostaglandin-endoperoxide synthase 2	*PTGS2*	13.95
PYD and CARD domain containing	*PYCARD*	13.92

3.8. Functional Enrichment Analysis with KEGG Pathways and GO Biological Process Databases

The potential functional pathways of ZP on gouty arthritis were predicted in the Kyoto Encyclopedia of Genes and Genomes (KEGG) pathway 2020 human [25] and Gene Ontology (GO) Process databases using the Cytoscape String App (ver. 3.9.1). A total of 8 overlapped genes between ZP and gouty arthritis were categorized by each database, and the number of matching genes was counted according to the FDR values. Eventually, a diagram image of the 'NOD-like receptor signaling pathway' was obtained from KEGG Pathways. The matched genes were marked in a red box.

3.9. Gas Chromatography-Mass Spectrometry Conditions

We conducted chromatographic analysis using a DB-5MS capillary column, 30 m length × 0.25 mm inner diameter, 0.25 μm film thickness; J&W Scientific (Folsom, CA, USA) combined with an Agilent 6890 N gas chromatograph system (Palo Alto, CA, USA). Helium gas served as the carrier gas with a flow rate of 1 mL/min. In split mode (5:1), we set the injection port temperature to 280 °C and the injection volume to 1 μL. The oven temperature was programmed as follows: initial temperature at 50 °C for 5 min, increased to 200 °C at a rate of 8 °C/min and held for 5 min, then increased to 320 °C at a rate of 30 °C/min and held for 4 min. We carried out mass spectrometric determination using an Agilent 5975C mass selective detector equipped with a quadrupole analyzer and electron ionization source. The mass spectrometer was optimized with the following conditions: electron ionization energy, 70 eV; mass range in scan mode, m/z 50–550; ion source temperature, 230 °C; and transfer line temperature, 280 °C.

3.10. Acute Gouty Arthritis Model

The mice were randomly divided into four groups, with five mice per group. Colchicine or deZP was orally administered in 0.1 mL of sterilized water. One hour after administration, MSU crystals were injected into the footpad of the mice. After 12 h, foot tissue was homogenized in T-PER, and the supernatants were collected for ELISA. Moreover, foot tissue was also homogenized in Trizol reagent for mRNA extraction. cDNA synthesis using the iScript cDNA synthesis kit obtained Bio-Rad (Hercules, CA, USA) according to the manufacturer. cDNA was synthesized by adjusting mRNA to a concentration of 1000 ng/μL. After synthesis, the purity and concentration of the cDNA were re-evaluated. The synthesized cDNA was then used to perform real-time PCR with the TB Green® Fast qPCR Mix. The primer sequences are as follows: IL-1b forward: 5′-GCAACTGTTCCTGAACTCAACT-3′; IL-1b reverse: 5′-ATCTTTTGGGGTCCGTCAACT-3′, NLRP3 forward: 5′-ATTACCCGCCCGAGAAAGG-3′; NLPR3 reverse: 5′-TCGCAGCAAAGATCCACACAG-3′, caspase-1 forward: 5′-ACAAGGCACGGGACCTATG-3′; caspase-1 reverse: 5′-TCCCAGTCAGTCCTGGAAATG-3′. The primers were obtained from Bioneer (Daejeon, Republic of Korea).

3.11. Western Blot

Footpad tissues were homogenized in 500 μL of T-PER buffer using a Bead Ruptor 12 purchased Omni Inc. (Kennesaw, GA, USA). The homogenate was centrifuged at 20,000× g for 10 min at 4 °C, and the clear upper phase was carefully transferred to a new tube. Protein concentration was determined using a BCA assay kit according to the manufacturer's instructions. Protein samples were denatured by boiling at 95 °C for 5 min and then loaded onto SDS-PAGE gels. Electrophoresis was performed at 80 V for 30 min, followed by 120 V for 1 h. Proteins were subsequently transferred onto nitrocellulose membranes.

To block nonspecific binding, the membranes were incubated with 5% BSA for 1 h. After overnight incubation with primary antibodies, membranes were treated with secondary antibodies conjugated to horseradish peroxidase for 1 h. The chemiluminescent reaction was initiated using ECL solution, and protein detection was conducted using the Amersham ImageQuant 800 system (Cytiva, Amersham, UK).

3.12. Statistics

Data were presented as the mean ± SD. Statistical analysis was conducted using GraphPad Prism 5 software (GraphPad). Data were analyzed by one-way ANOVA followed by Tukey's test. Statistical significance was set at $p < 0.05$.

4. Discussion

In this study, we demonstrated that deZP for pharmacopuncture reduced IL-1β levels and suggested that this effect may be mediated by the NLRP3 inflammasome signal pathway in a gouty arthritis mouse model. ZP is a summer green shrub from the Ructaceae family and is widely used in East Asia as a traditional spice and medicine due to its various active oil components [26]. Although ZP has been reported to have diverse therapeutic effects, including detoxifying, reducing hypertension, stroke, antibacterial and antioxidant, tyrosinase, and osteosarcoma proliferation control [27,28], its anti-inflammatory effect and therapeutic ability as a target for NLRP3 inflammasome have not been studied. Thus, we aimed to investigate the anti-inflammatory effect of deZP, a drug used in pharmacopuncture, on NLRP3 inflammasome-derived gouty arthritis via its effect on IL-1β regulation.

Firstly, we used network pharmacology to verify the association with 4-Terpinen-ol as the active compound of ZP-targeted genes and the gouty arthritis gene set, which could describe the complexity between compounds, target genes, and diseases, indicating the potential mechanism of action of multitarget medicinal herbs [29]. The target genes of Terpine-4-ol, a main compound of ZP, were overlapped with the gouty arthritis-related genes such as *NLRP3, PTGS2, CXCL8, IL6, TNF, IL1B, TLR4*, and *ALB*. Those genes constructed the network predicted to be related with the 'NOD-like receptor signaling pathway', 'NF-kappa B signaling pathway', 'Toll-like receptor signaling pathway', and others. Specifically, 'Pro-inflammatory cytokines', 'Chemokines', 'Toll-like receptor signaling pathway', and 'Pro-inflammatory effects' were predicted to associate with the potential target mechanisms of ZP.

Our study found that deZP administration reduced paw thickness, a symptom of acute gouty edema, and decreased IL-1β mRNA and protein levels in an MSU-induced gouty arthritis animal model. Previous studies have reported that ZP or 4-Terpinen-ol inhibits NF-κB and NLRP3 [15,30]. RNA sequencing analysis demonstrated a reduction in NLRP3 expression mediated by deZP. However, this result was not reproducible in the gouty model, where IL-1β mRNA expression was decreased. These findings suggest that deZP has the potential to regulate the mRNA expression of both NLRP3 and IL-1β. In gouty arthritis, MSU activates the NLRP3 inflammasome through the priming signal of NF-κB-dependent transcription of NLRP3 and pro-IL-1β, which triggers a downstream signal upon activation of the TLR 4 receptor [31,32]. This signal promotes the expression of inflammasome components such as NLRP3, pro-caspase-1, and pro-IL-1β. The second signal involved in the assembly and activation of the NLRP3 complex is induced by MSU. The ASC aggregate recruits CARD, which interacts with the CARD domain of pro-caspase-1 to promote caspase-1 activation and ASC specks [32,33]. Clustered pro-caspase-1 mediates self-cleavage and activation in the form of activated caspase-1, which cleaves precursors of IL-1β to generate activated forms of IL-1β. Once NLRP3 activation is initiated, the formation of ASC spec is considered an upstream indication of NLRP3 activation [34]. Although we did not directly investigate the expression level of direct biomarker proteins among NLRP3 pathways, such as NF-κB, ASC, and PYD, it was observed that deZP lowered IL-1β levels by regulating one of these pathways. Thus, additional studies are necessary to confirm the biomarker of NLRP3 based on single-component analysis of ZP.

In vitro experiments also demonstrated that deZP reduced inflammatory cytokines and factors in LPS-induced inflammation. LPS is one of the representative PAMPs that acts as a priming signal in NLRP3 inflammasome activation. Although LPS treatment alone does not activate the NLRP3 inflammasome, it was also confirmed that ZP has anti-inflammatory effects on other pathways, as reported in previous studies [9,12,26]. Thus, ZP is considered to have various anti-inflammatory effects that can be applied to the treatment of gouty arthritis.

In the TIC plot for deZP of Figure 7A, other minor peaks were observed around 10, 14, 17 and 20 min. However, these peaks could not be identified, since the mass spectra for these peaks could not be properly interpreted due to their low intensities and lack of match with the library database.

Numerous studies have aimed to find natural remedies that can treat gouty arthritis by regulating IL-1β through targeting the NLRP3 inflammasome [8]. This study highlights the potential of the ZP drug as an effective treatment that can be utilized in clinical practice. Moreover, further research on the safety and effectiveness of ZP in clinical studies, based on comprehensive mechanism research and preclinical studies, is necessary to establish it as an exceptional drug for gouty arthritis treatment.

Author Contributions: Conceptualization, J.U.K.; methodology, G.Y.; formal analysis, C.-J.L. and W.L.; investigation, T.H.Y.; resources, M.H.K. and S.H.J.; writing—original draft preparation, S.W.K.; writing—review and editing, G.Y.; visualization, S.W.K., S.H.J. and M.H.K.; supervision, G.Y. and T.H.Y.; funding acquisition, G.Y. and T.H.Y. All authors have read and agreed to the published version of the manuscript.

Funding: This study was supported by the National Research Foundation of Korea (NRF) (NRF2023R1A2C2005333) funded by the Korean government (Ministry of Science, ICT and Future Planning) and Woosuk University. G.Y. is supported by the Intra-mural Research Program of WOOSUK University.

Institutional Review Board Statement: The study was conducted in accordance with the Declaration of Helsinki, and approved by the Institutional Animal Care and Use Committee (IACUC) of Woosuk university of Korea (Permission no. WS2020-05).

Informed Consent Statement: Not applicable.

Data Availability Statement: The data presented in this study are available upon written and reasonable request directed at the corresponding authors.

Conflicts of Interest: The authors declare no conflicts of interest.

References

1. Dehlin, M.; Jacobsson, L.; Roddy, E. Global epidemiology of gout: Prevalence, incidence, treatment patterns and risk factors. *Nat. Rev. Rheumatol.* **2020**, *16*, 380–390. [CrossRef] [PubMed]
2. Sandler, D.P.; Burr, F.R.; Weinberg, C.R. Nonsteroidal anti-inflammatory drugs and the risk for chronic renal disease. *Ann. Intern. Med.* **1991**, *115*, 165–172. [CrossRef] [PubMed]
3. Sutterwala, F.S.; Haasken, S.; Cassel, S.L. Mechanism of NLRP3 inflammasome activation. *Ann. N. Y. Acad. Sci.* **2014**, *1319*, 82–95. [CrossRef] [PubMed]
4. Elliott, E.I.; Sutterwala, F.S. Initiation and perpetuation of NLRP3 inflammasome activation and assembly. *Immunol. Rev.* **2015**, *265*, 35–52. [CrossRef]
5. He, Y.; Hara, H.; Núñez, G. Mechanism and Regulation of NLRP3 Inflammasome Activation. *Trends Biochem. Sci.* **2016**, *41*, 1012–1021. [CrossRef]
6. Lamkanfi, M.; Dixit, V.M. Inflammasomes and their roles in health and disease. *Annu. Rev. Cell Dev. Biol.* **2012**, *28*, 137–161. [CrossRef]
7. Swanson, K.V.; Deng, M.; Ting, J.P.-Y. The NLRP3 inflammasome: Molecular activation and regulation to therapeutics. *Nat. Rev. Immunol.* **2019**, *19*, 477–489. [CrossRef]
8. Lee, J.H.; Kim, H.S.; Yang, G.; Kim, H.J. Natural Products as a Novel Therapeutic Strategy for NLRP3 Inflammasome-Mediated Gout. *Front. Pharmacol.* **2022**, *13*, 861399. [CrossRef]
9. Yun, S.-W.; Seo, Y.-J.; Kwon, J.E.; Park, D.W.; Lee, Y.-G.; Choe, T.H.; Kim, S.-K.; Lee, H.S.; Kim, H.; Kang, S.C. Preclinical evaluation of *Zanthoxylum piperitum* Benn., traditional muscle pain remedy, for joint inflammation. *J. Ethnopharmacol.* **2022**, *286*, 114921. [CrossRef]
10. Yang, S.Y.; Lee, S.-H.; Tai, B.H.; Jang, H.-D.; Kim, Y.H. Antioxidant and Anti-Osteoporosis Activities of Chemical Constituents of the Stems of *Zanthoxylum piperitum*. *Molecules* **2018**, *23*, 457. [CrossRef]

11. Lee, J.-H.; Jang, M.; Seo, J.; Kim, G.H. Antibacterial effects of natural volatile essential oil from *Zanthoxylum piperitum* A.P. DC. against foodborne pathogens. *J. Food Biochem.* **2012**, *36*, 667–674. [CrossRef]
12. Hwang, K.-A.; Kwon, J.E.; Noh, Y.; Park, B.; Jeong, Y.J.; Lee, S.-M.; Kim, S.-Y.; Kim, I.; Kang, S.C. Effects of *Zanthoxylum piperitum* ethanol extract on osteoarthritis inflammation and pain. *Biomed. Pharmacother.* **2018**, *105*, 481–490. [CrossRef] [PubMed]
13. Gwon, S.Y.; Ahn, J.Y.; Kim, T.W.; Ha, T.Y. *Zanthoxylum piperitum* DC ethanol extract suppresses fat accumulation in adipocytes and high fat diet-induced obese mice by regulating adipogenesis. *J. Nutr. Sci. Vitaminol.* **2012**, *58*, 393–401. [CrossRef]
14. Park, Y.S.; Park, Y.S.; Nam, G.H.; Nam, G.H.; Jo, K.J.; Jo, K.J.; Kawk, H.W.; Kawk, H.W.; Kim, S.Y.; Kim, S.Y.; et al. Extract from *Zanthoxylum piperitum* Induces Apoptosis of AGS Gastric Cancer Cells Through Akt/MDM2/p53 Signaling Pathway. *Chin. J. Integr. Med.* **2021**, *27*, 752–759. [CrossRef]
15. Zhang, Z.; Shen, P.; Lu, X.; Li, Y.; Liu, J.; Liu, B.; Fu, Y.; Cao, Y.; Zhang, N. In Vivo and In Vitro Study on the Efficacy of Terpinen-4-ol in Dextran Sulfate Sodium-Induced Mice Experimental Colitis. *Front. Immunol.* **2017**, *8*, 558. [CrossRef]
16. Yagnik, D.R.; Hillyer, P.; Marshall, D.; Smythe, C.D.W.; Krausz, T.; Haskard, D.O.; Landis, R.C. Noninflammatory phagocytosis of monosodium urate monohydrate crystals by mouse macrophages: Implications for the control of joint inflammation in gout. *Arthritis Rheum.* **2000**, *43*, 1779–1789. [CrossRef]
17. Landis, R.C.; Yagnik, D.R.; Florey, O.; Philippidis, P.; Emons, V.; Mason, J.C.; Haskard, D.O. Safe disposal of inflammatory monosodium urate monohydrate crystals by differentiated macrophages. *Arthritis Rheum.* **2002**, *46*, 3026–3033. [CrossRef]
18. Shin, O.S.; Isberg, R.R.; Akira, S.; Uematsu, S.; Behera, A.K.; Hu, L.T. Distinct roles for MyD88 and Toll-like receptors 2, 5, and 9 in phagocytosis of Borrelia burgdorferi and cytokine induction. *Infect. Immun.* **2008**, *76*, 2341–2351. [CrossRef]
19. Supabphol, R.; Tangjitjareonkun, J. Chemical Constituents and Biological Activities of *Zanthoxylum limonella* (Rutaceae): A Review. *Trop. J. Pharm. Res.* **2015**, *13*, 2119. [CrossRef]
20. Beier, R.C.; Byrd, J.A., II; Kubena, L.F.; Hume, M.E.; McReynolds, J.L.; Anderson, R.C.; Nisbet, D.J. Evaluation of linalool, a natural antimicrobial and insecticidal essential oil from basil: Effects on poultry. *Poult. Sci.* **2014**, *93*, 267–272. [CrossRef]
21. Cordeiro, L.; Figueiredo, P.; Souza, H.; Sousa, A.; Andrade-Júnior, F.; Medeiros, D.; Nóbrega, J.; Silva, D.; Martins, E.; Barbosa-Filho, J.; et al. Terpinen-4-ol as an Antibacterial and Antibiofilm Agent against *Staphylococcus aureus*. *Int. J. Mol. Sci.* **2020**, *21*, 4531. [CrossRef] [PubMed]
22. Huang, X.; Du, L.; Liu, T.; Ma, R.; Liu, X.; Yuan, H.; Liu, S. Insecticidal Activity of a Component, (-)-4-Terpineol, Isolated from the Essential Oil of *Artemisia lavandulaefolia* DC. against *Plutella xylostella* (L.). *Insects* **2022**, *13*, 1126. [CrossRef] [PubMed]
23. Loza-Tavera, H. Monoterpenes in essential oils: Biosynthesis and properties. *Adv. Exp. Med. Biol.* **1999**, *464*, 49–62. [CrossRef]
24. Woo, J.; Yang, H.; Yoon, M.; Gadhe, C.G.; Pae, A.N.; Cho, S.; Lee, C.J. 3-Carene, a Phytoncide from Pine Tree Has a Sleep-enhancing Effect by Targeting the $GABA_A$-benzodiazepine Receptors. *Exp. Neurobiol.* **2019**, *28*, 593–601. [CrossRef]
25. Kanehisa, M.; Goto, S. KEGG: Kyoto Encyclopedia of Genes and Genomes. *Nucleic Acids Res.* **2000**, *28*, 27–30. [CrossRef]
26. Kusuda, M.; Inada, K.; Ogawa, T.-O.; Yoshida, T.; Shiota, S.; Tsuchiya, T.; Hatano, T. Polyphenolic constituent structures of *Zanthoxylum piperitum* fruit and the antibacterial effects of its polymeric procyanidin on methicillin-resistant *Staphylococcus aureus*. *Biosci. Biotechnol. Biochem.* **2006**, *70*, 1423–1431. [CrossRef]
27. Lee, S.-J.; Lim, K.-T. Glycoprotein of *Zanthoxylum piperitum* DC has a hepatoprotective effect via anti-oxidative character in vivo and in vitro. *Toxicol. In Vitro* **2008**, *22*, 376–385. [CrossRef]
28. Lamkanfi, M.; Dixit, V.M. Mechanisms and functions of inflammasomes. *Cell* **2014**, *157*, 1013–1022. [CrossRef]
29. Hopkins, A.L. Network pharmacology: The next paradigm in drug discovery. *Nat. Chem. Biol.* **2008**, *4*, 682–690. [CrossRef]
30. Kim, M.H.; Lee, H.; Ha, I.J.; Yang, W.M. *Zanthoxylum piperitum* alleviates the bone loss in osteoporosis via inhibition of RANKL-induced c-fos/NFATc1/NF-κB pathway. *Phytomedicine* **2021**, *80*, 153397. [CrossRef]
31. Bauernfeind, F.G.; Horvath, G.; Stutz, A.; Alnemri, E.S.; MacDonald, K.; Speert, D.; Fernandes-Alnemri, T.; Wu, J.; Monks, B.G.; Fitzgerald, K.A.; et al. Cutting edge: NF-κB activating pattern recognition and cytokine receptors license NLRP3 inflammasome activation by regulating NLRP3 expression. *J. Immunol.* **2009**, *183*, 787–791. [CrossRef] [PubMed]
32. Masumoto, J.; Taniguchi, S.; Ayukawa, K.; Sarvotham, H.; Kishino, T.; Niikawa, N.; Hidaka, E.; Katsuyama, T.; Higuchi, T.; Sagara, J. ASC, a novel 22-kDa protein, aggregates during apoptosis of human promyelocytic leukemia HL-60 cells. *J. Biol. Chem.* **1999**, *274*, 33835–33838. [CrossRef] [PubMed]
33. Lu, A.; Magupalli, V.G.; Ruan, J.; Yin, Q.; Atianand, M.K.; Vos, M.R.; Schröder, G.F.; Fitzgerald, K.A.; Wu, H.; Egelman, E.H. Unified polymerization mechanism for the assembly of ASC-dependent inflammasomes. *Cell* **2014**, *156*, 1193–1206. [CrossRef] [PubMed]
34. Stutz, A.; Horvath, G.L.; Monks, B.G.; Latz, E. ASC speck formation as a readout for inflammasome activation. *Methods Mol. Biol.* **2013**, *1040*, 91–101. [CrossRef]

Disclaimer/Publisher's Note: The statements, opinions and data contained in all publications are solely those of the individual author(s) and contributor(s) and not of MDPI and/or the editor(s). MDPI and/or the editor(s) disclaim responsibility for any injury to people or property resulting from any ideas, methods, instructions or products referred to in the content.

pharmaceuticals

MDPI

Systematic Review

Efficacy and Key Materials of East Asian Herbal Medicine Combined with Conventional Medicine on Inflammatory Skin Lesion in Patients with *Psoriasis Vulgaris*: A Meta-Analysis, Integrated Data Mining, and Network Pharmacology

Hee-Geun Jo [1,2,†], Hyehwa Kim [3,†], Eunhye Baek [4], Donghun Lee [1,*] and Ji Hye Hwang [5,*]

1 Department of Herbal Pharmacology, College of Korean Medicine, Gachon University, 1342 Seongnamdae-ro, Sujeong-gu, Seongnam-si 13120, Gyeonggi-do, Republic of Korea; jho3366@hanmail.net
2 Naturalis Inc. 6, Daewangpangyo-ro, Bundang-gu, Seongnam-si 13549, Gyeonggi-do, Republic of Korea
3 KC Korean Medicine Hospital 12, Haeol 2-gil, Paju-si 10865, Gyeonggi-do, Republic of Korea; hh1635@hanmail.net
4 RexSoft Inc., 1 Gwanak-ro, Gwanak-gu, Seoul 08826, Republic of Korea
5 Department of Acupuncture and Moxibustion Medicine, College of Korean Medicine, Gachon University, 1342 Seongnamdae-ro, Sujeong-gu, Seongnam-si 13120, Gyeonggi-do, Republic of Korea
* Correspondence: dlee@gachon.ac.kr (D.L.); jhbori@gachon.ac.kr (J.H.H.)
† These authors contributed equally to this work.

Abstract: Psoriasis is a chronic inflammatory disease that places a great burden on both individuals and society. The use of East Asian herbal medicine (EAHM) in combination with conventional medications is emerging as an effective strategy to control the complex immune-mediated inflammation of this disease from an integrative medicine (IM) perspective. The safety and efficacy of IM compared to conventional medicine (CM) were evaluated by collecting randomized controlled trial literature from ten multinational research databases. We then searched for important key materials based on integrated drug data mining. Network pharmacology analysis was performed to predict the mechanism of the anti-inflammatory effect. Data from 126 randomized clinical trials involving 11,139 patients were used. Compared with CM, IM using EAHM showed significant improvement in the Psoriasis Area Severity Index (PASI) 60 (RR: 1.4280; 95% CI: 1.3783–1.4794; $p < 0.0001$), PASI score (MD: −3.3544; 95% CI: −3.7608 to −2.9481; $p < 0.0001$), inflammatory skin lesion outcome, quality of life, serum inflammatory indicators, and safety index of psoriasis. Through integrated data mining of intervention data, we identified four herbs that were considered to be representative of the overall clinical effects of IM: *Rehmannia glutinosa* (Gaertn.) DC., *Isatis tinctoria* subsp. *athoa* (Boiss.) Papan., *Paeonia × suffruticosa* Andrews, and *Scrophularia ningpoensis* Hemsl. They were found to have mechanisms to inhibit pathological keratinocyte proliferation and immune-mediated inflammation, which are major pathologies of psoriasis, through multiple pharmacological actions on 19 gene targets and 8 pathways in network pharmacology analysis. However, the quality of the clinical trial design and pharmaceutical quality control data included in this study is still not optimal; therefore, more high-quality clinical and non-clinical studies are needed to firmly validate the information explored in this study. This study is informative in that it presents a focused hypothesis and methodology for the value and direction of such follow-up studies.

Keywords: East Asian herbal medicine; psoriasis; systematic review; chronic inflammation; network pharmacology; integrative medicine; social network analysis; association rule mining

1. Introduction

Psoriasis is a chronic inflammatory disease of the skin that affects millions of people worldwide with a wide range of clinical symptoms [1]. A 2013 global epidemiological study found that adult prevalence ranged from 0.91 to 8.5%. The lifetime prevalence estimated by physicians was 6.3%, and a subsequent Danish cohort study supported this estimate [2,3]. Psoriasis is characterized by chronic inflammation resulting from the uncontrolled proliferation and differentiation of keratinocytes [4,5]. Inflammation of the skin on exposed areas of the body, such as the face and limbs, has a significant negative impact on the daily lives of the majority of people with psoriasis [6]. Many people with persistent psoriasis have a number of complications that might shorten their lifespans [7,8]. Recent studies have found a connection between psoriasis and other chronic diseases that may reduce life expectancy, such as psoriatic arthritis, hypertension, type 2 diabetes, dyslipidemia, myocardial infarction, and stroke [1,9–11]. In addition, a meta-analysis in 2022 revealed that the prevalence of psoriasis appears to increase the incidence of autoimmune thyroid disease [12]. According to these findings, psoriasis must be considered as a systemic disease that can increase the local skin condition of each patient's burden on society [13]. Therefore, to reduce the severe impact of psoriasis on physical, social, and psychological well-being, it is necessary to find a way to effectively treat psoriasis.

Psoriasis can be clinically classified into four main subtypes: erythrodermic, guttate, pustular, and plaque psoriasis [1]. Skin-related lesions make up the majority of psoriasis lesions, with *psoriasis vulgaris* accounting for approximately 90% of all cases [1,4,7]. Erythematous, itchy plaques coated in silvery plaques are features of *psoriasis vulgaris* [4,14]. The scalp, elbows, face, and lumbosacral area are the most common places for plaques to appear as scaly skin lesions. They can vary widely in width and thickness [1,4]. When these plaques cover only 3–5% of the body surface in mild cases, topical treatment or phototherapy is often helpful [15]. However, systemic oral medications are required for moderate-to-severe plaque psoriasis [1,15]. Since psoriasis is a chronic disease that requires long-term medication, people with psoriasis often require lifelong therapy [16]. Consequently, all treatment plans must adhere to strict standards for patient safety. Although many conventional medicines (CMs) are available, there are still certain limitations to systemic therapy for psoriasis that need to be addressed. For example, acitretin is contraindicated in women of reproductive age due to its teratogenicity and because adverse events (AEs) such as dose-dependent alopecia and xerosis have been observed. Meanwhile, the long-used drug methotrexate has side effects including hepatotoxicity and bone marrow suppression that can lead to cirrhosis [4,17]. On the other hand, although methotrexate has been used for a long time, the possibility of inducing liver cirrhosis along with side effects such as hepatotoxicity and bone marrow suppression is still a concern [1,18,19].

Natural products have been considered as promising candidates for the treatment of various chronic diseases worldwide because they are safer than novel synthetic drugs even after prolonged administration and with high patient compliance [14,20–22]. Of these trends, the most active area in the discovery of promising materials related to psoriasis is East Asian herbal medicine (EAHM) [23–29]. The term "EAHM" refers to herbal therapies approved for use as medicines in a number of East Asian countries, including China, Taiwan, Korea, and Japan [23–27,30–34]. EAHM is significantly different from natural resources in other parts of the world as many similar medicinal plants are commonly listed in the pharmacopeia of East Asian countries [24,32,35]. At the same time, integrative medicine (IM) studies of chronic disease management are often conducted in countries with a history of medicinal herb use [36–45]. IM is a comprehensive strategy that uses both complementary and conventional therapies simultaneously. As a subset of IM, the combination of CMs and natural products has shown superior therapeutic efficacy and safety compared to standard therapies for a number of ailments, including COVID-19, cancer, stroke, chronic pruritus, and rheumatoid arthritis [36,40,46–54]. Meanwhile, a recent study using a bioinformatics approach reported that the effectiveness of widely used EAHM prescriptions in psoriasis is related to the suppression of oxidative stress and the alleviation of the resulting inflammatory pathology [55].

Considering several previous studies on this topic, it is reasonable to assume that IM may improve the psoriasis area and severity index (PASI) and clinical symptoms while reducing AEs associated with CMs in psoriasis patients [56–60]. However, several issues must be resolved before identifying candidate materials for optimal IM utilization and making robust clinical decisions. First, EAHM was administered in the form of a polyherbal prescription tailored to the clinical findings of each patient, and the maximization of the synergistic effect expressed in the appropriate combination of these herbs was estimated as the key mechanism of action [30,41,61–63]. Therefore, it is challenging to identify candidate materials for optimal combination pharmacotherapy and sufficient discussion cannot be performed based only on the existing studies. Moreover, despite the considerable amount of evidence gathered, a detailed analysis of the dosage route and composition of diverse materials is lacking. These are important factors that prevent coherent conclusions from being drawn. Therefore, more research is needed to inform whether IM with EAHMs is a useful intervention with a benefit in psoriasis, and specifically which EAHMs should be used.

In recent years, "reliable data" and the "integration of a wide range of analytical methodologies" have been suggested as prerequisites for natural product-based drug research [64]. The authors established the following research objectives based on their understanding and expertise from previous studies: (1) a systematic review of randomized controlled clinical trials (RCTs) will determine whether IM using EAHM is worthy of investigation for the treatment of inflammatory skin lesions in plaque psoriasis, and (2) further data analysis of herbal prescription data collected through this review will lead to hypotheses regarding promising candidates for the best IM for plaque psoriasis. Through the above research, the authors have attempted to open a discussion on a multifaceted analysis method that can overcome the complex variable problem of EAHM and produce useful information that can be used in useful follow-up studies.

2. Materials and Methods

2.1. Research Workflow through Integrated Methodology

This study was carried out according to the following steps: (1) Clinical trial information was collected according to the systematic review methodology. At this stage, statistically valid evidence related to IM was obtained for efficacy and safety. Through a meta-analysis of the collected clinical trials, we first determined whether the IM approach using EAHM was a data pool worth searching for useful candidates. Once this condition was met, multiple data mining were performed as the second step, (2) deriving core materials through multiple data mining of drug information, was carried out. Through this, useful core materials predicted to exert the greatest weight in the pharmacology of EAHM theory were selected. Then, the appropriate dosage and duration of administration were investigated. (3) Prediction of the mechanism supporting the efficacy of the derived core material. The mechanism of action supporting the efficacy of the identified core herbs on inflammatory skin damage in psoriasis was predicted using systems biology methodology. Finally, based on the above steps, new IM information derived from clinical trial data was searched for crude drug usage patterns, doses, administration periods, compounds, targets, and pathways. This study was conducted as a process of building multidisciplinary-integrative-decision making-actual achievement-scientific creativity (M.I.D.A.S) research platform. The workflow of the methodology described above is summarized in Figure 1.

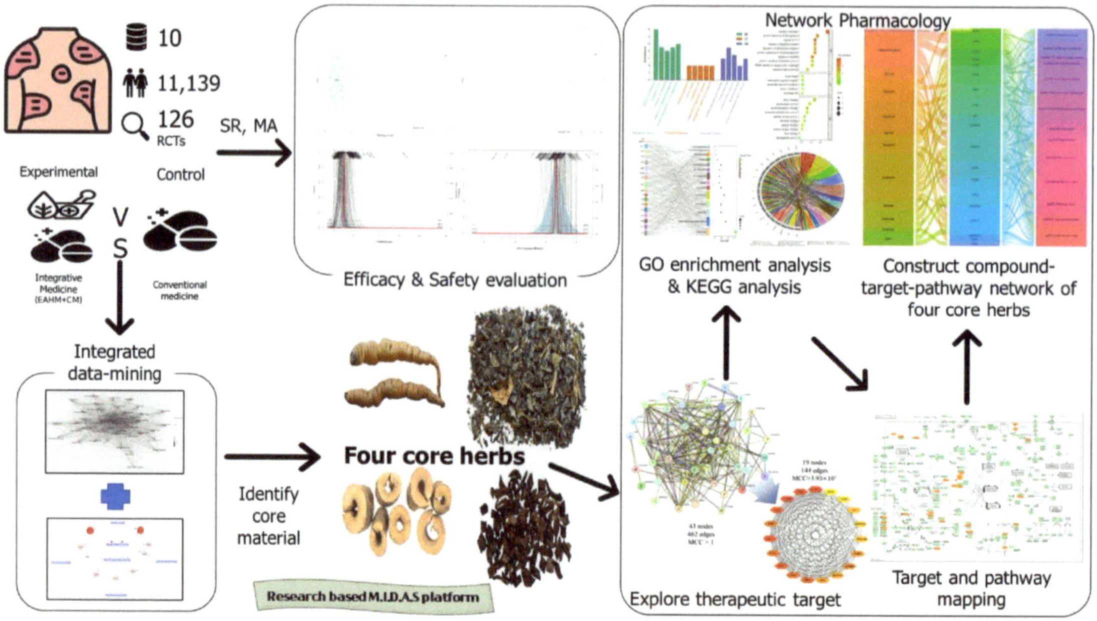

Figure 1. Study workflow of the multi-faceted analysis of IM for inflammatory skin lesion of psoriasis.

2.2. Data Sources and Search Strategy

The systematic exploration of clinical trial data for this study was conducted in accordance with the Preferred Reporting Items for Systematic Reviews and Meta-Analysis (PRISMA) 2020 statement [65], and the protocol was pre-registered in PROSPERO (registration number: CRD 42022296852, available from https://www.crd.york.ac.uk/PROSPERO/display_record.php?RecordID=296852). In addition, the protocol of this study has been formally published [66].

RCTs that evaluated the efficacy and safety of IM for *psoriasis vulgaris* were searched in the following 10 electronic databases from their inception until 29 July 2021: three English databases (PubMed, Cochrane Library, and Embase), four Korean databases (Korean Studies Information Service System, Research Information Service System, Oriental Medicine Advanced Searching Integrated System, and Korea Citation Index), two Chinese databases (Chinese National Knowledge Infrastructure Database, Wanfang data), and one Japanese database (Citation Information by National Institute of Informatics). The overall literature search procedure was carried out independently by two researchers (HGJ and HK). Detailed search strategies are presented in Supplementary Table S1.

2.3. Study Selection

2.3.1. Type of Studies

Only RCTs evaluating the efficacy and safety of IM for inflammatory skin lesion in plaque psoriasis were included. There were no restrictions on the language or publication time. Studies that meet the following criteria were excluded: (a) studies that are not RCTs or quasi RCTs; (b) studies not related to plaque psoriasis or related diseases; (c) primary intervention not related to IM; (d) no oral administration of medications; (e) not a clinical trial; (f) case reports or reviews; (g) studies not published in scientific peer-reviewed journals, including postgraduate theses or dissertations; and (h) studies in which the experimental intervention was not based on an IM approach, such as EAHM monotherapy.

2.3.2. Type of Participants

There were no restrictions on age, gender, or race, and studies were only eligible for inclusion if they were conducted in patients with a diagnosis of *psoriasis vulgaris*. Only studies that provided official or validated diagnostic criteria were included. Studies that included patients with additional psoriasis subtypes, such as psoriatic arthritis, guttate psoriasis, palmoplantar pulposus, and erythrodermic psoriasis, were excluded from the review because the focus of the review was on plaque psoriasis.

2.3.3. Type of Interventions

RCTs evaluating the active intervention of IM (EAHM combined with CMs) in the treatment group versus CMs alone in the control group were included. For inflammatory skin lesions in psoriasis, all dosage forms of IM intervention were considered, including decoction, granules, capsules, and tablets. There were no restrictions on dosage or duration of therapy, but oral ingestion was the only acceptable method of administration. Trials that included non-drug treatment, acupuncture, massage, or other complementary therapies only in the experimental group and not in the control group were excluded. Studies that could not confirm the composition of specific herbal ingredients included in the EAHM formula were excluded.

2.3.4. Type of Outcome Measures

The primary endpoint was the response rate of patients with a PASI improvement of 60% (PASI 60). The primary outcome was also the absolute difference in PASI scores between the groups. In the first set of secondary outcomes, an improved PASI of 70% (PASI 70), recurrence rate, dermatological quality of life index (DLQI), and visual analog scale (VAS) were adopted to measure the clinical response of inflammatory skin lesions in patients with psoriasis. As the second group of secondary outcomes, TNF-α, IL-8, IL-17, IL-22, IL-23, and IFN-γ were selected to evaluate changes in inflammation-related biomarkers that support improvement of inflammatory skin lesions. The safety of IM was assessed using the incidence rates of adverse events (AEs) in each group as the third set of secondary outcomes.

2.4. Data Extraction and Management

Two researchers (HGJ and HK) independently retrieved the titles and abstracts of potentially eligible articles using the search approach described above. The inclusion and exclusion criteria were then used to guide a full-text review. Two reviewers (HGJ and HK) independently retrieved data from the included studies. Two reviewers independently extracted the following data for the selected trials:

1. Publication information (title, first author, year of publication, and funding source).
2. Study characteristics (trial design, randomization method, sample size, treatment duration, and morbidity period).
3. Participants (age, sex, diagnostic criteria, and number of participants in each group).
4. Intervention (experimental intervention, comparator, ingredients, and detailed information on intervention frequency of medication, dosage, mode of delivery, and course of treatment).
5. Outcomes (primary and secondary outcomes, measurement point, blinding of outcome assessment, and AEs).

All disagreements were resolved through discussions with the researchers and the other author (DL).

2.5. Methodological Quality Assessment

The methodological quality of each included study was independently evaluated by two investigators (HGJ and HK) according to the revised version of risk of bias in randomized trials (RoB 2.0) [67]. Five areas of bias are addressed by RoB 2.0: bias arising from the randomization process, bias deviating from the intended intervention, bias due to the omission of outcome data, and bias in the selection of reported outcomes. There were three categories used to rate the study's methodological quality: "high risk of bias," "low risk of bias," and "some concerns.". Disagreements between investigators were resolved by consensus with the assistance of another author (DL).

2.6. Quality of Evidence according to Outcome Measures

The overall quality of evidence for each outcome was evaluated using the Grading of Recommendations Assessment, Development, and Evaluation (GRADE) pro framework [68]. GRADE evaluates the overall quality of evidence at four levels: very low, low, moderate, and high. The level of evidence is lowered according to factors such as the risk of bias, inconsistency, indirectness, imprecision, and publication bias.

2.7. Statistical Analysis

2.7.1. Data Synthesis of Clinical Outcomes

In the meta-analysis of the included data from clinical trials, the effect size and 95% confidence intervals (CI) were estimated using only the random-effects model. When either the $\chi 2$ test was <0.10 or I^2 was $\geq$50%, heterogeneity was considered statistically significant. Statistical synthesis of individual research results was performed using R software (version 4.1.2) and R studio program (Version 1.4.1106, Integrated Development for R. Rstudio, PBC, Boston, MA, USA) using the default settings of the "meta" and "metafor" packages [69].

Trials were categorized according to the type of intervention and comparator. The relative risk (RR) and 95% confidence interval (CI) were calculated for the PASI60, PASI70, and recurrence rates. The mean difference (MD) and 95% confidence interval (CI) were estimated for the PASI, DLQI, and VAS scores. For TNF-α, IL-8, IL-17, IL-22, IL-23, and IFN-γ, the standardized mean difference (SMD) and 95% confidence intervals (CIs) were computed to integrate the results of the indicators in different units related to the same measurement object. Due to the need to estimate a causal relationship, the odds ratio (OR) was used to quantify incidence rates of adverse events.

In addition to the forest plot, a drapery plot was used to better represent the effect size in this study, rather than relying solely on the primary outcome synthesis data with $p < 0.05$ as the level of significance [70]. Meanwhile, this study combined much more data than the previous meta-analysis. Forest plots cannot effectively represent such enormous amounts of data. After considering the results of more than 50 studies, an orchard plot was chosen to display the data instead of a forest plot [71].

When heterogeneity was identified in the primary outcomes of the meta-analysis, further analyses were performed to determine the explanation. First, a leave-one-out sensitivity analysis was performed to determine whether the included data were affected by outliers. If no outliers were identified, a subgroup analysis was performed after performing meta-regression analysis for the following seven pre-specified variables: (i) comparator drug, (ii) treatment duration, (iii) source of investigational medicine, (iv) formulation type, (v) sample size, (vi) overall risk of bias, and (vii) randomization method that caused a significant difference in results. A contour-enhanced funnel plot was used in the meta-analysis for the primary outcome of more than ten trials to distinguish publication bias [72]. Egger's and Begg's tests were performed to confirm the existence of publication bias for the asymmetry of the visually observed funnel plot [73,74].

2.7.2. Deriving Core Herbs Based on Data Mining Approach

In order to maximize the synergistic effect, the principle of use for EAHM differs from that of other natural medicines in that it uses a polyherbal formulation. "Gun-Shin-Jwa-Sa", also known as Sovereign-Minister-Assistant-Courier in the official nomenclature of the WHO, is the formulation theory for the ideal combination of EAHM [30]. The major pharmacological medication among them is referred to as the sovereign or monarch drug. EAHM prescriptions can be taken as a single dose due to the effects of this sovereign drug, while the other drugs both reduce the toxicity of the entire prescription and enhance the effects of the sovereign drugs [24,61,75]. Therefore, when data mining is used to identify core herbs that are likely to be sovereign drugs, their pharmacological effects can be considered representative of the overall clinical efficacy of EAHM. Based on this theoretical background, this review assumed that the improvement in psoriasis observed in the overall IM clinical data could be explained by the pharmacological effects of the core herbs, which were derived based on the following methodology.

First, the type of EAHM commonly used in more than 10% of the studies, the dose, and the duration of use of commonly prescribed EAHM were examined using a descriptive statistical approach.

Second, using social network analysis, we selected the EAHM component of the IM prescription construct, which plays a central role in the relationship between the different drugs. The social network analysis used in this study was divided into two parts. The network was assumed to be undirected and the degree distribution of the connections between the common EAHM substances utilized in each IM prescription was observed. The average degree of connection in this situation may be represented as follows because an undirected network is assumed:

$$A = \sum_{K=1}^{n} kP(k) = \frac{2E}{n}$$

where n is the number of nodes and E is the number of links.

By evaluating the power of specific materials on the association between frequently prescribed EAHMs, centrality was used to discover EAHM with relatively greater influence. The eigenvector centrality scale was used to analyze the association between each of the concurrently administered herbs.

$$Ci = \frac{1}{\lambda} \sum_{j \in N(i)} A_{ij} C_j$$

where N(i) represents the collection of herbs that are close to material i and λ is the eigenvalue of material i, a constant determined by the algorithm. If materials i and j are connected in the n $\times$ n-direction adjacency matrix A, Aij becomes "1"; otherwise, it becomes "0". Herb i and its neighbors constitute herb j, which is the eigenvector centrality value of Cj. Centrality measurements were performed on materials showing a frequency of use in more than 5% of the included trials.

Third, association rule mining was performed to discover a meaningful combination pattern among all materials included in the IM prescription [76]. The frequent combination pattern itself may be a relevant herb unit representing clinical tacit knowledge, as EAHM is administered in a combination to optimize synergy. Support, confidence, and lift were the primary measures used in association rule mining. The support of the itemset is the proportion of transactions in the dataset that contain it. Itemsets whose support exceeds a user-defined minimum support level are considered frequent. A rule X $\Rightarrow$ Y's confidence is expressed as conf (X $\Rightarrow$ Y) = support(X$\cup$Y)/support(X). Assuming that the transactions in question also contain the antecedent, this may be explained as an estimate of the probability P (Y | X) or the likelihood of finding the rule's consequent in those transactions. Another index called the lift can be represented by the following equation:

$$lift(X \Rightarrow Y) = \frac{supp(X \cup Y)}{supp(X)supp(Y)}$$

The lift can be explained as the deviation of all rules supports from the support anticipated under independence, given the support of the rule on both sides. Stronger associations are indicated by higher lift values.

Finally, through the above three steps, EAHM that simultaneously satisfies the three conditions of "frequent use in clinical practice", "central position within individual prescriptions", and "strong association with other drugs" was selected as the core herbs. The representative values of the dosing period and dose of these drugs are presented together by calculating the inter-quartile range.

2.7.3. Prediction of Anti-Inflammatory Mechanisms Based on Network Pharmacology

A network pharmacology analysis was performed to explore the anti-inflammatory mechanism of core herbs derived through the above data mining. The chemical ingredients in herbs were obtained from the TCMSP (Traditional Chinese medicine systems pharmacology database and analysis platform, https://therbsp-e.com/), TCMID (Traditional Chinese Medicine integrative database for herb molecular mechanism analysis, http://bidd.group/TCMID/), HERB (a high-throughput experiment-and reference-guided database of traditional Chinese medicine, http://herb.ac.cn/), and ETCM (an encyclopedia of traditional Chinese medicine, http://www.tcmip.cn/ETCM/) databases [77–80]. In this study, compounds with an oral bioavailability (OB) of 20% and a drug-like (DL) index of 0.1 were first screened according to the information provided by the TCMSP DB, and then a potential compound was finally selected by performing a second round of absorption, distribution, metabolism, and excretion (ADME) prediction using the SwissADME platform (http://www.swissadme.ch/). In the SwissADME platform, compounds were selected if at least three of the five ADME rules of Lipsinski, Ghose, Veber, Egan, and Muegge were evaluated as "yes" [81]. Potential target genes of selected active compounds were predicted by the SwissTargetPrediction platform (http://www.swisstargetprediction.ch) in the "*Homo sapiens*" setting and selected by the "Probability $\geq$ 0.1" criterion [82]. Using the "*Homo sapiens*" species filter in the Uniprot database (http://www.uniprot.org), the target information for active substances was standardized. Data on psoriasis-related target genes were retrieved from the GeneCards database (http://www.genecards.org) with "*psoriasis vulgaris*" as the keyword. For targets in GeneCards, only those with a score $\geq$ 10 were screened [83].

Using the "Bioinformatics & Evolutionary Genomics site" (https://bioinformatics.psb.ugent.be/webtools/Venn/), Venn diagrams for common targets between core herbs and psoriasis were created. To graphically represent the intricate interactions among chemicals and targets, a network comprising the elements of the core herbs and psoriasis targets was built using Cytoscape (version 3.10.0; https://cytoscape.org/). The protein classification "*Homo sapiens*" and the STRING protein analysis platform (12.0 beta; https://version-12-0.string-db.org/) were used to import the interaction gene targets of the core herbs and psoriasis [84]. The minimum required score to define a protein–protein interaction was set to "medium confidence (confidence score $\geq$ 0.4)". We constructed the PPI network, removed unnecessary protein nodes, and then loaded the data into Cytoscape using the Cytohubba plugin for topological analysis of the PPI network [85,86]. The selection of hub gene targets was based on the score calculated by cytoHubba's Maximum Clique Centrality (MCC) algorithm being greater than three fold of the median. The main method for characterizing the function of gene targets, including biological processes, cellular components, and molecular functions, was gene ontology (GO) functional analysis. The shared targets of the core herbs and psoriasis in signaling pathways were discovered using the Kyoto Encyclopedia of Genes and Genomes (KEGG) enrichment analysis. More than 40 gene function annotation datasets are included in the web tool for gene enrichment analysis Metascape (https://metascape.org/). For GO and KEGG analysis, the hub targets were uploaded to the Metascape platform [87]. The $p < 0.05$ data selection threshold was selected. The KEGG mapper (https://www.genome.jp/kegg/mapper/) was used to explore their underlying molecular mechanisms [88].

3. Results

3.1. Study Identification

By implementing the search strategy, the electronic search of the 10 databases identified 2434 potentially relevant articles. After removing 638 duplicate records, a total of 1796 records were collected. After screening for titles and abstracts, 1115 articles that met at least one of the exclusion criteria were excluded. The full-text assessment was performed on the remaining 460 studies, and 334 articles were excluded for the reasons listed in Figure 2. Finally, 126 eligible studies were included in this meta-analysis [89–214]. The screening process is summarized in the PRISMA 2020 flow diagram (Figure 2).

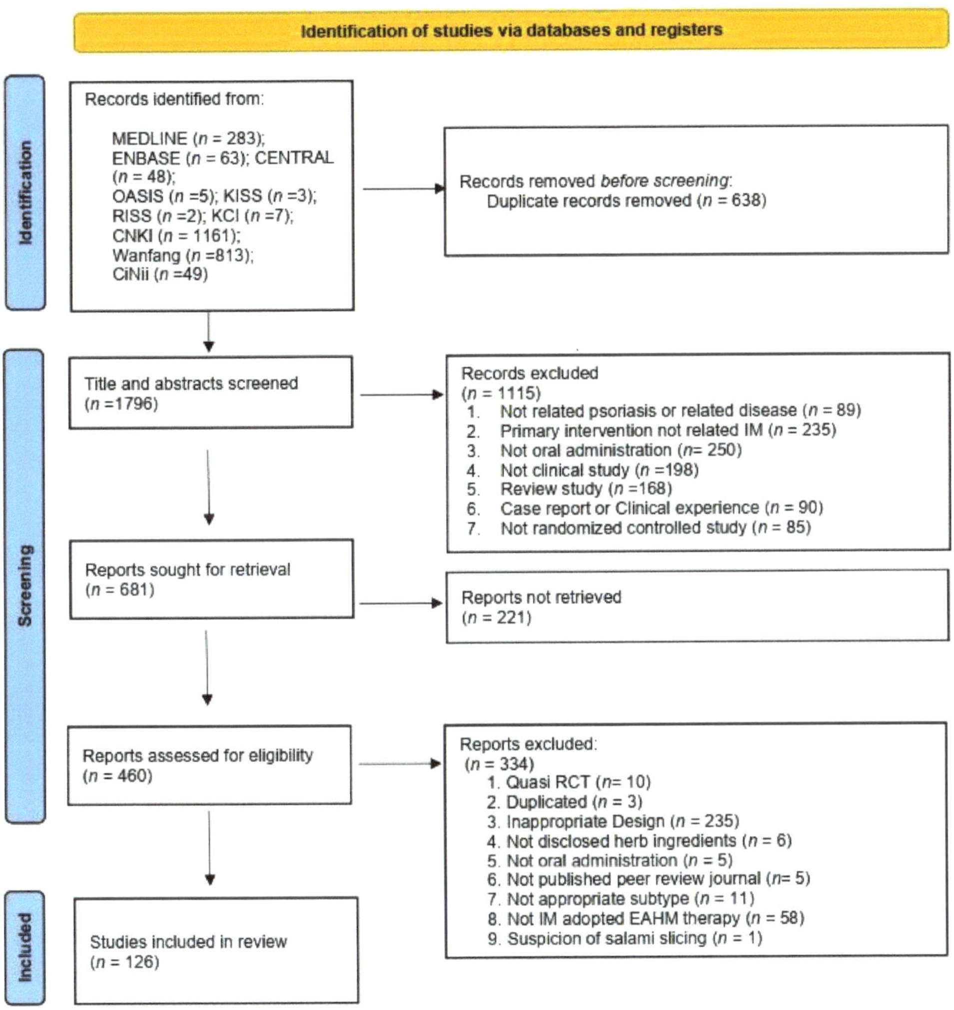

Figure 2. PRISMA 2020 flow diagram.

3.2. Study Characteristics

The basic characteristics of the 126 included studies are summarized in Table 1. In general, the sample size of the included studies ranged from 34 to 260, and 11,139 participants were divided into an experimental (n = 5624) and a control group (n = 5515). Acitretin (78 trials), topical corticosteroids (28 trials), immunosuppressants (8 trials), methotrexate (4 trials), topical urea (3 trials), eritretin (1 trial), topical retinoid (1 trial), topical pyrithione zinc (1 trial), topical boric acid (1 trial), and an ascorbate and pyridoxine combination (1 trial) were administered to the control group in all the included studies. PASI 60 was measured as the primary outcome in 96 studies. The PASI score, a primary outcome measure of the extent of inflammatory skin lesions, was used in 69 studies. In terms of inflammatory skin lesion-related secondary outcomes, PASI 70 was the score measured in 13 studies. Another secondary endpoint, the recurrence rate, was reported in 12 studies. A change in the DLQI was observed in ten studies. The VAS scores were observed in three studies. As a secondary outcome to measure laboratory findings related to inflammatory skin lesions, TNF-α, IL-8, IL-17, IL-22, IL-23, and IFN-γ were reported in twelve, eight, ten, four, five, and six studies, respectively. As a secondary outcome to evaluate the safety of IM versus CM, 75 trials reported information that could be used to compare the incidence rates of AEs in the experimental and control groups. All the included studies reported the treatment duration. It ranged from 2 to 16 weeks, with 31 studies adopting a treatment period of $\geq$12 weeks.

3.3. Risk of Bias

The methodological quality of the 126 included studies is summarized in Figure 3 and Table 2. The risk of bias in the studies was assessed using the Rob 2.0 tool [67]. In 101 studies, the overall risk of bias, in one or more domains, was evaluated to be of "some concern". In all other studies, the overall risk of bias was evaluated to be "high". Twenty-six studies reported detailed information on the randomization process, and only five studies explicitly revealed that a single-blind design was adopted [129,152,161,163,202]. In the remaining studies, it was difficult to obtain specific information other than the fact that randomization was performed, and the adoption of a blinded design was uncertain. In addition, a potential common factor that could affect the risk of bias was the absence of a preregistered protocol that could prevent selective outcome reporting.

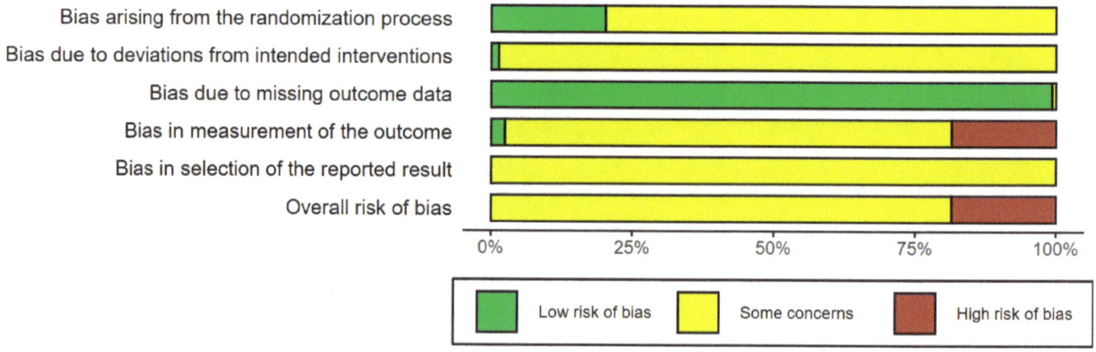

Figure 3. Risk of bias 2.0 summary: authors' judgements for each risk of bias domain across all included trials.

Table 1. Basic demographic data and intervention of studies included in the review.

Included Study (Reference)	Trial Design	Randomization Method	Number of Participants (Male/Female); Age (Mean ± SD)		Interventions		Morbidity Period (Mean ± SD or Range)		Outcome Index (Intergroup Differencies *p*-Value)	Course of Treat-ment	Adverse Events (Case/Symptom)
			Trial	Control	Trial	Control	Trial	Control			
Che 2004 [164]	Randomized; Single center; Parallel	NR	Both group 75 (41/34) 11–44 y Trial: 45	Both group 75 (41/34) 11–44 y Control: 30	1. Erdonghuoxue decoction (t.i.d.) 2. Immunosup-pressant (compound amino peptide tablets 5t, b.i.d.)	1. Immunosup-pressant (compound amino peptide tablets 5t, b.i.d.)	Both group 4 m–10 y	Both group 4 m–10 y	1. PASI 60 ($p < 0.05$)	4 w	Trial: 5 AEs Control: 22 AEs Including thirst, hyperhidrosis, pruritus, skin scale, dry lips, cheilitis, dry mouth, dry nose, nausea
Chen 2004 [211]	Randomized; Single center; Parallel	NR	30 (18/12) 32.4 y (16–65 y)	26 (16/10) 33.2 y (17–63 y)	1. Anti-psoriasis formula (q.d.) 2. Immunosup-pressant (compound amino peptide tablets 5t, b.i.d.)	1. Immunosup-pressant (compound amino peptide tablets 5t, b.i.d.)	6.2 y (3 m-28 y)	6.1 y (2 m-30 y)	1. PASI 60 ($p < 0.05$)	6 w	Trial: 29 AEs (6 xerostomia, 8 dry lips, 5 xeroderma, 4 pruritus, 6 skin scale) Control: 84 AEs (18 xerostomia, 20 dry lips, 17 xeroderma, 13 pruritus, 16 skin scale)
Xu 2005 [186]	Randomized; Single center; Parallel	NR	45 (27/18) Range 11–63 y	30 (18/12) Range 15–57 y	1. Liangxuejiedu decoction (b.i.d.) 2. Immunosup-pressant (compound amino peptide tablets 5t, b.i.d.)	1. Immunosup-pressant (compound amino peptide tablets 5t, b.i.d.)	3 w–22 y	3 w–22 y	1. PASI 60 ($p < 0.05$)	4 w	Trial: 5 AEs Control: 22 AEs Including skin scale, cheilitis, dry nasal cavity, thirst, hyperhidrosis, pruritus, nausea
Liu 2006 [137]	Randomized; Single center; Parallel	NR	Both group 60 (31/29) Trial: 40 (NR) 33.63 y (16–53 y)	Both group 60 (31/29) Control: 20 (NR) 32.7 y (21–64 y)	1. Jianpiyishen decoction (q.d.) 2. Immunosup-pressant (compound amino peptide tablets 5t, b.i.d.)	1. Immunosup-pressant (compound amino peptide tablets 5t, b.i.d.)	71.191 m (1–168 m)	70.95 m (3–120 m)	1. PASI 70 ($p < 0.05$)	8 w	Trial: 41 AEs (8 xerostomia, 14 xeroderma, 8 skin scale, 11 pruritus) Control: 71 AEs (17 xerostomia, 18 xeroderma, 18 skin scale, 18 pruritus)

Table 1. *Cont.*

Included Study (Reference)	Trial Design	Randomization Method	Number of Participants (Male/Female); Age (Mean ± SD)		Interventions		Morbidity Period (Mean ± SD or Range)		Outcome Index (Intergroup Differencies *p*-Value)	Course of Treat-ment	Adverse Events (Case/Symptom)
			Trial	Control	Trial	Control	Trial	Control			
Chen 2007 [92]	Randomized; Single center; Parallel Three arm trial	NR	Both group 222 (139/83) 33 y (19–58 y) Trial (IM): 74	Both group 222 (139/83) 33 y (19–58 y) Control: 74	1. Xiaoyin granule (3.5 g, t.i.d) 2. Acitretin capsule (10 mg, t.i.d.)	1. Acitretin capsule (10 mg, t.i.d.)	Both group 4.6 y (1–18 y)	Both group 4.6 y (1–18 y)	1. PASI 60 ($p < 0.05$)	16 w	Trial: 36 AEs (26 xeroderma, 3 pruritus, 1 skin poignant itch, 5 gastrointestinal discomfort, 1 hepatic dysfunction) Control: 29 AEs (21 xeroderma, 1 pruritus, 2 skin poignant itch, 4 gastrointestinal discomfort, 1 hepatic dysfunction)
Huang 2007 [129]	Randomized; Single center; Parallel; single blind	Simple ran-domization (envelope concealment method)	49 (30/19) 37 ± 9.12 y	49 (32/17) 38 ± 10.27 y	1. Yinxieling tablet (6t, t.i.d.) 2. Acitretin capsule (0.5 mg/kg/day, q.d.)	1. Acitretin capsule (0.5 mg/kg/day, q.d.)	3.36 ± 5.72 y	3.8 ± 5.44 y	1. PASI 60 ($p < 0.05$) 2. PASI score ($p < 0.01$)	8 w	Trial: 130 AEs (49 xerostomia, 13 xeroma, 21 xeroderma, 11 pruritus, 2 epistaxis, 14 folliculitis, 1 hepatic dysfunction, 19 hyperlipidemia) Control 238 AEs (49 xerostomia, 40 xeroma, 38 xeroderma, 20 pruritus, 8 epistaxis, 40 folliculitis, 8 hepatic dysfunctions, 35 hyperlipidemia)
Zeng 2009 [163]	Randomized; Single center; Parallel; single blind	Simple ran-domization	50 (30/20) 38.61 ± 13.12 y	50 (29/21) 39.41 ± 14.03 y	1. Xiaoyin granule (3.5 g, t.i.d) 2. Topical corticosteroid (clobetasol propionate ointment, b.i.d.)	1. Topical corticosteroid (clobetasol propionate ointment, b.i.d.)	7.68 ± 5.63 y	6.58 ± 5.92 y	1. PASI 60 ($p < 0.01$) 2. PASI score ($p < 0.01$)	4 w	Trial: 15 AEs Control: 18 AEs Including xeroderma, skin scale, pruritus, erythema, mild stabbing

Table 1. *Cont.*

Included Study (Reference)	Trial Design	Randomization Method	Number of Participants (Male/Female); Age (Mean ± SD)		Interventions		Morbidity Period (Mean ± SD or Range)		Outcome Index (Intergroup Differencies *p*-Value)	Course of Treat-ment	Adverse Events (Case/Symptom)
			Trial	Control	Trial	Control	Trial	Control			
Cao 2010 [157]	Randomized; Single center; Parallel Three arm trial	NR	40 (24/16) 35.3 y (18–64 y)	40 (25/15) 33.5 y (19–62 y)	1. Yangzhen decoction (200 mL, b.i.d.) 2. Acitretin capsule (0.5 mg/kg/day, q.d.)	1. Acitretin capsule (0.5 mg/kg/day, q.d.)	43.5 m (6 m–33 y)	40.5 m (6 m–30 y)	1. PASI 60 ($p < 0.01$)	12 w	Trial: 43 AEs (22 xeroderma and xerostomia, 9 pruritus, 11 hyperlipidemia, 2 hepatic dysfunction) Control: 77AEs (36 xeroderma and xerostomia, 27 pruritus, 13 hyperlipidemia, 1 hepatic dysfunction)
He 2010 [191]	Randomized; Single center; Parallel	NR	Both group 78 (36/42) 43.4 ± 6.2 y Trial: 39	Both group 78 (36/42) 43.4 ± 6.2 y Control: 39	1. Xiaoranqudan feng (200 mL, b.i.d) 2. Acitretin capsule (25 mg, q.d.)	1. Acitretin capsule (25 mg, q.d.)	Both group 3.6 ± 1.1 y	Both group 3.6 ± 1.1 y	1. PASI 60 ($p < 0.05$) 2. PASI score ($p < 0.01$)	8 w	Trial: 15 AEs (10 xerostomia, 5 headache) Control: 85 AEs (8 hepatic dysfunction, 30 xeroderma, 26 xerostomia, 7 headache, 14 gastrointestinal discomfort)
Hua 2010 [184]	Randomized; Single center; Parallel	NR	Both group 90 (50/40) 32.4 y (18–65 y) Trial: 46	Both group 90 (50/40) 32.4 y (18–65 y) Control: 44	1. Qingyinjiedu decoction (30 mg, q.d.) 2. Acitretin capsule (q.d.)	1. Acitretin capsule (q.d.)	Both group 4.2 y	Both group 4.2 y	1. PASI 60 ($p < 0.01$)	8 w	Trial: 15 AEs (8 skin scale and xerostomia, 5 hyperlipidemia, 2 hepatic dysfunction) Control: 30 AEs (17 skin scale and xerostomia, 10 hyperlipidemia, 3 hepatic dysfunction)
Luo 2010 [207]	Randomized; Single center; Parallel	NR	47 (25/22) 41.5 ± 9.8 y	47 (28/19) 39.7 ± 7.8 y	1. Piminxiao capsule (4c, t.i.d.) 2. Topical corticosteroid (Calcipotriol ointment, b.i.d.)	1. Topical corticosteroid (Calcipotriol ointment, b.i.d.)	9.3 ± 6.8 y	7.9 ± 5.7 y	1. PASI 70 ($p < 0.05$) 2. PASI score ($p < 0.01$) 3. Recurrence rate ($p < 0.05$)	12 w	Trial: No AEs Control: No AEs

Table 1. *Cont.*

Included Study (Reference)	Trial Design	Randomization Method	Number of Participants (Male/Female); Age (Mean ± SD)		Interventions		Morbidity Period (Mean ± SD or Range)		Outcome Index (Intergroup Differencies *p*-Value)	Course of Treatment	Adverse Events (Case/Symptom)
			Trial	Control	Trial	Control	Trial	Control			
Yu 2010 [135]	Randomized; Single center; Parallel	NR	40 (NR) 34.18 y (18–60 y)	40 (NR) 32.30 y (19–58 y)	1. Runzaozhiyang capsule (2 g, t.i.d.) 2. Acitretin capsule (30 mg, q.d.)	1. Acitretin capsule (30 mg, q.d.)	4.2 y (20 d–30 y)	3.8 y (1 m–29 y)	1. PASI 60 ($p < 0.01$) 2. PASI score ($p < 0.01$)	4 w	Trial: 5 AEs (5 nausea) Control: 6 AEs (6 hyperlipidemia)
Liu 2011 [165]	Randomized; Single center; Parallel	NR	Both group 68 (42/26) 35 y (18–65 y) Trial: 38	Both group 68 (42/26) 35 y (18–65 y) Control: 30	1. Fangfeng-tongsheng powder (6 g, b.i.d.) 2. Acitretin capsule (30 mg, q.d.)	1. Acitretin capsule (30 mg, q.d.)	Both group 5.7 y (2 m–12 y)	Both group 5.7 y (2 m–12 y)	1. PASI 60 ($p < 0.05$)	12 w	Trial: 17 AEs (detailed information NR) Control: 23 AEs (detailed information NR)
Lu 2011 [138]	Randomized; Single center; Parallel	NR	44 (28/16) 34.6 ± 3.82 y	44 (26/18) 33.96 ± 4.26 y	1. Qingrejiedu decoction (400 mL, b.i.d.) 2. Etretinate (0.5 mg, q.d.)	1. Etretinate (0.5 mg, q.d.)	5.63 ± 1.32 y	6.02 ± 1.50 y	1. PASI 60 ($p < 0.05$)	8 w	Trial: 1 AE (1 diarrhea) Control: 2 AEs (2 hepatic dysfunction)
Tian 2011 [197]	Randomized; Single center; Parallel	Simple randomization (random number table)	30 (12/18) 36.2 ± 9.8 y	30 (15/15) 34.5 ± 10.2 y	1. Qingfeil-iangxue decoction (100 mL, t.i.d.) 2. Vitamin C, Vitamin B6 (NR)	1. Vitamin C, Vitamin B6 (NR)	10.4 ± 7.6 y	10.4 ± 7.6 y	1. PASI 60 ($p < 0.05$) 2. PASI score ($p < 0.05$)	60 d	NR
Xu 2011 [208]	Randomized; Single center; Parallel	NR	39 (23/16) 34.5 y	39 (20/19) 36.5 y	1. Qingxuanyin (b.i.d.) 2. Acitretin capsule (20 mg, q.d.)	1. Acitretin capsule (20 mg, q.d.)	15 d–29 y	25 d–31 y	1. PASI 60 ($p < 0.01$) 2. PASI score ($p < 0.01$)	8 w	Trial: 8 AEs (8 diarrhea) Control: 10 AEs (10 xerostomia)
Yao 2011 [141]	Randomized; Single center; Parallel	Simple randomization (random number table)	70 (24/46) 36.5 y (15–62 y)	62 (34/28) 35.2 y (14–65 y)	1. Xiaoyin granule (3.5 g, t.i.d.) 2. Immunosuppressant (compound amino peptide tablets 5t, b.i.d.)	1. Immunosuppressant (compound amino peptide tablets 5t, b.i.d.)	7 d–45 y	10 d–47 y	1. PASI 60 ($p < 0.05$)	8 w	Detailed information NR

Table 1. *Cont.*

Included Study (Reference)	Trial Design	Randomization Method	Number of Participants (Male/Female); Age (Mean ± SD)		Interventions		Morbidity Period (Mean ± SD or Range)		Outcome Index (Intergroup Differencies *p*-Value)	Course of Treat-ment	Adverse Events (Case/Symptom)
			Trial	Control	Trial	Control	Trial	Control			
Zheng 2011 [91]	Randomized; Single center; Parallel	NR	60 (28/32) 42.1 ± 14.6 y	60 (26/34) 42.3 ± 15.4 y	1. Xiaoyinkeji decoction (100 mL, b.i.d.) 2. Acitretin capsule (20 mg, q.d.)	1. Acitretin capsule (20 mg, q.d.)	7.3 ± 0.9 y	7.5 ± 1.2 y	1. PASI 60 ($p < 0.05$) 2. PASI score ($p < 0.05$)	8 w	Trial: 8 AEs (6 gastrointestinal discomfort, 2 diarrhea) Control: 14 AEs (10 gastrointestinal discomfort with nausea or vomiting, 4 leukopenia)
Jiang 2012 [202]	Randomized; Single center; Parallel; single blind	Simple randomization (random number table)	30 (16/14) 33.37 ± 4.32 y	30 (17/13) 34.69 ± 5.01 y	1. Sendi particles (b.i.d.) 2. Acitretin capsule (10 mg, q.d.)	1. Acitretin capsule (10 mg, q.d.)	4.19 ± 2.77 y	3.98 ± 1.97 y	1. PASI 60 ($p < 0.05$) 2. PASI score ($p < 0.05$) 3. Recurrence rate ($p < 0.05$)	12 w	Trial: 3 AEs (2 loose stool, 1 hepatic dysfunction) Control: 1 AE (1 hepatic dysfunction)
Liu 2012 [143]	Randomized; Single center; Parallel	NR	42 (24/18) 40.00 ± 10.26 y	42 (23/19) 39.00 ± 10.50 y	1. Runzaozhiyang capsule (4c, t.i.d.) 2. Acitretin capsule (30 mg, q.d.)	1. Acitretin capsule (30 mg, q.d.)	3.0 ± 4.5 y	3.0 ± 4.6 y	1. PASI 60 ($p < 0.05$) 2. PASI score ($p < 0.05$)	8 w	Trial: 30 AEs (18 xeroderma, 5 xerostomia, 7 pruritus) Control: 98 AEs (32 xeroderma, 34 xerostomia, 32 pruritus)
Xie 2012 [183]	Randomized; Single center; Parallel	NR	Both group 94 (49/45) 35.5 ± 3.2 y Trial: 49	Both group 94 (49/45) 35.5 ± 3.2 y Control: 45	1. Runzaozhiyang capsule (4c, t.i.d.) 2. Acitretin capsule (20 mg, q.d.) 3. Topical corticosteroid (clobetasol propionate ointment, b.i.d.)	1. Acitretin capsule (20 mg, q.d.) 2. Topical corticosteroid (clobetasol propionate ointment, b.i.d.)	Both group 8.5 y (1 w–18 y)	Both group 8.5 y (1 w–18 y)	1. PASI 60 ($p < 0.05$)	4 w	Trial: 5 AEs (3 hyperlipidemia, 2 gastrointestinal discomfort) Control: 4 AEs (4 hyperlipidemia)

Table 1. *Cont.*

Included Study (Reference)	Trial Design	Randomization Method	Number of Participants (Male/Female); Age (Mean ± SD)		Interventions		Morbidity Period (Mean ± SD or Range)		Outcome Index (Intergroup Differencies *p*-Value)	Course of Treatment	Adverse Events (Case/Symptom)
			Trial	Control	Trial	Control	Trial	Control			
Zhang 2012 [127]	Randomized; Single center; Parallel	NR	40 (24/16) 36.65 ± 9.34 y	40 (27/13) 35.76 ± 10.26 y	1. Yinxie capsule (3c, t.i.d.) 2. Acitretin capsule (20 mg, q.d.)	1. Acitretin capsule (20 mg, q.d.)	16.43 ± 15.36 m	15.88 ± 16.48 m	1. PASI 60 ($p < 0.05$)	8 w	Trial: 19 AEs (detailed information NR) Control: 20 AEs (detailed information NR) Including xerostomia, dry lip, xeroma, pruritus, epistaxis
Zhou 2012a [124]	Randomized; Single center; Parallel	NR	70 (46/24) 38.3 y (19–65 y)	70 (44/26) 33.7 y (16–65 y)	1. Xiaoyin granule (3.5 g, t.i.d.) 2. Acitretin capsule (30 mg, q.d.) 3. 10% zing oxide ointment (b.i.d.)	1. Acitretin capsule (30 mg, q.d.) 2. 10% zing oxide ointment (b.i.d.)	5.6 y (3 m–12 y)	7.6 y (2 m–11 y)	1. PASI 60 ($p < 0.05$)	60 d	NR
Zhou 2012b [182]	Randomized; Single center; Parallel	NR	Both group 120 (49/71) 36.85 ± 6.32 y Trial: 49	Both group 120 (49/71) 36.85 ± 6.32 y Control: 45	1. Xiaoyin granule (3.5 g, t.i.d.) 2. Acitretin capsule (10 mg, q.d.)	1. Acitretin capsule (10 mg, q.d.)	Both group 6.70 ± 0.52 y	Both group 6.70 ± 0.52 y	1. PASI 60 ($p < 0.05$) 2. PASI score ($p < 0.01$) 3. Recurrence rate ($p > 0.05$)	12 w	Detailed information NR
Chen 2013 [153]	Randomized; Single center; Parallel	NR	30 (24/7) 31.2 ± 5.3 y	30 (24/7) 30.8 ± 6.1 y	1. Xiaoyin decoction (b.i.d.) 2. Acitretin capsule (10 mg, b.i.d.)	1. Acitretin capsule (10 mg, b.i.d.)	5.1 ± 2.0 y	5.5 ± 2.3 y	1. PASI 60 ($p < 0.05$) 2. PASI score ($p < 0.05$)	8 w	Trial: 2 AEs (1 xeroderma, 1 gastrointestinal discomfort) Control: 6 AEs (2 xeroderma, 1 xeroma, 1 xerostomia, 1 hyperlipidemia, 1 hepatic dysfunction)

Table 1. *Cont.*

Included Study (Reference)	Trial Design	Randomization Method	Number of Participants (Male/Female); Age (Mean ± SD)		Interventions		Morbidity Period (Mean ± SD or Range)		Outcome Index (Intergroup Differencies *p*-Value)	Course of Treat-ment	Adverse Events (Case/Symptom)
			Trial	Control	Trial	Control	Trial	Control			
Ding 2013 [146]	Randomized; Single center; Parallel	NR	30 (16/14) 52.32 ± 3.41 y	30 (12/18) 54.32 ± 2.15 y	1. NiupixuanII decoction (NR) 2. Acitretin capsule (0.3–1.0 mg/kg/day, NR)	1. Acitretin capsule (0.3–1.0 mg/kg/day, NR)	10.3 ± 2.2 y	11.1 ± 2.5 y	1. PASI score ($p < 0.05$)	4 w	NR
Mo 2013 [136]	Randomized; Single center; Parallel	NR	56 (33/23) 36.2 y (18–58 y)	54 (33/21) 35.9 y (20–60 y)	1. Total Glycosides of Paeoniae Alba capsule (2c, b.i.d.) 2. Immunosup-pressant (compound amino peptide tablets 5t, b.i.d.)	1. Immunosup-pressant (compound amino peptide tablets 5t, b.i.d.)	5.4 y (0.5 m–32 y)	5.6 y (1 m–40 y)	1. PASI 60 ($p < 0.05$)	8 w	Trial: No AEs Control No AEs
Song 2013 [90]	Randomized; Single center; Parallel	NR	43 (25/18) 34.5 ± 6.2 y	43 (26/17) 33.9 ± 6.0 y	1. Xiaoyin granule (3.5 g, t.i.d.) 2. Acitretin capsule (10 mg, t.i.d.)	1. Acitretin capsule (10 mg, t.i.d.)	8.1 ± 2.3 y	7.8 ± 2.4 y	1. PASI 60 ($p < 0.05$)	12 w	Trial: No AEs Control No AEs
Zhang 2013 [210]	Randomized; Single center; Parallel	NR	38 (20/18) 37.9 ± 5.9 y	38 (23/15) 33.9 ± 6.3 y	1. Xiaoyin capsule (5c, t.i.d.) 2. Acitretin capsule (10 mg, t.i.d.)	1. Acitretin capsule (10 mg, t.i.d.)	8.1 ± 2.7 y	7.8 ± 2.4 y	1. TNF-alpha ($p < 0.01$) 2. IL-8 ($p < 0.01$)	12 w	Trial: No AEs Control No AEs
Cheng 2014 [126]	Randomized; Single center; Parallel	NR	30 (17/13) 18–60 y	30 (16/14) 15–58 y	1. Yanghe decoction (b.i.d.) 2. Acitretin capsule (10 mg, b.i.d.)	1. Acitretin capsule (10 mg, b.i.d.)	1 w–5 y	2 w–6 y	1. PASI 60 ($p < 0.05$)	8 w	Detailed information NR

Table 1. *Cont.*

Included Study (Reference)	Trial Design	Randomization Method	Number of Participants (Male/Female); Age (Mean ± SD)		Interventions		Morbidity Period (Mean ± SD or Range)		Outcome Index (Intergroup Differencies p-Value)	Course of Treat-ment	Adverse Events (Case/Symptom)
			Trial	Control	Trial	Control	Trial	Control			
Du 2014 [125]	Randomized; Single center; Parallel	NR	70 (45/25) 39.6 ± 0.4 y	70 (45/25) 38.9 ± 0.5 y	1. Xiaoyin granule (3.5 g, t.i.d.) 2. Acitretin capsule (30 mg, q.d.)	1. Acitretin capsule (30 mg, q.d.)	6.3 ± 0.4 y	6.2 ± 0.4 y	1. PASI 60 ($p < 0.05$)	12 w	NR
Li 2014 [179]	Randomized; Single center; Parallel	NR	24 (14/10) 43.6 ± 10.78 y	24 (11/13) 45.3 ± 11.32 y	1. Huanglianjiedu decoction (200 mL, t.i.d.) 2. Methotrexate (2.5–5.0 mg, b.i.d.)	1. Methotrexate (2.5–5.0 mg, b.i.d.)	7.96 ± 4.41 y	7.49 ± 4.03 y	1. PASI 70 ($p < 0.05$)	4 w	Trial: 2 AEs Control: 5 AEs Including nausea, anorexia, hepatic dysfunction
Liang 2014 [172]	Randomized; Single center; Parallel	NR	39 (20/19) 38.43 ± 4.12 y	39 (19/20) 38.48 ± 4.15 y	1. Qinzhu-liangxue feng (200 mL, b.i.d.) 2. Acitretin capsule (10 mg, b.i.d.)	1. Acitretin capsule (10 mg, b.i.d.)	5.33 ± 1.05 y	5.38 ± 1.03 y	1. PASI 60 ($p < 0.05$) 2. PASI score ($p < 0.05$) 3. Recurrence rate ($p < 0.05$)	60 d	Trial: No AEs Control: 6 AEs (1 hepatic and renal dysfunction, 2 pruritus, 3 tinnitus)
Liu 2014 [115]	Randomized; Single center; Parallel	NR	47 (29/18) 28.9 ± 10.4 y	49 (26/21) 31.2 ± 12.6 y	1. Runzaozhiyang capsule (2 g, t.i.d) 2. Topical corticosteroid (compound flumetasone ointment, b.i.d.)	1. Topical corticosteroid (compound flumetasone ointment, b.i.d.)	2.3 ± 0.84 y	2.7 ± 0.91 y	1. PASI 60 ($p < 0.05$)	4 w	Trial: 3 AEs (2 mild abdominal discomfort, 1 mild burning sensation) Control: 1 AE (1 mild stabbing)
Qiu 2014 [99]	Randomized; Single center; Parallel	NR	49 (29/20) 43.7 ± 6.9 y	33 (22/11) 46.7 ± 7.1 y	1. EAHM prescription (q.d.) 2. Acitretin capsule (20–50 mg, q.d.)	1. Acitretin capsule (20–50 mg, q.d.)	9.5 ± 3.1 y	9.1 ± 3.9 y	1. PASI 60 ($p < 0.05$)	12 w	Trial: No AEs Control: 26 AEs (1 hypokalemia, 6 hyperlipidemia, 19 xeroderma and xerostomia)

Table 1. *Cont.*

Included Study (Reference)	Trial Design	Randomization Method	Number of Participants (Male/Female); Age (Mean ± SD)		Interventions		Morbidity Period (Mean ± SD or Range)		Outcome Index (Intergroup Differencies *p*-Value)	Course of Treat-ment	Adverse Events (Case/Symptom)
			Trial	Control	Trial	Control	Trial	Control			
Zhang 2014a [144]	Randomized; Single center; Parallel	NR	38 (26/12) 38.62 ± 6.11 y	38 (24/14) 36.74 ± 5.23 y	1. Yinxie capsule (3c, t.i.d.) 2. Acitretin capsule (20 mg, q.d.)	1. Acitretin capsule (20 mg, q.d.)	2–78 m	1–81 m	1. PASI 60 ($p < 0.05$)	8 w	Trial 15 AEs Control 16 AEs Including pruritus, xeroma, epistaxis, xerostomia, dry lib
Zhang 2014b [209]	Randomized; Single center; Parallel	NR	40 (26/14) 35.40 ± 2.83 y	30 (19/11) 34.69 ± 3.46 y	1. Ziyinyangx-uequfeng decoction (200 mL, b.i.d.) 2. Acitretin capsule (0.75 mg/kg/day, b.i.d.)	1. Acitretin capsule (0.75 mg/kg/day, b.i.d.)	6.35 ± 0.74 y	6.27 ± 0.68 y	1. PASI 60 ($p < 0.05$)	8 w	Trial: No AEs Control: No AEs
Zhang 2014c [196]	Randomized; Single center; Parallel	NR	36 (21/15) 33.5 y	36 (19/17) 35.3 y	1. Qingfeil-iangxue decoction (200 mL, b.i.d.) 2. Acitretin capsule (0.5 mg/kg/day, b.i.d.)	1. Acitretin capsule (0.5 mg/kg/day, b.i.d.)	1–28 y	1–26 y	1. PASI 60 ($p < 0.05$) 2. PASI score ($p < 0.05$)	2 w	NR
Cai 2015 [189]	Randomized; Single center; Parallel	NR	130 (72/58) 38.5 ± 12.3 y	130 (67/63) 41.8 ± 11.9 y	1. Xiaoyin feng (200 mL, b.i.d.) 2. Topical corticosteroid (Calcipotriol ointment, b.i.d.)	1. Topical corticosteroid (Calcipotriol ointment, b.i.d.)	4.83 ± 1.39 y	4.95 ± 1.12 y	1. PASI 70 ($p < 0.05$) 2. PASI score ($p < 0.05$)	12 w	Trial: 14 AEs (4 hematuria, 10 drug eruption) Control: 16 AEs (5 hematuria, 11 drug eruption)
Jin 2015a [132]	Randomized; Single center; Parallel	NR	50 (28/22) 44.38 ± 2.9 y	50 (30/20) 43.3 ± 2.5 y	1. Matrine capsule (2t, t.i.d.) 2. Acitretin capsule (b.i.d.)	1. Acitretin capsule (b.i.d.)	22–68 y	20–65 y	1. PASI 60 ($p < 0.05$)	8 w	Trial: 4 AEs (2 pruritus, 1 hyperlipidemia, 1 hepatic dysfunction) Control: 4 AEs (2 pruritus, 2 hyperlipidemia)

Table 1. *Cont.*

Included Study (Reference)	Trial Design	Randomization Method	Number of Participants (Male/Female); Age (Mean ± SD)		Interventions		Morbidity Period (Mean ± SD or Range)		Outcome Index (Intergroup Differencies *p*-Value)	Course of Treat-ment	Adverse Events (Case/Symptom)
			Trial	Control	Trial	Control	Trial	Control			
Jin 2015b [194]	Randomized; Single center; Parallel	Simple randomization (random number table)	43 (24/19) 41.33 ± 14.19 y	41 (26/15) 37.17 ± 11.30 y	1. Qinzhu-liangxue feng (200 mL, b.i.d.) 2. Acitretin capsule (10 mg, b.i.d.)	1. Acitretin capsule (10 mg, b.i.d.)	5.83 ± 1.60 y	7.16 ± 0.75 y	1. PASI 60 ($p < 0.01$) 2. PASI score ($p < 0.01$)	8 w	Trial: 21 AEs (18 xeroderma, 1 pruritus, 2 hepatic dysfunction) Control 42 AEs (27 xeroderma, 11 pruritus, 4 hepatic dysfuction)
Lu 2015a [102]	Randomized; Single center; Parallel	NR	Both group 62 (36/26) 29.5 ± 3.5 y Trial: 31	Both group 62 (36/26) 29.5 ± 3.5 y Control: 31	1. Qingying decoction (b.i.d.) 2. Acitretin capsule (30 mg, q.d.)	1. Acitretin capsule (30 mg, q.d.)	Both group 4.5 ± 1.2 y	Both group 4.5 ± 1.2 y	1. PASI 60 ($p < 0.05$) 2. PASI score ($p < 0.05$)	2 w	NR
Lu 2015b [118]	Randomized; Single center; Parallel	NR	54 (37/17) 21.2 ± 3.9 y	54 (33/21) 23.8 ± 2.1 y	1. Yangxierun-fuyin (q.d.) 2. Immunosup-pressant (compound amino peptide tablets 5t, b.i.d.)	1. Immunosup-pressant (compound amino peptide tablets 5t, b.i.d.)	Both group 3 m–12 y	Both group 3 m–12 y	1. PASI score ($p < 0.05$)	16 w	Trial: 2 AEs (2 burning sensation) Control: 1 AEs (1 burning sensation)
Ma 2015 [188]	Randomized; Single center; Parallel	NR	55 (28/27) 43.7 ± 7.6 y	55 (29/26) 43.2 ± 7.4 y	1. Liangxuerunfu decoction (q.d.) 2. Topical boric acid (q.d.)	1. Topical boric acid (q.d.)	NR	NR	1. PASI 60 ($p < 0.05$)	12 w	NR
Peng 2015 [168]	Randomized; Single center; Parallel	NR	40 (27/13) 36.5 y (18–65 y)	40 (25/15) 34.3 y (19–65 y)	1. Yinxiping pill (q.d.) 2. Acitretin capsule (10 mg, t.i.d.)	1. Acitretin capsule (10 mg, t.i.d.)	3 m–26 y	1 m–30 y	1. PASI 60 ($p < 0.05$) 2. PASI score ($p < 0.05$) 3. Recurrence rate	12 w	Detailed information NR

Table 1. *Cont.*

Included Study (Reference)	Trial Design	Randomization Method	Number of Participants (Male/Female); Age (Mean ± SD)		Interventions		Morbidity Period (Mean ± SD or Range)		Outcome Index (Intergroup Differencies *p*-Value)	Course of Treat-ment	Adverse Events (Case/Symptom)
			Trial	Control	Trial	Control	Trial	Control			
Sun 2015 [156]	Randomized; Single center; Parallel	NR	40 (21/19) 31.4 ± 2.8 y	40 (20/20) 30.9 ± 2.7 y	1. Qingrexiaoyin decoction (250 mL, b.i.d.) 2. Topical corticosteroid (Calcipotriol ointment, b.i.d.)	1. Topical corticosteroid (Calcipotriol ointment, b.i.d.)	4.6 ± 2.7 y	4.8 ± 2.4 y	1. PASI 60 ($p < 0.05$) 2. TNF alpha ($p < 0.05$) 3. IL-8 ($p < 0.05$)	12 w	NR
Wang 2015a [185]	Randomized; Single center; Parallel	NR	57 (30/27) 7.1 ± 2.5 y	57 (27/30) 7.2 ± 2.4 y	1. Compound glycyrrhizin (25 mg, t.i.d.) 2. Acitretin capsule (0.5 mg/kg/day, b.i.d.)	1. Acitretin capsule (0.5 mg/kg/day, b.i.d.)	4.3 ± 1.4 m	4.4 ± 1.5 m	1. PASI 60 ($p < 0.05$)	12 w	Trial: 3 AEs (1 hyperlipidemia, 2 hair loss) Control: 10 AEs (3 hyperlipidemia, 5 hair loss, 2 hepatic dysfunction)
Wang 2015b [140]	Randomized; Single center; Parallel	Simple ran-domization (random number table)	Both group 59 (38/21) 37.29 ± 10.24 y Trial: 30	Both group 59 (38/21) 37.29 ± 10.24 y Control: 29	1. Researcher prescription (q.d.) 2. Acitretin capsule (20 mg, q.d.)	2. Acitretin capsule (20 mg, q.d.)	Both group 7.21 ± 2.13 y	Both group 7.21 ± 2.13 y	1. PASI 60 ($p < 0.05$) 2. Recurrence rate ($p < 0.05$)	4 w	Trial: No AEs Control: No AEs
Wang 2015c [150]	Randomized; Single center; Parallel	NR	30 (18/12) 22.92 ± 3.08 y	30 (16/14) 23.08 ± 2.92 y	1. Danggui-yinzi granule (t.i.d.) 2. Topical urea ointment (t.i.d.)	1. Topical urea ointment (t.i.d.)	2.02 ± 0.79 y	1.98 ± 0.66 y	1. PASI 70 ($p < 0.05$)	4 w	NR
Yuan 2015 [169]	Randomized; Single center; Parallel	NR	Both group 80 (35/45) 32.5 ± 4.1 y Trial: 40	Both group 80 (35/45) 32.5 ± 4.1 y Control: 40	1. Qingrexiaoyin decoction (250 mL, b.i.d.) 2. Topical corticosteroid (Calcipotriol ointment, b.i.d.)	1. Topical corticosteroid (Calcipotriol ointment, b.i.d.)	Both group 2.4 ± 0.7 y	Both group 2.4 ± 0.7 y	1. PASI 60 ($p < 0.05$)	12 w	NR

Table 1. *Cont.*

Included Study (Reference)	Trial Design	Randomization Method	Number of Participants (Male/Female); Age (Mean ± SD)		Interventions		Morbidity Period (Mean ± SD or Range)		Outcome Index (Intergroup Differencies *p*-Value)	Course of Treat-ment	Adverse Events (Case/Symptom)
			Trial	Control	Trial	Control	Trial	Control			
Zhang 2015a [199]	Randomized; Single center; Parallel	Simple ran-domization (random number table)	90 (58/32) 44.6 ± 3.8 y	90 (64/26) 43.8 ± 3.4 y	1. Liangxuerunfu decoction (200 mL, b.i.d.) 2. Acitretin capsule (10 mg, b.i.d.)	1. Acitretin capsule (10 mg, b.i.d.)	7.8 ± 0.5 y	8.7 ± 0.5 y	1. PASI 60 ($p < 0.05$)	8 w	Trial: 12 AEs (6 pruritus, 3 xerostomia, 2 nausea, 1 headache) Control 13 AEs (5 pruritus, 4 xerostomia, 3 nausea, 1 headache)
Zhang 2015b [214]	Randomized; Single center; Parallel	Simple ran-domization (random number table)	65 (36/29) 35.5 ± 9.7 y	65 (38/27) 26.76 ± 7.34 y	1. Yinxie capsule (4c, t.i.d.) 2. Topical Pyrithione Zinc aerosol (t.i.d.)	1. Topical Pyrithione Zinc aerosol (t.i.d.)	60.7 ± 21.3 m	64.6 ± 22.5 m	1. PASI 60 ($p < 0.05$) 2. VAS ($p < 0.05$) 3. DLQI ($p < 0.05$) 4. TNF alpha ($p < 0.05$) 5. IL-8 ($p < 0.05$)	8 w	NR
Zhang 2015c [206]	Randomized; Single center; Parallel	Simple ran-domization (random number table)	63 (38/25) 31.29 ± 0.04 y	65 (38/27) 29.22 y (19–43 y)	1. Zinyinqin-grexiaofengsan (b.i.d.) 2. Acitretin capsule (20 mg, b.i.d.)	1. Acitretin capsule (20 mg, b.i.d.)	3 m–10 y	1–12 y	1. PASI 60 ($p < 0.05$)	8 w	Trial: 8 AEs (5 burning sensation, 2 erythema, 1 pruritus) Control: NR
Chen 2016 [96]	Randomized; Single center; Parallel	NR	35 (18/17) 37.5 ± 6.1 y	35 (19/16) 36.8 ± 6.0 y	1. Researcher prescription (b.i.d.) 2. Topical corticosteroid (Calcipotriol ointment, b.i.d.)	1. Topical corticosteroid (Calcipotriol ointment, b.i.d.)	9.2 ± 5.7 y	9.1 ± 5.4 y	1. PASI 70 ($p < 0.05$)	12 w	Trial: 0 AEs Control 0 AEs
He 2016a [110]	Randomized; Single center; Parallel	NR	Both group 90 (42/48) 40.1 ± 5.3 y Trial: 45	Both group 90 (42/48) 40.1 ± 5.3 y Control: 45	1. Liangxuerunfu decoction (b.i.d.) 2. Acitretin capsule (10 mg, b.i.d.)	1. Acitretin capsule (10 mg, b.i.d.)	Both group 11.8 ± 3.3 y	Both group 11.8 ± 3.3 y	1. PASI 60 ($p < 0.05$)	12 w	Trial: 4 AEs (4 gastrointestinal discomfort) Control: 12 AEs (3 gastrointestinal discomfort, 3 hyperlipdemia, 6 xerostomia)

Table 1. *Cont.*

Included Study (Reference)	Trial Design	Randomization Method	Number of Participants (Male/Female); Age (Mean ± SD)		Interventions		Morbidity Period (Mean ± SD or Range)		Outcome Index (Intergroup Differencies p-Value)	Course of Treatment	Adverse Events (Case/Symptom)
			Trial	Control	Trial	Control	Trial	Control			
He 2016b [93]	Randomized; Single center; Parallel	NR	34 (18/16) 39.21 ± 18.09 y	33 (17/16) 38.21 ± 17.68 y	1. Compound Qingdai pill (25 mg, t.i.d.) 2. Acitretin capsule (0.4 mg/kg/day, b.i.d.)	1. Acitretin capsule (0.4 mg/kg/day, b.i.d.)	4.85 ± 3.46 y	5.02 ± 3.96 y	1. PASI 60 ($p < 0.05$) 2. PASI score ($p < 0.05$)	8 w	Trial: 5 AEs (2 xerostomia, 2 hyperlipidemia, 1 gastrointestinal discomfort) Control: 4 AEs (3 xerostomia, 1 hyperlipidemia)
Jiang 2016 [131]	Randomized; Single center; Parallel	NR	40 (21/19) 34.62 ± 6.56 y	40 (23/17) 36.12 ± 5.44 y	1. Yinxiping pill (15 g, t.i.d.) 2. Topical corticosteroid (Calcipotriol ointment, b.i.d.)	1. Topical corticosteroid (Calcipotriol ointment, b.i.d.)	60.08 ± 41.03 m	59.45 ± 43.14 m	1. PASI 60 ($p < 0.05$) 2. PASI score ($p < 0.05$) 3. Recurrence rate ($p < 0.05$)	8 w	Trial: 5 AEs (2 burning sensation with skin rash, 1 gastrointestinal discomfort, 2 loose stool) Control: 3 AEs (burning sensation with skin rash)
Shan 2016 [154]	Randomized; Single center; Parallel	Simple randomization (random number table)	Both group 80 (45/35) 54.4 ± 10.4 y Trial: 40	Both group 80 (45/35) 54.4 ± 10.4 y Control: 40	1. Compound glycyrrhizin (50 mg, t.i.d.) 2. Acitretin capsule (50 mg, t.i.d.)	1. Acitretin capsule (50 mg, t.i.d.)	Both group 5.9 ± 3.2 y	Both group 5.9 ± 3.2 y	1. PASI 60 ($p < 0.05$) 2. PASI score ($p < 0.05$)	8 w	Trial: No AEs Control: 3 AEs (1 xerostomia, 1 xeroma, 1 xeroderma)
Wang 2016a [107]	Randomized; Single center; Parallel	NR	25 (16/9) 35.8 ± 7.6 y	25 (15/10) 37.1 ± 8.7 y	1. Xiaoyin granule (3.5 g, t.i.d.) 2. Acitretin capsule (10 mg, t.i.d.)	1. Acitretin capsule (10 mg, t.i.d.)	NR	NR	1. PASI 60 ($p < 0.05$)	16 w	NR
Wang 2016b [100]	Randomized; Single center; Parallel	NR	60 (35/25) 42.3 ± 6.9 y	60 (28/32) 39.5 ± 6.2 y	1. Piminxiao capsule (4c, t.i.d.) 2. Acitretin capsule (25 mg, q.d.)	1. Acitretin capsule (25 mg, q.d.)	NR	NR	1. PASI 60 ($p < 0.05$) 2. PASI score ($p < 0.05$)	8 w	NR

Table 1. *Cont.*

Included Study (Reference)	Trial Design	Randomization Method	Number of Participants (Male/Female); Age (Mean ± SD)		Interventions		Morbidity Period (Mean ± SD or Range)		Outcome Index (Intergroup Differencies p-Value)	Course of Treatment	Adverse Events (Case/Symptom)
			Trial	Control	Trial	Control	Trial	Control			
Wu 2016 [103]	Randomized; Single center; Parallel	NR	70 (43/27) 38.5 ± 2.6 y	70 (40/30) 39.1 ± 2.9 y	1. Xiaoyin granule (3.5 g, t.i.d.) 2. Acitretin capsule (25 mg, q.d.)	1. Acitretin capsule (25 mg, q.d.)	5.5 ± 1.4 y	5.9 ± 1.7 y	1. PASI 60 ($p < 0.05$)	8 w	Trial: 4 AEs Control: 13 AEs Including xerostomia, xeroderma, conjunctivitis, cheilitis
Xie 2016 [142]	Randomized; Single center; Parallel	Simple randomization (random number table)	52 (22/30) 39.1 ± 2.9 y	52 (24/28) 40.7 ± 9.5 y	1. Xiaoyin granule (3.5 g, t.i.d.) 2. Acitretin capsule (25–30 mg, q.d.)	1. Acitretin capsule (25–30 mg, q.d.)	5.32 ± 1.45 y	5.27 ± 1.42 y	1. PASI 60 ($p < 0.05$) 2. PASI score ($p < 0.05$) 3. Recurrence rate ($p < 0.05$) 4. IL-17 ($p < 0.05$)	4 w	Trial: 6 AEs (2 abdominal pain, 1 anorexia, 1 xerostomia, 1 dizziness, 1 conjunctivitis) Control: 5 AEs (1 pruritus, 1 tinnitus, 1 abdominal pain, 1 xeroma, 1 hepatic dysfunction)
Xu 2016 [195]	Randomized; Single center; Parallel	NR	Both group 114 (62/52) 40.5 ± 20.1 y Trial: 57	Both group 114 (62/52) 40.5 ± 20.1 y Control: 57	1. Compound glycyrrhizin (50 mg, t.i.d.) 2. Acitretin capsule (0.4 mg/kg/day, b.i.d.)	2. Acitretin capsule (0.4 mg/kg/day, b.i.d.)	NR	NR	1. PASI 60 ($p < 0.05$) 2. PASI score ($p < 0.05$)	6 w	Trial: 14 AEs Control: 30 AEs Including xeroderma, xeroma, xerostomia
Yang 2016a [178]	Randomized; Single center; Parallel	NR	40 (25/16) 39.33 ± 8.78 y	40 (30/10) 39.50 ± 9.37 y	1. Qinmei granule (b.i.d.) 2. Topical urea ointment (t.i.d.)	1. Topical urea ointment (t.i.d.)	NR	NR	1. PASI score ($p < 0.05$) 2. DLQI ($p < 0.05$)	12 w	Trial: 1 AE (1 Abnormal findings on urine test) Control: 1 AE (1 Abnormal findings on urine test)
Yang 2016b [104]	Randomized; Single center; Parallel	Simple randomization (random number generation)	23 (15/8) 30.5 y (25–54 y)	19 (12/7) 34.5 y (29–51 y)	1. Qingreliangxue decoction (b.i.d.) 2. Topical corticosteroid (Calcipotriol ointment, b.i.d.)	1. Topical corticosteroid (Calcipotriol ointment, b.i.d.)	7.62 m (3–26 m)	6.29 m (2–28 m)	1. PASI score ($p < 0.05$)	4 w	NR

Table 1. *Cont.*

Included Study (Reference)	Trial Design	Randomization Method	Number of Participants (Male/Female); Age (Mean ± SD)		Interventions		Morbidity Period (Mean ± SD or Range)		Outcome Index (Intergroup Differencies *p*-Value)	Course of Treat-ment	Adverse Events (Case/Symptom)
			Trial	Control	Trial	Control	Trial	Control			
Yu 2016 [201]	Randomized; Single center; Parallel	NR	Both group 40 (22/18) 40.3 y (18–74 y) Trial: 20	Both group 40 (22/18) 40.3 y (18–74 y) Control: 20	1. Qingreliangxue decoction (b.i.d.) 2. Topical corticosteroid (Calcipotriol ointment, b.i.d.) 3. Topical corticosteroid (Halometasone cream, b.i.d.)	1. Topical corticosteroid (Calcipotriol ointment, b.i.d.) 2. Topical corticosteroid (Halometa-sone cream, b.i.d.)	Both group 2.5 y (1 m–15 y)	Both group 2.5 y (1 m–15 y)	1. PASI score ($p < 0.05$)	8 w	NR
Cao 2017 [161]	Randomized; Single center; Parallel; single blind	Simple ran-domization (envelope concealment method)	30 (17/13) 36.02 ± 4.41 y	30 (18/12) 35.54 ± 4.36 y	1. Compound glycyrrhizin tablet (2t, t.i.d.) 2. Acitretin capsule (0.4 mg/kg/day, t.i.d.)	1. Acitretin capsule (0.4 mg/kg/day, t.i.d.)	4.66 ± 1.21 y	4.25 ± 1.02 y	1. PASI 60 ($p < 0.05$) 2. PASI score ($p < 0.05$)	8 w	Trial: 6 AEs (2 xeroderma, 3 xeroma, 1 hyperlipidemia) Control: 14 AEs (4 xeroderma, 5 xeroma, 3 hepatic dysfunction, 2 hyperlipidemia)
Cheng 2017 [94]	Randomized; Single center; Parallel	NR	27 (15/112) 38.4 ± 5.8 y	26 (15/11) 38.2 ± 5.3 y	1. Compound glycyrrhizin tablet (2–3t, t.i.d.) 2. Acitretin capsule (0.4 mg/kg/day, t.i.d.)	1. Acitretin capsule (0.4 mg/kg/day, t.i.d.)	5.6 ± 2.4 y	5.7 ± 2.5 y	1. PASI 60 ($p < 0.05$) 2. PASI score ($p < 0.05$)	8 w	Trial: 4AEs (4 xeroma, xerostomia, xeroderma) Control: 8 AEs (6 xeroma, xerostomia, xeroderma, 1 hair loss, 1 hyperlipidemia)
Ding 2017 [187]	Randomized; Single center; Parallel	Simple ran-domization (random number table)	40 (19/11) 36.15 ± 2.11 y	40 (22/18) 36.20 ± 2.07 y	1. Ziyinhuox-uerunzao decoction (200 mL, b.i.d.) 2. Acitretin capsule (20–50 mg, q.d.)	1. Acitretin capsule (20–50 mg, q.d.)	11.36 ± 1.00 y	11.41 ± 0.97	1. PASI 60 ($p < 0.05$) 2. PASI score ($p < 0.05$)	8 w	Trial: 1 AEs (1 focal pruritus with rash) Control: 2 AEs (2 focal pruritus with rash)

Table 1. *Cont.*

Included Study (Reference)	Trial Design	Randomization Method	Number of Participants (Male/Female); Age (Mean ± SD)		Interventions		Morbidity Period (Mean ± SD or Range)		Outcome Index (Intergroup Differencies *p*-Value)	Course of Treatment	Adverse Events (Case/Symptom)
			Trial	Control	Trial	Control	Trial	Control			
Du 2017 [114]	Randomized; Single center; Parallel	NR	32 (18/14) 32.2 ± 5.4 y	32 (21/11) 37.3 ± 5.2 y	1. Total Glycosides of Paeoniae Alba capsule (0.6 g, b.i.d.) 2. Topical corticosteroid (calcipotriol betamethasone ointment, q.d.)	1. Topical corticosteroid (calcipotriol betamethasone ointment, q.d.)	NR	NR	1. PASI 60 ($p < 0.05$)	4 w	Trial: 4 AEs (2 skin rash, 1 burning sensation, 1 folliculitis Control: No AE
Feng 2017 [130]	Randomized; Single center; Parallel	Simple randomization (random number table)	35 (21/14) 38.3 ± 4.1 y	35 (19/16) 35.7 ± 6.4 y	1. Yinxiping pill (15 g, t.i.d.) 2. Topical corticosteroid (Compound Flumetasone Ointment, b.i.d.)	1. Topical corticosteroid (Compound Flumetasone Ointment, b.i.d.)	5.9 ± 3.7 y	6.2 ± 3.3 y	1. PASI 60 ($p < 0.05$) 2. PASI score ($p < 0.05$)	8 w	Trial: 2 AEs (1 mild diarrhea, 1 mild skin rash) Control: 1 AEs (1 mild skin rash with pruritus)
Han 2017a [123]	Randomized; Single center; Parallel	NR	44 (25/19) 36.04 ± 7.15 y	44 (26/18) 35.69 ± 6.49 y	1. Xiaoyin granule (3.5 g, t.i.d.) 2. Acitretin capsule (30 mg, q.d.)	1. Acitretin capsule (30 mg, q.d.)	5.39 ± 2.48	5.21 ± 2.36	1. PASI 60 ($p < 0.05$) 2. Recurrence rate ($p < 0.05$) 3. IL-17 ($p < 0.05$)	12 w	Trial: 3 AEs (detailed information NR) Control 12 AEs (detailed information NR)
Han 2017b [162]	Randomized; Single center; Parallel	NR	48 (26/22) 45.72 ± 5.78 y	44 (27/21) 43.56 ± 4.43 y	1. Compound Qingdai capsule (4c, t.i.d.) 2. Acitretin capsule (20 mg, b.i.d.)	1. Acitretin capsule (20 mg, b.i.d.)	3.5 y (2 m–40 y)	4.2 y (3 m–36 y)	1. PASI 60 ($p < 0.05$) 2. PASI score ($p < 0.05$) 3. Recurrence rate ($p < 0.05$)	8 w	Detailed information NR

Table 1. *Cont.*

Included Study (Reference)	Trial Design	Randomization Method	Number of Participants (Male/Female); Age (Mean ± SD)		Interventions		Morbidity Period (Mean ± SD or Range)		Outcome Index (Intergroup Differencies *p*-Value)	Course of Treat-ment	Adverse Events (Case/Symptom)
			Trial	Control	Trial	Control	Trial	Control			
Li 2017 [203]	Randomized; Single center; Parallel	NR	30 (18/12) 36.4 ± 10.0 y	30 (15/15) 34.2 ± 12.7 y	1. Qingrejiedu decoction (6 g, b.i.d.) 2. Topical retinoid cream (b.i.d.)	1. Topical retinoid cream (b.i.d.)	5.73 ± 3.78 y	NR	1. PASI 60 ($p < 0.05$) 2. PASI score ($p < 0.05$) 3. TNF alpha ($p < 0.05$)	4 w	Trial: 3 AEs (3 diarrhea) Control: No AE
Liu 2017 [116]	Randomized; Single center; Parallel	NR	48 (27/21) 33.5 ± 6.5 y	52 (29/23) 33.8 ± 6.2 y	1. Yinxie capsule (4c, t.i.d.) 2. Acitretin capsule (20 mg, b.i.d.)	1. Acitretin capsule (20 mg, b.i.d.)	5.6 ± 7.2 y	5.2 ± 6.9 y	1. PASI 60 ($p < 0.05$) 2. PASI score ($p < 0.05$)	8 w	Trial: 65 AEs (13 xeroma, 16 xeroderma, 2 epistaxis, 14 folliculitis, 1 hepatic dysfunction, 19 hyperlipidemia) Control: 178 AEs (42 xeroma, 40 xeroderma, 9 epistaxis, 42 folliculitis, 8 hepatic dysfunction, 37 hyperlipidemia)
Luo 2017 [160]	Randomized; Single center; Parallel	NR	39 (26/13) 31.4 ± 2.3 y	36 (25/11) 32.7 ± 2.8 y	1. Compound glycyrrhizin (2–3t, t.i.d.) 2. Acitretin capsule (25–30 mg, t.i.d.)	1. Acitretin capsule (25–30 mg, t.i.d.)	NR	NR	1. PASI 60 ($p < 0.05$)	8 w	Trial: 5 AEs (2 pruritus, 1 xeroderma, 1 xerostomia, 1 xeroma) Control: 8 AEs (3 pruritus, 2 xeroderma, 2 xerostomia, 1 xeroma)
Pang 2017 [170]	Randomized; Single center; Parallel	NR	45 (22/23) 36.48 ± 14.21 y	45 (24/21) 37.02 ± 44.47 y	1. Compound Qingdai capsule (4c, t.i.d.) 2. Immunosup-pressant (compound amino peptide tablets 5t, t.i.d.)	1. Immunosup-pressant (compound amino peptide tablets 5t, t.i.d.)	47.68 ± 18.22 m	49.13 ± 18.80 m	1. PASI 70 ($p < 0.05$) 2. PASI score ($p < 0.01$) 3. DLQI ($p < 0.05$) 4. IL-8 ($p < 0.05$) 5. IFN gamma ($p < 0.01$)	8 w	NR

Table 1. *Cont.*

Included Study (Reference)	Trial Design	Randomization Method	Number of Participants (Male/Female); Age (Mean ± SD)		Interventions		Morbidity Period (Mean ± SD or Range)		Outcome Index (Intergroup Differencies *p*-Value)	Course of Treat-ment	Adverse Events (Case/Symptom)
			Trial	Control	Trial	Control	Trial	Control			
Shi 2017 [89]	Randomized; Single center; Parallel	NR	23 (12/11) 45.74 ± 8.43 y	23 (13/10) 45.72 ± 8.45 y	1. Liangxuexi-aofeng decoction (200 mL, b.i.d.) 2. Acitretin capsule (30 mg, q.d.)	1. Acitretin capsule (30 mg, q.d.)	7.29 ± 1.25 y	7.28 ± 1.28 y	1. PASI 60 ($p < 0.05$) 2. PASI score ($p < 0.01$) 3. IL-8 ($p < 0.05$)	8 w	Trial: 3 AEs (1 xerostomia, 1 xeroma, 1 xeroderma) Control: 2 AEs (1 xerostomia, 1 nausea)
Song 2017 [108]	Randomized; Single center; Parallel	NR	70 (32/28) 41.02 ± 5.39 y	70 (39/31) 40.76 ± 5.32 y	1. Yangxue-tongluo decoction (b.i.d.) 2. Acitretin capsule (10 mg, t.i.d.)	1. Acitretin capsule (10 mg, t.i.d.)	6.41 ± 1.00 y	6.49 ± 1.03 y	1. PASI 70 ($p < 0.05$) 2. VAS ($p < 0.05$) 3. DLQI ($p < 0.05$) 4. TNF alpha ($p < 0.05$)	8 w	Trial: 9 AEs (2 conjunctivitis, 4 xerostomia, 3 headache) Control: 8 AEs (3 conjunctivitis, 2 xerostomia, 2 headache, 1 muscular pain)
Wang 2017 [159]	Randomized; Single center; Parallel	NR	60 (32/28) 32.46 ± 6.25 y	60 (32/28) 33.08 ± 6.32 y	1. Yinxie capsule (4c, t.i.d.) 2. Acitretin capsule (20 mg, t.i.d.)	1. Acitretin capsule (20 mg, t.i.d.)	5.61 ± 7.32 y	5.29 ± 6.96 y	1. PASI 60 ($p < 0.05$) 2. PASI score ($p < 0.05$) 3. IL-8 ($p < 0.05$)	8 w	Trial: 54 AEs (15 xeroma, 18 xeroderma, 3 epistaxis, 16 folliculitis, 2 hepatic dysfunction, 15 hyperlipidemia) Control: 180 AEs (48 xeroma, 43 xeroderma, 11 epistaxis, 32 folliculitis, 9 hepatic dysfunction, 37 hyperlipidemia)

Table 1. *Cont.*

Included Study (Reference)	Trial Design	Randomization Method	Number of Participants (Male/Female); Age (Mean ± SD)		Interventions		Morbidity Period (Mean ± SD or Range)		Outcome Index (Intergroup Differencies *p*-Value)	Course of Treat-ment	Adverse Events (Case/Symptom)
			Trial	Control	Trial	Control	Trial	Control			
Wu 2017 [173]	Randomized; Single center; Parallel	NR	40 (26/14) 40.85 ± 15.48 y	40 (25/15) 41.64 ± 15.86 y	1. Shentongzhuyu decoction (200 mL, b.i.d.) 2. Methotrexate (10 mg, q.w.) 3. Sulfasalazine tablets (1.0 g, t.i.d.) 4. Diclofenac sodium extended-release tablet (0.1 g, q.d.) 5. Folic acid tablet (10 mg, q.d.)	1. Methotrexate (10 mg, q.w.) 2. Sulfasalazine tablets (1.0 g, t.i.d.) 3. Diclofenac sodium extended-release tablet (0.1 g, q.d.) 4. Folic acid tablet (10 mg, q.d.)	NR	NR	1. PASI 70 ($p < 0.01$) 2. PASI score ($p < 0.05$)	12 w	Trial: 9 AEs (2 leukopenia, 3 hepatic dysfunction, 1 hyperbilirubinemia, 3 nausea and vomiting) Control 10 AEs (3 leukopenia, 2 hepatic dysfunction, 1 hyperbilirubinemia, 4 nausea and vomiting)
Yang 2017 [213]	Randomized; Single center; Parallel	Simple ran-domization	40 (25/15) 30.6 ± 8.21 y	40 (24/16) 32.5 ± 7.10 y	1. Xiaoyin granule (3.5 g, t.i.d.) 2. Acitretin capsule (20 mg, q.d.)	1. Acitretin capsule (20 mg, q.d.)	4.51 ± 3.13 y	4.60 ± 3.02 y	1. PASI 60 ($p < 0.01$) 2. PASI score ($p < 0.05$) 3. TNF alpha ($p < 0.05$) 4. IL-17 ($p < 0.05$) 5. IL-23 ($p < 0.05$)	12 w	Trial: 9AEs (3 pruritus, 5 xerostomia, 1 xeroma) Control: 17 AEs (7 pruritus, 8 xerostomia, 1 dizzines, 1 xeroma)
Zeng 2017 [190]	Randomized; Single center; Parallel	NR	31 (21/11) 35.14 ± 0.15 y	31 (21/10) 35.29 ± 0.18 y	1. Compound glycyrrhizin capsule (2-3c, t.i.d.) 2. Acitretin capsule (10 mg, t.i.d.)	1. Acitretin capsule (10 mg, t.i.d.)	32.14 ± 1.25 m	33.45 ± 1.34 m	1. PASI 60 ($p < 0.05$)	8 w	Trial: 13 AEs (2 xerostomia, 3 xeroma, 2 pruritus, 4 nausea, 2 dizziness) Control: 8 AEs (1 xerostomia, 2 xeroma, 1 pruritus, 3 nausea, 1 dizziness)

Table 1. *Cont.*

Included Study (Reference)	Trial Design	Randomization Method	Number of Participants (Male/Female); Age (Mean ± SD)		Interventions		Morbidity Period (Mean ± SD or Range)		Outcome Index (Intergroup Differencies *p*-Value)	Course of Treat-ment	Adverse Events (Case/Symptom)
			Trial	Control	Trial	Control	Trial	Control			
Zhang 2017a [113]	Randomized; Single center; Parallel	NR	52 (27/25) 28.5 ± 5.2 y	52 (26/26) 30.1 ± 4.1 y	1. Taohongershao decoction (150 mL, t.i.d.) 2. Acitretin capsule (10 mg, t.i.d.)	1. Acitretin capsule (10 mg, t.i.d.)	7.3 ± 4.5 y	8.5 ± 4.9 y	1. PASI score ($p < 0.05$) 2. Recurrence rate ($p < 0.05$)	4 w	NR
Zhang 2017b [112]	Randomized; Single center; Parallel	NR	45 (21/24) 27.5 ± 7.4 y	45 (22/23) 28.0 ± 9.5 y	1. Taohongershao decoction (150 mL, t.i.d.) 2. Acitretin capsule (10 mg, t.i.d.)	1. Acitretin capsule (10 mg, t.i.d.)	10.5 ± 8.3 y	10.9 ± 8.0 y	1. PASI 60 ($p < 0.05$) 2. DLQI ($p < 0.05$) 3. IFN gamma ($p < 0.05$) 4. IL-17 ($p < 0.05$) 5. IL-23 ($p < 0.05$)	4 w	NR
Zhang 2017c [105]	Randomized; Single center; Parallel	NR	50 (23/27) 32.5 ± 5.9 y	50 (25/25) 30.9 ± 6.1 y	1. Taohongershao decoction (150 mL, t.i.d.) 2. Acitretin capsule (10 mg, t.i.d.)	1. Acitretin capsule (10 mg, t.i.d.)	12.6 ± 7.5 y	11.5 ± 6.9 y	1. PASI 60 ($p < 0.01$) 2. PASI score ($p < 0.05$)	8 w	NR
Zhang 2017d [121]	Randomized; Single center; Parallel	NR	17 (10/7) 46.47 ± 14.06	17 (9/8) 46.41 ± 18.45	1. Liangxuexiao feng (6 g, b.i.d.) 2. Acitretin capsule (10 mg, b.i.d.)	1. Acitretin capsule (10 mg, b.i.d.)	16.29 ± 10.49 y	16.08 ± 12.80 y	1. PASI 60 ($p < 0.05$) 2. PASI score ($p < 0.05$)	8 w	NR
Zhang 2017e [200]	Randomized; Single center; Parallel	NR	55 (24/31) 30.1 ± 4.4 y	55 (22/33) 29.8 ± 7.3 y	1. Taohongershao decoction (150 mL, t.i.d.) 2. Acitretin capsule (10 mg, t.i.d.)	1. Acitretin capsule (10 mg, t.i.d.)	7.5 ± 6.3 y	8.1 ± 6.4 y	1. PASI 60 ($p < 0.05$) 2. PASI score ($p < 0.05$) 3. IFN gamma ($p < 0.05$)	4 w	NR

Table 1. *Cont.*

Included Study (Reference)	Trial Design	Randomization Method	Number of Participants (Male/Female); Age (Mean ± SD)		Interventions		Morbidity Period (Mean ± SD or Range)		Outcome Index (Intergroup Differencies *p*-Value)	Course of Treat-ment	Adverse Events (Case/Symptom)
			Trial	Control	Trial	Control	Trial	Control			
Zhao 2017 [128]	Randomized; Single center; Parallel	NR	40 (23/17) 36.35 ± 2.09 y	40 (22/18) 36.25 ± 2.13 y	1. Yinxie capsule (3c, t.i.d.) 2. Acitretin capsule (20 mg, q.d.).	1. Acitretin capsule (20 mg, q.d.).	NR	NR	1. PASI 60 ($p < 0.05$)	8 w	Trial: 18 AEs (6 pruritus, 7 epistaxis, 5 xerostomia) Control: 19 AEs (7 pruritus, 7 epistaxis, 6 xerostomia)
Chai 2018 [158]	Randomized; Single center; Parallel	NR	46 (26/20) 45.3 ± 3.8 y	46 (25/21) 45.5 ± 3.6 y	1. Xiaoyin decoction (b.i.d.) 2. Topical corticosteroid (Calcipotriol ointment, b.i.d.)	1. Topical corticosteroid (Calcipotriol ointment, b.i.d.)	2.3 ± 0.5 y	2.1 ± 0.4 y	1. PASI 60 ($p < 0.05$) 2. IFN gamma ($p < 0.05$) 3. IL-8 ($p < 0.05$)	8 w	Trial: 2 AEs (1 skin rash, 1 mild gastrointestinal discomfort) Control: 3 AEs (2 skin rash, 1 pruritus)
Li 2018 [109]	Randomized; Single center; Parallel	Simple randomization (random number table)	44 (23/21) 35.01 ± 7.09 y	44 (24/20) 35.31 ± 7.29 y	1. Yinxie capsule (4c, t.i.d.) 2. Acitretin capsule (20 mg, b.i.d.)	1. Acitretin capsule (20 mg, b.i.d.).	7.69 ± 3.69 y	7.71 ± 3.46 y	1. PASI 60 ($p < 0.05$) 2. PASI score ($p < 0.01$)	8 w	Trial: 1 AEs (1 nausea) Control: 3 AEs (1 headache, 2 nausea)
Liu 2018 [147]	Randomized; Single center; Parallel Three arm trial	NR	25 (11/14) 37.21 ± 9.87 y	25 (13/12) 39.42 ± 9.23 y	1. Banzhilian decoction (200 mL, b.i.d.) 2. Topical corticosteroid (Calcipotriol ointment, b.i.d.).	1. Topical corticosteroid (Calcipotriol ointment, b.i.d.)	6.8 ± 5.1 y	6.2 ± 3.9 y	1. PASI 60 ($p < 0.05$) 2. PASI score ($p < 0.05$)	8 w	NR
Luo 2018 [212]	Randomized; Single center; Parallel	NR	50 (31/19) 32.46 ± 10.24 y	50 (29/21) 33.57 ± 10.82 y	1. Zicaohuoxue decoction (200 mL, bi.d.) 2. Acitretin capsule (30 mg, q.d.)	1. Acitretin capsule (30 mg, q.d.)	2.46 ± 1.24 y	2.52 ± 1.28 y	1. PASI 60 ($p < 0.05$) 2. TNF alpha ($p < 0.05$) 3. DLQI ($p < 0.05$)	8 w	Trial: 8 AEs (1 pruritus, 2 xeroma, 2 headache, 3 nausea) Control: 10 AEs (1 pruritus, 3 xeroma, 3 headache, 3 nausea)

Table 1. *Cont.*

Included Study (Reference)	Trial Design	Randomization Method	Number of Participants (Male/Female); Age (Mean ± SD)		Interventions		Morbidity Period (Mean ± SD or Range)		Outcome Index (Intergroup Differencies p-Value)	Course of Treat-ment	Adverse Events (Case/Symptom)
			Trial	Control	Trial	Control	Trial	Control			
Ma 2018a [192]	Randomized; Single center; Parallel	Simple randomization (random number table)	37 (21/16) 38.1 ± 4.2 y	42 (23/19) 37.4 ± 4.1 y	1. Xiaoyin granule (3.5 g, t.i.d.) 2. Topical corticosteroid (Calcipotriol ointment, b.i.d.) 3. Acitretin capsule (30 mg, q.d.)	1. Topical corticosteroid (Calcipotriol ointment, b.i.d.) 2. Acitretin capsule (30 mg, q.d.)	3.66 ± 1.01 y	3.7 ± 1.3 y	1. PASI 60 ($p < 0.05$)	4 w	NR
Ma 2018b [166]	Randomized; Single center; Parallel	NR	34 (19/15) 40.2 ± 6.9 y	34 (21/13) 41.6 ± 7.5 y	1. Qingrequshi decoction (b.i.d.) 2. Methotrexate (15 mg, q.w.)	1. Methotrexate (15 mg, q.w.)	8.7 ± 2.6 y	8.2 ± 2.4 y	1. PASI score ($p < 0.05$)	12 w	NR
Xiao 2018 [176]	Randomized; Single center; Parallel	Simple randomization (random number table)	21 (11/10) 27.5 ± 2.2 y	21 (15/6) 27.3 ± 1.2 y	1. Xiaoyin decoction (200 mL, b.i.d.) 2. Topical corticosteroid (Calcipotriol ointment, b.i.d.)	1. Topical corticosteroid (Calcipotriol ointment, b.i.d.)	NR	NR	1. TNF alpha ($p < 0.05$) 2. IL-17 ($p < 0.05$) 3. IL-22 ($p < 0.05$)	12 w	NR
Xie 2018 [106]	Randomized; Single center; Parallel	NR	60 (34/26) 36.5 y (19–65 y)	60 (29/31) 34.3 y (19–64 y)	1. Qingying decoction (b.i.d.) 2. Acitretin capsule (30 mg, q.d.)	1. Acitretin capsule (30 mg, q.d.)	6.4 y (3 m–26 y)	6.1 y (1 m–30 y)	1. PASI 60 ($p < 0.05$) 2. PASI score ($p < 0.05$)	8 w	Detailed information NR
Zhang 2018a [111]	Randomized; Single center; Parallel	NR	48 (25/23) 35.43 ± 0.16	48 (28/20) 35.28 ± 0.23	1. Compound glycyrrhizin (2–3t, t.i.d.) 2. Acitretin capsule (10 mg, t.i.d.)	1. Acitretin capsule (10 mg, t.i.d.)	33.12 ± 1.64 m	32.65 ± 1.14 m	1. PASI score ($p < 0.05$)	8 w	NR

Table 1. *Cont.*

Included Study (Reference)	Trial Design	Randomization Method	Number of Participants (Male/Female); Age (Mean ± SD)		Interventions		Morbidity Period (Mean ± SD or Range)		Outcome Index (Intergroup Differencies p-Value)	Course of Treatment	Adverse Events (Case/Symptom)
			Trial	Control	Trial	Control	Trial	Control			
Zhang 2018b [174]	Randomized; Single center; Parallel	Simple randomization	36 (20/16) 29.15 ± 6.24 y	36 (19/17) 29.36 ± 6.02	1. Xiaoyin granule (3.5 g, t.i.d.) 2. Topical corticosteroid (Calcipotriol ointment, b.i.d.)	1. Topical corticosteroid (Calcipotriol ointment, b.i.d.)	5.2 ± 1.3 y	5.4 ± 1.2 y	1. PASI 60 ($p < 0.05$) 2. PASI score ($p < 0.05$)	8 w	NR
Zhou 2018 [139]	Randomized; Single center; Parallel	NR	39 (17/22) 38.82 ± 1.29 y	39 (19/20) 38.71 ± 1.22 y	1. Keyin pills (10 mg, b.i.d.) 2. Acitretin capsule (10 mg, t.i.d.)	1. Acitretin capsule (10 mg, t.i.d.)	7.59 ± 0.78 y	7.46 ± 0.65 y	1. PASI 60 ($p < 0.05$) 2. PASI score ($p < 0.05$) 3. IFN gamma ($p < 0.05$) 4. IL-17 ($p < 0.05$)	8 w	Trial: 8 AEs (1 conjunctivitis, 1 skin rash, 2 arthralgia, 1 headache, 3 nausea and vomiting) Control: 11 AEs (1 conjunctivitis, 2 skin rash, 3 arthralgia, 1 headache, 4 nausea and vomiting)
Chen 2019 [198]	Randomized; Single center; Parallel	NR	62 (26/36) 37.62 ± 6.34 y	62 (28/34) 35.74 ± 5.54 y	1. Compound glycyrrhizin tablet (2–3t, t.i.d.) 2. Topical corticosteroid (Calcipotriol ointment, b.i.d.) 3. Acitretin capsule (25–30 mg, q.d.)	1. Topical corticosteroid (Calcipotriol ointment, b.i.d.) 2. Acitretin capsule (25–30 mg, q.d.)	6.92 ± 3.05 y	6.25 ± 2.47 y	1. PASI score ($p < 0.05$)	8 w	NR
Ge 2019 [145]	Randomized; Single center; Parallel	NR	40 (21/19) 27 ± 2.4 y	40 (22/18) 26 ± 2.3 y	1. Compound glycyrrhizin (50 mg, t.i.d.) 2. Acitretin capsule (20 mg, t.i.d.)	1. Acitretin capsule (20 mg, t.i.d.)	3 m–31 y	3 m–31.5 y	1. PASI 60 ($p < 0.05$) 2. PASI score ($p < 0.05$)	8 w	Trial: 5 AEs (including xerostomia, xeroma, xeroderma) Control: 8 AEs (including xerostomia, xeroma, xeroderma, 2 hair loss, 1 hyperlipidemia)

Table 1. *Cont.*

Included Study (Reference)	Trial Design	Randomization Method	Number of Participants (Male/Female); Age (Mean ± SD)		Interventions		Morbidity Period (Mean ± SD or Range)		Outcome Index (Intergroup Differencies *p*-Value)	Course of Treatment	Adverse Events (Case/Symptom)
			Trial	Control	Trial	Control	Trial	Control			
Han 2019 [97]	Randomized; Single center; Parallel	NR	Both group 80 (59/21) 37.37 ± 9.48 y	Both group 80 (59/21) 37.37 ± 9.48 y	1. Qinzhu-liangxue feng (b.i.d.) 2. Acitretin capsule (10 mg, b.i.d.)	1. Acitretin capsule (10 mg, b.i.d.)	Both group 5.29 ± 1.44 y	Both group 5.29 ± 1.44 y	1. PASI 60 ($p < 0.05$) 2. PASI score ($p < 0.05$)	4 w	NR
Hu 2019 [119]	Randomized; Single center; Parallel	Simple randomization	62 (34/28) 36.8 ± 8.1 y	59 (33/26) 37.8 ± 9.4 y	1. Liangxuexiaobi decoction (150 mL, b.i.d.) 2. Topical corticosteroid (Calcipotriol betamethasone ointment, q.d.)	1. Topical corticosteroid (Calcipotriol betamethasone ointment, q.d.)	45.4 ± 12.5 m	46.1 ± 11.1 m	1. PASI score ($p < 0.05$)	8 w	Trial: 2 AEs (2 loose stool) Control: 1AE (1 skin rash)
Lu 2019 [152]	Randomized; Single center; Parallel; single blind	Simple randomization (envelope concealment method)	30 (15/15) 21–66 y	30 (16/14) 22–65 y	1. Yangxuequfeng granule (b.i.d.) 2. Topical urea ointment (t.i.d.)	1. Topical urea ointment (t.i.d.)	NR	NR	1. PASI 70 ($p < 0.05$) 2. DLQI ($p < 0.05$)	4 w	NR
Xun 2019 [134]	Randomized; Single center; Parallel	NR	52 (24/28) 40–50 y	52 (22/30) 40–51 y	1. Yinxie capsule (15 g, t.i.d.) 2. Acitretin capsule (25–50 mg, t.i.d.)	1. Acitretin capsule (25–50 mg, t.i.d.)	NR	NR	1. Recurrence rate (*p*-value NR)	4 w	NR
Yang 2019 [149]	Randomized; Single center; Parallel	NR	41 (24/17) 38.24 ± 4.19 y	41 (25/16) 38.57 ± 4.03 y	1. Compound Qingdai capsule (4c, t.i.d.) 2. Topical corticosteroid (Calcipotriol betamethasone ointment, q.d.)	1. Topical corticosteroid (Calcipotriol betamethasone ointment, q.d.)	14.39 ± 2.78 y	14.79 ± 1.93 y	1. PASI 60 ($p < 0.05$) 2. TNF alpha ($p < 0.05$) 3. IL-17 ($p < 0.05$) 4. IL-23 ($p < 0.05$)	8 w	Trial: 4 AEs (1 xeroderma, 1 pruritus, 1 hyperlipidemia, 1 hepatic dysfunction) Control: 6 AEs (1 skin scale with edema, 1 xeroderma, 2 pruritus, 1 hyperlipidemia, 1 hepatic dysfunction)

Table 1. *Cont.*

Included Study (Reference)	Trial Design	Randomization Method	Number of Participants (Male/Female); Age (Mean ± SD)		Interventions		Morbidity Period (Mean ± SD or Range)		Outcome Index (Intergroup Differencies p-Value)	Course of Treat-ment	Adverse Events (Case/Symptom)
			Trial	Control	Trial	Control	Trial	Control			
Yao 2019 [180]	Randomized; Single center; Parallel	Simple randomization (random number table)	52 (29/23) 45.37 ± 6.12 y	52 (27/25) 45.13 ± 6.08 y	1. Liangxuerunzao decoction (250 mL, b.i.d.) 2. Topical corticosteroid (Calcipotriol ointment, b.i.d.)	1. Topical corticosteroid (Calcipotriol ointment, b.i.d.)	9.77 ± 1.30 y	9.86 ± 1.35 y	1. PASI 70 ($p < 0.05$) 2. PASI score ($p < 0.01$) 3.IL-17 ($p < 0.01$) 4.IL-22 ($p < 0.01$) 5.IL-23 ($p < 0.01$)	8 w	NR
Zhong 2019 [122]	Randomized; Single center; Parallel	NR	46 (22/24) 36.72 ± 6.21 y	46 (23/23) 37.23 ± 5.78 y	1. Piminxiao capsule (4c, t.i.d.) 2. Acitretin capsule (30 mg, q.d.)	1. Acitretin capsule (30 mg, q.d.)	5.10 ± 1.76 y	5.25 ± 1.28 y	1. PASI 60 ($p < 0.05$) 3. IL-17 ($p < 0.05$)	12 w	NR
Chen 2020 [155]	Randomized; Single center; Parallel	NR	47 (25/22) 36.9 ± 5.3 y	47 (24/23) 38.2 ± 5.1 y	1. Compound glycyrrhizin capsule (3c, t.i.d.) 2. Acitretin capsule (25–30 mg, q.d.)	2. Acitretin capsule (25–30 mg, q.d.)	NR	NR	1. PASI 60 ($p < 0.05$) 2. PASI score ($p < 0.05$)	8 w	Trial: 7 AEs (1 nausea, 3 hyperesthesia, 2 xeroderma, 1 xerostomia) Control: 5 AEs (2 nausea, 1 hyperesthesia 1 xeroderma, 1 xerostomia)
Hao 2020 [120]	Randomized; Single center; Parallel	Simple randomization (random number table)	30 (22/8) 42 ± 13 y	30 (21/9) 41 ± 12 y	1. Liangxuexiaobi pill (6c, t.i.d.) 2. Topical corticosteroid (Calcipotriol ointment, b.i.d.)	1. Topical corticosteroid (Calcipotriol ointment, b.i.d.)	3.71 ± 3.26 y	4.28 ± 2.96 y	1. PASI 60 ($p < 0.05$) 2. PASI score ($p < 0.05$) 3. DLQI ($p < 0.05$)	4 w	Trial: 1 AE (1 skin rash) Control: 4 AEs (2 erythema, 1 skin rash with burning sensation, 1 pruritus)
Ji 2020 [204]	Randomized; Single center; Parallel	NR	25 (15/10) 37.8 y (25.3–49.2 y)	25 (17/8) 37.5 y (27.1–48.2 y)	1. Compound Qingdai capsule (6 g, t.i.d.) 2. Methotrexate (detailed dosage NR)	1. Methotrexate (detailed dosage NR)	NR	NR	1. PASI score ($p < 0.05$) 2. DLQI ($p < 0.05$)	8 w	NR

Table 1. *Cont.*

Included Study (Reference)	Trial Design	Randomization Method	Number of Participants (Male/Female); Age (Mean ± SD)		Interventions		Morbidity Period (Mean ± SD or Range)		Outcome Index (Intergroup Differencies *p*-Value)	Course of Treatment	Adverse Events (Case/Symptom)
			Trial	Control	Trial	Control	Trial	Control			
Liu 2020a [181]	Randomized; Single center; Parallel	NR	30 (18/12) 41.3 ± 5.1 y	30 (17/13) 42.3 ± 6.2 y	1. Compound glycyrrhizin tablet (2t, t.i.d.) 2. Acitretin capsule (10 mg, b.i.d.)	1. Acitretin capsule (10 mg, b.i.d.)	5.23 ± 1.13 y	5.31 ± 1.23 y	1. PASI score ($p < 0.05$) 2. DLQI ($p < 0.05$)	8 w	Trial: 5 AEs (2 xerostomia, 2 xeroma, 1 xeroderma) Control: 11 AEs (3 xerostomia, 3 xeroma, 2 xeroderma, 2 hair loss, 1 hyperlipidemia)
Liu 2020b [167]	Randomized; Single center; Parallel	NR	92 (46/46) 54.18 ± 4.19 y	92 (45/47) 54.81 ± 4.33 y	1. Xiaoyin decoction (6c, t.i.d.) 2. Topical corticosteroid (Calcipotriol ointment, b.i.d.)	1. Topical corticosteroid (Calcipotriol ointment, b.i.d.)	8.43 ± 2.28 y	8.79 ± 2.30 y	1. IL-8 ($p < 0.05$)	8 w	NR
Lu 2020 [95]	Randomized; Single center; Parallel	NR	56 (31/25) 44.58 ± 7.12 y	56 (29/27) 45.01 ± 6.93 y	1. Total Glycosides of Paeoniae Alba capsule (0.6 g, t.i.d.) 2. Acitretin capsule (25–30 mg, q.d.)	2. Acitretin capsule (25–30 mg, q.d.)	7.63 ± 1.83 y	7.79 ± 1.72 y	1. PASI 60 ($p < 0.05$) 2. PASI score ($p < 0.05$)	8 w	Detailed information NR
Qu 2020 [177]	Randomized; Single center; Parallel	Simple randomization (computer assisted random assignment)	46 (25/21) 40.44 ± 6.74 y	46 (26/20) 40.13 ± 6.48 y	1. Yinxie capsule (4c, t.i.d.) 2. Acitretin capsule (25–30 mg, q.d.)	2. Acitretin capsule (25–30 mg, q.d.)	7.75 ± 3.75 y	7.45 ± 3.46 y	1. PASI 60 ($p < 0.05$)	8 w	NR

Table 1. *Cont.*

Included Study (Reference)	Trial Design	Randomization Method	Number of Participants (Male/Female); Age (Mean ± SD)		Interventions		Morbidity Period (Mean ± SD or Range)		Outcome Index (Intergroup Differencies *p*-Value)	Course of Treat-ment	Adverse Events (Case/Symptom)
			Trial	Control	Trial	Control	Trial	Control			
Ren 2020 [151]	Randomized; Single center; Parallel	NR	51 (27/24) 38.19 ± 2.14 y	51 (26/25) 38.62 ± 2.37 y	1. Xiaoyin granule (10 g, t.i.d.) 2. Acitretin capsule (10 mg, t.i.d.)	2. Acitretin capsule (10 mg, t.i.d.)	2.25 ± 0.95 y	2.34 ± 0.83 y	1. PASI 60 ($p < 0.05$) 2. PASI score ($p < 0.05$)	12 w	Trial: 6 AEs (2 xerostomia, 1 pruritus, 2 gastrointestinal discomfort, 1 hepatic dysfunction) Control: 18 AEs (5 xerostomia, 6 pruritus, 4 gastrointestinal discomfort, 3 hepatic dysfunction)
Shen 2020 [133]	Randomized; Single center; Parallel	NR	31 (15/16) 36 ± 14 y	32 (16/16) 35 ± 11 y	1. Liangxuejiedu decoction (200 mL, b.i.d.) 2. Topical corticosteroid (Calcipotriol ointment, q.d.) 3. Topical corticosteroid (Mometasone furoate cream, q.d.)	1. Topical corticosteroid (Calcipotriol ointment, q.d.) 2. Topical corticosteroid (Mometasone furoate cream, q.d.)	6.67 ± 6.30 y	6.42 ± 6.13 y	1. PASI 60 ($p < 0.05$) 2. PASI score ($p < 0.05$) 3. DLQI ($p < 0.05$)	8 w	NR
Wu 2020 [193]	Randomized; Single center; Parallel	NR	50 (28/22) 35.4 ± 7.9 y	50 (27/23) 34.4 ± 7.6 y	1. Compound glycyrrhizin tablet (10 g, t.i.d.) 2. Acitretin capsule (10 mg, t.i.d.)	2. Acitretin capsule (10 mg, t.i.d.)	4.6 ± 3.1 y	4.5 ± 2.9 y	1. PASI 70 ($p < 0.05$)	8 w	Trial: 25 AEs (21 xerostomia, xeroma, xeroderma, 2 hepatic dysfunction, 2 hyperlipidemia) Control: 47 AEs (35 xerostomia, xeroma, xeroderma, 8 hepatic dysfunction, 4 hyperlipidemia)

Table 1. *Cont.*

Included Study (Reference)	Trial Design	Randomization Method	Number of Participants (Male/Female); Age (Mean ± SD)		Interventions		Morbidity Period (Mean ± SD or Range)		Outcome Index (Intergroup Differencies p-Value)	Course of Treat-ment	Adverse Events (Case/Symptom)
			Trial	Control	Trial	Control	Trial	Control			
Yang 2020 [101]	Randomized; Single center; Parallel	NR	78 (44/34) 39.24 ± 6.51 y	78 (45/33) 39.17 ± 6.48 y	1. Xiaoyin granule (3.5 g, t.i.d.) 2. Acitretin capsule (30 mg, q.d.)	1. Acitretin capsule (30 mg, q.d.)	7.71 ± 2.51 y	7.69 ± 2.53 y	1. PASI 60 ($p < 0.05$)	8 w	NR
Zheng 2020 [205]	Randomized; Single center; Parallel	NR	32 (16/16) 29.8 ± 6.2 y	32 (14/18) 30.5 ± 5.8 y	1. Jueyin granule (b.i.d.) 2. Acitretin capsule (20 mg, q.d.) 3. Topical corticosteroid (Compound flumetasone cream, b.i.d.)	1. Acitretin capsule (20 mg, q.d.) 2. Topical corticosteroid (Compound flumetasone cream, b.i.d.)	NR	NR	1. PASI 60 ($p < 0.05$) 2. PASI score ($p < 0.05$)	8 w	NR
Jin 2021 [148]	Randomized; Single center; Parallel	NR	50 (24/26) 42.06 ± 4.37 y	32 (14/18) 41.32 ± 4.93 y	1. Xiaoyin decoction (200 mL, b.i.d.) 2. Topical corticosteroid (Calcipotriol ointment, b.i.d.)	1. Topical corticosteroid (Calcipotriol ointment, b.i.d.)	3.47 ± 1.40 y	3.32 ± 1.03 y	1. TNF alpha ($p < 0.05$) 2. IL-17 ($p < 0.05$) 3. IL-22 ($p < 0.05$)	4 w	NR
Lan 2021 [117]	Randomized; Single center; Parallel	NR	40 (25/15) 46.21 ± 8.07 y	40 (23/17) 46.05 ± 8.69 y	1. Dangguiyinzi (150 mL, b.i.d.) 2. Acitretin capsule (10 mg, b.i.d.)	2. Acitretin capsule (10 mg, b.i.d.)	6.25 ± 3.00 y	6.18 ± 1.95 y	1. PASI 60 ($p < 0.05$) 2. PASI score ($p < 0.05$)	12 w	Trial: 5 AEs (1 hyperlipidemia, 2 xeroma, 2 xerostomia) Control: 21 AEs (7 hyperlipidemia, 4 xeroma, 8 xerostomia, 2 cheilitis)

Table 1. *Cont.*

Included Study (Reference)	Trial Design	Randomization Method	Number of Participants (Male/Female); Age (Mean ± SD)		Interventions		Morbidity Period (Mean ± SD or Range)		Outcome Index (Intergroup Differencies *p*-Value)	Course of Treatment	Adverse Events (Case/Symptom)
			Trial	Control	Trial	Control	Trial	Control			
Le 2021 [175]	Randomized; Single center; Parallel	NR	43 (24/19) 42.17 ± 2.75 y	42 (21/21) 42.34 ± 2.66 y	1. Tianxian decoction (t.i.d.) 2. Acitretin capsule (30 mg, q.d.)	2. Acitretin capsule (30 mg, q.d.)	4.68 ± 0.91 y	4.71 ± 0.88 y	1. PASI score ($p < 0.05$) 2. DLQI ($p < 0.05$)	12 w	NR
Tang 2021 [171]	Randomized; Single center; Parallel	Simple randomization (random number table)	36 (19/17) 34.17 ± 1.75 y	36 (20/16) 33.25 ± 1.67 y	1. Qingreyangxuejiedu decoction (200 mL, b.i.d.) 2. Topical corticosteroid (Calcipotriol betamethasone ointment, b.i.d.)	1. Topical corticosteroid (Calcipotriol betamethasone ointment, b.i.d.)	1.96 ± 0.76 y	1.75 ± 0.49 y	1.P ASI 60 ($p < 0.05$) 2. PASI score ($p < 0.05$) 3. IL-17 ($p < 0.05$) 4. IL-23 ($p < 0.05$)	8 w	Trial: 1 AE (1 gastrointestinal discomfort) Control: 2 AEs (1 telangiectasia, 1 folliculitis)
Wang 2021 [98]	Randomized; Single center; Parallel	NR	42 (22/20) 46.75 ± 21.23 y	41 (21/20) 45.91 ± 20.89 y	1. Qingreliangxue decoction (100 mL, b.i.d.) 2. Acitretin capsule (20 mg, q.d.) 3. Topical corticosteroid (Mometasone furoate cream, q.d.)	1. Acitretin capsule (20 mg, q.d.) 2. Topical corticosteroid (Mometasone furoate cream, q.d.)	12.74 ± 5.23 y	11.76 ± 5.48 y	1. PASI 60 ($p < 0.05$) 3. Recurrence rate ($p < 0.05$)	8 w	Trial: 9AEs (2 xerostomia, 5 xeroderma with pruritus, 2 diarrhea) Control: 10 AEs (2 xerostomia, 6 xeroderma with pruritus, 2 diarrhea)

AEs: adverse events; b.i.d: bis in die; c: capsules; d: days; DLQI: dermatology life quality index; g: grams; IFN: interferon; IL: interleukin; m: months; mg: milligrams; NR: not reported; p: packs; PASI: psoriasis area severity index; q.d: quaque die; SD: standard deviation; t: tablets; t.i.d: ter in die; TNF: tumor necrosis factor; w: weeks; y: years; μg: micrograms.

Table 2. Methodological quality of the included studies according to the risk of bias 2.0.

Author	Year	D1	D2	D3	D4	D5	Overall
Che	2004	Sc	Sc	L	H	Sc	H
Chen	2004	Sc	Sc	L	Sc	Sc	Sc
Xu	2005	Sc	Sc	L	Sc	Sc	Sc
Liu	2006	Sc	Sc	L	H	Sc	H
Chen	2007	Sc	Sc	L	H	Sc	H
Huang	2007	L	L	L	L	Sc	Sc
Zeng	2009	L	Sc	L	Sc	Sc	Sc
Cao	2010	Sc	Sc	L	Sc	Sc	Sc
He	2010	Sc	Sc	L	H	Sc	H
Hua	2010	Sc	Sc	L	H	Sc	H
Luo	2010	Sc	Sc	L	Sc	Sc	Sc
Yu	2010	Sc	Sc	L	H	Sc	H
Liu	2011	Sc	Sc	L	H	Sc	H
Lu	2011	Sc	Sc	L	Sc	Sc	Sc
Tian	2011	L	Sc	L	Sc	Sc	Sc
Xu	2011	Sc	Sc	L	Sc	Sc	Sc
Yao	2011	L	Sc	L	Sc	Sc	Sc
Zheng	2011	Sc	Sc	L	Sc	Sc	Sc
Jiang	2012	L	L	L	L	Sc	Sc
Liu	2012	Sc	Sc	L	Sc	Sc	Sc
Xie	2012	Sc	Sc	L	H	Sc	H
Zhang	2012	Sc	Sc	L	Sc	Sc	Sc
Zhou	2012a	Sc	Sc	L	Sc	Sc	Sc
Zhou	2012b	Sc	Sc	L	H	Sc	H
Chen	2013	Sc	Sc	L	Sc	Sc	Sc
Ding	2013	Sc	Sc	L	Sc	Sc	Sc
Mo	2013	Sc	Sc	L	Sc	Sc	Sc
Song	2013	Sc	Sc	L	Sc	Sc	Sc
Zhang	2013	Sc	Sc	L	Sc	Sc	Sc
Cheng	2014	Sc	Sc	L	Sc	Sc	Sc
Du	2014	Sc	Sc	L	Sc	Sc	Sc
Li	2014	Sc	Sc	L	Sc	Sc	Sc
Liang	2014	Sc	Sc	L	Sc	Sc	Sc
Liu	2014	Sc	Sc	L	Sc	Sc	Sc
Qiu	2014	Sc	Sc	L	Sc	Sc	Sc
Zhang	2014a	Sc	Sc	L	Sc	Sc	Sc
Zhang	2014b	Sc	Sc	L	Sc	Sc	Sc
Zhang	2014c	Sc	Sc	L	Sc	Sc	Sc
Cai	2015	Sc	Sc	L	Sc	Sc	Sc

Table 2. *Cont.*

Author	Year	D1	D2	D3	D4	D5	Overall
Jin	2015a	Sc	Sc	L	Sc	Sc	Sc
Jin	2015b	Sc	Sc	L	Sc	Sc	Sc
Lu	2015a	Sc	Sc	L	H	Sc	H
Lu	2015b	Sc	Sc	L	Sc	Sc	Sc
Ma	2015	Sc	Sc	L	Sc	Sc	Sc
Peng	2015	Sc	Sc	L	Sc	Sc	Sc
Sun	2015	Sc	Sc	L	Sc	Sc	Sc
Wang	2015a	Sc	Sc	L	Sc	Sc	Sc
Wang	2015b	L	Sc	L	H	Sc	H
Wang	2015c	Sc	Sc	L	Sc	Sc	Sc
Yuan	2015	Sc	Sc	L	H	Sc	H
Zhang	2015a	L	Sc	L	Sc	Sc	Sc
Zhang	2015b	L	Sc	L	Sc	Sc	Sc
Zhang	2015c	L	Sc	L	Sc	Sc	Sc
Chen	2016	Sc	Sc	L	Sc	Sc	Sc
He	2016a	Sc	Sc	L	H	Sc	H
He	2016b	Sc	Sc	L	Sc	Sc	Sc
Jiang	2016	Sc	Sc	L	Sc	Sc	Sc
Shan	2016	L	Sc	L	H	Sc	H
Wang	2016a	Sc	Sc	L	H	Sc	H
Wang	2016b	Sc	Sc	L	H	Sc	H
Wu	2016	Sc	Sc	L	Sc	Sc	Sc
Xie	2016	L	Sc	L	Sc	Sc	Sc
Xu	2016	Sc	Sc	L	H	Sc	H
Yang	2016a	Sc	Sc	Sc	H	Sc	H
Yang	2016b	L	Sc	L	Sc	Sc	Sc
Yu	2016	Sc	Sc	L	H	Sc	H
Cao	2017	L	Sc	L	L	Sc	Sc
Cheng	2017	Sc	Sc	L	Sc	Sc	Sc
Ding	2017	L	Sc	L	Sc	Sc	Sc
Du	2017	Sc	Sc	L	Sc	Sc	Sc
Feng	2017	L	Sc	L	Sc	Sc	Sc
Han	2017a	Sc	Sc	L	Sc	Sc	Sc
Han	2017b	Sc	Sc	L	Sc	Sc	Sc
Li	2017	Sc	Sc	L	H	Sc	H
Liu	2017	Sc	Sc	L	Sc	Sc	Sc
Luo	2017	Sc	Sc	L	H	Sc	H
Pang	2017	Sc	Sc	L	Sc	Sc	Sc
Shi	2017	Sc	Sc	L	Sc	Sc	Sc
Song	2017	Sc	Sc	L	Sc	Sc	Sc

Table 2. *Cont.*

Author	Year	D1	D2	D3	D4	D5	Overall
Wang	2017	Sc	Sc	L	Sc	Sc	Sc
Wu	2017	Sc	Sc	L	H	Sc	H
Yang	2017	L	Sc	L	Sc	Sc	Sc
Zeng	2017	Sc	Sc	L	Sc	Sc	Sc
Zhang	2017a	Sc	Sc	L	Sc	Sc	Sc
Zhang	2017b	Sc	Sc	L	Sc	Sc	Sc
Zhang	2017c	Sc	Sc	L	Sc	Sc	Sc
Zhang	2017d	Sc	Sc	L	Sc	Sc	Sc
Zhang	2017e	Sc	Sc	L	Sc	Sc	Sc
Zhao	2017	Sc	Sc	L	H	Sc	H
Chai	2018	Sc	Sc	L	Sc	Sc	Sc
Li	2018	L	Sc	L	Sc	Sc	Sc
Liu	2018	Sc	Sc	L	Sc	Sc	Sc
Luo	2018	Sc	Sc	L	Sc	Sc	Sc
Ma	2018a	L	Sc	L	Sc	Sc	Sc
Ma	2018b	Sc	Sc	L	Sc	Sc	Sc
Xiao	2018	L	Sc	L	Sc	Sc	Sc
Xie	2018	Sc	Sc	L	Sc	Sc	Sc
Zhang	2018a	Sc	Sc	L	Sc	Sc	Sc
Zhang	2018b	L	Sc	L	Sc	Sc	Sc
Zhou	2018	Sc	Sc	L	Sc	Sc	Sc
Chen	2019	Sc	Sc	L	Sc	Sc	Sc
Ge	2019	Sc	Sc	L	Sc	Sc	Sc
Han	2019	Sc	Sc	L	H	Sc	H
Hu	2019	L	Sc	L	Sc	Sc	Sc
Lu	2019	L	Sc	L	L	Sc	Sc
Xun	2019	Sc	Sc	L	Sc	Sc	Sc
Yang	2019	Sc	Sc	L	Sc	Sc	Sc
Yao	2019	L	Sc	L	Sc	Sc	Sc
Zhong	2019	Sc	Sc	L	Sc	Sc	Sc
Chen	2020	Sc	Sc	L	H	Sc	H
Hao	2020	L	Sc	L	Sc	Sc	Sc
Ji	2020	Sc	Sc	L	Sc	Sc	Sc
Liu	2020a	Sc	Sc	L	Sc	Sc	Sc
Liu	2020b	Sc	Sc	L	Sc	Sc	Sc
Lu	2020	Sc	Sc	L	Sc	Sc	Sc
Qu	2020	L	Sc	L	Sc	Sc	Sc
Ren	2020	Sc	Sc	L	Sc	Sc	Sc
Shen	2020	Sc	Sc	L	Sc	Sc	Sc
Wu	2020	Sc	Sc	L	Sc	Sc	Sc

Table 2. *Cont.*

Author	Year	D1	D2	D3	D4	D5	Overall
Yang	2020	Sc	Sc	L	Sc	Sc	Sc
Zheng	2020	Sc	Sc	L	Sc	Sc	Sc
Jin	2021	Sc	Sc	L	Sc	Sc	Sc
Lan	2021	Sc	Sc	L	Sc	Sc	Sc
Le	2021	Sc	Sc	L	Sc	Sc	Sc
Tang	2021	L	Sc	L	Sc	Sc	Sc
Wang	2021	Sc	Sc	L	Sc	Sc	Sc

D1–D5: 5 domain criteria; D1, bias arising from the randomization process; D2, bias due to deviations from intended interventions; D3, bias due to missing outcome data; D4, bias in the measurement of the outcome; D5, bias in the selection of the reported results; H: high risk of bias; L: low risk of bias; Sc: some concerns.

3.4. Primary Outcomes

3.4.1. PASI 60

A meta-analysis was performed on 96 studies that reported PASI 60. The combined effect of IM on the response rate was significantly better than that of the CM control (96 trials, n = 8367; RR: 1.4280; 95% CI: 1.3783–1.4794; p < 0.0001, heterogeneity chi-square = 118.79, df = 95, I^2 = 20.0%; p = 0.0498; Figure 4A,B).

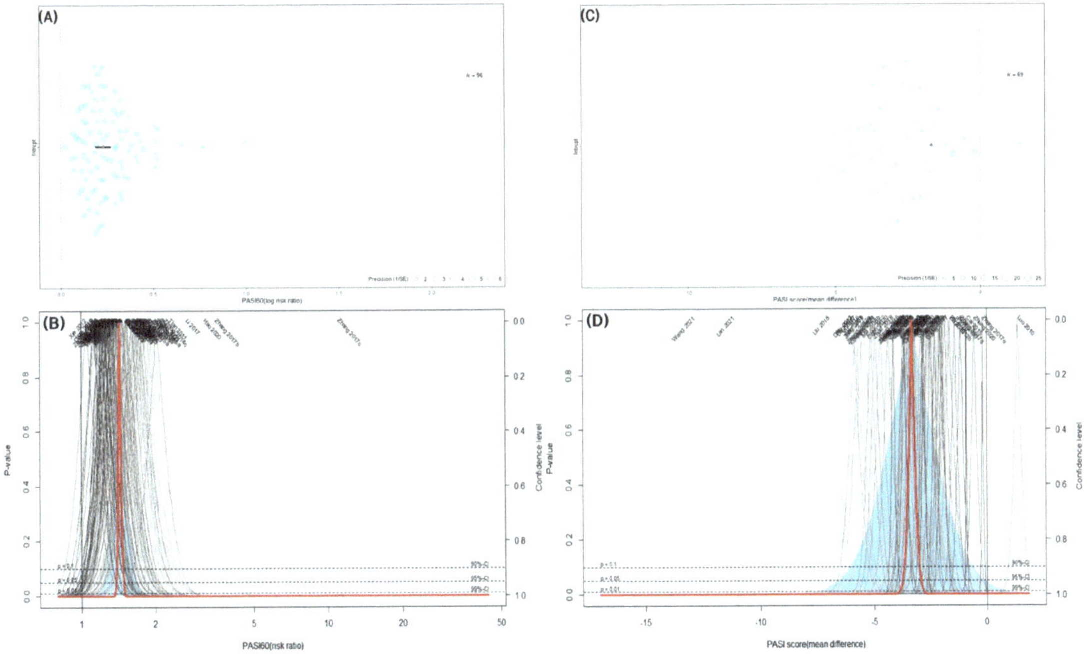

Figure 4. (**A**) Orchard plot of the trials that compared IM with CM for the PASI 60; (**B**) Drapery plot of the trials that compared IM with CM for the PASI 60; (**C**) Orchard plot of the trials that compared IM with CM for the PASI score; (**D**) Drapery plot of the trials that compared IM with CM for the PASI score. The red line in the drapery plot is the P-value curve of the pooled estimates; the light blue region in the drapery plot is the prediction region.

3.4.2. PASI Score

In the 69 studies comparing the effect of IM with that of the CM control, IM exhibited a significantly improved PASI score compared to the CM control (69 trials, n = 5801; MD: −3.3544; 95% CI: −3.7608 to −2.9481; $p < 0.0001$, heterogeneity chi-square = 3712.58, df = 68, I^2 = 98.2%; $p < 0.0001$; Figure 4C,D).

3.5. Secondary Outcomes Group 1: Assessment of Patient Outcome Related to Inflammatory Skin Lesion

In the 13 studies comparing the effect of IM with that of the CM control, IM exhibited a significantly improved PASI 70 compared to the CM control (13 trials, n = 1276; RR: 1.4994; 95% CI: 1.3208–1.7021; $p < 0.0001$; heterogeneity chi-square = 19.82, df = 12, I^2 = 39.5%; p = 0.0705; Figure 5A). Twelve studies compared IM with CM controls regarding the recurrence rate. The combined effect of IM on the recurrence rate was significantly lower than that of the CM control (12 trials, n = 692; RR: 0.3095; 95% CI: 0.2256–0.4245; $p < 0.0001$, heterogeneity chi-square = 4.97, df = 11, I^2 = 0%; p = 0.9325; Figure 5B). In the ten studies comparing the effect of IM with the CM control, IM significantly improved the DLQI compared to the CM control (10 trials, n = 811; MD: −2.6072; 95% CI: −3.7100 to −1.5043; $p < 0.0001$, heterogeneity chi-square = 198.18, df = 9, I^2 = 95.5%; $p < 0.0001$; Figure 5C). VAS scores were reported in three studies. The reduction in VAS score was significantly greater in the IM group than in the CM control (three trials, n = 273; MD: −0.8888; 95% CI: −1.4769 to −0.3008; p = 0.0031, heterogeneity chi-square = 25.04, df = 2, I^2 = 92.0%; $p < 0.0001$; Figure 5D).

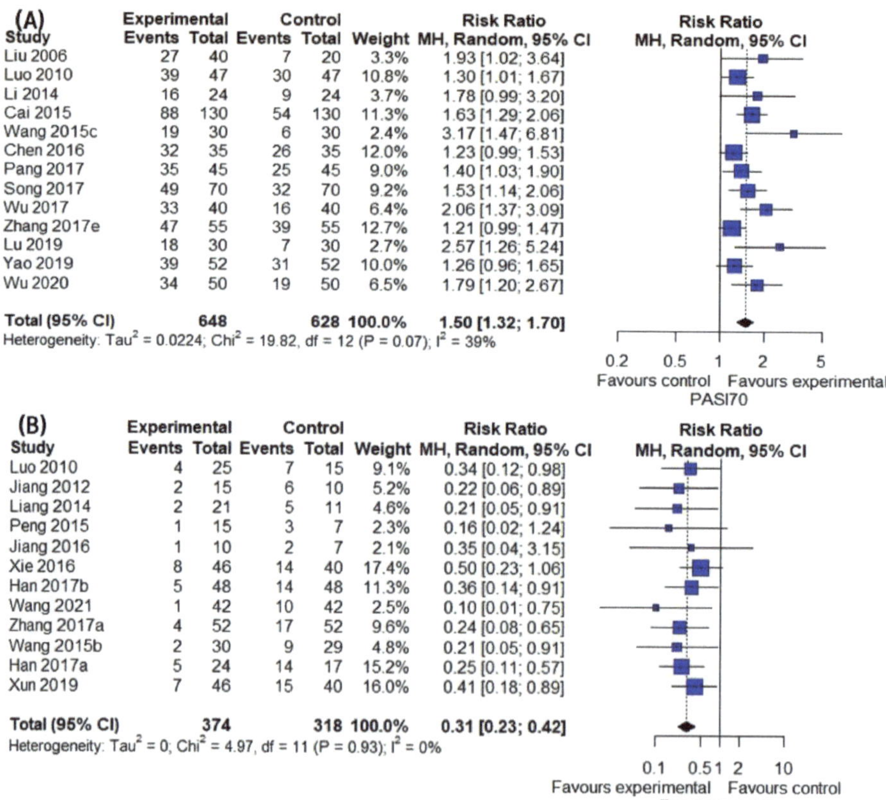

Figure 5. *Cont.*

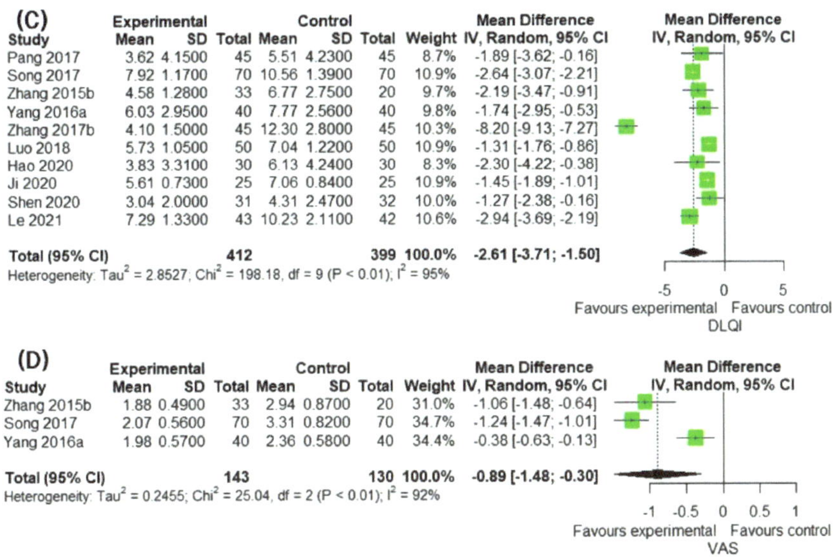

Figure 5. Forest plot of the trials that compared IM with CM for (**A**) PASI 70; (**B**) Recurrence rate; (**C**) DLQI; (**D**) VAS.

3.6. Secondary Outcomes Group 2: Assessment of Laboratory Biomarkers Related to Inflammatory Skin Lesion

The TNF-α levels were measured in 12 studies. TNF-α levels in the IM group were significantly lower than in the CM control (12 trials, n = 997; SMD: −1.9948; 95% CI: −2.5964–−1.3932; $p < 0.0001$, heterogeneity chi-square = 154.77, df = 11, I^2 = 92.9%; p = 0.3361; Figure 6A). IM remarkably decreased IL-8 levels compared to the CM control (seven trials, n = 657; SMD: −1.0752; 95% CI: −1.9647 to −0.1856; p = 0.0178, heterogeneity chi-square = 52.93, df = 6, I^2 = 92.9%; $p < 0.0001$; Figure 6B). In the ten studies comparing the effect of IM with that of the CM control, IM significantly decreased IL-17 levels compared to the CM control (10 trials, n = 842; SMD: −2.0009; 95% CI: −3.2927 to −0.7091; p = 0.0024, heterogeneity chi-square = 127.74, df = 9, I^2 = 93.0%; $p < 0.0001$; Figure 6C). The effect of IM on IL-22 was reported in four studies. A significant improvement in IL-22 by IM was identified by the CM control (four trials, n = 348; SMD: −3.1874; 95% CI: −6.0441 to −0.3307; p = 0.0288, heterogeneity chi-square = 113.40, df = 3, I^2 = 97.4%; $p < 0.0001$; Figure 6D). IL-23 was reported in five studies. The reduction in IL-23 was significantly greater in the IM group than in the CM control (five trials, n = 428; SMD: −1.7398; 95% CI: −2.6139 to −0.8657; $p < 0.0001$, heterogeneity chi-square = 56.86, df = 4, I^2 = 93.0%; $p < 0.0001$; Figure 6E). IFN-γ levels were reported in six studies. Compared with CM controls, IM significantly decreased IFN-γ levels (six trials, n = 524; SMD: −2.4211; 95% CI: −3.2187 to −1.6235; $p < 0.0001$, heterogeneity chi-square = 59.15, df = 5, I^2 = 91.5%; $p < 0.0001$; Figure 6F).

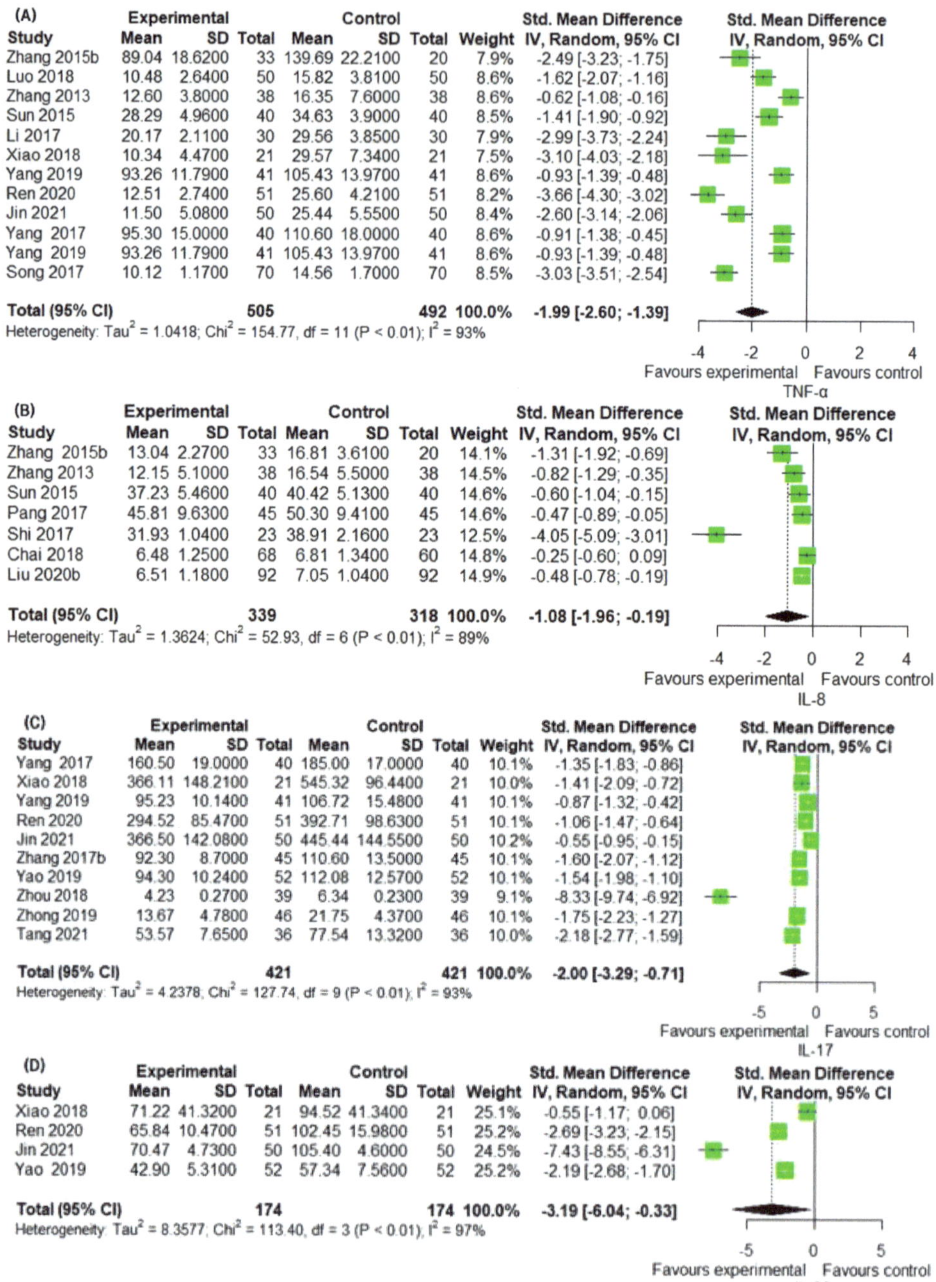

Figure 6. *Cont.*

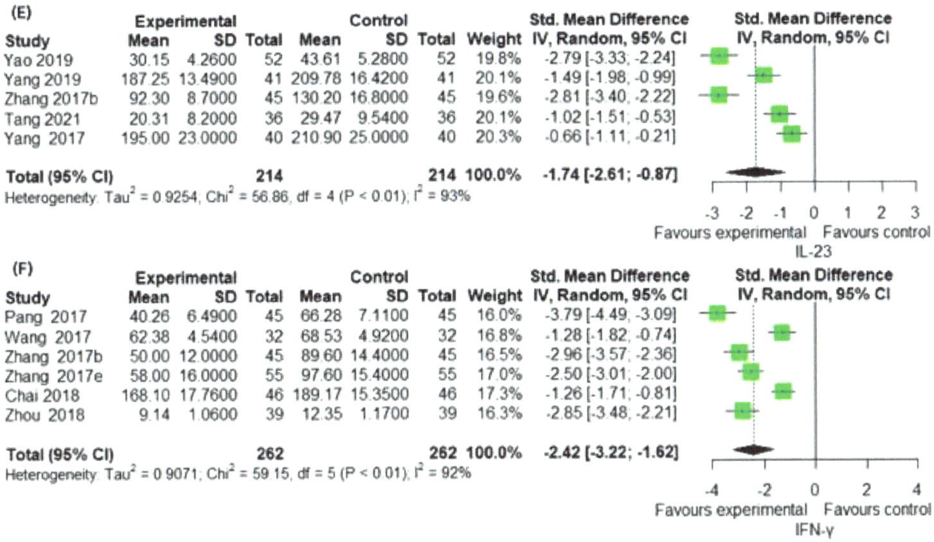

Figure 6. Forest plot of the trials that compared IM with CM for (**A**) TNF-α; (**B**) IL-8; (**C**) IL-17; (**D**) IL-22; (**E**) IL-23; (**F**) IFN-γ.

3.7. Secondary Outcomes Group 3: Safety Assessment

A total of 79 trials (79/126, 62.7%) reported detailed information on AEs in the IM and CM control groups. In four trials, detailed information for each group of AEs was not reported [95,106,126,162]. Therefore, a safety assessment was performed by extracting AEs data from 75 trials. The individual symptoms of the reported AEs for each trial are presented in Table 1. AEs that occurred during the treatment of psoriasis could be classified into the following four categories: drug-induced liver injury, including findings of impaired hepatic function and elevated blood aspartate transaminase levels; cutaneous symptoms suggesting worsening of skin findings, such as xerostomia, xeroma, and xeroderma; alimentary symptoms, including abdominal discomfort and diarrhea after taking medication; and metabolic disorders reported primarily in the form of hyperlipidemia. Hence, the incidence rates were compared between the groups by dividing the reported AEs into the four categories mentioned above and one category that included other symptoms that could not be classified into these categories in this review.

A meta-analysis of 19 studies showed that IM significantly reduced the incidence of drug-induced liver injury compared to the CM control (19 trials, n = 1728; OR: 0.4219; 95% CI: 0.2482–0.7170; p = 0.0014, heterogeneity chi-square =11.90, df = 18, I^2 = 0%; p = 0.8523; Figure 7A). The pooled results of 31 studies comparing IM and CM controls suggested that the incidence of AEs was significantly reduced by IM administration. A total of 57 trials evaluated the effect of IM on reducing the incidence rate of cutaneous AEs compared to the CM control. The meta-analysis revealed a significant reducing effect of IM on AEs (57 trials, n = 2723; OR: 0.3578; 95% CI: 0.2687 to 0.4763; $p < 0.0001$, heterogeneity chi-square = 205.79, df = 56, I^2 = 72.8%; $p < 0.0001$; Figure 7B). The incidence rate of alimentary AEs in IM was similar to that of the CM controls, and there was no statistically significant difference between the two groups, according to a meta-analysis (31 trials, n =2723; OR: 0.9552; 95% CI: 0.6516 to 1.4002; p = 0.8143, heterogeneity chi-square = 30.70, df = 30, I^2 = 2.3%; p = 0.4305; Figure 7C). In contrast, the pooled results of 21 studies comparing IM and the CM control suggested that the incidence of metabolic AEs was significantly reduced by IM (21 trials, n = 1793; OR: 0.3288; 95% CI: 0.2369–0.4562; $p < 0.0001$, heterogeneity chi-square = 13.61, df = 20, I^2 = 0%; p = 0.8497; Figure 7D). In the 21 studies comparing the effect of IM with the CM control, IM significantly decreased the incidence rate compared to the CM control (21 trials, $n = 2152$; OR: 0.5098; 95% CI: 0.3488–0.7450; $p = 0.0005$, heterogeneity chi-square = 15.11, df = 20, $I^2 = 0\%$; $p = 0.7701$; Figure 7E).

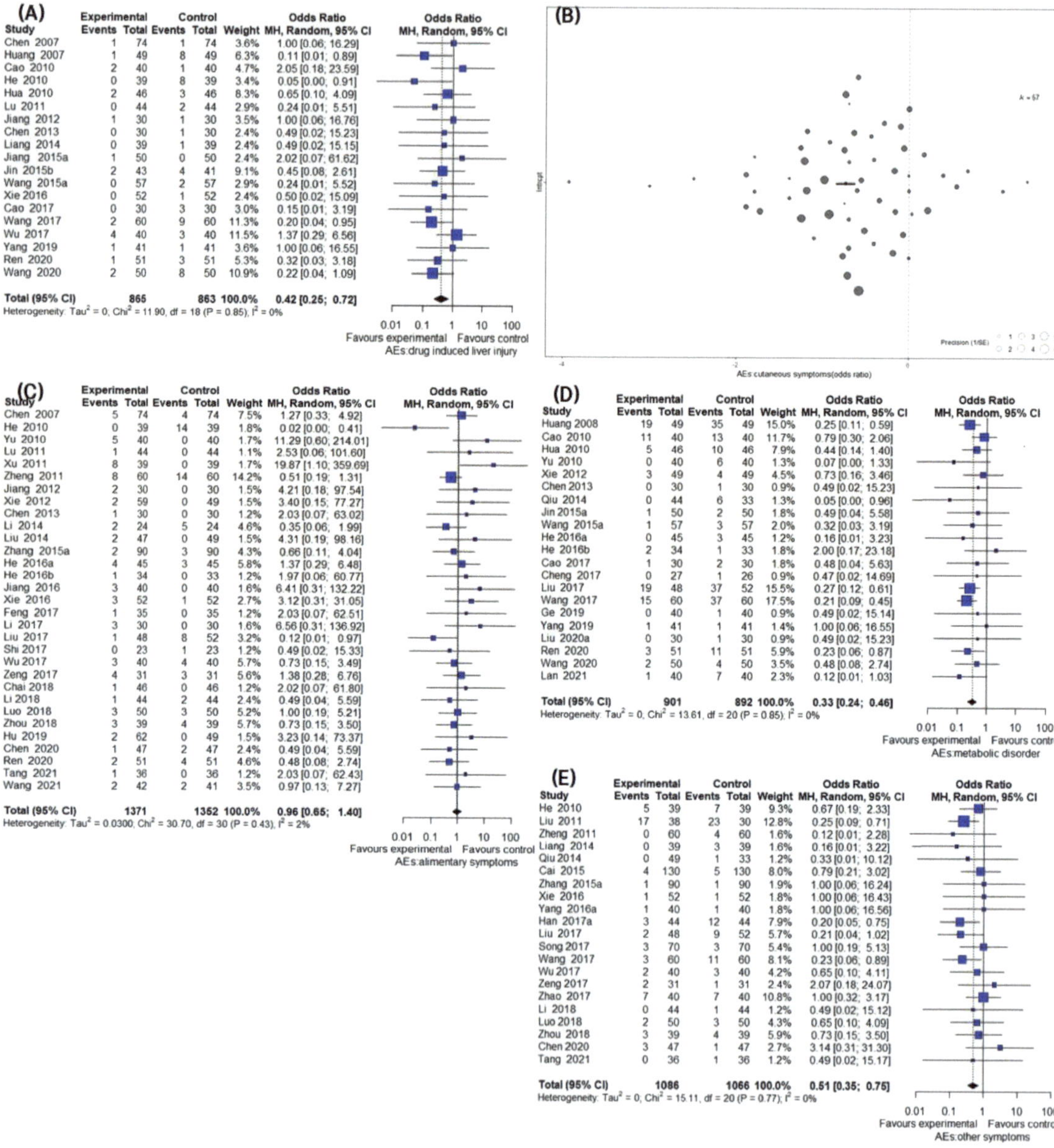

Figure 7. Forest plot of the incidence rates of reported adverse events: (**A**) drug induced liver injury; (**B**) cutaneous symptoms; (**C**) alimentary symptoms; (**D**) metabolic disorder; (**E**) other symptoms.

3.8. Assessing Heterogeneity

3.8.1. Sensitivity Analysis

Considerable heterogeneity was found in the synthesis of the trial data for PASI scores between the two primary outcomes of this study. In this respect, trials estimated to be potential outliers were also observed in the drapery plot. To ascertain whether particular research matching these outliers was the cause of heterogeneity, a sensitivity analysis using the leave-one-out method was performed. Sensitivity analysis revealed that no single study's omission considerably affected the change in heterogeneity (Figure 8A,B).

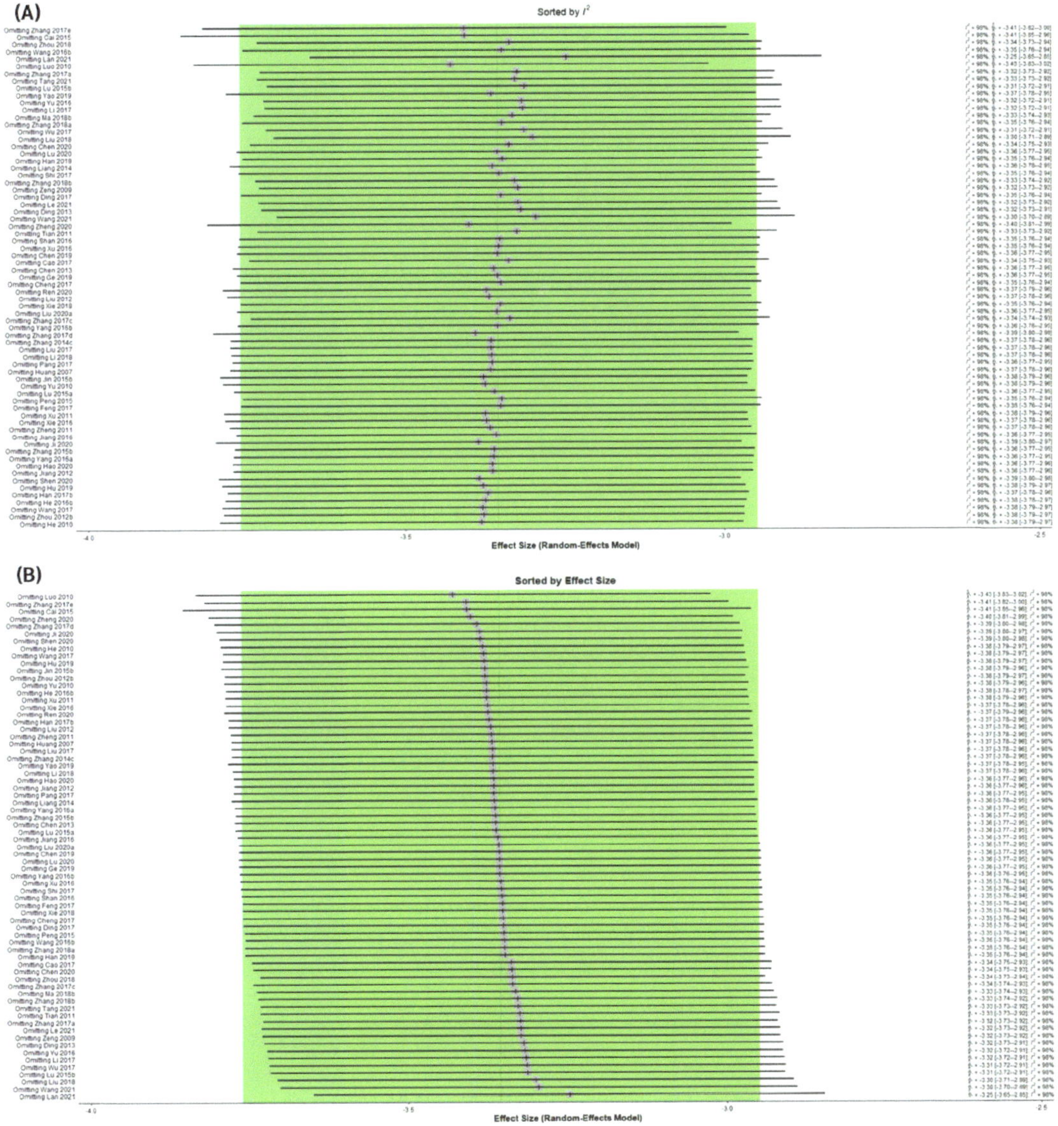

Figure 8. Forest plot of the sensitivity analysis ordered by heterogeneity for (**A**) PASI 60 and (**B**) PASI score.

3.8.2. Meta-Regression and Subgroup Analysis

Sensitivity analysis demonstrated that specific individual studies corresponding to outliers did not affect heterogeneity. Consequently, a meta-regression analysis was carried out on moderators predicted to have an impact on the results to identify other possible sources of heterogeneity. The following moderators were examined in this study: (i) comparator drug, (ii) treatment duration, (iii) source of investigational medicine, (iv) formulation type, (v) sample size, (vi) overall risk of bias, and (vii) randomization method. Meta-regression analysis revealed that the variables "source of investigational medicine" (p = 0.0240) and "formulation type" (p = 0.0145) had a statistically noticeable influence on the pooled results (Figure 9A,B). However, the results of the meta-analysis were not significantly influenced by the comparator drug (p = 0.6215, Supplementary Figure S1A), treatment duration (p = 0.5563, Supplementary Figure S1B), sample size (p = 0.6098, Supplementary Figure S1C), overall risk of bias (p = 0.5615, Supplementary Figure S1D), or randomization method (p = 0.9857, Supplementary Figure S1E). Based on the meta-regression results, a subgroup analysis was performed on the source of investigational medicine and formulation type. However, subgroup analysis showed that neither moderator had a significant influence on the magnitude of the effect or degree of heterogeneity (Table 3).

Table 3. Subgroup analysis of the trials that compared IM with CM for PASI score.

	k	MD	95% CI	Heterogeneity (I^2)	$P_{subgroup}$
Source of investigational medicine					**0.0240**
Research institute prescription	39	−3.7420	−4.2580 to −3.2261	98.4%	
Other conventional medicine	30	−2.8494	−3.4276 to −2.2712	95.9%	
Formulation type					**0.0116**
Decoction	32	−2.8951	−3.3769 to −2.4132	95.2%	
Other types	37	−3.8594	−4.4331 to −2.2856	98.6%	

IM: Integrative medicine; CM: conventional medicine; PASI: psoriasis area severity index.

3.9. Assessing Publication Bias

A contour-enhanced funnel plot and Egger's and Begg's tests were used to assess the potential publication bias of the primary outcomes in this meta-analysis. Asymmetric shapes were observed in the contour-enhanced funnel plot for PASI 60, suggesting potential bias (Figure 10A). The publication bias was statistically significant in the Egger's and Begg's tests for PASI 60 (Egger's test: t = 10.67, df = 94, $p < 0.0001$; Begg's test: z = 7.48, $p < 0.0001$). Unlike PASI 60, a distinct asymmetric shape suggestive of publication bias was not observed in the contour-enhanced funnel plot for the PASI score (Figure 10B). The publication bias of the continuous PASI score was significant in Egger's test, but no significant bias was confirmed in Begg's test (Egger's test: t = −6.58, df = 67, $p < 0.0001$; Begg's test: z = 1.19, p = 0.2335).

3.10. Summary of Evidence according to Outcome Measures

For all outcome measures in the IM compared to CM trials, the overall quality of the evidence ranged from very low to moderate. Table 4 shows the results of the GRADE assessments.

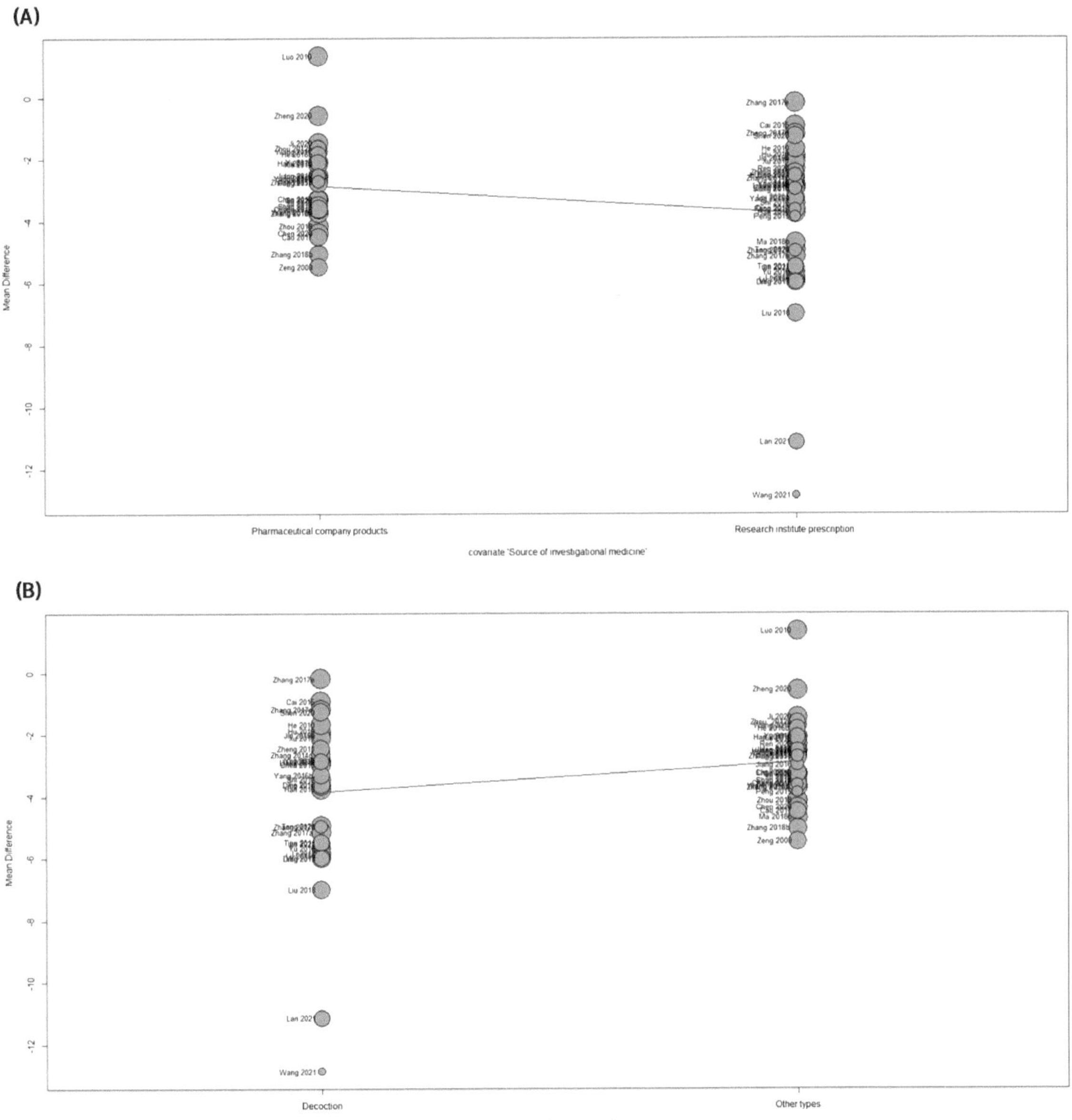

Figure 9. Bubble plot of the meta-regression analysis for (**A**) source of investigational medicine and (**B**) formulation type.

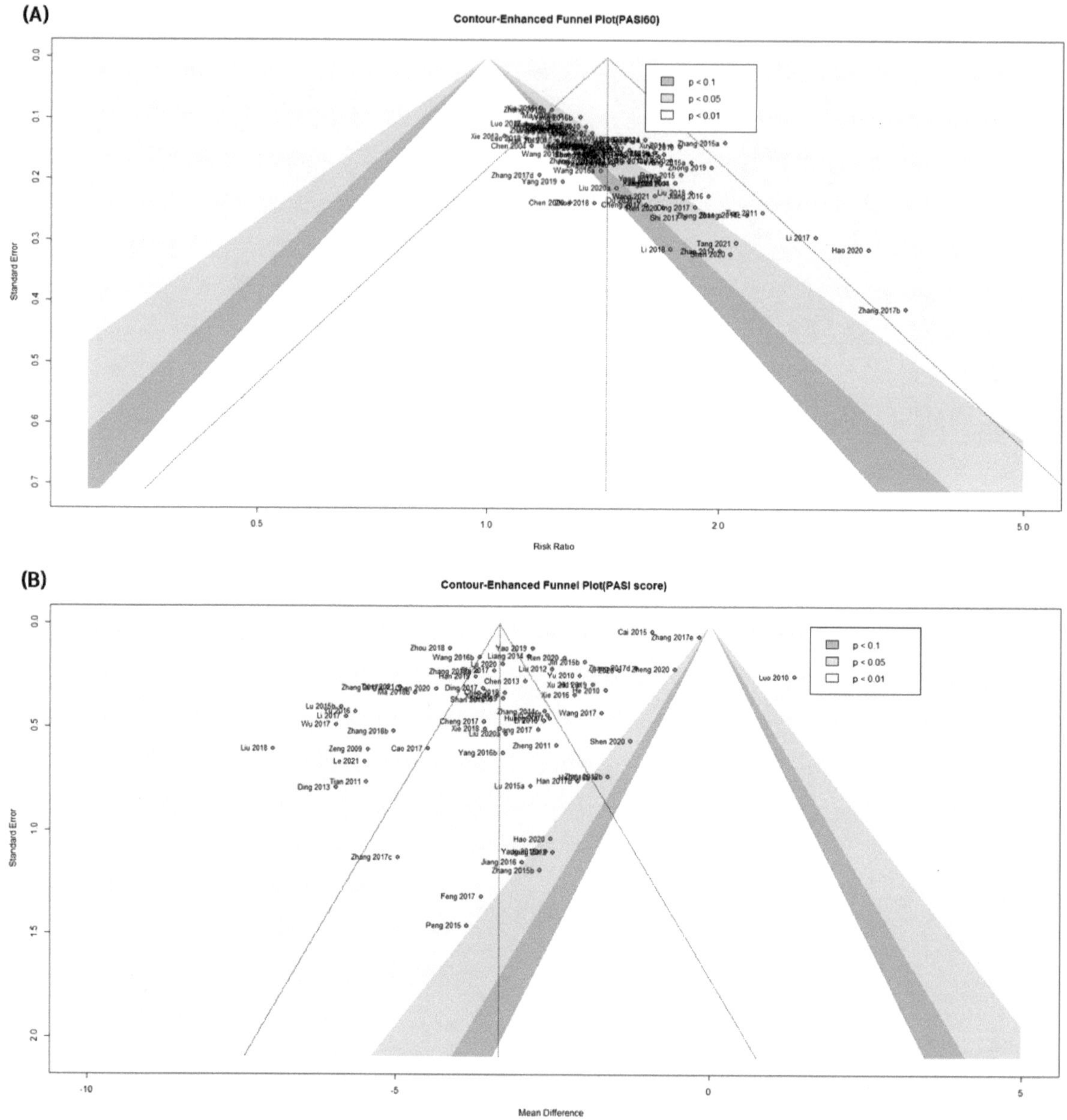

Figure 10. Contour-enhanced funnel plot of (**A**) PASI 60 and (**B**) PASI scores.

Table 4. Summary of findings for studies in this meta-analysis.

Intervention and Comparator Intervention	Outcomes	Number of Participants (Studies)	Anticipated Absolute Effects (95% CI)	Quality of the Evidence (GRADE)
IM compared to CM for inflammatory pain of rheumatoid arthritis	PASI60	8367 (96)	224 more per 1000 (from 139 more to 318 more)	⊕⊕⊕○ MODERATE [a]
	PASI score	5801 (69)	MD 3.3544 PASI score lower (3.7608 lower to 2.9481 lower)	⊕⊕⊕○ MODERATE [b]
	PASI70	1276 (13)	239 more per 1000 (From 154 more to 337 more)	⊕⊕⊕○ MODERATE [a]
	Recurrence rate	692 (12)	252 fewer per 1000 (From 282 fewer to 210 fewer)	⊕⊕⊕○ MODERATE [a]
	DLQI	811 (10)	MD 2.6072 DLQI lower (3.71 lower to 1.5043 lower)	⊕⊕⊕○ MODERATE [a]
	VAS	273 (3)	MD 0.89 VAS lower (1.4769 lower to 0.3008 lower)	⊕⊕○○ LOW [a,b]
	TNF-α	997 (12)	SMD 1.9948 SD lower (2.5964 lower to 1.3932 lower)	⊕⊕○○ LOW [a,b]
	IL-8	657 (7)	SMD 1.0752 SD lower (1.9647 lower to 0.1856 lower)	⊕⊕○○ LOW [a,b]
	IL-17	842 (10)	SMD 2.0009 SD lower (3.2927 lower to 0.7091 lower)	⊕⊕○○ LOW [a,b]
	IL-22	348 (4)	SMD 3.1874 SD lower (6.0441 lower to 0.3307 lower)	⊕⊕○○ LOW [a,b]
	IL-23	428 (5)	SMD 1.7398 SD lower (2.6139 lower to 0.8657 lower)	⊕⊕○○ LOW [a,b]
	INF-γ	524 (6)	SMD 2.4211 SD lower (3.2187 lower to 1.6235 lower)	⊕⊕○○ LOW [a,b]

CM, conventional medicine; DLQI, dermatology life quality index; IM, integrative medicine; MD, mean difference; RR, risk ratio; RCT, randomized clinical trial; SD, standardized difference; SMD, standardized mean difference; TNF-α, tumor necrosis factor-α; [a]: substantial concerns of publication bias; [b]: the confidence intervals are less overlapping.

3.11. Core Herbs Discovery Based on Data Mining

3.11.1. Detailed Information on Investigational Medicine Ingredients

A total of 137 EAHMs were used as components of IM in the 126 clinical trials included in this review. Supplementary Table S2 provides comprehensive details on the herbal components of IM. The following 26 individual herbs, listed in descending order of frequency, were among those used in 10% of IM prescriptions: *Rehmannia glutinosa* (Gaertn.) DC., *Glycyrrhiza uralensis* Fisch., *Dictamnus dasycarpus* Turcz., *Paeonia* × *suffruticosa* Andrews, *Paeonia anomala* subsp. *veitchii* (Lynch) D.Y.Hong and K.Y.Pan, *Smilax glabra* Roxb., *Angelica sinensis* (Oliv.) Diels, *Lonicera japonica Thunb.*, *Saposhnikovia divaricata* (Turcz.) Schischk., *Arnebia euchroma* (Royle) I.M.Johnst., *Sophora flavescens Aiton*, *Isatis tinctoria* subsp. *athoa* (Boiss.) Papan., *Scrophularia ningpoensis* Hemsl., *Salvia miltiorrhiza Bunge*, *Cryptotympana dubia* (Haupt), *Carthamus tinctorius* L., *Styphnolobium japonicum* (L.) *Schott*, *Spatholobus suberectus* Dunn, *Arctium lappa* L., *Scleromitrion diffusum (Willd.) R.J.Wang*, *Paeonia lactiflora* Pall., *Ligusticum striatum* DC., *Ophiopogon japonicus* (Thunb.) Ker Gawl., *Scutellaria baicalensis* Georgi, *Scutellaria Nepeta tenuifolia* Benth., and *Taraxacum mongolicum* Hand.-Mazz. The relative frequencies of the top 26 herbal materials ranged from 10.16% to 62.5%. Table 5 provides detailed information on the frequency distributions of EAHM materials.

Table 5. Characters of top 26 commonly prescribed herbs utilized with relatively frequencies exceeding 10% inclusion trials and its prestige centrality.

No	Herbal Material (Latin Name)	Frequency of Prescription	Relative Frequency (%)	PageRank Centrality	Eigenvector Centrality
1	*Rehmannia glutinosa* (Gaertn.) DC.	80	62.5%	0.0284	1
2	*Glycyrrhiza uralensis* Fisch.	57	44.53%	0.0278	0.982
3	*Dictamnus dasycarpus* Turcz.	54	42.19%	0.0272	0.968
4	*Paeonia* × *suffruticosa* Andrews	53	41.41%	0.0271	0.974
5	*Paeonia anomala subsp. veitchii* (Lynch) D.Y.Hong and K.Y.Pan	51	39.84%	0.0278	0.982
6	*Smilax glabra* Roxb.	47	8.59%	0.0277	0.988
7	*Angelica sinensis* (Oliv.) Diels	44	34.38%	0.0271	0.974
8	*Lonicera japonica* Thunb.	39	30.47%	0.0266	0.953
9	*Saposhnikovia divaricata* (Turcz.) Schischk.	39	30.47%	0.0266	0.945
10	*Arnebia euchroma* (Royle) I.M.Johnst.	36	28.13%	0.0265	0.959
11	*Sophora flavescens* Aiton	36	28.13%	0.0254	0.914
12	*Isatis tinctoria subsp. athoa* (Boiss.) Papan.	35	27.34%	0.0253	0.92
13	*Scrophularia ningpoensis* Hemsl.	33	25.78%	0.0247	0.907
14	*Salvia miltiorrhiza* Bunge	30	23.44%	0.0277	0.988
15	*Cryptotympana dubia* (Haupt)	28	21.88%	0.0253	0.923
16	*Carthamus tinctorius* L.	27	21.09%	0.0224	0.823
17	*Styphnolobium japonicum* (L.) Schott	21	16.41%	0.0224	0.824
18	*Spatholobus suberectus* Dunn	20	15.63%	0.0253	0.92
19	*Arctium lappa* L.	18	14.06%	0.0128	0.442
20	*Scleromitrion diffusum* (Willd.) R.J.Wang	17	13.28%	0.022	0.787
21	*Paeonia lactiflora* Pall.	17	13.28%	0.023	0.843
22	*Ligusticum striatum* DC.	14	10.94%	0.0219	0.793
23	*Ophiopogon japonicus (Thunb.)* Ker Gawl.	14	10.94%	0.0235	0.871
24	*Scutellaria baicalensis* Georgi	14	10.94%	0.0248	0.895
25	*Scutellaria Nepeta tenuifolia* Benth.	13	10.16%	0.0242	0.868
26	*Taraxacum mongolicum* Hand.-Mazz.	13	10.16%	0.0218	0.804

3.11.2. Social Network Analysis

Social network analysis was conducted on materials that were utilized regularly in more than 5% of trials to examine the correlation and centrality among herbal medicines prescribed to patients, as well as to identify the candidate sovereign herbs (Figure 11). In more than 5% of the trials, a total of 45 herbal medicines were included in the analysis. Centrality analysis showed that the range of the eigenvector centrality index of all materials was in the range of 0.338 to 1, and the range of the centrality index according to the PageRank method was found to be 0.0106 to 0.0284 (Table 5 and Table S3). The core materials that exceeded 0.9 in the eigenvector centrality were as follows: *Rehmannia glutinosa* (Gaertn.) DC., *Salvia miltiorrhiza* Bunge, *Smilax glabra* Roxb., *Glycyrrhiza uralensis* Fisch., *Paeonia anomala* subsp. *veitchii (Lynch)* D.Y.Hong and K.Y.Pan, *Angelica sinensis* (Oliv.) Diels,

Paeonia × suffruticosa Andrews, *Dictamnus dasycarpus* Turcz., *Arnebia euchroma* (Royle) I.M.Johnst., *Lonicera japonica* Thunb., *Saposhnikovia divaricata* (Turcz.) Schischk., *Cryptotympana dubia* (Haupt), *Isatis tinctoria* subsp. *athoa* (Boiss.) Papan., *Spatholobus suberectus* Dunn, *Sophora flavescens* Aiton, and *Scrophularia ningpoensis* Hemsl. Only these materials exceeded 0.026 in the PageRank centrality, and there was no change in the ranking trend. These materials have the strongest reciprocity based on the properties of the prestige centrality. Therefore, they were considered as candidate sovereign herbs in this review.

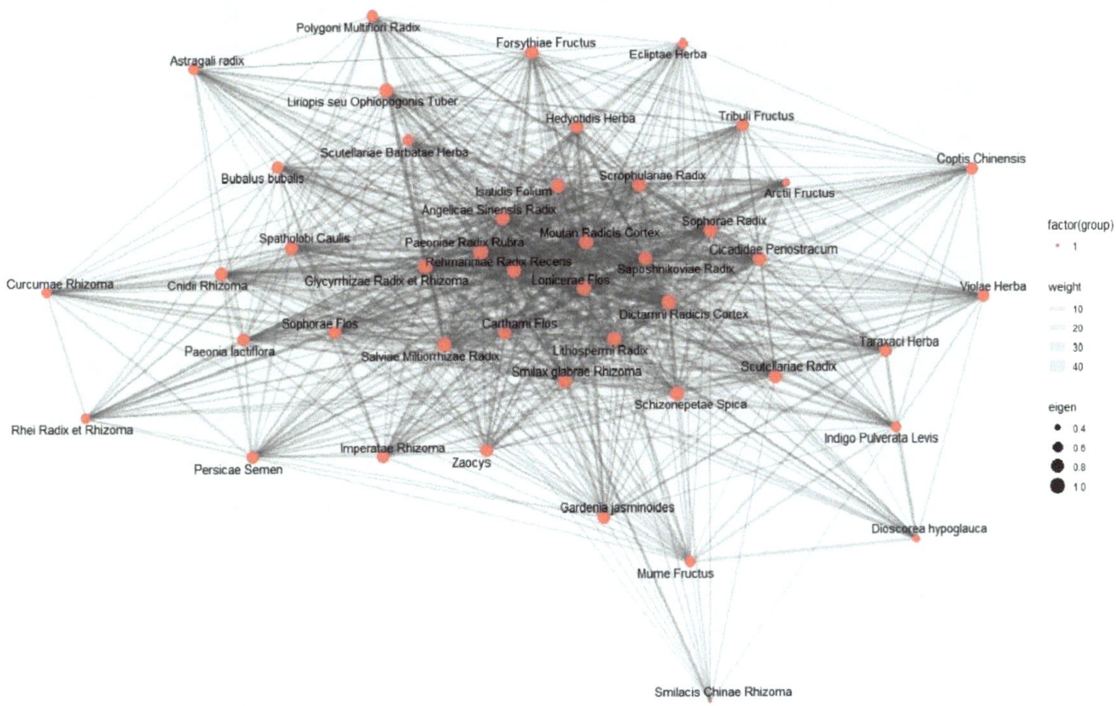

Figure 11. IM herbal materials network used in more than 5% of trials for inflammatory skin lesion of psoriasis.

3.11.3. A Priori Algorithm-Based Association Rule Analysis

Nine association rules were discovered via a priori algorithm analysis of the ingredient data for the 137 herbs included in this review (Table 6). Among the investigated association rules, the following three patterns showed the highest confidence value: # 1 {*Scrophularia ningpoensis* Hemsl.} => {*Rehmannia glutinosa* (Gaertn.) DC.} (confidence:0.9697); # 7 {*Isatis tinctoria* subsp. *athoa* (Boiss.) Papan., *Paeonia × suffruticosa* Andrews} => {*Rehmannia glutinosa* (Gaertn.) DC.}; # 8 {*Isatis tinctoria* subsp. *athoa* (Boiss.) Papan., *Rehmannia glutinosa* (Gaertn.) DC.} => {*Paeonia × suffruticosa* Andrews}. The following two association rules had lift values higher than 2: # 2 {*Isatis tinctoria* subsp. *athoa* (Boiss.) Papan.} => {*Paeonia × suffruticosa* Andrews} (Lift:2.277) and # 8 {*Isatis tinctoria* subsp. *athoa* (Boiss.) Papan., *Rehmannia glutinosa* (Gaertn.) DC.} => {*Paeonia × suffruticosa* Andrews} (Lift:2.342). Therefore, four association rules with excellent characteristics, as described above, were selected as the most important patterns that can be confirmed among IM prescriptions. The relationship diagram of all the association rules is presented as a separate network graph (Figure 12).

Table 6. A priori algorithm-based association rules in the IM component herbs for inflammatory skin lesion in psoriasis.

No	Associations Rules	Support	Confidence	Lift
1	{*Scrophularia ningpoensis* Hemsl.} => {*Rehmannia glutinosa* (Gaertn.) DC.}	0.2500000	0.9696970	1.551515
2	{*Isatis tinctoria* subsp. *athoa* (Boiss.) Papan.} => {*Paeonia* × *suffruticosa* Andrews}	0.2578125	0.9428571	2.277089
3	{*Isatis tinctoria* subsp. *athoa* (Boiss.) Papan.} => {*Rehmannia glutinosa* (Gaertn.) DC.}	0.2578125	0.9428571	1.508571
4	{*Saposhnikovia divaricata* (Turcz.) Schischk.} => {*Rehmannia glutinosa* (Gaertn.) DC.}	0.2812500	0.9230769	1.476923
5	Paeonia *anomala* subsp. *veitchii* (Lynch) D.Y.Hong and K.Y.Pan} => {*Rehmannia glutinosa* (Gaertn.) DC.}	0.3671875	0.9215686	1.474510
6	{*Paeonia* × *suffruticosa* Andrews} => {*Rehmannia glutinosa* (Gaertn.) DC.}	0.3828125	0.9245283	1.479245
7	{*Isatis tinctoria* subsp. *athoa* (Boiss.) Papan., *Paeonia* × *suffruticosa* Andrews} => {*Rehmannia glutinosa* (Gaertn.) DC.}	0.2500000	0.9696970	1.551515
8	{*Isatis tinctoria* subsp. *athoa* (Boiss.) Papan., *Rehmannia glutinosa* (Gaertn.) DC.} => {*Paeonia* × *suffruticosa* Andrews}	0.2500000	0.9696970	2.341910
9	{*Paeonia* × *suffruticosa* Andrews, *Paeonia anomala* subsp. *veitchii* (Lynch) D.Y.Hong and K.Y.Pan} => {*Rehmannia glutinosa* (Gaertn.) DC.}	0.2734375	0.9210526	1.473684

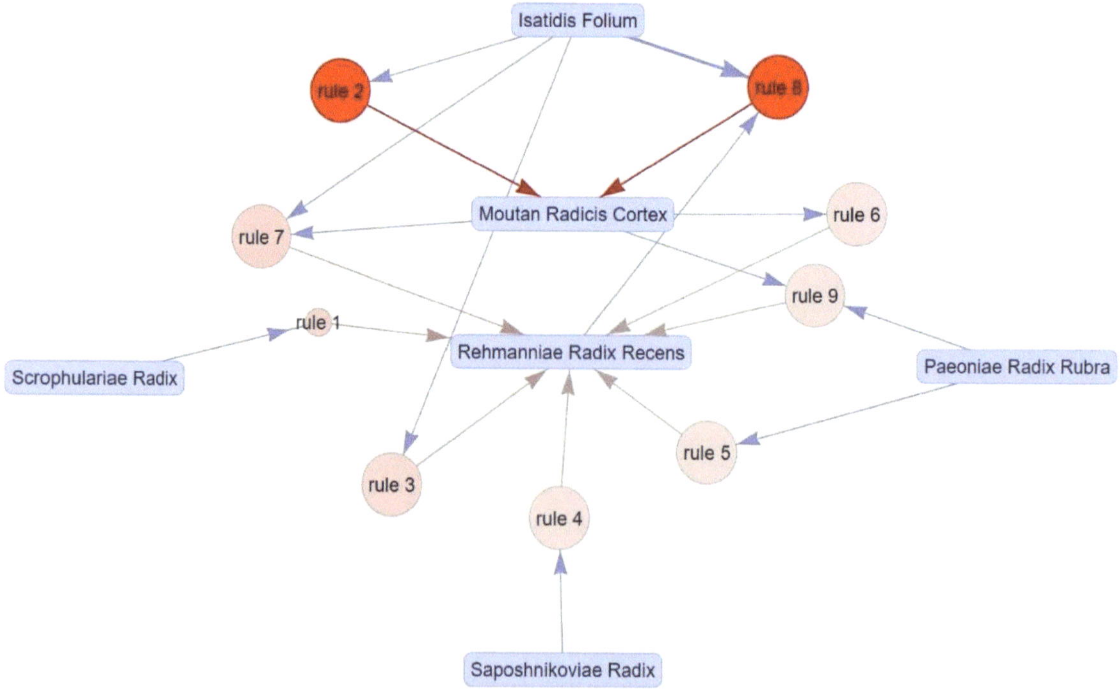

Figure 12. Network graph of the core association rule in the IM component herbs prescribed for inflammatory skin lesion in psoriasis.

3.11.4. Derivation of Core Herbs

Based on the above three-step data mining, four core herbs, *Rehmannia glutinosa* (Gaertn.) DC., *Isatis tinctoria* subsp. *athoa* (Boiss.) Papan., *Paeonia* × *suffruticosa* Andrews, and *Scrophularia ningpoensis Hemsl.*, were selected to simultaneously satisfy the three conditions defined in the research method section. Table 7 shows the results calculated using the inter-quartile range, with the extreme values removed for the range of dose and administration period of each material based on the data of the included trial.

Table 7. Main information of four core herbs identified through this review.

Four Core Herbs	Interquartile Range of Dosage (g/Day)	Interquartile Range of Treatment Duration (Week)
Rehmannia glutinosa (Gaertn.) DC.	15–20 g	4–12 w
Isatis tinctoria subsp. *athoa* (Boiss.) Papan.	10–30 g	8–12 w
Paeonia × *suffruticosa* Andrews	10–15 g	8–12 w
Scrophularia ningpoensis Hemsl.	10–15 g	7–12 w

g: gram; w: weeks.

3.12. *Analysis of Four Core Herbs through Network Pharmacology*

3.12.1. Active Ingredients and Anti-Psoriasis Targets of Four Core Herbs

The chemical ingredients of four core herbs were retrieved from a traditional Chinese medicine system pharmacology database (TCMSP), an encyclopedia of traditional Chinese medicine (ETCM), a high-throughput experiment-and-reference-guided database of traditional Chinese medicine (HERB), and a traditional Chinese medicine integrative database (TCMID). For the above compounds collected from multiple databases, after completing the first screening by utilizing OB and DL information from TCMSP DB, we conducted a second screening by performing an ADME prediction for each compound and finally derived a list of 22 potential active compounds. (Table 8) We then performed target prediction by compound on the SwissTargetPrediction platform and obtained pieces of 473 compound–target relationship data after removing duplicates (Supplementary Table S4). Additionally, the GeneCards database contained information on 584 human target genes associated with *psoriasis vulgaris* (Supplementary Table S5). After intersection mapping, 43 consensus genes were identified as potential therapeutic targets for four core herbs against psoriasis (Figure 13A).

3.12.2. PPI Network Construction

Using the STRING 11.5 platform, we imported common targets and constructed the PPI network model (minimum required interaction score: 0.4) by restricting the organism to "*homo sapiens*". A total of 41 nodes with 462 edges were obtained with an average degree of 11.268. Two targets (CA3, CHRNA7) were excluded from the PPI network because they did not interact with other targets. The MCC algorithm was used by cytoHubba to screen the main hub genes. The outcome was the identification of 19 nodes as hub gene targets, each of which had an MCC score at least three times higher than the median value 3.93×10^5). Table 9 lists the 19 hub gene targets according to their degree centrality. Figure 13B shows the PPI network of hub targets.

Table 8. Major active compounds of four core herbs.

Molecule Number	Molecule Name	Chemical Structure	ADME Evaluation				
			Lipsinski	Ghose	Veber	Egan	Muegge
1	Dihydrochelerythrine	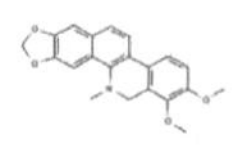	Yes	Yes	Yes	Yes	Yes
2	Sitogluside	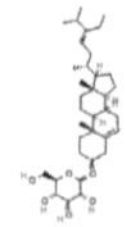	Yes	No	Yes	Yes	No
3	Glycyrol	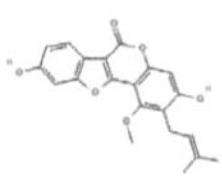	Yes	Yes	Yes	Yes	Yes
4	Hesperetin	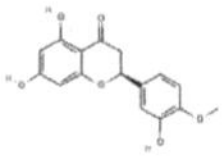	Yes	Yes	Yes	Yes	Yes
5	Salutaridine	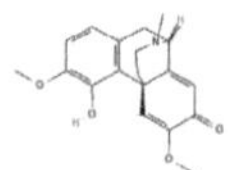	Yes	Yes	Yes	Yes	Yes
6	indirubin	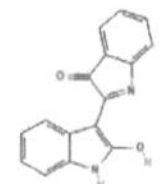	Yes	Yes	Yes	Yes	Yes
7	qingdainone	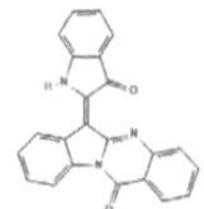	Yes	Yes	Yes	Yes	Yes
8	indigo	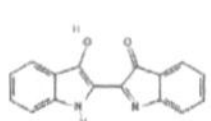	Yes	Yes	Yes	Yes	Yes
9	Quindoline	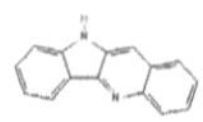	Yes	Yes	Yes	Yes	Yes

Table 8. *Cont.*

Molecule Number	Molecule Name	Chemical Structure	ADME Evaluation				
			Lipsinski	Ghose	Veber	Egan	Muegge
10	Indicaxanthin	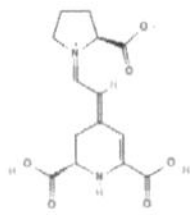	Yes	No	Yes	Yes	Yes
11	Paeoniflorgenin	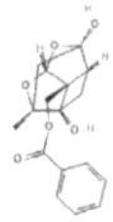	Yes	Yes	Yes	Yes	Yes
12	6-Deglucosyl-3-O-methylpaeoniflorin	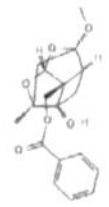	Yes	Yes	Yes	Yes	Yes
13	(2R,3R)-2-(3,5-dihydroxyphenyl)-3,5,7-trihydroxy-2,3-dihydrochromen-4-one	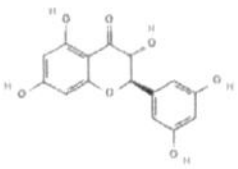	Yes	Yes	Yes	Yes	Yes
14	Columbianetin acetate	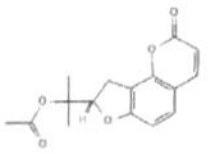	Yes	Yes	Yes	Yes	Yes
15	Cianidanol	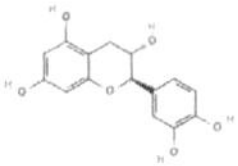	Yes	Yes	Yes	Yes	Yes
16	Triptolide	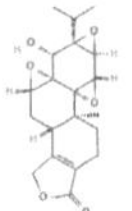	Yes	Yes	Yes	Yes	Yes
17	quercetin	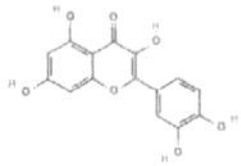	Yes	Yes	Yes	Yes	Yes
18	5-[5-(4-Methoxy-phenyl)-furan-2-ylmethylene]-pyrimidine-2,4,6-trione	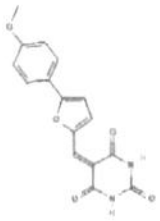	Yes	Yes	Yes	Yes	Yes

Table 8. *Cont.*

Molecule Number	Molecule Name	Chemical Structure	ADME Evaluation				
			Lipsinski	Ghose	Veber	Egan	Muegge
19	kaempferol	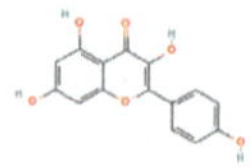	Yes	Yes	Yes	Yes	Yes
20	Sugiol	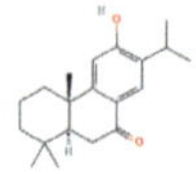	Yes	Yes	Yes	Yes	No
21	Imperatorin	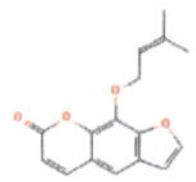	Yes	Yes	Yes	Yes	Yes
22	Palmitoleic acid	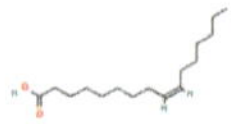	Yes	Yes	No	Yes	No

ADME: absorption, distribution, metabolism, and excretion.

3.12.3. GO and KEGG Pathway Enrichment Analyses

The results of GO and KEGG analyses of the top 19 hub targets are shown in Figure 13A–E. A total of 312 biological processes were identified, including the response to hormones, response to UV, response to estradiol, response to lipopolysaccharide, positive regulation of cell migration, the regulation of inflammatory response, positive regulation of cytokine production, response to glucocorticoid, cellular response to an organic cyclic compound, and the positive regulation of hydrolase activity (Figure 14A,B). A total of 21 molecular functions were identified, including protein phosphatase binding, kinase binding, protein kinase activity, protease binding, protein homodimerization activity, cytokine receptor binding, chromatin binding, enzyme activator activity, heme binding, and endopeptidase activity (Figure 14A,B). Five cellular components were identified, including nuclear membrane, vesicle lumen, membrane raft, transcription regulator complex, perinuclear region of cytoplasm (Figure 14A,B). A total of 104 pathways were identified using KEGG pathway analysis. The results suggested that the mechanisms of four core herbs were mainly linked to pathways in cancer, the AGE-RAGE signaling pathway in diabetic complications, pancreatic cancer, small cell lung cancer, MicroRNAs in cancer, Endocrine resistance, Leishmaniasis, the IL-17 signaling pathway, HIF-1 signaling pathway, MAPK signaling pathway, Chagas disease, Acute myeloid leukemia, Thyroid hormone signaling pathway, and the NF-kappa B signaling pathway (Figure 14C–E).

Table 9. The 19 hub gene targets.

Node Rank	Gene Target	MCC Centrality	Degree Centrality	Betweenness Centrality	Closeness Centrality
1	STAT3	1.01×10^9	24	75.10296	31.58333
2	CASP3	1.01×10^9	19	12.27081	29.08333
3	PTGS2	1.01×10^9	27	284.44838	33.08333
4	BCL2	1.01×10^9	24	102.82056	31.58333
5	MMP9	1.01×10^9	24	144.3609	31.58333
6	EGFR	1.01×10^9	22	69.20533	30.58333
7	ESR1	1.01×10^9	21	305.04296	30.33333
8	CCND1	1.01×10^9	18	12.97538	28.58333
9	STAT1	1.01×10^9	21	42.05261	30.08333
10	NFKB1	1.01×10^9	22	39.82745	30.58333
11	MAPK8	9.59×10^8	16	8.18678	27.58333
12	CASP9	9.58×10^8	15	5.20558	27.08333
13	JAK2	5.23×10^8	19	123.91169	29.08333
14	MAPK14	4.90×10^8	16	5.94696	27.25
15	KIT	4.04×10^7	14	8.86641	26.58333
16	PDGFRB	1.09×10^7	17	30.99454	28.08333
17	SERPINE1	7,297,946	14	34.3894	26.25
18	CCR2	453,610	15	58.75556	26.61667
19	NOS2	413,280	12	3.37798	24.95

3.12.4. Construction of Compounds–Target-Pathway Network of Four Core Herbs against Psoriasis

To intuitively illustrate the compound–target-pathway relationships between the four core herbs and psoriasis, we used a network visualization method, Alubian plots. As shown in Figure 15, the network contained 45 nodes (including 12 compound nodes, 19 target nodes, and 14 pathway nodes) and 652 edges. The compounds in the network, qingdainone, Dihydrochelerythrine, Imperatorin, Hesperetin, kaempferol, Salutaridine, Sugiol, Glycyrol, quercetin, 5-[5-(4-Methoxy-phenyl)-furan-2-ylmethylene]-pyrimidine-2,4,6-trione, indirubin, and Sitogluside, showed degree centrality values of 69, 51, 42, 32, 29, 25, 22, 20, 15, 8, 7, and 6, respectively. Among therapeutic targets, PTGS, MAPK8, and NOS2 were identified as important nodes with high-degree centrality values of 80, 80, and 64, respectively. In the case of pathways, pathways in cancer (hsa05200) and the AGE-RAGE pathway in diabetic complications (hsa04933) showed high values of 39 and 42, respectively.

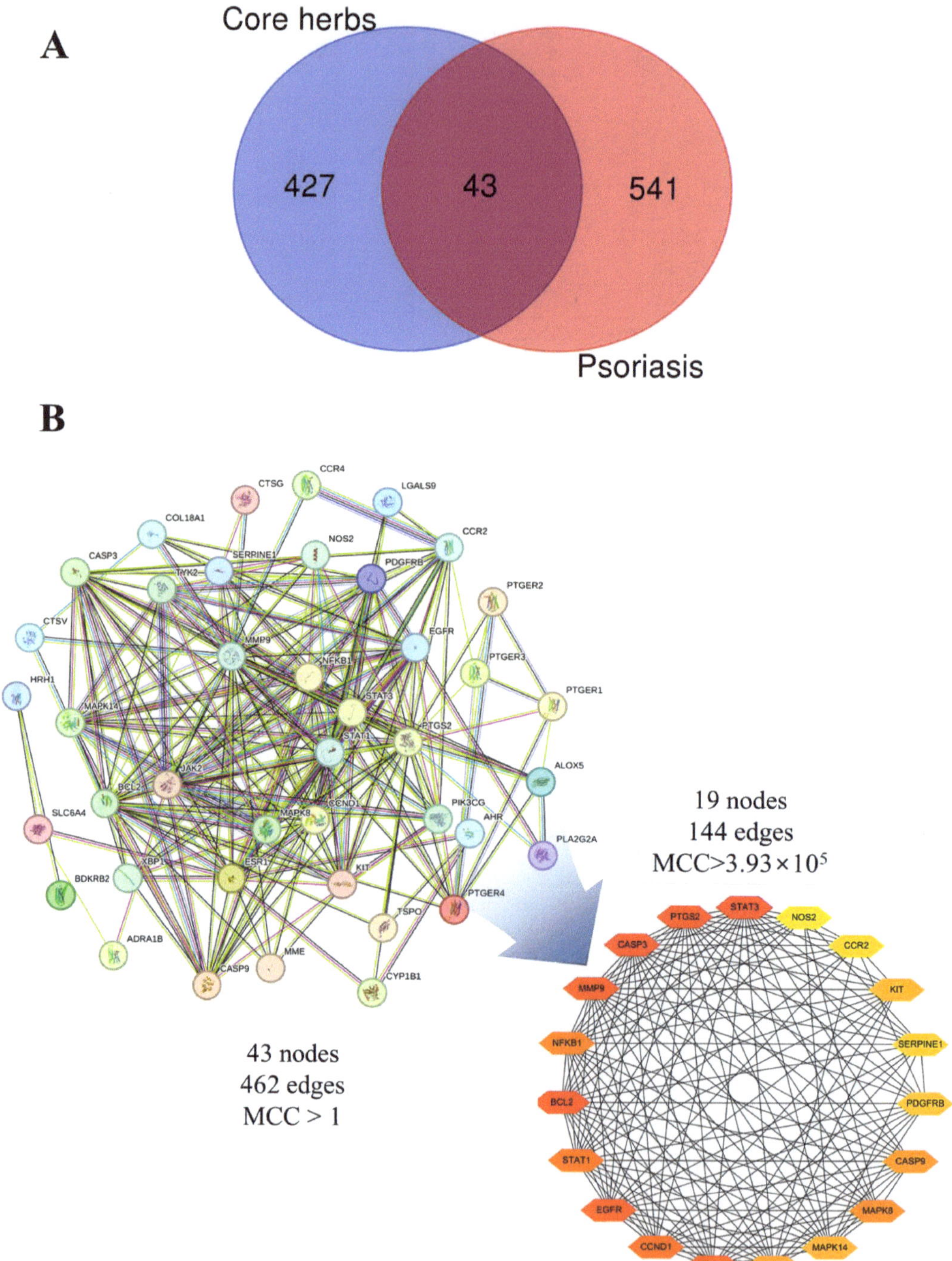

Figure 13. (**A**) Venn diagram of targets of the four core herbs against psoriasis; (**B**) PPI network construction sequence of four core herbs against psoriasis gene targets by MCC algorithm; MCC: Maximum Clique Centrality.

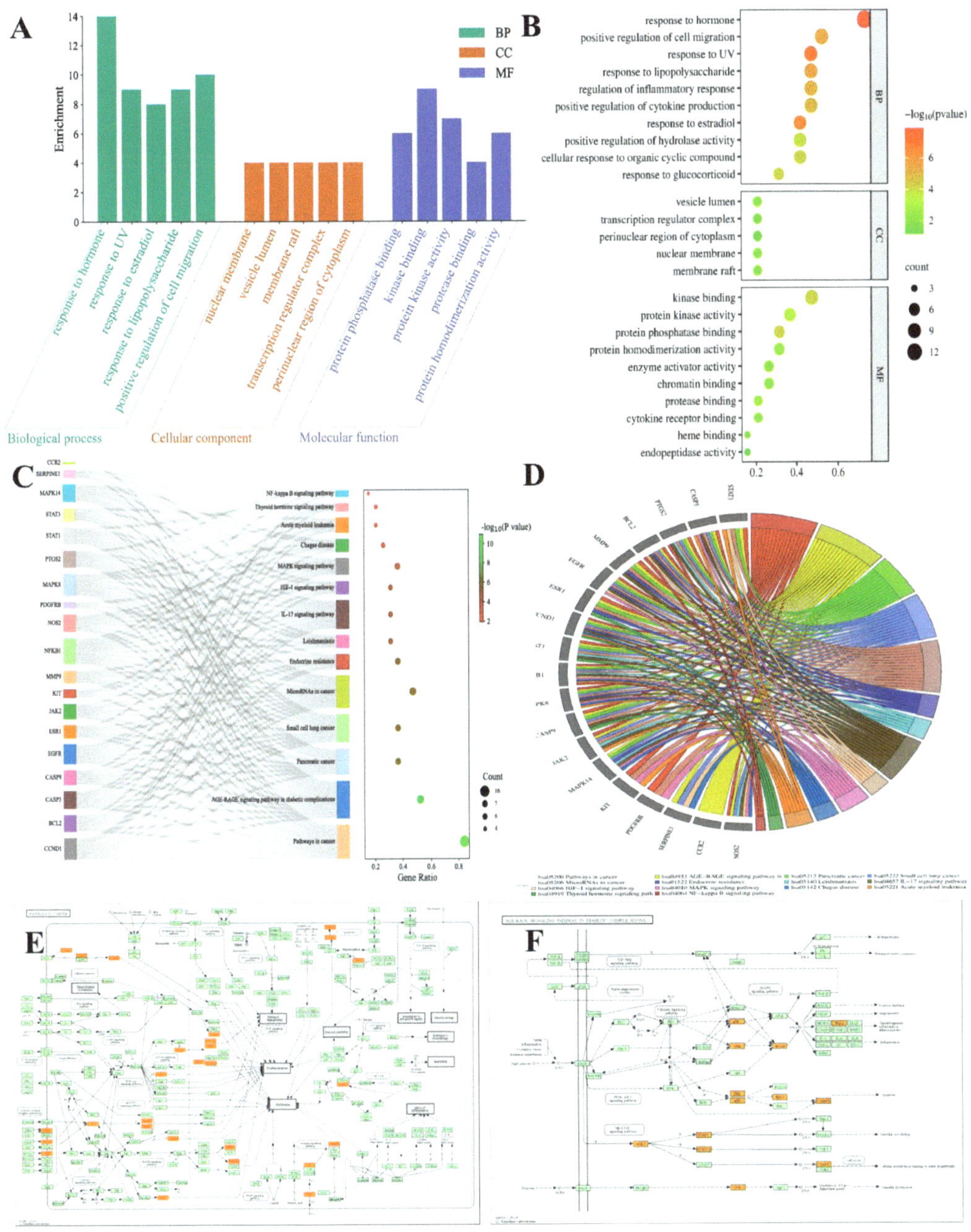

Figure 14. (**A**) Top 5 of GO enrichment analysis for biological process, cellular components, and molecular functions.; (**B**) Bubble plot of GO enrichment; (**C**) Sankey and dot plot of KEGG pathway enrichment analysis illustrating 8 enriched pathways; (**D**) Gene ontology chord diagram of KEGG pathway analysis; (**E**) Pathways in cancer were colored using the KEGG mapper. (**F**) AGE-RAGE signaling pathway in diabetic complications were colored using the KEGG mapper. Orange represents the therapeutic targets in this pathway where the four core herbs act to alleviate psoriasis.

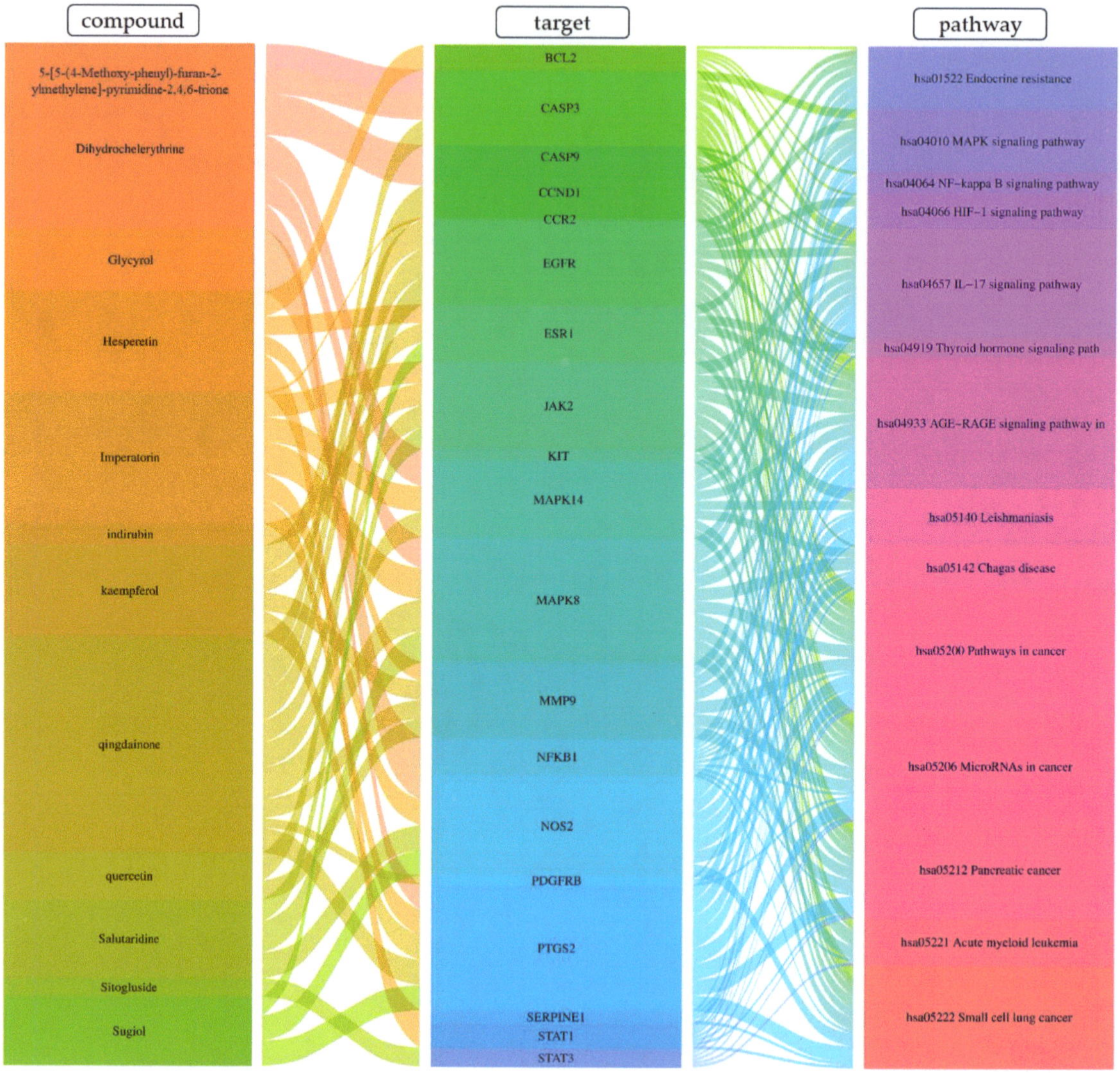

Figure 15. Alluvial plot showing the compound–target-pathway network for the therapeutic mechanism of psoriasis from four core herbs.

4. Discussion

4.1. Summary of the Main Findings and Comparision with Previous Research

According to the clinical trial data analyzed in this study, IM can have a superior effect compared to CM in the improvement of skin damage as measured by PASI as well as inflammation-related biomarkers including TNF-α, IL-8, IL-17, IL-22, IL-23, and IFN-γ. Through integrated data mining to explore the core herbs that contribute most to this effect, *Rehmannia glutinosa* (Gaertn.) DC., *Isatis tinctoria* subsp. *athoa* (Boiss.) Papan., *Paeonia* × *suffruticosa* Andrews, and *Scrophularia ningpoensis* Hemsl. were identified as the four core herbs. To explore their therapeutic mechanisms for psoriasis from a holistic perspective, we further conducted network pharmacology analysis and found that the multiple bioactive compounds contributed to the clinical effects, mainly based on their actions on 19 gene targets and 14 signaling pathways.

The above results can be compared to previous studies we have conducted. Previously, we analyzed data from clinical trials comparing the efficacy of CM with the oral administration of EAHM alone and explored the core herbal materials through frequency analysis and network analysis [215]. In that study, EAHM showed superiority in PASI 70, PASI 60, and continuous PASI scores compared to the CM group. It also showed significant results in immune-mediated inflammatory markers such as IL-17, IL-23, and TNF-α, and in the incidence of adverse events. This supports the clinical benefit of IM identified in this study. On the other hand, the data mining used to extract the core drugs in the previous study is different from the methodology of this study, and no separate pharmacopredictive analysis was performed, making a direct comparison difficult. However, *Rehmannia glutinosa* (Gaertn.) DC. *and Paeonia* × *suffruticosa* Andrews were classified as core drugs in EAHM monotherapy. *Isatidis radix*, a plant of the same origin as *Isatis tinctoria subsp. athoa* (Boiss.) Papan. in this study, but administered at different sites and with almost similar traditional indications, was also a key drug in EAHM monotherapy [216]. Despite the completely different nature of EAHM monotherapy and its combination with CM, the findings of previous studies that similar drugs are important in psoriasis provide some support for the conclusion that the four core herbs in this study exert important pharmacological effects that represent the overall clinical efficacy of IM.

Compared to other relevant previous studies, the favorable effect of IM on psoriasis demonstrated in this study reveals a consistent conclusion [56–60]. However, the existing systematic reviews for simply evaluating the effectiveness of IM or EAHM monotherapy performed an analysis of the overall clinical endpoints related to psoriasis. Additionally, these previous studies did not limit the administration route and application form of the interventional drugs. Considering this, here we evaluated whether EAHM could be a useful candidate for a new IM drug, with a specific target, by setting strict criteria for the administration route and whether to use it as a combined treatment. As for the evaluation index, only the range that specifically supports the indication of "inflammatory skin lesion" was adopted, and not overall psoriasis. The results of this perspective analysis provided a favorable answer to the question of whether the whole EAHM used as an IM, including core herbs, may be a promising candidate material for inflammatory skin lesions caused by psoriasis.

On the other hand, owing to the nature of psoriasis, which has a complex underlying mechanism and a high incidence of comorbid diseases, safety concerns due to polypharmacy are important topics to be addressed [217,218]. Additionally, a detailed evaluation of the relationship between this dosage form and certain adverse effects, particularly drug-induced liver damage, is required owing to the nature of IM, which requires combination therapy with CM. In this study, taking into consideration the aforementioned issues, the incidence rates of AEs were grouped and compared by the lineage in which the symptoms developed. The findings of the analysis demonstrate that IM can be a safer option than CM in the treatment of psoriasis by reducing the incidence of several AEs, such as drug-induced liver damage, cutaneous symptoms, and metabolic disorders. In addition, this justifies the development of a new EAHM drug for patients with poor adherence to CM.

Overall, based on the information from this study, EAHM can be considered worthy to continue its research, as a more effective and safer IM drug candidate for inflammatory skin lesions in psoriasis.

4.2. What Are the Limitations and Future Tasks of This Study?

The long history of EAHM administration on humans in numerous countries is perhaps the most substantial advantage of using them to identify potential new drug candidates. If the source data for drug discovery are confined to the classical literature or the experience of individual clinicians, the reproducibility of this information should be questioned. Encouragingly, the number of clinical trials using EAHM has increased significantly over the past decade. Our study reflects the quantitative growth of IM-related clinical and pharmacological data on psoriasis. However, due to data quality issues and methodological limitations not fully addressed in this study, considerable caution should be taken in interpreting the results.

The recent literature reflects the advances that have been made in research methodologies and suggests several requirements for optimized herbal medicine research [219]. Among them, the reproducibility/consistency of the results of the investigation is presented as an important factor. With regard to this requirement, the present study has limitations that cannot be overlooked. Despite the increase in the number of clinical trials identified in this review, there is still a significant risk of bias due to issues such as lack of pre-protocol enrollment, uncertainty in randomization methods, and selective outcome reporting. In addition, the lack of chemical and quality control information in all studies, which is essential for natural medicine trials, is an important reason to question the reproducibility of inferred IM efficacy. However, this is exactly the problem that this study aims to address. There are currently hundreds of EAHMs with pharmaceutical status, and it is unclear which individual herb or herbal preparations are worthy candidates for further study in psoriasis. Therefore, selecting more promising candidates and focusing resources on them according to the principle of selection and concentration to obtain higher quality and reproducible clinical evidence is a key challenge for future research and the aim of this manuscript. In the same sense, it is appropriate to accept the information in this study only at the level of a useful research hypothesis proposal that is relatively consistent with subsequent studies.

Second, the literature recommends the use of network pharmacology in herbal medicine research, but the results should be validated by in vivo experiments or clinical trials. The present study partially satisfies the above conditions in that it derived herbs that are believed to play a key role in a number of clinical trials and performed network pharmacology analysis on them. Although the four core herbs derived in this study showed a pattern of superiority in the data science characteristics based on clinical utilization information compared to other drugs, their pharmacological activity against psoriasis needs to be further verified by separate follow-up animal experiments to show their superiority over other drugs. Meanwhile, the methodology of network pharmacology itself needs to be improved. For example, during the course of this study, we recognized the risk that compound screening methodologies that rely solely on information provided by specific network pharmacology databases may not be updated or may erroneously lead to assays based on compounds that are not actually druggable. This would render the overall predictive results of Network Pharmacology unreliable. Therefore, future network pharmacology studies should be conducted in a manner that recognizes the potential for such errors in the information provided by existing databases and has measures in place to compensate for them.

Third, the specification of synergistic or antagonistic interactions of multi-component preparations was also presented as a very important condition for herbal medicine research. In particular, in the case of EAHM, which is the subject of this study, maximizing the synergistic effects between drugs in a multi-herb combination is a theoretical principle of clinical practice, and most clinical trials have adopted such a combination as an intervention. However, there are limitations due to the scope of the design of this study that prevented us from exploring these interaction effects in more depth. Furthermore, this study examined combination therapy with CM as an intervention, not EAHM monotherapy. The most appropriate CM that can be used in combination with core herbs could not be identified, and various expected interactions could not be considered. The most important issue to be addressed in relation to combined pharmacotherapy is the safety issue due to drug–drug interactions. Overall, this study's meticulous examination of all types of adverse events concluded that IM would benefit patients with psoriasis in terms of safety. Further studies of IM will require a more detailed analysis of the interactions between specific CMs and the candidate EAHM materials that will make up the actual IM.

Fourth, the four core herbs and their predicted pharmacodynamics in this study do not fully represent the full spectrum of EAHM efficacy for psoriasis. This approach was chosen to address some of the weaknesses of systematic reviews of EAHM efficacy in general, including a lack of information on the mechanisms of action and on key drug identification. This allowed us to filter out drugs that were likely to support the concept of a sovereign herb in EAHM, at least in terms of drug prescription data that reflected clinical implications, and that had prominent data science characteristics compared to many other individual herbs. We also used network pharmacology techniques to confirm that these potential sovereign herbs do indeed have pharmacological activities consistent with the efficacy of EAHM for psoriasis. However, this study cannot answer the question of whether herbs that exhibit prominent patterns and frequencies in data mining are likely to be actual theoretical EAHMs. Furthermore, due to the nature of psoriasis as an immune-mediated inflammatory disease, the targets and pathways involved in the pathology are far beyond the scope of modulation by the four core herbs in this study. Therefore, the methodology attempted in this study should be further developed through continuous follow-up and validation, and the results should not be interpreted as more than a high-quality research hypothesis.

In addition to the BLT4s discussed above, other factors important in the pathogenesis of psoriasis are being studied, including cytoplasmic phospholipase A2, glycerophosphodiester phosphodiesterase domain containing 3, arachidonate 12-lipoxygenase R type, phospholipase B-like 1, sphingomyelin phosphodiesterase 3, ganglioside GM2 activator, and serine palmitoyltransferase long chain subunit have been suggested as important gene targets to consider for future studies.

4.3. Implications of the Four Core Herbs That Emerged from This Study

The notable feature of this study is the discovery of four core herbs as a promising candidate material for further investigation using clinical trial data. It is anticipated that this idea can be employed as an effective research hypothesis in subsequent experiments for the following reasons.

EAHM is different from the experience of using other natural medicinal products in that the principle of use is to incorporate a polyherbal formulation that can maximize the synergistic effect. "Gun-Shin-Jwa-Sa" theory, which in WHO standard nomenclature is known as a sovereign–minister–assistant–courier, is the formulation theory for the ideal combination of EAHM [30]. Among them, the drug responsible for the main pharmacology is called a sovereign or monarch drug. Due to the impact of this sovereign drug, EAHM prescriptions may be taken as a single medication, and the remaining drugs both minimize the overall prescription's toxicity and enhance the activity of sovereign drugs [24,61,75].

The sovereign drug concept can be explained in two ways. First, identifying a sovereign drug in the EAHM formula, in which several materials are combined, is to screen candidates that represent the overall pharmacology of the prescription, so it is possible to effectively narrow down the dimension of the hypothesis for actual drug discovery. In addition, based on the premise that it is a sovereign drug that represents the mechanism of action of EAHM for a specific disease, an in-depth exploration of the mechanism of action of this material can provide more refined information for subsequent research.

However, the criteria used to identify this sovereign herb are ambiguous and the general system depends on the clinician's tacit knowledge. For example, the material with the highest capacity may not be a sovereign herb; conversely, the material used in a very small amount may be a sovereign herb. Consequently, the scientific research methodology to address this problem has not yet reached its full potential.

As a result, we focused on how to recognize such four core herbs as the most essential hypothesis for IM drug discovery in this work. In the authors' previous study, four core herbs was selected only by social network analysis of the frequently prescribed drug group and the centrality index calculated from it [220]. In contrast, this study offered four core herbs of a more specifically compressed range compared to previous research by combining several data-mining approaches and choosing materials that exhibit essential features across all procedures as intersections.

4.4. Possible Pharmacological Mechanisms of Four Core Herbs aginst Psoriasis

4.4.1. Pharmacology of Individual Crude Herbs to Support Efficacy

The greatest strength of EAHM research is that it is currently used as a medicine in many East Asian countries. In other words, because crude EAHM is used as a drug, drug data such as human dose, administration period, and prescription trend by the disease can be obtained in a more direct form, giving it a comparative advantage in terms of safety as a candidate material for new drug research. In addition, the effects of polysaccharides and other phytochemicals with low bioavailability, which are difficult to predict using only compound-based pharmacological studies because of the nature of natural products, cannot be overlooked. Therefore, the development of botanical drugs using crude materials in EAHM research remains one of the most important topics.

The materials derived from the core herbs in this study were *Rehmannia glutinosa* (Gaertn.) DC., *Isatis tinctoria* subsp. *athoa* (Boiss.) Papan., *Paeonia* × *suffruticosa* Andrews, and *Scrophularia ningpoensis* Hemsl., an EAHM that has been widely used for inflammatory and autoimmune diseases for a long time owing to its broad immunomodulatory effects. It can also reduce oxidative stress and inhibit apoptotic responses [221,222]. One study regarding the topical application of *Rehmannia glutinosa* (Gaertn.) DC. extracts to an atopic dermatitis animal model reported that skin inflammation was suppressed through the inhibition of dermal infiltration by inflammatory cells and the suppression of chemokine production by keratinocytes [223]. *Isatis tinctoria* subsp. *athoa* (Boiss.) Papan. has recently been attracting attention for its antiviral activity against various viral infections, including COVID-19, and its related effects, such as a wide range of anti-endotoxin, anti-inflammatory, and immune regulation activities, have also been confirmed [224,225]. Recently, it was reported that the anti-aging effect of this material is based on its antioxidant and anti-inflammatory effects [226].

Paeonia × *suffruticosa* Andrews is the most widely used material for various skin inflammatory reactions and is expressed in redness and dryness [227]. As for its direct effect on psoriasis, it has been reported that the improvement of psoriasis-like skin lesions in a mouse model occurs through the inhibition of the mRNA expression of IL-23 and the activation of dendritic cells via actions on the toll-like receptor 7/8 signaling pathway [228]. Furthermore, it is well recognized that the *Paeonia* × *suffruticosa* Andrews has a variety of benefits for atopic dermatitis-related inflammation, dryness, and UV-induced skin aging [229–231]. *Scrophularia ningpoensis* Hemsl. is a material known for its role in various inflammatory diseases based on its pharmacological effects across anti-inflammatory, immune-enhancing, anti-apoptotic, and anti-allergic effects [232,233]. *Scrophularia ningpoensis* Hemsl. extract effectively inhibited inflammatory cell infiltration and the expression of IL-5, IL-13, IL-17, and immunoglobulin E, as well as reduced NF-κB phosphorylation, thereby exhibiting an anti-inflammatory effect [234].

As the four core herbs have a wide range of anti-inflammatory and immunomodulatory properties, it may be a promising candidate material for successfully reducing a variety of skin inflammatory consequences. At the same time, the assumption of the four core herbs' identification in this study will be a methodology to identify materials that play a key role in the manifestation of clinical effects found throughout IM clinical trial data which are also convincing.

4.4.2. The Key Therapeutic Targets and Pathways Associated with the Four Core Herbs

It is widely known that the mechanism of action of natural products, including East Asian herbal medicine, is mainly based on the action of multi-compound multi-target/signaling pathways [235–237]. In addition, the synergistic effect generated in this action process further enhances the potential of natural product pharmacology [238]. However, it is not easy to interpret and predict such a wide range of action unless a separate research methodology is used for such multilayered and multifaceted pharmacology. Therefore, in this study, a network pharmacology technique was additionally used for an exploratory analysis of the range of mechanisms of action for psoriasis of the derived four core herbs. The active compounds of the four core herbs that treat psoriasis were predicted to act on 19 gene targets, including STAT3, CASP3, PTGS2, BCL2, MMP9, EGFR, ESR1, CCND1, and STAT1, by PPI analysis. Based on this, as a result of KEGG pathway enrichment analysis, eight pathways were identified to be related to the psoriasis treatment effect of four core herbs. Among them, the pathways expected to be most relevant to important therapeutic gene targets are the "pathways in cancer" and "AGE-RAGE signaling pathway in diabetic complications". These include a number of very complex partial signaling pathways. Thus, the detailed relationship between the target and the pathway was additionally explored using the KEGG mapper to more specifically clarify where the psoriasis treatment effect occurs.

The major gene targets of four core herbs, MMP9, BCL2, STAT1, and STAT3, each as a subset of pathways in cancer, were confirmed to be involved in the MAPK pathway, the mTOR pathway, and cytokine–cytokine receptor interaction. These are all pathways involved in proliferation. Pharmacological action on these pathways may be related to the alleviation of psoriasis. For example, previous studies in which metformin was applied to the MAPK and mTOR pathways showed the significant inhibition of cell proliferation and inflammatory responses in human immortalized keratinocytes [239,240]. In view of this, it is not difficult to predict that various natural products that act on the MAPK pathway will show significant psoriasis treatment effects, and consistent progress and multiple previous studies have reported this [241,242]. Several recent studies on the mTOR signaling pathway have identified it as an important regulatory target for the treatment of the epidermal and functional structure in dermatologic diseases, including psoriasis, and relatedly, there are reports that natural products that can modulate this signaling pathway have shown good efficacy in ameliorating animal models of psoriasis with dyslipidemia [243,244]. Meanwhile, STAT1 and STAT3 have been shown to be involved in the cytokine–cytokine receptor interaction pathway, and indeed, these targets are involved in suppressing the pathology of immune-mediated inflammation in psoriasis. For example, blocking the STAT1 function is known to suppress the activity of pathogenic Th1 cells and induce the hyperactivation of Th17 cells, thereby suppressing psoriasis caused by them [245]. On the other hand, in the case of STAT3, previous research reports have been made stating it to be a mediating pathway of alleviating skin inflammatory damage of psoriasis through the anti-inflammatory and antioxidant activities of natural products [246]. CASP3 and CASP9 have been shown to be involved in the regulation of the apoptotic pathway, and given that impaired apoptosis is an important contributor to keratinocyte proliferation abnormalities in psoriasis pathology, it is conceivable that the inhibition of these pathways may be related to similar mechanisms of action as in proliferative pathologies [247]. Based on the above, it is possible to interpret that the activity of the four core herbs on pathways in cancer is mainly exerted through the mechanism of the inhibition of the abnormal proliferation of keratinocytes. This is also the pharmacological mechanism of many natural products that have been shown to alleviate psoriasis, a finding that is relatively consistent with the results of the clinical trial data [248].

The major gene targets of the four herbs in relation to AGE-RAGE pathways include JNK, BLC2, CASP3, NFKB1, JAK2, STAT1, and STAT3, which have been implicated in the regulation of PI3K-AKT, JAK/STAT, and MAPK pathways as partial signaling pathways. Among these, the PI3K-AKT pathway is important as it is involved in the development of general immune-mediated inflammatory diseases. Notably, mTOR kinase, a downstream effect of this pathway, is over-activated during the course of psoriasis pathology, promoting keratinocyte proliferation and suppressing differentiation [249]. Additionally, when the immune-mediated inflammatory state of psoriasis is dysregulated, mTOR signaling is activated in keratinocytes. Hence, the PI3K-AKT-mTOR pathway is an important component of psoriasis pathology in which immune and inflammatory pathologies are linked to keratin hyperplasia. In this context, among the natural products and phytochemicals effective in psoriasis, a number of cases showing a correlation with the PI3K-AKT-mTOR pathway have already been reported in previous studies [250]. The JAK/STAT pathway has also been implicated in the pathophysiology of psoriasis, involving the proliferation of Th17 cells, keratinocytes, and γ- δ T cells [251]. It is similar to the PI3K-AKT pathway above in that it is activated by cytokines such as IL-17 and IL-23 and serves as a link between immune-mediated inflammatory pathology and the overproduction of epidermal keratinocytes. Therefore, the JAK/STAT pathway can be considered as one of the other major targets for the treatment of psoriasis with natural products [252].

In summary, the multiple pathways of action of the four core herbal targets extracted from the herbal prescription data of 126 clinical trials are all closely related to the actual pathogenesis of psoriasis, and their effects appear to be centered on modulating pathologic keratinocyte proliferation and inhibiting immune-mediated inflammation. This is meaningful because it is consistent with the inhibitory effects of PASI and various inflammatory cytokines seen in clinical trial results. Therefore, it may be worthwhile to conduct intensive non-clinical and clinical follow-up studies on the multi-component, multi-pathway mechanisms of the four core herbs in suppressing psoriasis.

4.4.3. Promising Therapeutic Targets in Psoriasis Unexplained by the Pharmacology of Four Core Herbs

As described above, the four core herbs in this study exert modulatory effects on key pathologies of psoriasis through broad multi-target, multi-pathway pharmacological modulation. However, recent studies have reported promising therapeutic targets for psoriasis beyond the scope of the four core herbs.

Notable among these are the leukotriene B4 (LTB4) receptors, BLT1 and BLT2. BLT1 blockade has been shown to treat immune-mediated inflammatory diseases such as rheumatoid arthritis and asthma, and should be strongly considered in the treatment of psoriasis [253]. Recent studies in this regard include a report on the most widely used EAHM with potent anti-inflammatory activity, *Coptis chinensis* Franch. [254]. This study showed that *Coptis chinensis* Franch. can inhibit the pathology of immunosuppressive cytokine secretion by tumor-associated macrophages that cause immune checkpoint blockade (ICB) treatment for cancer to fail. In particular, the blockade of LTB4 signaling and inhibition of ICB were observed as one of the important mechanisms, and this study confirmed that the combination of *Coptis chinensis* Franch. and anti-cancer immunotherapy may lead to better results. Although this herb is not used in the drug data included in this study, it can be considered as a promising candidate for psoriasis treatment considering the above mechanisms of action. On the other hand, BLT2 is a low affinity receptor and agonists against it are considered to be able to treat a number of skin diseases. On a related note, another EAHM, branches of *Morus alba* L. (Mori ramulus), has been reported to exert antioxidant and anti-inflammatory activity by blocking BLT2-dependent NADPH oxidase 1 and inhibiting the production of IL-6 [255]. It is worth noting that this is a mechanism that may also be useful for inhibiting inflammation-mediated skin damage in psoriasis, and indeed, the drug data in this study include a prescription for the leaves of *Morus alba* L. (Mori folium) of the same plant. Actually, a number of Korean references also report its use for skin conditions such as hyperpigmentation [256].

In addition to the BLT4s discussed above, other factors important in the pathogenesis of psoriasis are being studied, including cytoplasmic phospholipase A2, glycerophosphodiester phosphodiesterase domain containing 3, arachidonate 12-lipoxygenase R type, phospholipase B-like 1, sphingomyelin phosphodiesterase 3, ganglioside GM2 activator, and serine palmitoyltransferase long chain subunit, which have been suggested as important gene targets to consider for future studies [257]. Immunomodulatory and anti-inflammatory activities via these targets have been reported in EAHMs, but to date, there appears to be a paucity of data. However, based on the above discussion, two directions for further research can be suggested. First, as we have seen, the more complex the disease, the more important the action on key gene targets and pathways involved in the pathology, so it is necessary to conduct efficacy studies for specific indications of EAHM by targeting modulatory actions on newly identified important pathologies. In this case, the multi-target, multi-pathway effects of EAHM, a characteristic and advantage of EAHM, can be reduced in advance by actively using the network pharmacology methodology. In the present study, there were 26 drugs that showed significant patterns in terms of prescription frequency for psoriasis. Further research could be conducted to determine the extent to which these drugs are involved in the important targets of psoriasis discussed above and which pathological endotypes they modulate, which would be a worthwhile endeavor from a drug discovery perspective. Overall, the value of this study is that it provides insights that can be used to narrow down hypotheses in this large EAHM dataset for further research.

5. Conclusions

The present study showed that the combination therapy of EAHM and CM for the improvement of inflammatory skin lesions in psoriasis patients is a potentially more effective and safer option compared to CM monotherapy. Further integrated data mining and network pharmacology analysis of the drug data in this study revealed that the four core herbs, *Rehmannia glutinosa* (Gaertn.) DC., *Isatis tinctoria* subsp. *athoa* (Boiss.) Papan., *Paeonia × suffruticosa* Andrews, *and Scrophularia ningpoensis* Hemsl., exerted curative effects on psoriasis by inhibiting keratinocyte proliferation and immune-mediated inflammation based on their effects on multiple targets and pathways directly related to psoriasis. However, the quality of the clinical trial data underlying this study is considered suboptimal in terms of both study design and quality control of the intervention. Therefore, the results should be interpreted with caution, bearing in mind that reproducibility is an issue. Further preclinical efficacy studies of the four core herbs with promising pharmacological properties, as well as data from clinical trials with substantially improved designs that meet CONSORT criteria, are needed to validate the information explored in this study. In this respect, this study may be of value in providing summary hypotheses regarding the value and direction of conducting such follow-up studies.

Supplementary Materials: The following supporting information can be downloaded at: https://www.mdpi.com/article/10.3390/ph16081160/s1, Figure S1: (A) Bubble plot of the meta-regression analysis for type of comparator. (B) Bubble plot of the meta-regression analysis for treatment duration. (C) Bubble plot of the meta-regression analysis for sample size. (D) Bubble plot of the meta-regression analysis for overall risk of bias. (E) Bubble plot of the meta-regression analysis for randomization method; Table S1: Search terms used in each databases; Table S2: Detailed information on East Asian herbal medicine utilized as a component of integrative medicine interventions; Table S3: Prestige centrality of each of the East Asian herbs used in the clinical trials of 5 percent or more; Table S4: Potential gene targets for bioactive compounds from four core herbs; Table S5: Detailed information on potential psoriasis targets from the GeneCard database.

Author Contributions: H.-G.J.: Conceptualization, Methodology, Software, Validation, Formal analysis, Investigation, Resources, Data curation, Writing—original draft preparation, Writing—review and editing, Visualization, Project administration. H.K.: Conceptualization, Methodology, Validation, Formal analysis, Investigation, Resources, Writing—review and editing. E.B.: Conceptualization, Methodology, Validation, Writing—original draft preparation, Writing—review and editing. D.L.: Conceptualization, Methodology, Writing—review and editing, Supervision, Project administration, Funding acquisition. J.H.H.: Conceptualization, Resources, Writing—review and editing, Supervision, Project administration. All authors have read and agreed to the published version of the manuscript.

Funding: This work was supported by the National Research Foundation of Korea (NRF, https://www.nrf.re.kr/eng/index) grant funded by the Korean government (MSIT) (NRF-2022R1A2C1013518). This research was supported by the Basic Science Research Program through the National Research Foundation of Korea (NRF), funded by the Ministry of Education (RS-2023-00247663).

Institutional Review Board Statement: Not applicable.

Informed Consent Statement: Not applicable.

Data Availability Statement: Data sharing not applicable.

Conflicts of Interest: The authors declare that the research was conducted in the absence of any commercial or financial relationships that could be construed as a potential conflict of interest.

References

1. Griffiths, C.E.M.; Armstrong, A.W.; Gudjonsson, J.E.; Barker, J.N.W.N. Psoriasis. *Lancet* **2021**, *397*, 1301–1315. [CrossRef]
2. Parisi, R.; Symmons, D.P.M.; Griffiths, C.E.M.; Ashcroft, D.M. Identification and Management of Psoriasis and Associated ComorbidiTy (IMPACT) project team. Global Epidemiology of Psoriasis: A Systematic Review of Incidence and Prevalence. *J. Investig. Dermatol.* **2013**, *133*, 377–385. [CrossRef]
3. Egeberg, A.; Andersen, Y.M.F.; Thyssen, J.P. Prevalence and Characteristics of Psoriasis in Denmark: Findings from the Danish Skin Cohort. *BMJ Open* **2019**, *9*, e028116. [CrossRef]
4. Armstrong, A.W.; Read, C. Pathophysiology, Clinical Presentation, and Treatment of Psoriasis: A Review. *JAMA* **2020**, *323*, 1945–1960. [CrossRef] [PubMed]
5. Calabrese, E.J.; Dhawan, G.; Kapoor, R.; Agathokleous, E.; Calabrese, V. Hormesis: Wound Healing and Keratinocytes. *Pharmacol. Res.* **2022**, *183*, 106393. [CrossRef] [PubMed]
6. Strober, B.; Greenberg, J.D.; Karki, C.; Mason, M.; Guo, N.; Hur, P.; Zhao, Y.; Herrera, V.; Lin, F.; Lebwohl, M. Impact of Psoriasis Severity on Patient-Reported Clinical Symptoms, Health-Related Quality of Life and Work Productivity among US Patients: Real-World Data from the Corrona Psoriasis Registry. *BMJ Open* **2019**, *9*, e027535. [CrossRef]
7. Mattei, P.L.; Corey, K.C.; Kimball, A.B. Psoriasis Area Severity Index (PASI) and the Dermatology Life Quality Index (DLQI): The Correlation between Disease Severity and Psychological Burden in Patients Treated with Biological Therapies. *J. Eur. Acad. Dermatol. Venereol.* **2014**, *28*, 333–337. [CrossRef] [PubMed]
8. Singh, S.; Taylor, C.; Kornmehl, H.; Armstrong, A.W. Psoriasis and Suicidality: A Systematic Review and Meta-Analysis. *J. Am. Acad. Dermatol.* **2017**, *77*, 425–440.e2. [CrossRef]
9. Semenov, Y.R.; Herbosa, C.M.; Rogers, A.T.; Huang, A.; Kwatra, S.G.; Cohen, B.; Anadkat, M.J.; Silverberg, J.I. Psoriasis and Mortality in the United States: Data from the National Health and Nutrition Examination Survey. *J. Am. Acad. Dermatol.* **2021**, *85*, 396–403. [CrossRef]
10. Amin, M.; Lee, E.B.; Tsai, T.-F.; Wu, J.J. Psoriasis and Co-Morbidity. *Acta Derm. Venereol.* **2020**, *100*, adv00033. [CrossRef]
11. Talotta, R.; Atzeni, F.; Sarzi-Puttini, P.; Masala, I.F. Psoriatic Arthritis: From Pathogenesis to Pharmacologic Management. *Pharmacol. Res.* **2019**, *148*, 104394. [CrossRef] [PubMed]
12. Zhang, X.; Zhang, S.; Wu, R.; Li, S.; Su, Y.; Zhang, P. Prevalence of Autoimmune Thyroid Disease in Patients with Psoriasis: A Meta-Analysis. *BMJ Open* **2022**, *12*, e055538. [CrossRef]
13. Korman, N.J. Management of Psoriasis as a Systemic Disease: What Is the Evidence? *Br. J. Dermatol.* **2020**, *182*, 840–848. [CrossRef] [PubMed]
14. Dabholkar, N.; Rapalli, V.K.; Singhvi, G. Potential Herbal Constituents for Psoriasis Treatment as Protective and Effective Therapy. *Phytother. Res.* **2020**, *35*, 2429–2444. [CrossRef] [PubMed]
15. Nast, A.; Smith, C.; Spuls, P.I.; Avila Valle, G.; Bata-Csörgö, Z.; Boonen, H.; De Jong, E.; Garcia-Doval, I.; Gisondi, P.; Kaur-Knudsen, D.; et al. EuroGuiDerm Guideline on the Systemic Treatment of Psoriasis Vulgaris—Part 1: Treatment and Monitoring Recommendations. *J. Eur. Acad. Dermatol. Venereol.* **2020**, *34*, 2461–2498. [CrossRef] [PubMed]
16. Lv, F.; Hu, S.; Lin, C.; Cai, X.; Zhu, X.; Ji, L. Association between Biologic Therapy and Fracture Incidence in Patients with Selected Rheumatic and Autoimmune Diseases: A Systematic Review and Meta-Analysis of Randomized Controlled Trials. *Pharmacol. Res.* **2022**, *181*, 106278. [CrossRef]
17. Ortiz, N.E.G.; Nijhawan, R.I.; Weinberg, J.M. Acitretin. *Dermatol. Ther.* **2013**, *26*, 390–399. [CrossRef]
18. Wang, W.; Zhou, H.; Liu, L. Side Effects of Methotrexate Therapy for Rheumatoid Arthritis: A Systematic Review. *Eur. J. Med. Chem.* **2018**, *158*, 502–516. [CrossRef]
19. An, J.; Zhang, D.; Wu, J.; Li, J.; Teng, X.; Gao, X.; Li, R.; Wang, X.; Xia, L.; Xia, Y. The Acitretin and Methotrexate Combination Therapy for Psoriasis Vulgaris Achieves Higher Effectiveness and Less Liver Fibrosis. *Pharmacol. Res.* **2017**, *121*, 158–168. [CrossRef]
20. Hussain, Z.; Thu, H.E.; Shuid, A.N.; Kesharwani, P.; Khan, S.; Hussain, F. Phytotherapeutic Potential of Natural Herbal Medicines for the Treatment of Mild-to-Severe Atopic Dermatitis: A Review of Human Clinical Studies. *Biomed. Pharmacother.* **2017**, *93*, 596–608. [CrossRef]
21. Tasneem, S.; Liu, B.; Li, B.; Choudhary, M.I.; Wang, W. Molecular Pharmacology of Inflammation: Medicinal Plants as Anti-Inflammatory Agents. *Pharmacol. Res.* **2019**, *139*, 126–140. [CrossRef]
22. Ren, J.-L.; Yang, L.; Qiu, S.; Zhang, A.-H.; Wang, X.-J. Efficacy Evaluation, Active Ingredients, and Multitarget Exploration of Herbal Medicine. *Trends Endocrinol. Metab.* **2023**, *34*, 146–157. [CrossRef] [PubMed]

23. Yin, C.S.; Ko, S.-G. Introduction to the History and Current Status of Evidence-Based Korean Medicine: A Unique Integrated System of Allopathic and Holistic Medicine. *Evid. Based Complement. Altern. Med.* **2014**, *2014*, 740515. [CrossRef]
24. Shim, J.-M.; Kim, J. Cross-National Differences in the Holistic Use of Traditional East Asian Medicine in East Asia. *Health Promot. Int.* **2018**, *33*, 536–544. [CrossRef]
25. Shim, J.-M.; Lee, Y.-S. The Association between the Use of Biomedical Services and the Holistic Use of Traditional East Asian Medicine: A National Survey of Outpatients in South Korea. *BMJ Open* **2017**, *7*, e018414. [CrossRef] [PubMed]
26. Shim, J.-M. The Relationship Between the Use of Complementary and Alternative Medicine and the Use of Biomedical Services: Evidence from East Asian Medical Systems. *Asia Pac. J. Public Health* **2016**, *28*, 51–60. [CrossRef]
27. de Seabra Rodrigues Dias, I.R.; Lo, H.H.; Zhang, K.; Law, B.Y.K.; Nasim, A.A.; Chung, S.K.; Wong, V.K.W.; Liu, L. Potential Therapeutic Compounds from Traditional Chinese Medicine Targeting Endoplasmic Reticulum Stress to Alleviate Rheumatoid Arthritis. *Pharmacol. Res.* **2021**, *170*, 105696. [CrossRef] [PubMed]
28. Li, Z.; Lu, J.; Ou, J.; Yu, J.; Lu, C. Effect of Chinese Herbal Medicine Injections for Treatment of Psoriasis Vulgaris: A Systematic Review and Meta-Analysis. *Front Pharmacol* **2023**, *14*, 1148445. [CrossRef]
29. Jo, H.-G.; Baek, E.; Lee, D. Comparative Efficacy of East Asian Herbal Formulae Containing Astragali Radix–Cinnamomi Ramulus Herb-Pair against Diabetic Peripheral Neuropathy and Mechanism Prediction: A Bayesian Network Meta-Analysis Integrated with Network Pharmacology. *Pharmaceutics* **2023**, *15*, 1361. [CrossRef]
30. Kim, H.U.; Ryu, J.Y.; Lee, J.O.; Lee, S.Y. A Systems Approach to Traditional Oriental Medicine. *Nat. Biotechnol.* **2015**, *33*, 264–268. [CrossRef]
31. Cha, W.-S.; Oh, J.-H.; Park, H.-J.; Ahn, S.-W.; Hong, S.-Y.; Kim, N.-I. Historical Difference between Traditional Korean Medicine and Traditional Chinese Medicine. *Neurol. Res.* **2007**, *29* (Suppl. 1), 5–9. [CrossRef] [PubMed]
32. Weng, S.-W.; Chen, B.-C.; Wang, Y.-C.; Liu, C.-K.; Sun, M.-F.; Chang, C.-M.; Lin, J.-G.; Yen, H.-R. Traditional Chinese Medicine Use among Patients with Psoriasis in Taiwan: A Nationwide Population-Based Study. *Evid. Based Complement. Altern. Med.* **2016**, *2016*, 3164105. [CrossRef] [PubMed]
33. Su, Y.; Qin, W.; Wu, L.; Yang, B.; Wang, Q.; Kuang, H.; Cheng, G. A Review of Chinese Medicine for the Treatment of Psoriasis: Principles, Methods and Analysis. *Chin. Med.* **2021**, *16*, 138. [CrossRef] [PubMed]
34. Tai, C.-J.; El-Shazly, M.; Tsai, Y.-H.; Csupor, D.; Hohmann, J.; Wu, Y.-C.; Tseng, T.-G.; Chang, F.-R.; Wang, H.-C. Uncovering Modern Clinical Applications of Fuzi and Fuzi-Based Formulas: A Nationwide Descriptive Study with Market Basket Analysis. *Front. Pharmacol.* **2021**, *12*, 641530. [CrossRef] [PubMed]
35. Coyle, M.E.; Yu, J.J.; Zhang, A.L.; Jones, L.; Xue, C.C.; Lu, C. Patient Experiences of Using Chinese Herbal Medicine for Psoriasis Vulgaris and Chronic Urticaria: A Qualitative Study. *J. Dermatol. Treat.* **2020**, *31*, 352–358. [CrossRef]
36. Park, M.; Hunter, J.; Kwon, S. Evaluating Integrative Medicine Acute Stroke Inpatient Care in South Korea. *Health Policy* **2018**, *122*, 373–379. [CrossRef]
37. Alschuler, L.; Chiasson, A.M.; Horwitz, R.; Sternberg, E.; Crocker, R.; Weil, A.; Maizes, V. Integrative Medicine Considerations for Convalescence from Mild-to-Moderate COVID-19 Disease. *Explore (NY)* **2022**, *18*, 140–148. [CrossRef]
38. Kim, D.; Shin, J.-S.; Moon, Y.-J.; Ryu, G.; Shin, W.; Lee, J.; Lim, S.; Jeon, H.A.; Seo, J.-Y.; Wang, W.H.; et al. Long-Term Follow-Up of Spinal Stenosis Inpatients Treated with Integrative Korean Medicine Treatment. *J. Clin. Med.* **2020**, *10*, 74. [CrossRef]
39. Yu, H.-H.; Hsieh, C.-J. Integrative Therapy Combining Chinese Herbal Medicines with Conventional Treatment Reduces the Risk of Cardiovascular Disease among Patients with Systemic Lupus Erythematosus: A Retrospective Population-Based Cohort Study. *Front. Pharmacol.* **2021**, *12*, 737105. [CrossRef]
40. Parvizi, M.M.; Salami, M.H.; Moini Jazani, A.; Javaheri, R.; Jaladat, A.M.; Handjani, F. Complementary and Integrative Remedies in the Treatment of Chronic Pruritus: A Review of Clinical Trials. *J. Cosmet. Dermatol.* **2022**, *21*, 5360–5369. [CrossRef]
41. Melby, M.K.; Yoshino, T.; Tonob, D.; Horiba, Y.; Watanabe, K. Differences in Demographics and Complementary and Alternative Medicine Use between Patients Attending Integrative Kampo versus Biomedical Clinics in Japan. *Complement. Ther. Med.* **2019**, *46*, 202–209. [CrossRef]
42. Motoo, Y.; Yukawa, K.; Hisamura, K.; Arai, I. Physician Perspectives on Traditional, Complementary, and Integrative Medicine and the National Evidence-Based Japanese Integrative Medicine Information Website: A Mixed-Method Study. *Integr. Med. Res.* **2021**, *10*, 100454. [CrossRef]
43. Heo, I.; Hwang, M.-S.; Hwang, E.-H.; Cho, J.-H.; Ha, I.-H.; Shin, K.-M.; Lee, J.-H.; Kim, N.-K.; Son, D.-W.; Shin, B.-C. Electroacupuncture as a Complement to Usual Care for Patients with Non-Acute Low Back Pain after Back Surgery: A Pilot Randomised Controlled Trial. *BMJ Open* **2018**, *8*, e018464. [CrossRef]
44. Leung, E.L.-H.; Wu, Q.-B. Concurrent Use of Herbal Products with Prescription Drugs Is a Double-Edged Sword and Evidence-Based Medicine Contributes to Reshaping the Practice. *Pharmacol. Res.* **2019**, *141*, 609–610. [CrossRef] [PubMed]
45. Lee, J.H.; Jo, H.G.; Min, S.Y. East Asian Herbal Medicine Combined with Conventional Therapy for Children with Autism Spectrum Disorder: A Systematic Review and Meta-Analysis. *Explore* **2022**, *18*, 646–656. [CrossRef] [PubMed]
46. Liu, Y.; May, B.H.; Zhang, A.L.; Guo, X.; Lu, C.; Xue, C.C.; Zhang, H. Integrative Herbal Medicine for Chemotherapy-Induced Peripheral Neuropathy and Hand-Foot Syndrome in Colorectal Cancer: A Systematic Review and Meta-Analysis. *Integr. Cancer Ther.* **2019**, *18*, 1534735418817833. [CrossRef]
47. Chen, L.; Michalsen, A. Management of Chronic Pain Using Complementary and Integrative Medicine. *BMJ* **2017**, *357*, j1284. [CrossRef]

48. Ang, L.; Song, E.; Lee, M.S. Randomized Controlled Trials of Traditional, Complementary, and Integrative Medicine-Based Interventions for Coronavirus Disease 2019 (COVID-19): A Bibliometric Analysis and Review of Study Designs. *Integr. Med. Res.* **2021**, *10*, 100777. [CrossRef] [PubMed]
49. Fang, J.; Chen, L.; Ma, R.; Keeler, C.L.; Shen, L.; Bao, Y.; Xu, S. Comprehensive Rehabilitation with Integrative Medicine for Subacute Stroke: A Multicenter Randomized Controlled Trial. *Sci. Rep.* **2016**, *6*, 25850. [CrossRef]
50. Wang, C.; Lin, K.Y.-H.; Wu, M.-Y.; Lin, C.-L.; Lin, J.-G.; Chang, C.Y.-Y.; Lin, W.-C.; Yen, H.-R. Adjunctive Chinese Herbal Medicine Treatment Is Associated with an Improved Survival Rate in Patients with Cervical Cancer in Taiwan: A Matched Cohort Study. *Integr. Cancer Ther.* **2021**, *20*, 15347354211061752. [CrossRef]
51. Xiong, Y.; Gao, M.; van Duijn, B.; Choi, H.; van Horssen, F.; Wang, M. International Policies and Challenges on the Legalization of Traditional Medicine/Herbal Medicines in the Fight against COVID-19. *Pharmacol. Res.* **2021**, *166*, 105472. [CrossRef] [PubMed]
52. Lee, J.H.; Jo, H.G.; Min, S.Y. East Asian Herbal Medicine for the Treatment of Children with Attention Deficit Hyperactivity Disorder: A Systematic Review and Meta-Analysis. *Explore* **2023**, *19*, 330–355. [CrossRef]
53. Jo, H.-G.; Seo, J.; Choi, S.; Lee, D. East Asian Herbal Medicine to Reduce Primary Pain and Adverse Events in Cancer Patients: A Systematic Review and Meta-Analysis with Association Rule Mining to Identify Core Herb Combination. *Front. Pharmacol.* **2021**, *12*, 800571. [CrossRef]
54. Jo, H.-G.; Lee, D. Oral Administration of East Asian Herbal Medicine for Peripheral Neuropathy: A Systematic Review and Meta-Analysis with Association Rule Analysis to Identify Core Herb Combinations. *Pharmaceuticals* **2021**, *14*, 1202. [CrossRef]
55. Zhou, L.; Zhang, L.; Tao, D. Investigation on the Mechanism of Qubi Formula in Treating Psoriasis Based on Network Pharmacology. *Evid. Based Complement. Altern. Med.* **2020**, *2020*, 4683254. [CrossRef] [PubMed]
56. Zhang, J.; Yu, Q.; Peng, L.; Qin, Y.; Jing, M.; Huang, D.; Guo, J.; Xiao, M.; Chen, M. Benefits and Safety of Chinese Herbal Medicine in Treating Psoriasis: An Overview of Systematic Reviews. *Front. Pharmacol.* **2021**, *12*, 680172. [CrossRef] [PubMed]
57. Zhang, C.S.; Yang, L.; Zhang, A.L.; May, B.H.; Yu, J.J.; Guo, X.; Lu, C.; Xue, C.C. Is Oral Chinese Herbal Medicine Beneficial for Psoriasis Vulgaris? A Meta-Analysis of Comparisons with Acitretin. *J. Altern. Complement. Med.* **2016**, *22*, 174–188. [CrossRef]
58. Deng, S.; May, B.H.; Zhang, A.L.; Lu, C.; Xue, C.C.L. Topical Herbal Formulae in the Management of Psoriasis: Systematic Review with Meta-Analysis of Clinical Studies and Investigation of the Pharmacological Actions of the Main Herbs. *Phytother. Res.* **2014**, *28*, 480–497. [CrossRef]
59. Parker, S.; Zhang, C.S.; Yu, J.J.; Lu, C.; Zhang, A.L.; Xue, C.C. Oral Chinese Herbal Medicine versus Placebo for Psoriasis Vulgaris: A Systematic Review. *J. Dermatol. Treat.* **2017**, *28*, 21–31. [CrossRef]
60. Luo, Y.; Chen, J.; Kuai, L.; Zhang, Y.; Ding, X.; Luo, Y.; Ru, Y.; Xing, M.; Li, H.; Sun, X.; et al. Chinese Herbal Medicine for Psoriasis: Evidence from 11 High-Quality Randomized Controlled Trials. *Front. Pharmacol.* **2020**, *11*, 599433. [CrossRef]
61. Zhou, X.; Seto, S.W.; Chang, D.; Kiat, H.; Razmovski-Naumovski, V.; Chan, K.; Bensoussan, A. Synergistic Effects of Chinese Herbal Medicine: A Comprehensive Review of Methodology and Current Research. *Front. Pharmacol.* **2016**, *7*, 201. [CrossRef]
62. Bae, K.-H.; Lee, Y.; Park, K.-H.; Yoon, Y.; Mun, S.; Lee, S. Perception of Cold and Heat Pattern Identification in Diseases: A Survey of Korean Medicine Doctors. *Integr. Med. Res.* **2017**, *6*, 26–32. [CrossRef]
63. Lee, M.S.; Lee, J.A.; Alraek, T.; Bian, Z.-X.; Birch, S.; Goto, H.; Jung, J.; Kao, S.-T.; Moon, S.-K.; Park, B.; et al. Current Research and Future Directions in Pattern Identification: Results of an International Symposium. *Chin. J. Integr. Med.* **2016**, *22*, 947–955. [CrossRef] [PubMed]
64. Wainwright, C.L.; Teixeira, M.M.; Adelson, D.L.; Braga, F.C.; Buenz, E.J.; Campana, P.R.V.; David, B.; Glaser, K.B.; Harata-Lee, Y.; Howes, M.-J.R.; et al. Future Directions for the Discovery of Natural Product-Derived Immunomodulating Drugs: An IUPHAR Positional Review. *Pharmacol. Res.* **2022**, *177*, 106076. [CrossRef]
65. Page, M.J.; McKenzie, J.E.; Bossuyt, P.M.; Boutron, I.; Hoffmann, T.C.; Mulrow, C.D.; Shamseer, L.; Tetzlaff, J.M.; Akl, E.A.; Brennan, S.E.; et al. The PRISMA 2020 Statement: An Updated Guideline for Reporting Systematic Reviews. *BMJ* **2021**, *372*, n71. [CrossRef]
66. Kim, H.; Jo, H.-G.; Hwang, J.-H.; Lee, D. Integrative Medicine (East Asian Herbal Medicine Combined with Conventional Medicine) for Psoriasis: A Protocol for Systematic Review and Meta-Analysis. *Medicine* **2023**, *102*, e32360. [CrossRef] [PubMed]
67. Sterne, J.A.C.; Savović, J.; Page, M.J.; Elbers, R.G.; Blencowe, N.S.; Boutron, I.; Cates, C.J.; Cheng, H.-Y.; Corbett, M.S.; Eldridge, S.M.; et al. RoB 2: A Revised Tool for Assessing Risk of Bias in Randomised Trials. *BMJ* **2019**, *366*, l4898. [CrossRef] [PubMed]
68. Guyatt, G.H.; Oxman, A.D.; Vist, G.E.; Kunz, R.; Falck-Ytter, Y.; Alonso-Coello, P.; Schünemann, H.J.; GRADE Working Group. GRADE: An Emerging Consensus on Rating Quality of Evidence and Strength of Recommendations. *BMJ* **2008**, *336*, 924–926. [CrossRef]
69. Lortie, C.J.; Filazzola, A. A Contrast of Meta and Metafor Packages for Meta-Analyses in R. *Ecol. Evol.* **2020**, *10*, 10916–10921. [CrossRef]
70. Rücker, G.; Schwarzer, G. Beyond the Forest Plot: The Drapery Plot. *Res. Synth. Methods* **2021**, *12*, 13–19. [CrossRef]
71. Nakagawa, S.; Lagisz, M.; O'Dea, R.E.; Rutkowska, J.; Yang, Y.; Noble, D.W.A.; Senior, A.M. The Orchard Plot: Cultivating a Forest Plot for Use in Ecology, Evolution, and Beyond. *Res. Synth. Methods* **2021**, *12*, 4–12. [CrossRef]
72. Peters, J.L.; Sutton, A.J.; Jones, D.R.; Abrams, K.R.; Rushton, L. Contour-Enhanced Meta-Analysis Funnel Plots Help Distinguish Publication Bias from Other Causes of Asymmetry. *J. Clin. Epidemiol.* **2008**, *61*, 991–996. [CrossRef]

73. Egger, M.; Davey Smith, G.; Schneider, M.; Minder, C. Bias in Meta-Analysis Detected by a Simple, Graphical Test. *BMJ* **1997**, *315*, 629–634. [CrossRef] [PubMed]
74. Begg, C.B.; Mazumdar, M. Operating Characteristics of a Rank Correlation Test for Publication Bias. *Biometrics* **1994**, *50*, 1088–1101. [CrossRef] [PubMed]
75. Castañeda, R.; Natarajan, S.; Jeong, S.Y.; Hong, B.N.; Kang, T.H. Traditional Oriental Medicine for Sensorineural Hearing Loss: Can Ethnopharmacology Contribute to Potential Drug Discovery? *J. Ethnopharmacol.* **2019**, *231*, 409–428. [CrossRef]
76. Agrawal, R.; Imieliński, T.; Swami, A. Mining Association Rules between Sets of Items in Large Databases. In Proceedings of the 1993 ACM SIGMOD International Conference on Management of Data, Washington, DC, USA, 25–28 May 1993; pp. 207–216. [CrossRef]
77. Ru, J.; Li, P.; Wang, J.; Zhou, W.; Li, B.; Huang, C.; Li, P.; Guo, Z.; Tao, W.; Yang, Y.; et al. TCMSP: A Database of Systems Pharmacology for Drug Discovery from Herbal Medicines. *J. Cheminformatics* **2014**, *6*, 13. [CrossRef] [PubMed]
78. Xue, R.; Fang, Z.; Zhang, M.; Yi, Z.; Wen, C.; Shi, T. TCMID: Traditional Chinese Medicine Integrative Database for Herb Molecular Mechanism Analysis. *Nucleic Acids Res.* **2013**, *41*, D1089–D1095. [CrossRef]
79. Fang, S.; Dong, L.; Liu, L.; Guo, J.; Zhao, L.; Zhang, J.; Bu, D.; Liu, X.; Huo, P.; Cao, W.; et al. HERB: A High-Throughput Experiment- and Reference-Guided Database of Traditional Chinese Medicine. *Nucleic Acids Res.* **2021**, *49*, D1197–D1206. [CrossRef]
80. Xu, H.-Y.; Zhang, Y.-Q.; Liu, Z.-M.; Chen, T.; Lv, C.-Y.; Tang, S.-H.; Zhang, X.-B.; Zhang, W.; Li, Z.-Y.; Zhou, R.-R.; et al. ETCM: An Encyclopaedia of Traditional Chinese Medicine. *Nucleic Acids Res.* **2019**, *47*, D976–D982. [CrossRef]
81. Daina, A.; Michielin, O.; Zoete, V. SwissADME: A Free Web Tool to Evaluate Pharmacokinetics, Drug-Likeness and Medicinal Chemistry Friendliness of Small Molecules. *Sci. Rep.* **2017**, *7*, 42717. [CrossRef]
82. Daina, A.; Michielin, O.; Zoete, V. SwissTargetPrediction: Updated Data and New Features for Efficient Prediction of Protein Targets of Small Molecules. *Nucleic Acids Res.* **2019**, *47*, W357–W364. [CrossRef] [PubMed]
83. Stelzer, G.; Rosen, N.; Plaschkes, I.; Zimmerman, S.; Twik, M.; Fishilevich, S.; Stein, T.I.; Nudel, R.; Lieder, I.; Mazor, Y.; et al. The GeneCards Suite: From Gene Data Mining to Disease Genome Sequence Analyses. *Curr. Protoc. Bioinform.* **2016**, *54*, 1.30.1–1.30.33. [CrossRef]
84. Szklarczyk, D.; Kirsch, R.; Koutrouli, M.; Nastou, K.; Mehryary, F.; Hachilif, R.; Gable, A.L.; Fang, T.; Doncheva, N.T.; Pyysalo, S.; et al. The STRING Database in 2023: Protein-Protein Association Networks and Functional Enrichment Analyses for Any Sequenced Genome of Interest. *Nucleic Acids Res.* **2023**, *51*, D638–D646. [CrossRef] [PubMed]
85. Shannon, P.; Markiel, A.; Ozier, O.; Baliga, N.S.; Wang, J.T.; Ramage, D.; Amin, N.; Schwikowski, B.; Ideker, T. Cytoscape: A Software Environment for Integrated Models of Biomolecular Interaction Networks. *Genome Res.* **2003**, *13*, 2498–2504. [CrossRef] [PubMed]
86. Chin, C.-H.; Chen, S.-H.; Wu, H.-H.; Ho, C.-W.; Ko, M.-T.; Lin, C.-Y. CytoHubba: Identifying Hub Objects and Sub-Networks from Complex Interactome. *BMC Syst. Biol.* **2014**, *8*, S11. [CrossRef] [PubMed]
87. Zhou, Y.; Zhou, B.; Pache, L.; Chang, M.; Khodabakhshi, A.H.; Tanaseichuk, O.; Benner, C.; Chanda, S.K. Metascape Provides a Biologist-Oriented Resource for the Analysis of Systems-Level Datasets. *Nat. Commun.* **2019**, *10*, 1523. [CrossRef]
88. Kanehisa, M.; Sato, Y.; Kawashima, M. KEGG Mapping Tools for Uncovering Hidden Features in Biological Data. *Protein Sci.* **2022**, *31*, 47–53. [CrossRef]
89. Shi, B. 23 Cases of Blood Heat Type Psoriasis Treated with Liangxuexiaofeng Decoction. *Guangming J. Chin. Med.* **2017**, *32*, 3122–3124. [CrossRef]
90. Song, X. 43 Cases of Psoriasis Vulgaris Treated with Xiaoyin Granules. *China Pharm.* **2013**, *22*, 82.
91. Zheng, X. 60 Cases of Psoriasis Vulgaris Treated with Xiaoyinkeji Decoction. *Henan Tradit. Chin. Med.* **2011**, *31*, 383–384. [CrossRef]
92. Chen, B. 74 Cases of Psoriasis Vulgaris Treated by Xiaoyin Granules Combined with Awei A Capsules. *Mod. Tradit. Chin. Med.* **2007**, *27*, 29–30.
93. He, Y.; Yang, Y.; Yong, M.; Li, C.; Xiong, X. A Clinical Study of Compound Qingdai Pill Combined with Acitretin and Calcipotriol in the Treatment of Patients with Moderate to Severe Psoriasis Vulgaris. *Chin. J. Dermatovenereology* **2016**, *30*, 769–770. [CrossRef]
94. Cheng, J. Analysis of Curative Effect and Safety of Compound Glycyrrhizin and Acitretin Capsules in Treatment of Plaque Psoriasis. *China Foreign Med. Treat.* **2017**, *24*, 126–128. [CrossRef]
95. Lu, X. Analysis of Curative Effect of Total Glucosides of Peony Root in the Treatment of Psoriasis. *Dermatol. Venereol.* **2020**, *42*, 689–691.
96. Chen, X. Analysis of the Effect of Combined Use of Calcipotriol Ointment and Traditional Chinese Medicine in the Treatment of Psoriasis. *Contemp. Med. Symp.* **2016**, *14*, 17–18.
97. Han, J. Analysis of the Effect of Qinzhuliangxue Feng Combined with Awei A Capsules in the Treatment of Psoriasis Vulgaris. *Guide China Med.* **2019**, *17*, 170. [CrossRef]
98. Wang, X.; Wang, K.; Shi, X.; Yu, X. Analysis of the Short-Term Effect of Qingreliangxue Decoction in the Treatment of Progressive Psoriasis Vulgaris. *Dermatol. Venereol.* **2021**, *43*, 212–213.
99. Qiu, S.; Lin, A. Avi A Combined Traditional Chinese Medicine Treatment of Skin Diseases, Psoriasis Clinical Experience. *China Health Ind.* **2014**, *12*, 5–6. [CrossRef]
100. Wang, Y. Clinical Analysis of Awei A Capsule Combined with Piminxiao Capsule in the Treatment of Plaque Psoriasis. *All Health* **2016**, *10*, 137–138.

101. Yang, S.; Gao, X. Clinical Analysis of Xiaoyin Granule Combined with Avia Capsule in the Treatment of 78 Cases of Psoriasis. *Electron. J. Clin. Med. Lit.* **2020**, *7*, 4–5. [CrossRef]
102. Lu, Z. Clinical Application Effect of Qingying Decoction in the Treatment of 62 Cases of Psoriasis. *Nei Mong. J. Tradit. Chin. Med.* **2015**, *30*, 19. [CrossRef]
103. Wu, W. Clinical Effect Analysis of Xiaoyin Granules Combined with Awei A Capsules in the Treatment of Psoriasis Vulgaris. *World Latest Med. Inf.* **2016**, *16*, 132.
104. Yang, X.; Sun, D.; Zeng, W.; He, Q.; Ma, L.; Tao, M. Clinical Effect and Regulation of Peripheral-Blood Mir-21 Expression of Qingre Liangxue Decoction on Psoriasis Patients with Blood Heat Syndrome. *China J. Tradit. Chin. Med. Pharm.* **2016**, *31*, 2299–2301.
105. Zhang, X.; Yuan, X.; Liu, C.; Li, H. Clinical Effect Observation of Taohong Ershao Decoction in Treating Psoriasis Vulgaris of Blood Stasis Type. *Clin. J. Chin. Med.* **2017**, *9*, 79–80.
106. Xie, J.; Xi, J.; Li, X.; Xie, W. Clinical Effect of Formulated Granules of Qingying Decoction Combined with Acitretin Capsules in Treatment of Psoriasis Vulgaris with Blood-Heat Syndrome: An Analysis of 60 Cases. *Hunan J. Tradit. Chin. Med.* **2018**, *34*, 18–20. [CrossRef]
107. Wang, H. Clinical Effect of Xiaoyin Granules Combined with Awei A Capsules in the Treatment of Psoriasis Vulgaris. *Womens Health Res.* **2016**, *1*, 212.
108. Song, Y.; Li, F.; Li, Y.; Liu, X.; Yang, F. Clinical Effect of Yangxuetongluo Decoction in the Treatment of Plaque Psoriasis and Its Influence on the Levels of IL-6,TNF-α and VEGF. *Mod. J. Integr. Tradit. Chin. West. Med.* **2017**, *26*, 2298–2301.
109. Li, D.; Zhao, J.; Su, Q. Clinical Effect of Yinxie Capsule Combined with Acitretin in Treatment of Patients with Psoriasis Vulgaris and Its Effect on Immune Function of T Lymphocytes in Peripheral Blood. *Clin. Misdiagnosis Mistherapy* **2018**, *31*, 98–101. [CrossRef]
110. He, G. Clinical Efficacy and Significance of Liangxuerunfu Decoction in the Treatment of Psoriasis Vulgaris. *Nei Mong. J. Tradit. Chin. Med.* **2016**, *31*, 23. [CrossRef]
111. Zhang, J. Clinical Efficacy of Compound Glycyrrhizin Combined with Avi A Capsules in the Treatment of Psoriasis. *World Latest Med. Inf.* **2018**, *18*, 107. [CrossRef]
112. Zhang, X.; Liu, C.; Yuan, X.; Li, H. Clinical Efficacy of the Taohong Ershao Decoction on Psoriasis of the Xueyu Type and the Influence on Serum Immune Cells. *Clin. J. Chin. Med.* **2017**, *9*, 26–28. [CrossRef] [PubMed]
113. Zhang, X.; Yuan, X.; Liu, C.; Li, H. Clinical Efficacy of the Taohongershao Decoction on Psoriasis of the Xueyu Type and the Influence on Serum Immune Cells. *Clin. J. Chin. Med.* **2017**, *9*, 14–16.
114. Du, A.; Pan, L.; Li, B.; Zhang, H. Clinical Experience of Calcipotriol Betamethasone Ointment Combined with Paeony Total Glycosides Capsules in the Treatment of Psoriasis Vulgaris. *Nei Mong. J. Tradit. Chin. Med.* **2017**, *33*, 79. [CrossRef]
115. Liu, H.; Li, H. Clinical Observation of 47 Psoriasis Vulgaris Treated by Runzaozhiyang Capsule Combined with Compound Flumetasone Ointment. *Chin. J. Dermatovenereology* **2014**, *28*, 654–655. [CrossRef]
116. Liu, S.; Zhang, S. Clinical Observation of Acitretin Capsule Combined with Yin-Xie Capsule in Treating 48 Psoriasis with Blood Deficiency and Wind-Dryness Syndrome. *Chin. J. Dermatovenereology* **2017**, *31*, 474–476. [CrossRef]
117. Lan, X. Clinical Observation of Awei A Capsule Combined with Dangguiyinzi in the Treatment of Psoriasis Vulgaris. *Chin. J. Mod. Drug Appl.* **2021**, *15*, 230–232. [CrossRef]
118. Lu, X. Clinical Observation of Diyin Tablet Combined with Traditional Chinese Medicine in the Treatment of Psoriasis Vulgaris. *Intern. Med. China* **2015**, *10*, 197–198.
119. Hu, R.; Zhou, F.; Zhou, X. Clinical Observation of Liangxuexiaobi Decoction Combined Calcipotriol Betamethasone Ointment in the Treatment of Psoriasis Vulgaris. *World Clin. Drugs* **2019**, *40*, 790–794. [CrossRef]
120. Hao, J.; Xi, J.; Li, L.; Shang, S. Clinical Observation of Liangxuexiaobi Pills Combined with Calcipotriol Ointment in Treating Bloodheat Syndrome of Psoriasis Vulgaris. *Clin. J. Tradit. Chin. Med.* **2020**, *32*, 2335–2338. [CrossRef]
121. Zhang, X.; Liu, D.; Nie, Z. Clinical Observation of Liangxuexiaofeng Decoction Combined with Acitretin in the Treatment of 34 Cases of Psoriasis with Blood Heat Type. *Chin. J. Dermatovenereology Integr. Tradit. West. Med.* **2017**, *16*, 59–60.
122. Zhong, J. Clinical Observation of Piminxiao Capsule Combined with Awei A Capsule in the Treatment of Psoriasis Vulgaris. *World Latest Med. Inf.* **2019**, *19*, 189.
123. Han, K. Clinical Observation of Xiaoyin Granule Combined with Acitretin Capsule for Treating Psoriasis Vulgaris in 44 Cases. *China Pharm.* **2017**, *26*, 83–85.
124. Zhou, L. Clinical Observation of Xiaoyin Granules Combined with Awei A Capsules in Treating 70 Cases of Psoriasis Vulgaris. *J. Aerosp. Med.* **2012**, *23*, 1352–1353.
125. Du, Q. Clinical Observation of Xiaoyin Granules Combined with Awei A Capsules in Treating 70 Cases of Psoriasis Vulgaris. *China Foreign Med. Treat.* **2014**, *21*, 121–122. [CrossRef]
126. Cheng, X.; Zhou, M.; Ban, X. Clinical Observation of Yanghe Decoction Combined with Awei A Capsule in the Treatment of 30 Cases of Pustular Psoriasis. *Nei Mong. J. Tradit. Chin. Med.* **2014**, *30*, 1. [CrossRef]
127. Zhang, J.; Yu, J. Clinical Observation of Yinxie Capsule Combined with Acitretin Capsule in the Treatment of Psoriasis Vulgaris. *Chin. J. Dermatovenereology* **2012**, *26*, 279–280.
128. Zhao, X. Clinical Observation of Yinxie Capsule Combined with Avi A Capsule in the Treatment of Psoriasis Vulgaris. *Med. Forum* **2017**, *21*, 2201–2202. [CrossRef]

129. Huang, Y.; Wu, Y.; Liao, L.; Wang, L. Clinical Observation of Yinxieling Tablet Combinate with Acitretin Capsule in the Treatment of Psoriasis Vulgaris. *Chin. J. Dermatovenereology* **2007**, *21*, 643–644.
130. Feng, R.; Xi, J.; Li, X.; Qi, L.; Xie, W. Clinical Observation of Yinxieping Pill Combined with Compound Flumetasone Ointment in the Treatment of Psoriasis Vulgaris. *Chin. J. Dermatovenereology Integr. Tradit. West. Med.* **2017**, *2017*, 57–59.
131. Jiang, N.; Xi, J.; Li, X.; Qi, L.; Xie, W. Clinical Observation of Yinxieping Pills Combined with Calcipotriol Ointment on Treatment of Psoriasis Vulgaris in Resting Stage. *J. Hunan Univ. Chin. Med.* **2016**, *36*, 55–57.
132. Jin, X. Clinical Observation on 50 Cases of Psoriasis Vulgaris Treated by Awei A Capsule Combined with Matrine Tablets. *Chin. J. Ethnomedicine Ethnopharmacy* **2015**, *22*, 72.
133. Shen, W.; Ma, L.; Fang, Y.; Chen, C.; Ding, H.; Shen, S. Clinical Observation on Liangxuejiedu Decoction Combined with External Application of Western Medicine for Psoriasis Vulgaris with Blood Heat Syndrome. *J. New Chin. Med.* **2020**, *52*, 47–49. [CrossRef]
134. Xun, Y. Clinical Observation on Psoriasis Vulgaris Treated with Psoriasis Granules and Avian Capsules. *Electron. J. Clin. Med. Lit.* **2019**, *6*, 6–7.
135. Yu, X.; Wu, L.; Xiong, W. Clinical Observation on the Treatment of Psoriasis Vulgaris with Acitretin Combined with Runzaozhiyang Capsules. *Chin. J. Dermatovenereology Integr. Tradit. West. Med.* **2010**, *9*, 111–112.
136. Mo, Y. Clinical Observation on the Treatment of Psoriasis Vulgaris with Total Glycosides of Paeoniae Alba Combined with Compound Aminopeptide Tablets. *Asia Pac. Tradit. Med.* **2013**, *9*, 159–160.
137. Liu, Z.; Liu, H. Clinical Observation on Treatment of Quiescent Psoriasis Vulgaris with Jianpi Yishen Decoction. *Xinjiang J. Tradit. Chin. Med.* **2006**, *24*, 46–47.
138. Lu, Y. Clinical Research on Psoriasis Treated with Qingrejiedu Decoction Combined with Etretinate, Tigason. *China J. Chin. Med.* **2011**, *26*, 1001–1002. [CrossRef]
139. Zhou, S.; Huang, H. Clinical Study on Keyin Pills Combined with Acitretin Capsules in Treatment of Psoriasis. *Drugs Clin.* **2018**, *33*, 101–104.
140. Wang, Y.; Qi, Y.; Zhang, Y. Clinical Study on the Treatment of Psoriasis Vulgaris with Researcher Prescribed Traditional Chinese Medicine Decoction Combined with Acitretin. *Asia Pac. Tradit. Med.* **2015**, *11*, 115–116.
141. Yao, H.; Xu, K.; Shan, B. Clinical Study on Xiaoyin Granules in Treating Psoriasis Vulgaris. *Chin. J. Mod. Drug Appl.* **2011**, *5*, 66–67. [CrossRef]
142. Xie, C.; Fu, M. Clinical Study on Xiaoying Granules Combined with Acitretin in Treatment of Psoriasis Vulgaris. *Drugs Clin.* **2016**, *31*, 1834–1837.
143. Liu, J.; Liu, Y.; Li, J.; Zhang, Y. Clinical Treatment of Psoriasis Vulgaris with Acitretin and Runzaozhiyang Capsule. *Chin. J. Dermatovenereology Integr. Tradit. West. Med.* **2012**, *11*, 32–33.
144. Zhang, N. Comparative Analysis of Combined and Single Use of Avi A Capsules in the Treatment of Psoriasis Vulgaris. *Clin. Med.* **2014**, *34*, 43–44.
145. Ge, M.; Lu, H. Curative Effect of Combination of Compound Glycyrrhizin and Acitretin A Capsule in Treatment of Psoriasis Vulgaris. *J. Med. Forum* **2019**, *40*, 29–31.
146. Ding, Y.; Li, T.; Wang, Z.; Liu, Y.; Chen, Q.; Cheng, X.; Lin, S. Effect of NiupixuanII Decoction on TGF-B1 of Lymphocytes in Patients with Psoriasis. *J. Dermatol. Venereol.* **2013**, *12*, 144–146.
147. Liu, Y.; Sun, D.; Yan, X. Effect of Banzhilian Decoction on Psoriasis Vulgaris Patients with Blood-Heat Syndrome. *Guid. J. Tradit. Chin. Med. Pharm.* **2018**, *24*, 81–83. [CrossRef]
148. Jin, J. Effect of Xiaoyin Decoction on Immune Function and Serum Levels of IL-17,IL-22 and TNF-α in Patient with Psoriasis Vulgaris. *Guangming J. Chin. Med.* **2021**, *36*, 1390–1392.
149. Yang, C.; Lin, J.; Wang, W.; Wang, Y.; Han, P.; Wang, B. Effects of Compound Qingdai Capsule Combined with Calcipotriol Betamethasone Cream on Serum IL-17, TNF-α and IL-23 Levels in Patients with Psoriasis. *Chin. J. Dermatovenereology Integr. Tradit. West. Med.* **2019**, *18*, 602–604.
150. Wang, H.; Huang, P.; Yang, Z. Effects of Danggui-Yinzi Granules on the Skin Barrier Functions of Psoriasis Patients with Blood-Deficiency and Wind-Dryness. *J. Hunan Univ. Chin. Med.* **2015**, *35*, 41–43.
151. Ren, F.; Ni, X.; Liu, J. Effects of Xiaoyin Granules on TNF-α and VEGF Levels in Patients with Blood Heat Type Psoriasis. *Mod. Med. Health Res.* **2020**, *4*, 123–124.
152. Lu, H.; Liu, X. Effects of Yangxuequfeng Granules Combined with Moisturizing Cream on Skin Barrier Function and Quality of Life of Patients with Blood Deficiency and Dryness Psoriasis. *Clin. J. Tradit. Chin. Med.* **2019**, *31*, 547–550. [CrossRef]
153. Chen, S.; Dou, J.; Zhang, Z. Efficacy Analysis of Awei A Combined with Xiaoyin Decoction in the Treatment of Psoriasis Vulgaris. *China Pract. Med.* **2013**, *8*, 169–170.
154. Shan, K.; Li, L. Efficacy Analysis of Compound Glycyrrhizin Combined with Awei A Capsules in the Treatment of Psoriasis. *J. North Pharm.* **2016**, *13*, 44–45.
155. Chen, J. Efficacy Analysis of Compound Glycyrrhizin in Combination with Awei A Capsule in the Treatment of Psoriasis. *Sci. Tech. Inf. Gansu* **2020**, *49*, 27–29.
156. Sun, Y.; Hu, X.; Zhen, Z.; Liu, H.; Chao, D. Efficacy Analysis of Qingrexiaoyin Decoction in the Treatment of Psoriasis Vulgaris with Blood Heat and Wind. *Mod. J. Integr. Tradit. Chin. West. Med.* **2015**, *24*, 297–298, 331.
157. Cao, Y.; Li, B.; Wang, Y. Efficacy Analysis of Researcher Prescribed Yangzhen Decoction Combined with Awei A Capsules in the Treatment of Psoriasis Vulgaris. *Chin. J. Misdiagnostics* **2010**, *10*, 1553–1554.

158. Chai, R. Efficacy Analysis of Xiaoyin Decoction Combined with Calcipotriol in the Treatment of Psoriasis. *J. Pract. Tradit. Chin. Med.* **2018**, *34*, 1235–1236.
159. Wang, J.; Zhang, B.; Gao, L.; Yao, Y. Efficacy and Effect of Serum IL-4, IL-10 and INF-γ of Yinxie Capsule Combined with Acitretin Capsule in the Treatment of Psoriasis Vulgaris. *Chin. J. Dermatovenereology Integr. Tradit. West. Med.* **2017**, *16*, 116–118.
160. Luo, M. Efficacy and Safety Evaluation of Compound Glycyrrhizin Combined with Avi A Capsules in the Treatment of Psoriasis. *Med. Forum* **2017**, *21*, 565–566. [CrossRef]
161. Cao, Z. Efficacy and Safety of Psoriasis Compound Glycyrrhizin Combined Treatment with Avi A Capsules. *China Foreign Med. Treat.* **2017**, *24*, 8–10. [CrossRef]
162. Han, X.; Yin, D.; Tian, K.; Guo, X.; Yue, Z. Efficacy Observation of Awei A Capsule Combined with Compound Qingdai Capsule in the Treatment of Progressive Psoriasis. *Chin. J. Dermatovenereology Integr. Tradit. West. Med.* **2017**, *16*, 66–67.
163. Zeng, W.; Wang, L. Efficacy Observation of Compound Clobetasol Propionate Ointment Combined with Xiaoyin Granules in the Treatment of Psoriasis Vulgaris. *Chin. J. Dermatovenereology Integr. Tradit. West. Med.* **2009**, *8*, 90–91.
164. Che, X.; Zhang, Z. Efficacy Observation of Diyin Tablet Combined with Erdong Huoxue Decoction in the Treatment of Psoriasis. *J. Emerg. Tradit. Chin. Med.* **2004**, *13*, 217.
165. Liu, B.; Miao, G.; Li, Z. Efficacy Observation of Fangxi Capsule Combined with Fangfengtongsheng Powder in the Treatment of Psoriasis Vulgaris. *Shandong Med. J.* **2011**, *51*, 106–107.
166. Feng, Y. Efficacy Observation of Qingrequshi Decoction Combined with Methotrexate in the Treatment of Psoriatic Arthritis. *J. Pract. Tradit. Chin. Med.* **2018**, *34*, 567–568.
167. Liu, A. Efficacy Observation of Xiaoyin Decoction Combined with Calcipotriol Ointment in the Treatment of Patchy Psoriasis. *J. Dermatol. Venereol.* **2020**, *42*, 271–272. [CrossRef]
168. Peng, D.; Xi, J.; Qi, L.; Li, X. Efficacy Observation of Yinxiping Pill Combined with Awei A Capsule in the Treatment of Psoriasis Vulgaris. *Chin. J. Dermatovenereology Integr. Tradit. West. Med.* **2015**, *14*, 41–42.
169. Yuan, C.; Li, L.; Zheng, K.; Ke, J.; Ke, X. Efficacy of Calcipotriol Ointment Combined with Qingrexiaoyin Decoction in Treatment of Progressive Psoriasis Vulgaris. *J. Hubei Polytech. Univ.* **2015**, *31*, 51–55. [CrossRef]
170. Pang, J.; Ye, T.; Shu, Z. Efficacy of Compound Qingdai Capsule in the Treatment of Psoriasis Vulgaris and Its Effect on Serum Inflammatory Factors. *World Chin. Med.* **2017**, *12*, 1298–1301. [CrossRef]
171. Tang, J.; Fei, L. Efficacy of Qingreyangxuejiedu Decoction Combined with Calcipotriol Betamethasone for Psoriasis Vulgaris of Blood-Dryness Type and the Influence on Peripheral Blood IL-17,IL-23. *Chin. J. Dermatovenereology Integr. Tradit. West. Med.* **2021**, *20*, 146–149.
172. Liang, Y.; Li, L.; Yang, Y. Efficacy of Qinzhuliangxue Feng Combined with Awei A Capsules in the Treatment of Psoriasis Vulgaris with Blood Heat Syndrome. *Guangdong Med. J.* **2014**, *35*, 768–769. [CrossRef]
173. Wu, X.; Wang, X.; Li, Y. Efficacy of Shentongzhuyu Decoction Combined with Western Medicine in Patients with Blood Stasis of Psoriatic Arthritis. *Intern. Med.* **2017**, *12*, 480–483. [CrossRef]
174. Zhang, N. Efficacy of the Xiaoyin Granules plus Calcipotriene on Psoriasis Vulgaris. *Clin. J. Chin. Med.* **2018**, *10*, 116–117.
175. Le, Q.; Sun, P.; Liu, Y. Efficacy of Tianxian Decoction in the Treatment of Psoriasis Vulgaris and Its Promoting Effect on Skin Healing. *J. Sichuan Tradit. Chin. Med.* **2021**, *39*, 169–171.
176. Xiao, H. Efficacy of Xiaoyin Decoction Combined with Calcipotriol Ointment in the Treatment of Advanced Blood-Heat Type Mild/Moderate (PASI Score < 10) Psoriasis Vulgaris and Its Influence on the Levels of IL-17, IL-22 and TNF-α. *J. Clin. Med. Lit.* **2018**, *5*, 158–159. [CrossRef]
177. Qu, L.; Liu, L.; Lu, X. Efficacy of Yinxie Capsules Combined with Acitretin in the Treatment of Psoriasis Vulgaris and Its Effect on Immune Function of T Lymphocytes. *J. Dermatol. Venereol.* **2020**, *42*, 374–375. [CrossRef]
178. Yang, D.; Zheng, Q.; Wu, J.; Liu, J. Evaluation of Clinical Effect of Qinmei Granules Combined with Compound Xinwei Cream on Psoriasis Vulgaris. *Chin. Tradit. Pat. Med.* **2016**, *38*, 1472–1476.
179. Li, Y.; Zhou, J. Huanglianjiedu Decoction Combined with Methotrexate in the Treatment of 48 Cases of Psoriasis Vulgaris. *Henan Tradit. Chin. Med.* **2014**, *34*, 101–102. [CrossRef]
180. Yao, L.; Wang, X. Liangxuerunzao Decoction Combined with Calcipotriol Ointment in Treatment of Psoriasis Vulgaris with Blood Heat and Wind Dryness Syndrome. *Chin. J. Dermatovenereology Integr. Tradit. West. Med.* **2019**, *18*, 309–312.
181. Liu, G.; Chen, D. Observation of Clinical Efficacy of Compound Glycyrrhizin Combined with Acitretin Capsules in Treatment of Psoriasis. *China Mod. Dr.* **2020**, *58*, 105–107.
182. Zhou, Y.; Yu, R. Observation of Combined Acitretin Capsule and Xiaoyin Granule Therapy for 60 Patients with Psoriasis Vulgaris. *Jilin Med. J.* **2012**, *33*, 6504–6505.
183. Xie, B. Observation of Curative Effect of Acitretin Combined with Runzaozhiyang Capsules in the Treatment of Psoriasis Vulgaris. *Guide China Med.* **2012**, *10*, 499. [CrossRef]
184. Hua, L.; Fan, J. Observation of Curative Effect of Acitretin Combined with Traditional Chinese Medicine Qingyinjiedu Decoction in the Treatment of Psoriasis Vulgaris. *J. Chin. Med. Mater.* **2010**, *33*, 1026–1027. [CrossRef]
185. Wang, S.; Li, L.; Yuan, S.; Wang, S. Observation of Curative Effect of Compound Glycyrrhizin Combined with Avi A Capsules in the Treatment of Children with Psoriasis. *China Pract. Med.* **2015**, *10*, 149–150. [CrossRef]
186. Xu, W.; Wang, H.; Wang, Y. Observation of Curative Effect of Diyin Tablet Combined with Liangxuejiedu Decoction in the Treatment of Psoriasis Vulgaris. *J. Ningxia Med. Coll.* **2005**, *27*, 412–413.

187. Ding, R. Observation of Curative Effect of Traditional Chinese Medicine Ziyinhuoxuerunzao Decoction Combined with Western Medicine in the Treatment of Plaque Psoriasis. *Mod. J. Integr. Tradit. Chin. West. Med.* **2017**, *26*, 2030–2032, 2043.
188. Ma, M. Observation on Curative Effect of Liangxuerunfu Decoction Combined with Western Medicine in Treating Psoriasis Vulgaris. *Shanxi J. Tradit. Chin. Med.* **2015**, *31*, 48.
189. Cai, Y.; Zhang, J.; Yan, Z.; Zhang, L.; Yin, J.; Ren, Y.; Zhang, L.; Yin, Z.; Li, H. Observation on the Clinical Effect of Researcher Prescribed Xiaoyin Feng in the Treatment of Patients with Psoriasis Vulgaris. *Chin. J. Clin. Ration. Drug Use* **2015**, *8*, 105–106.
190. Zeng, M.; Li, Q.; Huang, W.; Chen, J.; Liu, J.; Wang, C.; Dai, H.; Li, J. Observation on the Clinical Efficacy of Compound Glycyrrhizin Combined with Awei A Capsules in the Treatment of Psoriasis. *Qingdao Med. J.* **2017**, *49*, 120–122.
191. He, D. Observation on the Curative Effect of Xiaoranqudan Feng and Awei A Capsules in the Treatment of Psoriasis Vulgaris. *Youjiang Med. J.* **2010**, *38*, 434–435.
192. Ma, L. Observation on the Effect of Xiaoyin Granule in the Adjuvant Treatment of 37 Cases of Psoriasis Vulgaris. *J. Dermatol. Venereol.* **2018**, *40*, 840–842.
193. Wu, S.; Wen, C. Observation on the Efficacy and Adverse Reactions of Compound Glycyrrhizin Combined with Acitretin Capsule in the Treatment of Psoriasis Vulgaris. *Chin. Community Dr.* **2020**, *36*, 89–90.
194. Jin, X. Observation on the Efficacy of Researcher Prescribed Liangxuequfeng Powder Combined with Acitretin in the Treatment of Psoriasis Vulgaris. *Gansu Sci. Technol.* **2015**, *31*, 103–104.
195. Xu, Y. Preliminary Evaluation on the Combined Treatment of Psoriasis with Compound Glycyrrhizin and Avi A Capsules. *China Rural Health* **2016**, *10*, 77–78.
196. Zhang, B. Qingfeiliangxue Decoction and Acitretin in Treating 36 Cases of Psoriasis Vulgaris. *West. J. Tradit. Chin. Med.* **2014**, *27*, 106–107.
197. Tian, Y. Qingfeiliangxue Decoction Clinical Study of Psoriasis Vulgaris. *J. Liaoning Univ. Tradit. Chin. Med.* **2011**, *13*, 187–189. [CrossRef]
198. Chen, J. Study on the Effect of Calcipotriol Ointment, Avitamin A Capsule Combined with Compound Glycyrrhizin Tablets in the Treatment of Psoriasis. *Contemp. Med. Symp.* **2019**, *17*, 165–166.
199. Zhao, C.; Ma, L.; Xia, D. The Clinical Effect Observation of Avermectin A Combined with Traditional Chinese Medicine in the Treatment of Patients with Psoriasis Vulgaris. *Chin. J. Clin. Ration. Drug Use* **2015**, *8*, 17–18.
200. Zhang, X.; Liu, C.; Yuan, X.; Li, H. The Curative Effect of Taohong Ershao Decoction in the Treatment of Blood Stasis Psoriasis Vulgaris and Its Influence on Serum Levels of TH1/TH2. *Chin. Med. Mod. Distance Educ. China* **2017**, *15*, 56–57.
201. Yu, Y.; He, Q.; Liu, Y.; Sun, D.; Cao, Y. The Effect of Peripheral Blood Parathyroid Hormone (PTH) Level of Qingreliangxue Decoction on Psoriasis Patients with Blood-Heat Syndrome. *Chin. J. Dermatovenereology* **2016**, *30*, 841–843.
202. Jiang, Q.; Liu, W.; Chen, H. The Efficacy and Serum TNF-α and IL-8 Content in the Treatment of Blood Heat Psoriasis Vulgaris by Sendi Particles Combined Acitretin Capsules. *Asia Pac. Tradit. Med.* **2012**, *8*, 57–58.
203. Li, J.; Li, H.; Zhang, J. The Efficacy of Qingrejiedu Decoction in the Treatment of Psoriasis Vulgaris with the Pattern of Blood-Heat and the Influence on Level of TNF-α in the Serum. *Chin. J. Dermatovenereology* **2017**, *31*, 554–556. [CrossRef]
204. Ji, T.; Chen, J.; Meng, Y. Therapeutic Effect and Mechanism of Compound Qingdai Capsule Combined with MTX in the Treatment of Psoriasis. *Lab. Med.* **2020**, *35*, 120–124.
205. Zheng, T.; Xiao, J.; Zhou, W.; Feng, D.; Zhu, P.; Niu, F.; Guo, F.; Liu, J.; Wang, S. Therapeutic Effect of Jueyin Granule on Psoriasis Vulgaris with Blood-Heat Syndrome. *Gansu Med. J.* **2020**, *39*, 27–28, 31.
206. Zhang, T.; Jia, Z. To Observe the Effect of Nourishing Yin and Zinyinqingrexiaofengsan Efficacy and Safety in the Treatment of Psoriasis Vulgaris. *Clin. J. Chin. Med.* **2015**, *7*, 89–90.
207. Luo, C.; Ma, L. Topical Calcipotriol Ointment Therapy Combination with Oral Piminxiao Capsule Treatment in Plaque Psoriasis Vulgaris. *Mod. Hosp.* **2010**, *10*, 44–46.
208. Xu, H. Treatment of 39 Cases of Psoriasis Vulgaris with Qingxuanyin Combined with Awei A Capsules. *China Pharm.* **2011**, *20*, 72.
209. Zhang, G. Treatment of 40 Cases of Psoriasis Due to Blood Deficiency and Wind-Dryness with Ziyinyangxuequfeng Decoction. *Henan Tradit. Chin. Med.* **2014**, *34*, 1558–1559. [CrossRef]
210. Zhang, X. Treatment of 76 Cases of Psoriasis Vulgaris with Xiaoyin Capsules. *J. Shaanxi Coll. Tradit. Chin. Med.* **2013**, *36*, 69–70. [CrossRef]
211. Chen, B.; Tan, X. Treatment of Psoriasis Vulgaris with Diyin Tablets Combined with Anti-Psoriasis Formula. *Guangdong Med. J.* **2004**, *25*, 727–728. [CrossRef]
212. Luo, W.; Wang, T.; Wang, C. Treatment of Psoriasis Vulgaris with Zicaohuoxue Decoction Combined with Acitretin. *Acta Chin. Med.* **2018**, *33*, 2462–2465. [CrossRef]
213. Yang, L.; Wu, X.; Ma, J. Xiaoying Granules Combined with A Citretin Capsule in the Treatment of Psoriasis Clinical Analysis. *China Med. Cosmetol.* **2017**, *7*, 61–64.
214. Zhang, J.; Chen, X.; Li, F.; Li, X. Yinxie Capsule Combined with Pyrithione Zinc Aerosol in Treating 65 Cases of Psoriasis Vulgaris. *Chin. J. Exp. Tradit. Med. Formulae* **2015**, *21*, 169–172. [CrossRef]
215. Jo, H.-G.; Kim, H.; Lee, D. Oral Administration of East Asian Herbal Medicine for Inflammatory Skin Lesions in Plaque Psoriasis: A Systematic Review, Meta-Analysis, and Exploration of Core Herbal Materials. *Nutrients* **2022**, *14*, 2434. [CrossRef] [PubMed]
216. Chen, J.; Zhu, Z.; Gao, T.; Chen, Y.; Yang, Q.; Fu, C.; Zhu, Y.; Wang, F.; Liao, W. Isatidis Radix and Isatidis Folium: A Systematic Review on Ethnopharmacology, Phytochemistry and Pharmacology. *J. Ethnopharmacol.* **2022**, *283*, 114648. [CrossRef] [PubMed]

217. Di Caprio, R.; Caiazzo, G.; Cacciapuoti, S.; Fabbrocini, G.; Scala, E.; Balato, A. Safety Concerns with Current Treatments for Psoriasis in the Elderly. *Expert Opin. Drug Saf.* **2020**, *19*, 523–531. [CrossRef]
218. George, C.J.; Verghese, J.; Izzetoglu, M.; Wang, C.; Holtzer, R. The Effect of Polypharmacy on Prefrontal Cortex Activation during Single and Dual Task Walking in Community Dwelling Older Adults. *Pharmacol. Res.* **2019**, *139*, 113–119. [CrossRef]
219. Panossian, A. Challenges in Phytotherapy Research. *Front. Pharmacol.* **2023**, *14*, 1199516. [CrossRef]
220. Jo, H.-G.; Seo, J.; Lee, D. Clinical Evidence Construction of East Asian Herbal Medicine for Inflammatory Pain in Rheumatoid Arthritis Based on Integrative Data Mining Approach. *Pharmacol. Res.* **2022**, *185*, 106460. [CrossRef]
221. Li, M.; Jiang, H.; Hao, Y.; Du, K.; Du, H.; Ma, C.; Tu, H.; He, Y. A Systematic Review on Botany, Processing, Application, Phytochemistry and Pharmacological Action of Radix Rehmnniae. *J. Ethnopharmacol.* **2022**, *285*, 114820. [CrossRef]
222. Sun, M.; Shen, X.; Ma, Y. Rehmannioside A Attenuates Cognitive Deficits in Rats with Vascular Dementia (VD) through Suppressing Oxidative Stress, Inflammation and Apoptosis. *Biomed. Pharmacother.* **2019**, *120*, 109492. [CrossRef]
223. Sung, Y.-Y.; Yoon, T.; Jang, J.Y.; Park, S.-J.; Kim, H.K. Topical Application of Rehmannia Glutinosa Extract Inhibits Mite Allergen-Induced Atopic Dermatitis in NC/Nga Mice. *J. Ethnopharmacol.* **2011**, *134*, 37–44. [CrossRef] [PubMed]
224. Kuchta, K.; Cameron, S.; Lee, M.; Cai, S.-Q.; Shoyama, Y. Which East Asian Herbal Medicines Can Decrease Viral Infections? *Phytochem. Rev.* **2022**, *21*, 219–237. [CrossRef] [PubMed]
225. Zhou, W.; Zhang, X.-Y. Research Progress of Chinese Herbal Medicine *Radix isatidis* (Banlangen). *Am. J. Chin. Med.* **2013**, *41*, 743–764. [CrossRef]
226. Kim, J.H.; Gao, D.; Jeong, W.S.; Kim, C.T.; Cho, C.W.; Kim, H.M.; Kang, J.S. Anti-Wrinkle Effect of Isatis Indigotica Leaf Extract: Evaluation of Antioxidant, Anti-Inflammation, and Clinical Activity. *Antioxidants* **2021**, *10*, 1339. [CrossRef] [PubMed]
227. Wang, Z.; He, C.; Peng, Y.; Chen, F.; Xiao, P. Origins, Phytochemistry, Pharmacology, Analytical Methods and Safety of Cortex Moutan (*Paeonia suffruticosa* Andrew): A Systematic Review. *Molecules* **2017**, *22*, 946. [CrossRef]
228. Meng, Y.; Wang, M.; Xie, X.; Di, T.; Zhao, J.; Lin, Y.; Xu, X.; Li, N.; Zhai, Y.; Wang, Y.; et al. Paeonol Ameliorates Imiquimod-Induced Psoriasis-like Skin Lesions in BALB/c Mice by Inhibiting the Maturation and Activation of Dendritic Cells. *Int. J. Mol. Med.* **2017**, *39*, 1101–1110. [CrossRef]
229. Sun, Z.; Du, J.; Hwang, E.; Yi, T.-H. Paeonol Extracted from *Paeonia suffruticosa* Andr. Ameliorated UVB-Induced Skin Photoaging via DLD/Nrf2/ARE and MAPK/AP-1 Pathway. *Phytother. Res.* **2018**, *32*, 1741–1749. [CrossRef]
230. Wu, X.X.; Siu, W.S.; Wat, C.L.; Chan, C.L.; Koon, C.M.; Li, X.; Cheng, W.; Ma, H.; Tsang, M.S.M.; Lam, C.W.-K.; et al. Effects of Topical Application of a Tri-Herb Formula on Inflammatory Dry-Skin Condition in Mice with Oxazolone-Induced Atopic Dermatitis. *Phytomedicine* **2021**, *91*, 153691. [CrossRef]
231. Wang, W.; Li, Q.; Zhao, Z.; Liu, Y.; Wang, Y.; Xiong, H.; Mei, Z. Paeonol Ameliorates Chronic Itch and Spinal Astrocytic Activation via CXCR3 in an Experimental Dry Skin Model in Mice. *Front. Pharmacol.* **2021**, *12*, 805222. [CrossRef]
232. Lee, H.-J.; Kim, H.-L.; Lee, D.-R.; Choi, B.-K.; Yang, S.-H. Scrophulariae Radix: An Overview of Its Biological Activities and Nutraceutical and Pharmaceutical Applications. *Molecules* **2021**, *26*, 5250. [CrossRef] [PubMed]
233. Ren, D.; Shen, Z.-Y.; Qin, L.-P.; Zhu, B. Pharmacology, Phytochemistry, and Traditional Uses of Scrophularia Ningpoensis Hemsl. *J. Ethnopharmacol.* **2021**, *269*, 113688. [CrossRef]
234. Shin, N.-R.; Lee, A.Y.; Song, J.-H.; Yang, S.; Park, I.; Lim, J.-O.; Jung, T.-Y.; Ko, J.-W.; Kim, J.-C.; Lim, K.S.; et al. Scrophularia Buergeriana Attenuates Allergic Inflammation by Reducing NF-KB Activation. *Phytomedicine* **2020**, *67*, 153159. [CrossRef] [PubMed]
235. Panossian, A.; Abdelfatah, S.; Efferth, T. Network Pharmacology of Ginseng (Part II): The Differential Effects of Red Ginseng and Ginsenoside Rg5 in Cancer and Heart Diseases as Determined by Transcriptomics. *Pharmaceuticals* **2021**, *14*, 1010. [CrossRef]
236. Panossian, A.; Abdelfatah, S.; Efferth, T. Network Pharmacology of Ginseng (Part III): Antitumor Potential of a Fixed Combination of Red Ginseng and Red Sage as Determined by Transcriptomics. *Pharmaceuticals* **2022**, *15*, 1345. [CrossRef]
237. Panossian, A.; Abdelfatah, S.; Efferth, T. Network Pharmacology of Red Ginseng (Part I): Effects of Ginsenoside Rg5 at Physiological and Sub-Physiological Concentrations. *Pharmaceuticals* **2021**, *14*, 999. [CrossRef]
238. Panossian, A.; Seo, E.-J.; Wikman, G.; Efferth, T. Synergy Assessment of Fixed Combinations of Herba Andrographidis and Radix Eleutherococci Extracts by Transcriptome-Wide Microarray Profiling. *Phytomedicine* **2015**, *22*, 981–992. [CrossRef]
239. Liu, Y.; Yang, F.; Ma, W.; Sun, Q. Metformin Inhibits Proliferation and Proinflammatory Cytokines of Human Keratinocytes in Vitro via MTOR-Signaling Pathway. *Pharm. Biol.* **2016**, *54*, 1173–1178. [CrossRef] [PubMed]
240. Li, W.; Ma, W.; Zhong, H.; Liu, W.; Sun, Q. Metformin Inhibits Proliferation of Human Keratinocytes through a Mechanism Associated with Activation of the MAPK Signaling Pathway. *Exp. Ther. Med.* **2014**, *7*, 389–392. [CrossRef] [PubMed]
241. Zhu, X.; Xu, Q.; Liu, Z.; Cao, Z.; Li, M.; Wei, J.; Du, L.; Han, C.; Zhang, C. Qingre Lishi Decoction Ameliorates Imiquimod-Induced Psoriasis-like Skin Lesions in SKH-1 Mice by Regulating the Treg-DC-Th17 Axis and Inhibiting MAPK-Mediated DC Maturation. *J. Ethnopharmacol.* **2023**, *318*, 116931. [CrossRef]
242. Xu, Q.; Sheng, L.; Zhu, X.; Liu, Z.; Wei, G.; Zhang, T.; Du, H.; Yang, A.; Yao, J.; Zhang, G.; et al. Jingfang Granules Exert Anti-Psoriasis Effect by Targeting MAPK-Mediated Dendritic Cell Maturation and PPARγ-Mediated Keratinocytes Cell Cycle Progression in Vitro and in Vivo. *Phytomedicine* **2023**, *117*, 154925. [CrossRef] [PubMed]
243. Wang, J.; Cui, B.; Chen, Z.; Ding, X. The Regulation of Skin Homeostasis, Repair and the Pathogenesis of Skin Diseases by Spatiotemporal Activation of Epidermal MTOR Signaling. *Front. Cell Dev. Biol.* **2022**, *10*, 950973. [CrossRef] [PubMed]

244. Xie, X.; Zhang, L.; Li, X.; Liu, W.; Wang, P.; Lin, Y.; Han, X.; Li, P. Liangxue Jiedu Formula Improves Psoriasis and Dyslipidemia Comorbidity via PI3K/Akt/MTOR Pathway. *Front. Pharmacol.* **2021**, *12*, 591608. [CrossRef] [PubMed]
245. Park, J.; Son, M.-J.; Ho, C.-C.; Lee, S.-H.; Kim, Y.; An, J.; Lee, S.-K. Transcriptional Inhibition of STAT1 Functions in the Nucleus Alleviates Th1 and Th17 Cell-Mediated Inflammatory Diseases. *Front. Immunol.* **2022**, *13*, 1054472. [CrossRef] [PubMed]
246. Lin, P.; Shi, H.-Y.; Lu, Y.-Y.; Lin, J. Centella Asiatica Alleviates Psoriasis through JAK/STAT3-Mediated Inflammation: An in Vitro and in Vivo Study. *J. Ethnopharmacol.* **2023**, *317*, 116746. [CrossRef] [PubMed]
247. Elango, T.; Thirupathi, A.; Subramanian, S.; Ethiraj, P.; Dayalan, H.; Gnanaraj, P. Methotrexate Treatment Provokes Apoptosis of Proliferating Keratinocyte in Psoriasis Patients. *Clin. Exp. Med.* **2017**, *17*, 371–381. [CrossRef] [PubMed]
248. Huang, T.-H.; Lin, C.-F.; Alalaiwe, A.; Yang, S.-C.; Fang, J.-Y. Apoptotic or Antiproliferative Activity of Natural Products against Keratinocytes for the Treatment of Psoriasis. *Int. J. Mol. Sci.* **2019**, *20*, 2558. [CrossRef]
249. Yu, J.; Zhao, Q.; Wang, X.; Zhou, H.; Hu, J.; Gu, L.; Hu, Y.; Zeng, F.; Zhao, F.; Yue, C.; et al. Pathogenesis, Multi-Omics Research, and Clinical Treatment of Psoriasis. *J. Autoimmun.* **2022**, *133*, 102916. [CrossRef]
250. Roy, T.; Boateng, S.T.; Uddin, M.B.; Banang-Mbeumi, S.; Yadav, R.K.; Bock, C.R.; Folahan, J.T.; Siwe-Noundou, X.; Walker, A.L.; King, J.A.; et al. The PI3K-Akt-MTOR and Associated Signaling Pathways as Molecular Drivers of Immune-Mediated Inflammatory Skin Diseases: Update on Therapeutic Strategy Using Natural and Synthetic Compounds. *Cells* **2023**, *12*, 1671. [CrossRef]
251. Gómez-García, F.; Gómez-Arias, P.J.; Montilla-López, A.; Hernández-Parada, J.; Sanz-Cabanillas, J.L.; Ruano, J.; Parra-Peralbo, E. A Scoping Review on Use of Drugs Targeting the JAK/STAT Pathway in Psoriasis. *Front. Med.* **2022**, *9*, 754116. [CrossRef]
252. Kadagothy, H.; Nene, S.; Amulya, E.; Vambhurkar, G.; Rajalakshmi, A.N.; Khatri, D.K.; Singh, S.B.; Srivastava, S. Perspective Insights of Small Molecules, Phytoconstituents and Biologics in the Management of Psoriasis: A Focus on Targeting Major Inflammatory Cytokine Pathways. *Eur. J. Pharmacol.* **2023**, *947*, 175668. [CrossRef] [PubMed]
253. Nakamura, M.; Shimizu, T. Therapeutic Target of Leukotriene B4 Receptors, BLT1 and BLT2: Insights from Basic Research. *Biochimie* 2023, *in press*. [CrossRef] [PubMed]
254. Yan, J.; Zhu, J.; Li, X.; Yang, R.; Xiao, W.; Huang, C.; Zheng, C. Blocking LTB4 Signaling-Mediated TAMs Recruitment by Rhizoma Coptidis Sensitizes Lung Cancer to Immunotherapy. *Phytomedicine* **2023**, *119*, 154968. [CrossRef]
255. Park, G.-S.; Kim, J.-K.; Kim, J.-H. Anti-Inflammatory Action of Ethanolic Extract of Ramulus Mori on the BLT2-Linked Cascade. *BMB Rep.* **2016**, *49*, 232–237. [CrossRef] [PubMed]
256. Joh, B.; Jeon, E.S.; Lim, S.H.; Park, Y.L.; Park, W.; Chae, H. Intercultural Usage of Mori Folium: Comparison Review from a Korean Medical Perspective. *Evid. Based Complement. Altern. Med.* **2015**, *2015*, 379268. [CrossRef]
257. Burger, B.; Sagiorato, R.N.; Cavenaghi, I.; Rodrigues, H.G. Abnormalities of Sphingolipids Metabolic Pathways in the Pathogenesis of Psoriasis. *Metabolites* **2023**, *13*, 291. [CrossRef]

Disclaimer/Publisher's Note: The statements, opinions and data contained in all publications are solely those of the individual author(s) and contributor(s) and not of MDPI and/or the editor(s). MDPI and/or the editor(s) disclaim responsibility for any injury to people or property resulting from any ideas, methods, instructions or products referred to in the content.

pharmaceuticals

MDPI

Article

Salvia miltiorrhiza and Its Compounds as Complementary Therapy for Dyslipidemia: A Meta-Analysis of Clinical Efficacy and In Silico Mechanistic Insights

Min-Seong Lee [1], Han-Young Lee [1], Seung-Hyun Oh [1], Chang-Bum Kim [1], Ji-Han Kim [1], Seung-Hoon Yoo [1], Yeon-Joo Yoo [1], Su-Yeon Lee [2] and Byung-Cheol Lee [1,*]

[1] Department of Clinical Korean Medicine, Graduate School, Kyung Hee University, 26, Kyungheedae-ro, Dongdaemun-gu, Seoul 02447, Republic of Korea; min6691@khu.ac.kr (M.-S.L.); hanyoung2357@gmail.com (H.-Y.L.); sshyunbiz@naver.com (S.-H.O.); kinkcb@gmail.com (C.-B.K.); ilpc21@naver.com (J.-H.K.); ysh9306@khu.ac.kr (S.-H.Y.); alice98113@khu.ac.kr (Y.-J.Y.)

[2] College of Korean Medicine, Kyung Hee University, Seoul 02447, Republic of Korea; suyeon427@khu.ac.kr

* Correspondence: hydrolee@khu.ac.kr; Tel.: +82-2-958-9182

Abstract: Background/Objectives: Dyslipidemia is a significant risk factor for atherosclerotic cardiovascular disease (ASCVD), a leading cause of death worldwide. *Salvia miltiorrhiza* Burge is widely used in East Asia for cardiovascular health, showing potential benefits in lowering cholesterol and reducing inflammation. **Methods:** This study systematically reviewed and conducted a meta-analysis of randomized controlled trials (RCTs) to assess the clinical effectiveness of *Salvia miltiorrhiza* in treating dyslipidemia. Moreover, network pharmacology and molecular docking analyses were performed to explore the mechanisms underlying the effects of *Salvia miltiorrhiza*. **Results:** The meta-analysis revealed that when *Salvia miltiorrhiza* is combined with statin therapy, it significantly enhances lipid profiles, including reductions in total cholesterol, low-density lipoprotein cholesterol (LDL-C), and triglycerides and improvements in high-density lipoprotein cholesterol (HDL-C), compared to statin therapy alone. The in silico analyses indicated that *Salvia miltiorrhiza* may influence key biological pathways, such as the PI3K/Akt, JAK/STAT, and HMGCR pathways, which are involved in inflammation, lipid metabolism, and the development of atherosclerosis. **Conclusions:** *Salvia miltiorrhiza* shows potential as a complementary therapy for dyslipidemia, offering additional lipid-lowering and anti-inflammatory benefits.

Keywords: dyslipidemia; *Salvia miltiorrhiza*; meta-analysis; network pharmacology

1. Introduction

Dyslipidemia is a key risk factor for atherosclerotic cardiovascular disease (ASCVD), a leading cause of mortality worldwide. Elevated low-density lipoprotein cholesterol (LDL-C) levels in the bloodstream lead to endothelial damage, monocyte recruitment, and plaque formation, culminating in atherosclerotic lesions [1]. Furthermore, oxidized LDL exacerbates this process by impairing nitric oxide production, increasing oxidative stress, and activating nuclear factor kappa B (NF-κB), which promotes inflammation, cell apoptosis, and foam cell development in vascular smooth muscle cells [2]. Thus, the proper management of dyslipidemia, particularly by addressing LDL, is crucial for reducing ASCVD risk [3].

Cholesterol is primarily synthesized in the liver, either through the absorption of lipoproteins from the blood or through de novo synthesis, a process regulated by 3-hydroxy-3-methyl-glutaryl-coenzyme A reductase (HMGCR) [4]. Statins, which inhibit HMGCR, remain the cornerstone of dyslipidemia treatment. However, statin therapy is associated with several side effects, including muscle symptoms such as myalgia and an increased risk

https://doi.org/10.3390/ph17111426

of developing diabetes in some patients [5]. Additionally, a significant number of patients on statins do not achieve optimal LDL-C targets, highlighting the need for alternative or additional lipid-lowering therapies [6].

Salvia miltiorrhiza Burge, commonly known as Danshen, has been used in traditional East Asian medicine to improve blood circulation and relieve blood stasis [7]. Recent studies have indicated that Tanshinone IIa, one of the major constituents of Danshen, has shown a wide range of beneficial effects, including anti-inflammatory and anti-atherosclerotic properties [8,9]. Although Danshen and its formulations are commonly used in East Asia including Korea and China for treating dyslipidemia and ASCVD, and while some studies have reported favorable effects of Danshen on dyslipidemia, no comprehensive meta-analysis has yet been conducted to evaluate its therapeutic efficacy. Furthermore, the precise mechanisms by which components of Danshen exert their effects on dyslipidemia have been explored but not fully elucidated.

In this study, we conducted a systematic review and meta-analysis of randomized controlled trials (RCTs) to assess the clinical effectiveness of Danshen in the treatment of dyslipidemia. Additionally, we performed mechanism analyses through network pharmacology and molecular docking to investigate the underlying biological pathways involved in Danshen's action.

2. Results

2.1. Results of Meta-Analysis

2.1.1. Description of Studies

A total of 166 relevant studies were identified, with 159 from CNKI and 7 from PubMed. After removing 16 duplicate studies, 150 articles were screened, of which 94 were excluded for the following reasons: animal studies (n = 38), acute pancreatitis-related (n = 12), diabetes-related (n = 3), case studies (n = 21), Danshen injection (n = 16), network pharmacology (n = 2), meta-analysis/protocol (n = 2), cohort study (n = 1), and reports not retrieved (n = 2). The remaining 53 studies were assessed for eligibility (n = 56), among which 20 were further excluded for the following reasons: interventions combined with other herbal medicine (n = 13), did not include any lipid profiles as an outcome (n = 2), reported lipid profile on a 10-point scale (n = 1), only suggested effectiveness/ineffectiveness (n = 3), and reported only the results of the intervention group (n = 1). Finally, 33 studies were included in the analysis. The process is illustrated in Figure 1.

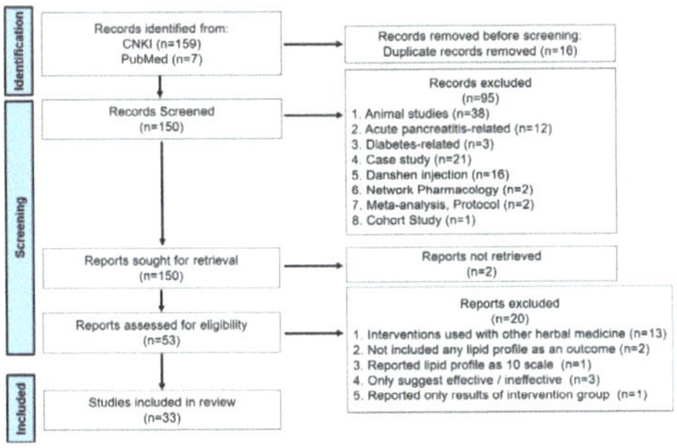

Figure 1. Meta-analysis flow diagram.

Among the 33 RCTs, 32 were conducted in China and 1 was conducted in the Netherlands. The characteristics of each study are summarized in Table 1. The trials assessed the following four indicators: total cholesterol (TC) (n = 33), triglycerides (TGs) (n = 29), LDL-C (n = 28), and high-density lipoprotein cholesterol (HDL-C) (n = 25).

Table 1. Characteristics of each experiment.

First Author	Participants	Number of Cases/Age, Years		Intervention Methods		Observation Period	Outcome Indicators
		Trial Group	Control Group	Trial Group	Control Group		
Chen, 2022 [10]	Patients aged 51–83 years	63.97 ± 5.48	64.38 ± 5.79	Standard treatment (anticoagulants, antiplatelet aggregation agents, blood sugar-lowering agents, antihypertensive drugs) with Danshen dripping tablets 10 T tid	Standard treatment (anticoagulants, antiplatelet aggregation agents, blood sugar-lowering agents, antihypertensive drugs) with Atorvastatin 20 mg qd	12 weeks	Blood lipid levels via TC, TG, HDL-C, LDL-C
Deng Chaolan, 2015 [11]	Patients aged 41–80 years	49/61.58 ± 7.30	48/61.16 ± 7.15	Compound Danshen soft capsules 27 mg tid	Atorvastatin 10 mg qd	12 weeks	Blood lipid levels via TC, TG, HDL-C, LDL-C
Gong Xingping, 2013 [12]	Patients aged 34–76 years	77/ 45.7 ± 8.2	73/ 46.2 ± 8.6	Simvastatin 20 mg qd with compound Danshen dripping tablet 10 T tid	Simvastatin 20 mg qd	12 weeks	Blood lipid levels via TC, TG, HDL-C, LDL-C
Guo Hongmei, 2016 [13]	76 patients	38/57.4 ± 5.2	38/56.1 ± 4.8	Simvastatin 10 mg qd with compound Danshen dripping tablets 10 T tid	Simvastatin 10 mg qd	12 weeks	Blood lipid levels via TC, TG,
Jin Ji-bin, 2019 [14]	Patients aged 30–65 years	64/45.8 ± 8.5	64/46.7 ± 7.9	Simvastatin 5 mg qd with compound Danshen dripping tablets 10 T tid	Simvastatin 5 mg qd	12 weeks	Blood lipid levels via TG, TC, LDL-C
Li Shujia, 2017 [15]	Patients aged 45–85 years	50/62.5 ± 5.1	50/61.9 ± 3.6	Rosuvastatin 10 mg/20 mg/40 mg qd with compound Danshen dripping tablets 10 T tid	Rosuvastatin 10 mg/20 mg/40 mg qd	8 weeks	Blood lipid levels via TG, TC, LDL-C
Li Si-si, 2019 [16]	Patients average age of 56.4 ± 7.9 years	not mentioned	not mentioned	Baseline medications (enteric-coated aspirin 100 mg qd and isosorbide mononitrate 40 mg qd) with compound Danshen dripping tablets 270 mg tid	Baseline medications (enteric-coated aspirin 100 mg qd and isosorbide mononitrate 40 mg qd) with Rosuvastatin 10 mg qd	24 weeks	Blood lipid levels via TG, LDL, HDL, TC
Li Tao, 2017 [17]	Patients aged 42–81 years	48/59.2 ± 6.1	48/60.4 ± 5.4	Capsules containing compound Danshen dripping capsules 10 C tid	Simvastatin 20 mg/day	12 weeks	Blood lipid levels via TC, TG, HDL-C, LDL-C

Table 1. *Cont.*

First Author	Participants	Number of Cases/Age, Years		Intervention Methods		Observation Period	Outcome Indicators
		Trial Group	Control Group	Trial Group	Control Group		
Liu Jun, 2015 [18]	80 patients	40/54.8 ± 11.6	40/53.8 ± 9.4	Simvastatin 20 mg qd with compound Danshen dripping pill 10 C tid	Simvastatin 20 mg qd	8 weeks	Blood lipid levels via TG, TC, LDL-C
Liu Qingan, 2012 [19]	102 patients	52/average age 51.7	50/average age 59.8	Simvastatin 10 mg qd with complex Danshen preparation 25 mg tid	Simvastatin 10 mg qd	20 weeks	Blood lipid levels via TG, LDL-C
Liu Yongchun, 2014 [20]	90 patients	45/65.4 ± 5.1	45/64.8 ± 3.7	Rosuvastatin calcium 10 mg qd with compound Danshen dripping pill 270 mg tid	Rosuvastatin 10 mg qd	12 weeks	Blood lipid levels via TG, TC, LDL-C
Rong Chunlan, 2014 [21]	Patients aged 45–77 years	40/57.33 ± 3.29	40/57.33 ± 3.29	Standard treatment with compound Danshen dripping capsule 10 C tid	Standard treatment (β-blockers, calcium channel blockers, antiplatelet agents, statins)	4 weeks	Blood lipid levels via TC, TG, HDL-C, LDL-C
Wei Ting, 2016 [22]	Patients aged 45–86 years	58/62.8 ± 11.5	58/64.7 ± 12.8	Simvastatin 10 mg qd with compound Danshen dripping capsules 10 C tid	Simvastatin 10 mg qd	12 weeks	Blood lipid levels via TG, TC, LDL-C
Wu Zichun, 2020 [23]	Patients aged 52–81 years	65.42 ± 2.44	66.42 ± 3.21 years	Conventional treatment with Danshen dripping capsules 10 C tid	Conventional treatment with Rosuvastatin Calcium 10 mg qd	8 weeks	Blood lipid levels via TC, TG, HDL-C, LDL-C
Xu Ling, 2014 [24]	182 patients	91 (randomized)	91 (randomized)	Rosuvastatin calcium 10 mg qd with compound Danshen dripping capsule 10 C tid	Rosuvastatin 10 mg qd	8 weeks	Blood lipid levels via TC, TG, HDL-C, LDL-C
Yang Di 2022 [25]	Patients aged 36–82 years	63. 18 ± 4. 10	62. 38 ± 4. 28	Danshen dripping tablets 270 mg tid	Atorvastatin 20 mg qd	12 weeks	Blood lipid levels via TC, TG, HDL-C, LDL-C
Ye Sha, 2019 [26]	Patients aged <75 years	41/59.2 ± 6.1	40/60.4 ± 5.4	Compound Danshen dripping capsules 10 C tid	Trimetazidine 20 mg tid	12 weeks	Blood lipid levels via TC, TG, HDL-C, LDL-C
Zhang, 2019 [27]	Patients aged 52–89 years	65.8 ± 9.7	67.5 ± 7.6	Danshen dripping pills 27 mg bid	Atorvastatin 10 mg qd	12 weeks	Blood lipid levels via TG, TC, LDL-C

Table 1. *Cont.*

First Author	Participants	Number of Cases/Age, Years		Intervention Methods		Observation Period	Outcome Indicators
		Trial Group	Control Group	Trial Group	Control Group		
Zhang Mingyan, 2011 [28]	Patients aged 31–69	37/56 ± 11.45	35/55 ± 10.78	Lovastatin 20 mg qd with compound *Salvia miltiorrhiza* dripping tablets 10 T tid	Lovastatin 20 mg qd	8 weeks	Blood lipid levels via TG, TC, LDL-C, HDL-C
Zhao Ming, 2018 [29]	Patients aged 46–83 years	46/61.82 ± 4.90	46/61.95 ± 4.92	Rosuvastatin 10–40 mg qd with compound Danshen driping tablets 10 T tid	Rosuvastatin 10–40 mg qd	8 weeks	Blood lipid levels via TG, LDL-C, HDL-C
Zheng Lewei, 2014 [30]	100 patients aged 60~86(72 ± 2.6)	50 (randomized)	50 (randomized)	Atorvastatin calcium 10 mg qd with compound Danshen dripping capsules 10 C tid	Atorvastatin Calcium 20 mg qd	8 weeks	TC, TG, HDL-C
Li Xiaoyang, 2003 [31]	Patients average age under 60	20	20	Complex Salvia ginseng preparation 30 C tid	Inositol nicotinate 0.6~1.2 g tid	8 weeks	Blood lipid levels via TG, TC,
Pleun, 2015 [32]	Patients aged 40–70 years	Randomized double-blind placebo-controlled crossover study *n*= 20		Danshen capsules 500 mg (4 capsules) tid	Placebo capsules 500 mg (4 capsules) tid	4 weeks	Blood lipid levels via TG, TC, LDL-C, HDL-C
Bei Guangmin, 2010 [33]	Patients aged 42–65	92/52.9 ± 7	93/54.3 ± 6.4	Candesartan ester capsules 8 mg qd, Hydrochlorothiazide 10 mg qd, long-acting nifedipine 10 mg bid with Danshen tablets 4 T tid	Candesartan ester capsules 8 mg qd, Hydrochlorothiazide 10 mg qd, Long-acting nifedipine 10 mg bid, Atorvastatin 10 mg qd	24 weeks	Blood lipid levels via TG, TC, LDL-C, HDL-C
Zhang Huina, 2020 [34]	No information	63.74 ± 3.28	63.52 ± 3.79	Danshen dripping capsules 10 mg tid	Conventional symptomatic treatment (Betaloc (metoprolol), aspirin, lifestyle adjustment, low-sodium, low-fat diet, metformin, nifedipine, simvastatin 10 mg qd)	4 weeks	Blood lipid levels via TC, TG, HDL-C, LDL-C
Zhang Shijun, 2007 [35]	Patients aged 26–63	40/38 ± 10.2	41/39 ± 11.8	Composite Salviae dropping pills 10 C tid	Simvastatin 20 mg qd	12 weeks	Blood lipid levels via TG, TC, LDL-C, HDL-C

Table 1. *Cont.*

First Author	Participants	Number of Cases/Age, Years		Intervention Methods		Observation Period	Outcome Indicators
		Trial Group	Control Group	Trial Group	Control Group		
Han Qinghua, 2000 [36]	Patients aged 50–70	30/64.2 ± 0.3	28/63.8 ± 0.3	Compound Danshen dropping capsules 10 C tid	No treatment	4 weeks	Blood lipid levels via TG, TC, LDL-C
Li Yangyuan, 2013 [37]	73 patients	35/60.7 ± 13.6	38/59.2 ± 11.4	Atorvastatin 10 mg qd with compound Danshen dripping capsules 10 C tid	Atorvastatin 10 mg qd	4 weeks	Blood lipid levels via TG, TC, LDL-C, HDL-C
Pan Xiaojian, 2015 [38]	98 patients	49/67.3 ± 3.2	49/67.1 ± 4.2	Atorvastatin 10 mg qd with compound Danshen tablets 10 T tid	Atorvastatin 10 mg qd	20 weeks	Blood lipid levels via TG, TC, LDL-C, HDL-C
Yin Xiangshi, 2012 [39]	80 patients aged 58–75	40	40	Standard treatment (aspirin, β-blockers, calcium channel blockers, nitrates with compound Danshen preparation 10 T tid	Standard treatment (aspirin, β-blockers, calcium channel blockers, nitrates)	4 weeks	Blood lipid levels via TG, TC, LDL-C, HDL-C
Ma Jianying, 1998 [40]	Patients aged 40–45 years	40	20	Complex Salvia ginseng preparation 10 T tid	Sodium alginate double ester capsule 50 mg tid	8 weeks	Blood lipid levels via TG, TC, LDL-C, HDL-C
Dai Xiangdong, 2002 [41]	No information	20/56.3 ± 8.5	20/58.7 ± 7.5	Danshen capsules 10 C tid	Xuezhikang 0.6 g bid	4 weeks	Blood lipid levels via TG, TC, LDL-C, HDL-C
Cui Shikui, 2002 [42]	Patients aged 36–79	69/58	44/51	Compound Danshen preparation 10 T tid	Zhibitu capsule 1 C bid	4 weeks	Blood lipid levels via TG, TC, LDL-C

The characteristics of each experiment are summarized in Table 1.

2.1.2. Risk of Bias in the Included Studies

A risk of bias table was constructed based on the following criteria: Random Sequence Generation (selection bias), Allocation Concealment (selection bias), the Blinding of Participants and Personnel (performance bias), the Blinding of Outcome Assessment (detection bias), Incomplete Outcome Data (attrition bias), Selective Reporting (reporting bias), and Other Bias. Low risk is indicated in green, uncertain risk in yellow, and high risk in red. The risk of bias assessments for each study are presented in Figure 2 below.

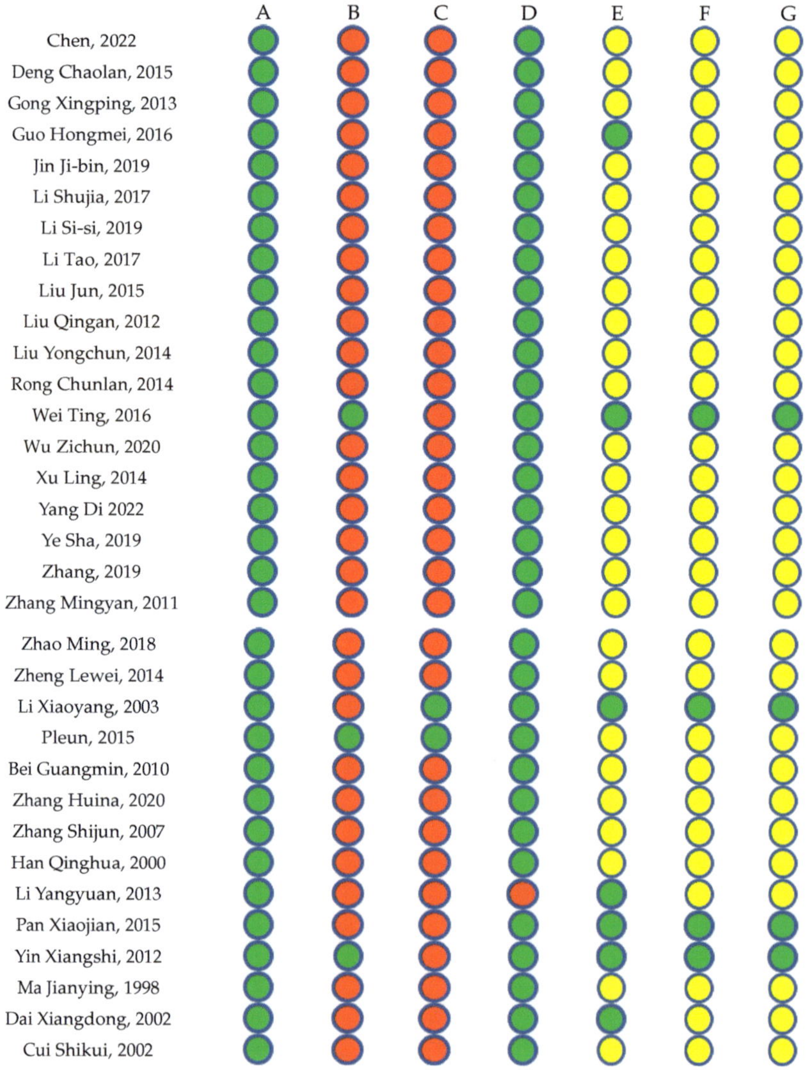

Figure 2. A: Random Sequence Generation (selection bias), B: Allocation Concealment (selection bias), C: Blinding of Participants and Personnel (performance bias), D: Blinding of Outcome Assessment (detection bias), E: Incomplete Outcome Data (attrition bias), F: Selective Reporting (reporting bias), G: Other Bias. Red: high risk, Yellow: uncertain risk, Green: low risk [10–42].

2.1.3. Effective Rate

Total Cholesterol

Twenty-one trials compared the concurrent use of Danshen and statins versus statins alone. Two trials compared Danshen alone versus a placebo. Three trials compared Danshen alone versus statins alone, and four trials compared Danshen alone versus the observation. Additionally, the number of trials comparing Danshen alone versus sodium alginate, Xuezhikang (XZK), or Zhibituo tablets as control groups was one each. The results are presented in Table 2 and Figure 3.

Table 2. Effective rate of total cholesterol.

	Std. Mean Difference	Heterogeneity
Danshen + Statin vs. Statin (n = 21)	−0.65 [−0.92, −0.38]	$p < 0.00001$, $I^2 = 89\%$
Danshen vs. Placebo (n = 2)	0.41 [−1.01, 1.84]	$p = 0.0004$, $I^2 = 92\%$
Danshen vs. Statin (n = 3)	0.08 [−0.13, 0.28]	$p = 0.48$, $I^2 = 0\%$
Danshen vs. Observation (n = 4)	−0.78 [−1.01, −0.55]	$p = 0.95$, $I^2 = 0\%$
Danshen vs. Sodium Alginate (n = 1)/ Xuezhikang (n = 1)/ Zhibituo Tablet (n = 1)		
Total	−0.55 [−0.77, −0.33]	$p < 0.00001$, $I^2 = 89\%$

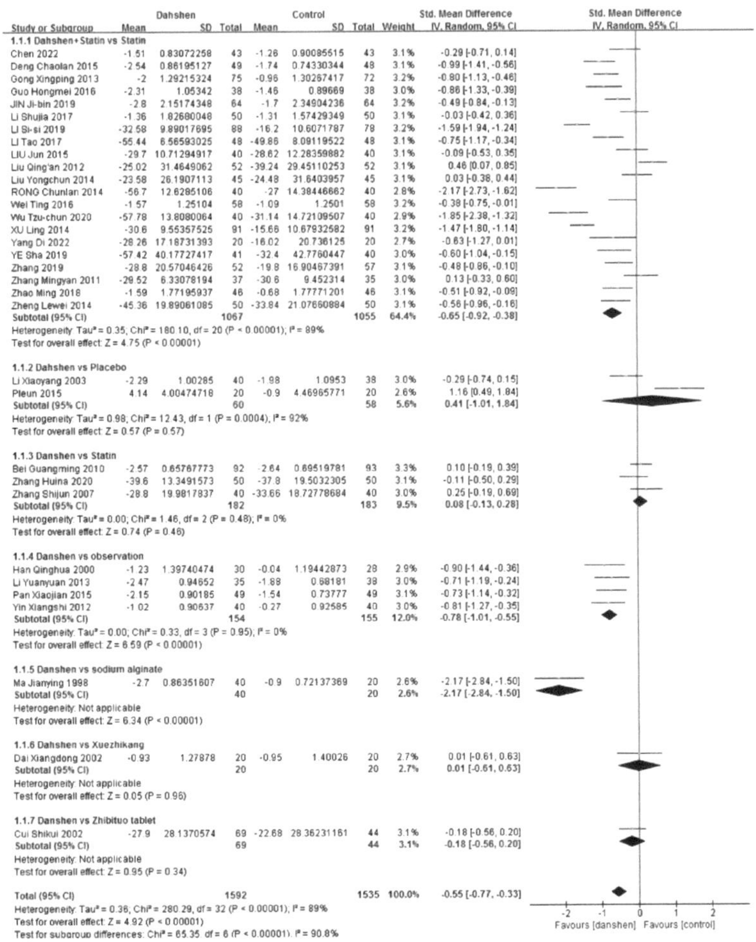

Figure 3. Complete data of total cholesterol [10–42].

Triglycerides

Seventeen trials compared the concurrent use of Danshen and statins versus statins alone. Two trials compared Danshen alone versus a placebo. Three trials compared Danshen alone versus statins alone, and four trials compared Danshen versus an observation. Additionally, there were trials each comparing Danshen alone versus sodium alginate, Xuezhikang (XZK), or Zhibituo tablets as control groups. The results are presented in Table 3, and the complete data are presented in Figure 4.

Table 3. Effective rate of triglycerides.

	Std. Mean Difference	Heterogeneity
Danshen + Statin vs. Statin (n = 17)	−0.67 [−0.80, −0.54]	$p = 0.02$, $I^2 = 46\%$
Danshen vs. Placebo (n = 2)	−0.32 [−0.68, 0.04]	$p = 0.97$, $I^2 = 0\%$
Danshen vs. Statin (n = 3)	−1.13 [−3.06, 0.79]	$p < 0.00001$, $I^2 = 98\%$
Danshen vs. Observation (n = 4)	−1.44 [−2.53, −0.35]	$p < 0.00001$, $I^2 = 94\%$
Danshen vs. Sodium Alginate (n = 1)/ Xuezhikang (n = 1)/Zhibituo Tablet (n = 1)		
Total	−0.73 [−0.96, −0.50]	$p < 0.00001$, $I^2 = 88\%$

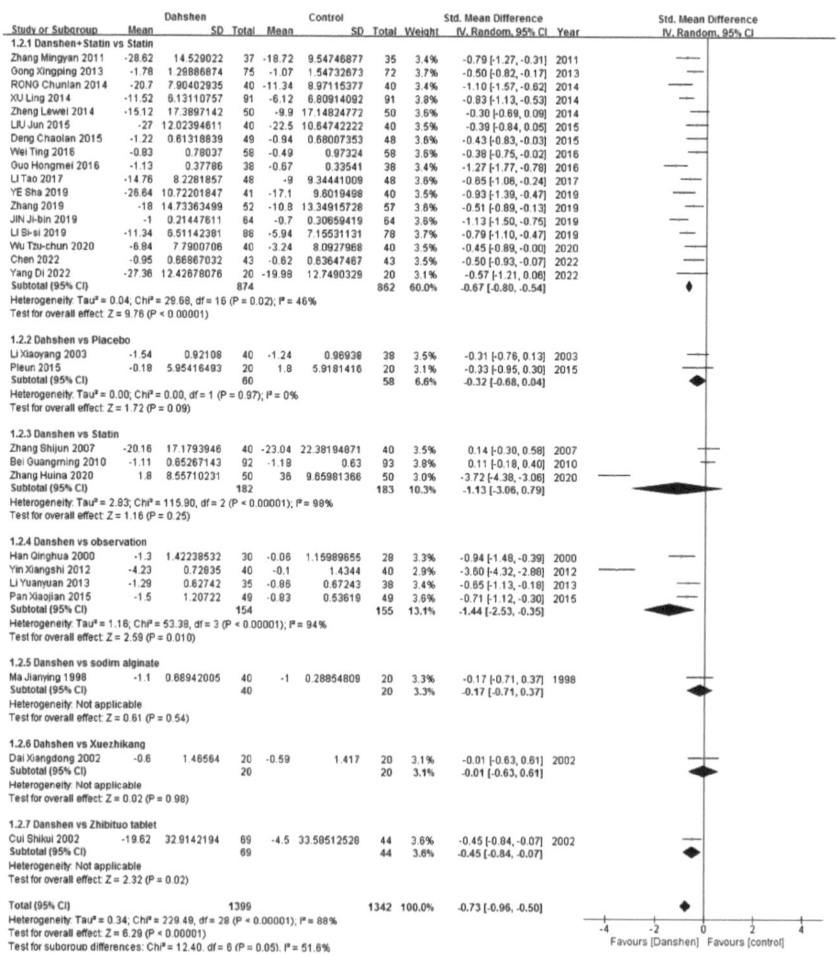

Figure 4. Complete data of triglycerides [10–14,16–18,21–28,30–42].

LDL-Cholesterols

Nineteen trials compared the concurrent use of Danshen and statins versus statins alone. Two trials compared Danshen alone versus statins alone. Four trials compared Danshen alone versus the observation. Additionally, there were trials each comparing Danshen alone versus sodium alginate, Xuezhikang (XZK), or placebo as control groups. The results are presented in Table 4, and the complete data are presented in Figure 5.

Table 4. Effective rate of LDL-cholesterol.

	Std. Mean Difference	Heterogeneity
Danshen + Statin vs. Statin (n = 19)	−0.56 [−0.78, −0.34]	$p < 0.00001$, $I^2 = 82\%$
Danshen vs. Statin (n = 2)	0.18 [−0.07, 0.43]	$p = 0.31$, $I^2 = 5\%$
Danshen vs. Observation (n = 4)	−1.39 [−2.51, −0.28]	$p < 0.00001$, $I^2 = 95\%$
Danshen vs. Sodium Alginate (n = 1)/ Xuezhikang (n = 1)/Placebo (n = 1)		
Total	−0.58 [−0.83, −0.33]	$p < 0.00001$, $I^2 = 89\%$

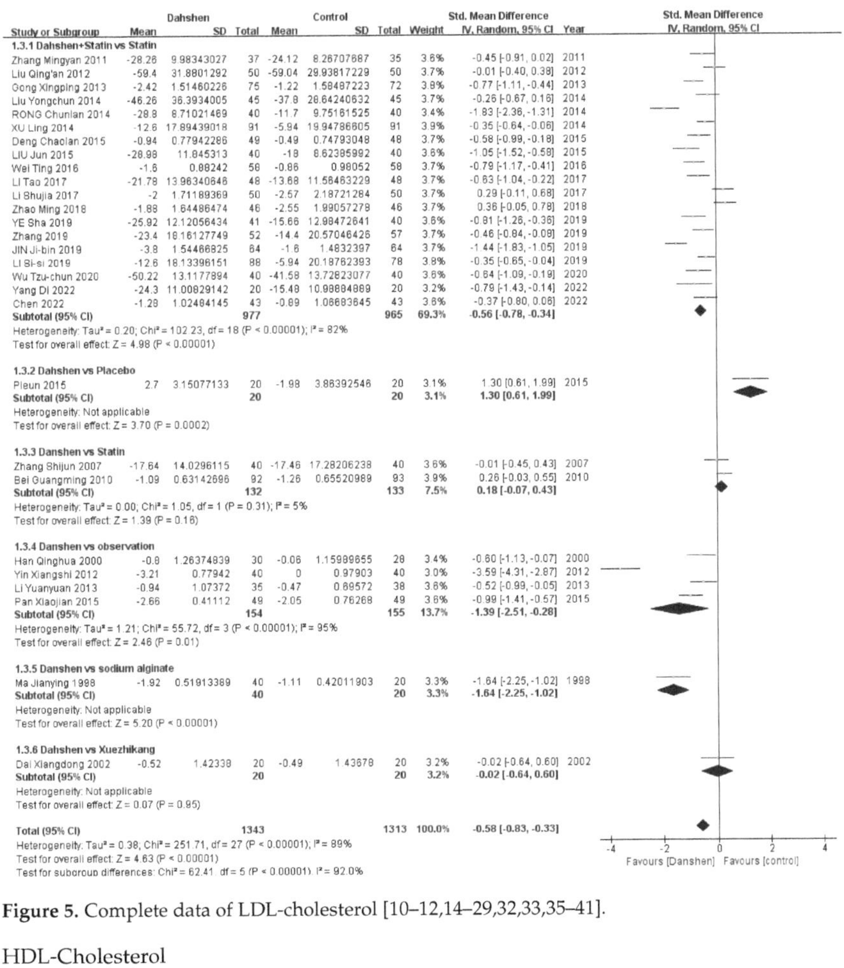

Figure 5. Complete data of LDL-cholesterol [10–12,14–29,32,33,35–41].

HDL-Cholesterol

Sixteen trials compared the concurrent use of Danshen and statins versus statins alone. Two trials compared Danshen alone versus statins alone. Three trials compared Danshen alone versus the observation. Additionally, there were trails each comparing Danshen alone versus sodium alginate, Xuezhikang (XZK), Zhibituo tablet or placebo as control groups. The results are presented in Table 5, and the complete data are presented in Figure 6.

Table 5. Effective rate of HDL-cholesterol.

	Std. Mean Difference	Heterogeneity
Danshen + Statin vs. Statin (n = 16)	0.70 [0.41, 0.98]	$p < 0.00001$, $I^2 = 87\%$
Danshen vs. Statin (n = 2)	−0.05 [−0.37, 0.28]	$p = 0.20$, $I^2 = 39\%$
Danshen vs. Observation (n = 3)	−2.13 [−4.88, 0.62]	$p < 0.00001$, $I^2 = 99\%$
Danshen vs. Sodium Alginate (n = 1)/ Xuezhikang (n = 1)/ Zhibituo Tablet (n = 1)/Placebo (n = 1)		
Total	0.34 [0.03, 0.64]	$p < 0.00001$, $I^2 = 92\%$

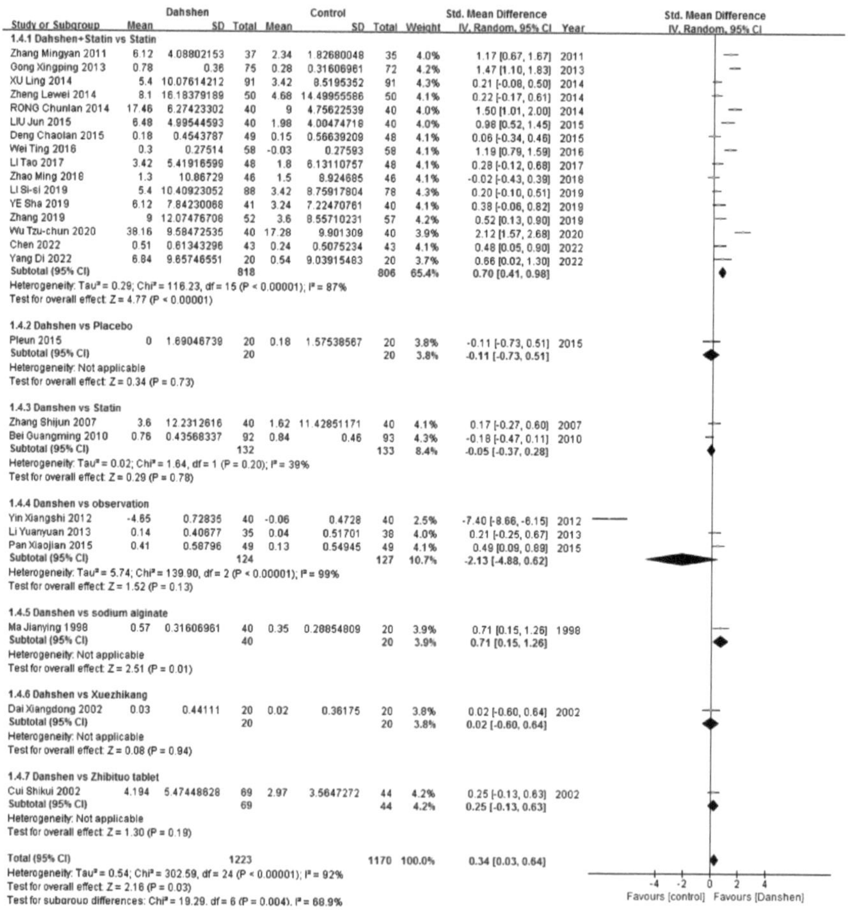

Figure 6. Complete data of HDL-cholesterol [10–12,16–18,21–30,32,33,35,37–42].

2.2. Results of In Silico Network Construction and Analysis

2.2.1. ADME of Active Compounds of *Salvia miltiorrhiza*

Among the various compounds identified from *Salvia miltiorrhiza*, five key constituents—Tetrahydrotanshinone, Cryptotanshinone, Dihydrotanshinone I, Tanshinone IIa, and Tanshinone I—were selected for evaluation based on their pharmacokinetic and ADME (absorption, distribution, metabolism, and excretion) properties and HPLC fingerprints results of *Salvia miltiorrhiza* [43]. The molecular weights of these compounds ranged from 276.3 to

296.39, with Tanshinone I having the lowest molecular weight (276.3) and Cryptotanshinone the highest (296.39). Oral bioavailability (OB) values varied between 29.27% (Tanshinone I) and 52.34% (Cryptotanshinone), indicating moderate to high bioavailability across the selected constituents. The drug likeness (DL) scores ranged from 0.36 to 0.4, suggesting favorable drug-like properties. Notably, all five compounds complied with Lipinski's rule, showing no violations, which underscores their potential as orally active therapeutic agents (Table 6).

Table 6. Physicochemical properties, ADME profiles, and toxicity Evaluation of selected *Salvia miltiorrhiza* constituents.

Molecular Name		Tetrahydrotanshinone	Cryptotanshinone	Dihydrotanshinone I	Tanshinone IIa	Tanshinone I
MW		280.34	296.39	278.32	294.37	276.3
OB (%)		38.75	52.34	45.04	49.89	29.27
DL		0.36	0.4	0.36	0.4	0.36
Lipinski		Yes; 0 violations	Yes; 0 violations	Yes; 0 violations	Yes; 0 violations	Yes; 0 violations
Absorption	Caco-2	0.96	0.95	0.95	1.05	1.05
Distribution	PPB	82.42	88.37	86.71	89.13	87.91
	BBB	0.39	0.51	0.43	0.7	0.53
Metabolism	CYP1A2 inhibitor	Yes	Yes	Yes	Yes	Yes
	CYP2C19 inhibitor	Yes	Yes	Yes	Yes	Yes
	CYP2C9 inhibitor	Yes	Yes	Yes	Yes	No
	CYP2D6 inhibitor	No	No	Yes	Yes	No
	CYP3A4 inhibitor	Yes	Yes	Yes	Yes	Yes
Excretion	T 1/2	1.729	1.886	1.919	2.084	2.145
Toxicity	Human hepatoxicity	0.758	0.73	0.826	0.776	0.83
	LD50	2.583	2.605	2.562	2.712	2.525

MW, molecular weight; OB, oral bioavailability; DL, drug likeness; Lipinski, Lipinski's rule of five compliance; Caco-2, Caco-2 permeability; PPB, plasma protein binding; BBB, blood–brain barrier penetration; CYP, cytochrome P450 enzyme inhibition status, indicating potential drug–drug interactions; T 1/2, half-Life; LD50, median lethal dose.

The ADME properties of these selected Danshen constituents were also comprehensively analyzed. The absorption potential, as indicated by the Caco-2 permeability values, ranged from 0.95 to 1.05, with Tanshinone IIa and Tanshinone I exhibiting the highest permeability. Plasma protein binding (PPB) values were relatively high across all compounds, ranging from 82.424% to 89.131%, suggesting an extensive distribution. Blood–brain barrier (BBB) penetration varied, with Tanshinone IIa showing the highest potential (0.7) and Tetrahydrotanshinone the lowest (0.39) (Table 6).

Metabolically, all compounds were predicted to inhibit key cytochrome P450 enzymes (CYP1A2, CYP2C19, CYP2C9, and CYP3A4), with the exception of Tanshinone I, which did not inhibit CYP2C9, and Tetrahydrotanshinone and Cryptotanshinone, which did not inhibit CYP2D6. The half-life (T 1/2) of the compounds ranged from 1.729 h (Tetrahydrotanshinone) to 2.145 h (Tanshinone I), indicating moderate elimination rates. In terms of toxicity, human hepatotoxicity values were within acceptable ranges, from 0.73 to 0.83, and the LD50 values, representing acute toxicity, ranged from 2.525 to 2.712, indicating a relatively low toxicity profile (Table 6).

Collectively, these results highlight the favorable pharmacokinetic and safety profiles of the selected Danshen compounds, reinforcing their potential as candidates for further drug development and clinical evaluation.

2.2.2. Identifying Overlapping Genes of Compounds and Dyslipidemia

A Venn diagram analysis was conducted to identify the overlapping genes between five key compounds from Danshen (Tetrahydrotanshinone, Cryptotanshinone, Dihydrotanshinone I, Tanshinone IIa, and Tanshinone I) and dyslipidemia-related genes. The results demonstrated varying degrees of overlap, indicating the potential relevance of these compounds in modulating dyslipidemia pathways. Specifically, Tetrahydrotanshinone showed 33 overlapping genes (1.7%), Cryptotanshinone had 40 overlapping genes (2.1%), and Dihydrotanshinone I presented 34 overlapping genes (1.8%) with dyslipidemia. Similarly, Tanshinone IIa and Tanshinone I exhibited 36 (1.9%) and 38 (2.0%) overlapping genes, respectively (Figure 7).

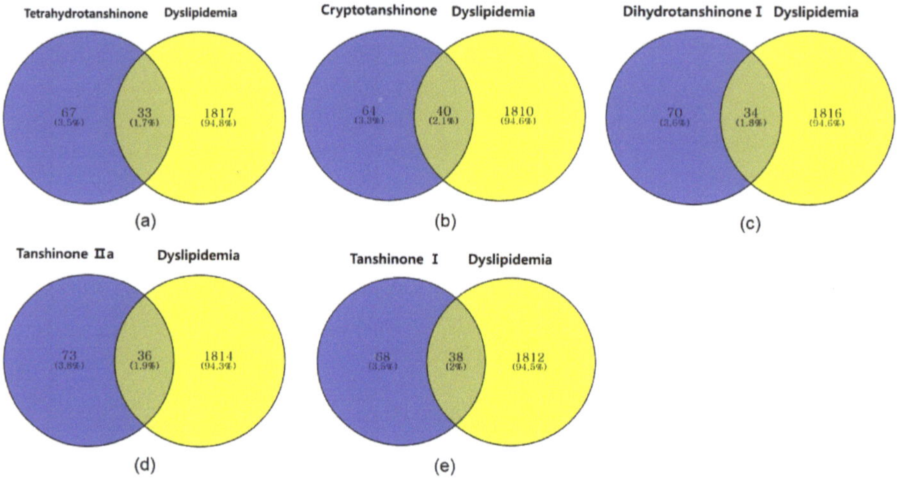

Figure 7. Overlapping genes between *Salvia miltiorrhiza* compounds and dyslipidemia. (**a**) Venn diagram of Tetrahydrotanshinone and dyslipidemia-related genes; (**b**) Venn diagram of Cryptotanshinone and dyslipidemia-related genes; (**c**) Venn diagram of Dihydrotanshinone I and dyslipidemia-related genes; (**d**) Venn diagram of Tanshinone IIa and dyslipidemia-related genes; (**e**) Venn diagram of Tanshinone I and dyslipidemia-related genes.

2.2.3. Network Description of Compounds and Dyslipidemia

A network analysis was conducted to investigate the interaction between the key constituents of Danshen and dyslipidemia-associated genes. The first visualization, generated using Cytoscape, illustrates the interaction network between the five selected compounds—Tetrahydrotanshinone, Cryptotanshinone, Dihydrotanshinone I, Tanshinone IIa, and Tanshinone I—and their target genes. This network reveals the complex interconnections between the compounds and multiple dyslipidemia-related targets, underscoring the multi-target therapeutic potential of these compounds (Figure 8a).

Subsequently, a protein–protein interaction (PPI) network was constructed using the STRING database, encompassing the broader network of interactions among the identified target proteins. This PPI network highlights the interconnected nature of the dyslipidemia-related pathways and the central role of several key proteins, suggesting their importance in mediating the therapeutic effects of Danshen constituents (Figure 8b).

To further refine the analysis, we focused on the top 16 genes with the highest degree of connectivity (degree ≥ 4) within the PPI network. This refined network demonstrates the core targets that may play crucial roles in the therapeutic modulation of dyslipidemia. Key genes such as phosphatidylinositol-4,5-bisphosphate 3-kinase catalytic subunit alpha (PIK3CA), Akt Serine/Threonine Kinase 1 (AKT1), and signal transducer and activator of transcription 3 (STAT3) were identified as central nodes, suggesting their potential as critical mediators of the observed pharmacological effects. These findings provide valuable insights into the molecular mechanisms through which Danshen compounds exert their effects on dyslipidemia, offering a foundation for future experimental validation and drug development efforts (Figure 8c).

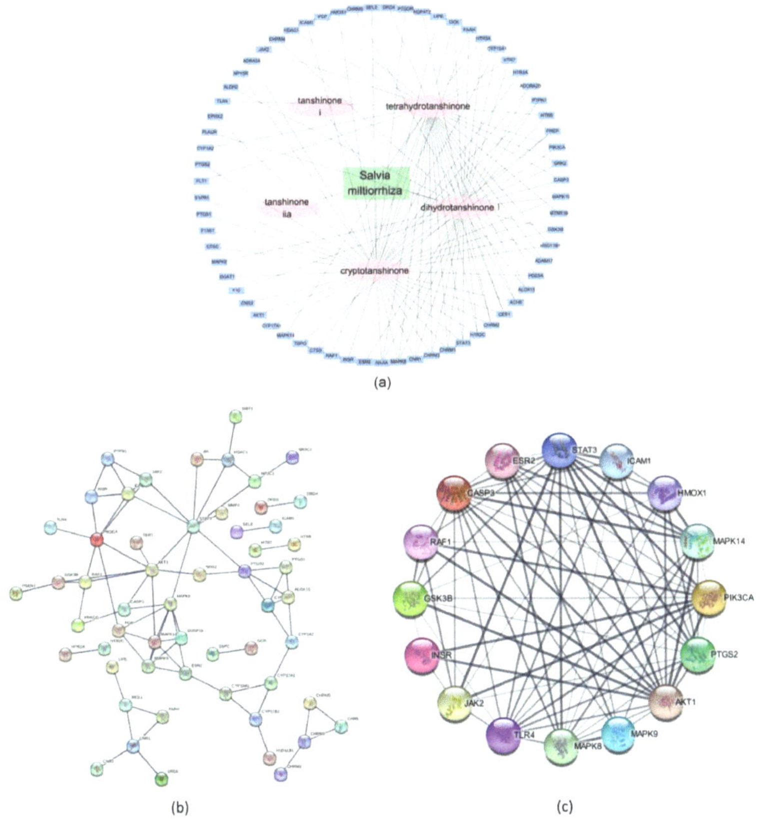

Figure 8. Network analysis of *Salvia miltiorrhiza* constituents and their interaction with dyslipidemia-related genes. (**a**) Interaction network of *Salvia miltiorrhiza* compounds and target genes; (**b**) protein–protein interaction network of target genes; (**c**) refined network of top 16 genes.

2.2.4. GO and KEGG Analyses

A ClueGO analysis was performed to elucidate the biological pathways associated with the targets of Danshen compounds. The analysis revealed that the most enriched pathways were related to lipid and atherosclerosis (82.46%) and insulin resistance (17.54%), indicating that these pathways are significantly involved in the pharmacological effects of the compounds. The network visualization demonstrated a highly interconnected structure among the identified pathways, with lipid metabolism and insulin signaling being central nodes (Figure 9a–c).

Gene Ontology (GO) and Kyoto Encyclopedia of Genes and Genomes (KEGG) pathway enrichment analyses were conducted to explore the functional implications of the Danshen compounds' targets. GO analysis revealed that the most enriched molecular functions were protein binding, adenosine 5′-triphosphate (ATP) binding, and enzyme binding, while the top cellular components included the plasma membrane, cytoplasm, and nucleoplasm. In terms of biological processes, signal transduction and phosphorylation were prominently enriched. The KEGG pathway analysis identified significant enrichment in pathways related to lipid and atherosclerosis, insulin resistance, and Phosphoinositide 3-kinase (PI3K)-Akt signaling (Figure 9d,e).

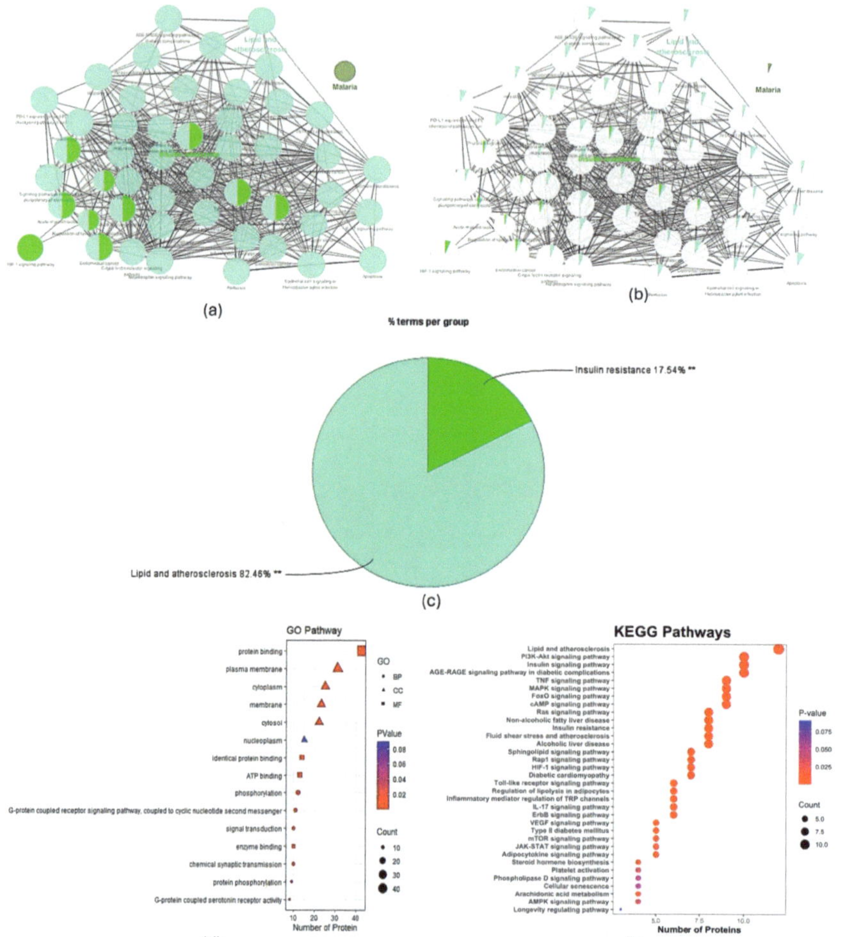

Figure 9. ClueGO, GO, and KEGG pathway enrichment analysis of Danshen compounds' targets. (**a**) ClueGO network visualization of selected pathways; (**b**) proportional representation of pathways in the network; (**c**) KEGG pathway distribution with filtered terms; (**d**) GO top 5 BP, CC, and MF pathways; (**e**) KEGG analysis identifying key pathways. ** p-value<0.001.

2.2.5. Molecular Docking

The molecular docking analysis revealed that the five Danshen compounds—Cryptotanshinone, Dihydrotanshinone I, Tanshinone I, Tanshinone IIa, and Tetrahydrotanshinone—exhibited strong binding affinities across a panel of lipid and atherosclerosis-related protein targets, which are class I HMGCR, class II HMGCR, low-density lipoprotein receptor (LDLR), peroxisome proliferator-activated receptor-α (PPARA), janus kinase2 (JAK2), and receptor for advanced glycation end products (RAGE), with binding energy values consistently lower than -7.4 kcal/mol. Binding affinities with STAT3, macrophage and tumor necrosis factor (TNF), which are insulin resistance-related proteins, demonstrated significant interactions. Proteins related to PI3K-Akt pathways also exhibited intimate interactions, which were lower than −7.7 kcal/mol. Molecular docking targets included key proteins of the PPI network including JAK2, STAT3, PIK3CA, and AKT1, suggesting their potential as effective modulators of these targets (Table 7 and Figure 10).

Table 7. Binding affinities of Danshen compounds to dyslipidemia-related protein targets.

Target Name	PBD ID	Binding Affinity (kcal/mol)				
		Crypto-Tanshinone	Dihydro-Tanshinone I	Tanshinone I	Tanshinone IIa	Tetrahydro-Tanshinone
Lipid and atherosclerosis-related proteins						
HMGCR	1dqa	−8.2	−8.4	−8.2	−8.1	−8.2
Class II HMGCR	1t02	−8.3	−8.5	−7.9	−7.9	−8
LDLR	3p5b	−8.2	−7.4	−8.6	−7.4	−7.3
PPARA	1i7g	−8.8	−8.6	−8.8	−9.1	−8.8
JAK2	4fvq	−9.1	−10.3	−10.1	−9.5	−10.4
RAGE	3o3u	−9.2	−8.4	−10.2	−9.1	−8.2
Insulin resistance-related proteins						
STAT3	1bg1	−7.9	−8	−8.1	−8	−7.3
Macrophage	1gd0	−7.3	−7.8	−7.3	−7.2	−7.2
TNF	2az5	−9.8	−10.3	−10.4	−10	−10.2
PI3K-Akt-related proteins						
PIK3CA	8tsa	−8.4	−8.6	−9	−9	−8.9
AKT1	4gv1	−8	−7.7	−9.2	−8.1	−8.8

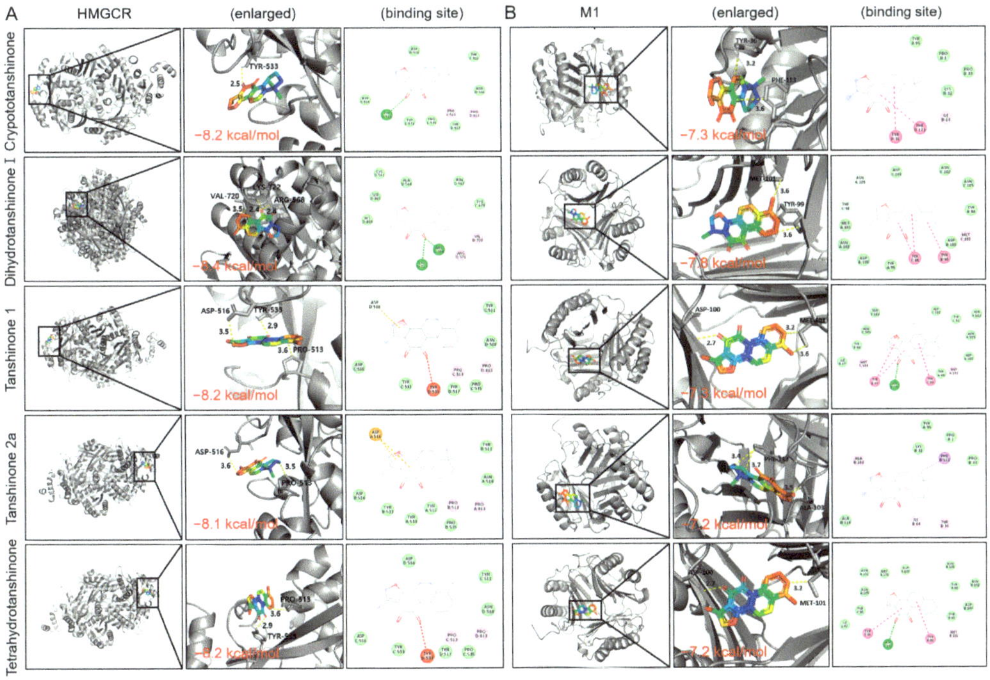

Figure 10. *Cont.*

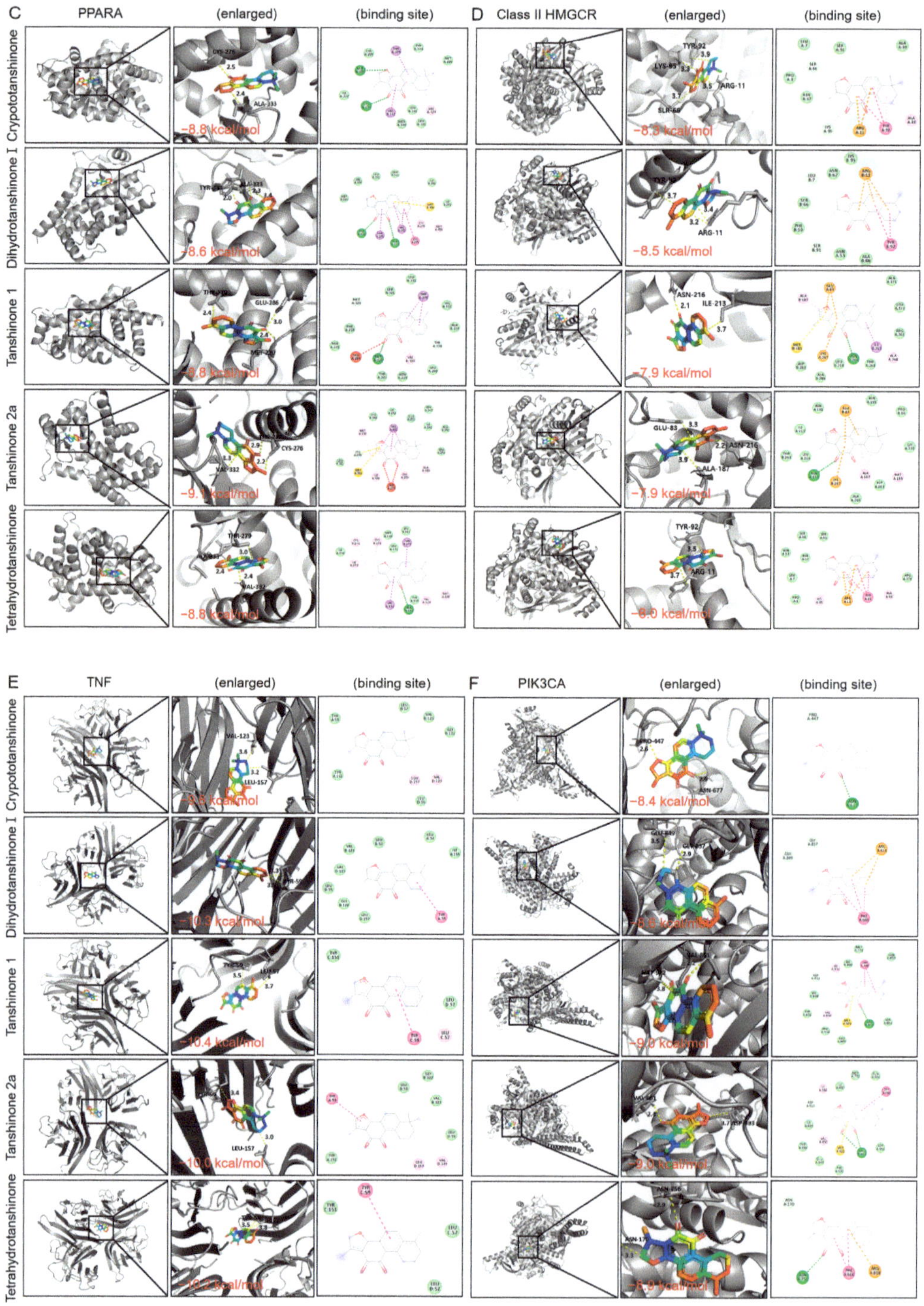

Figure 10. *Cont.*

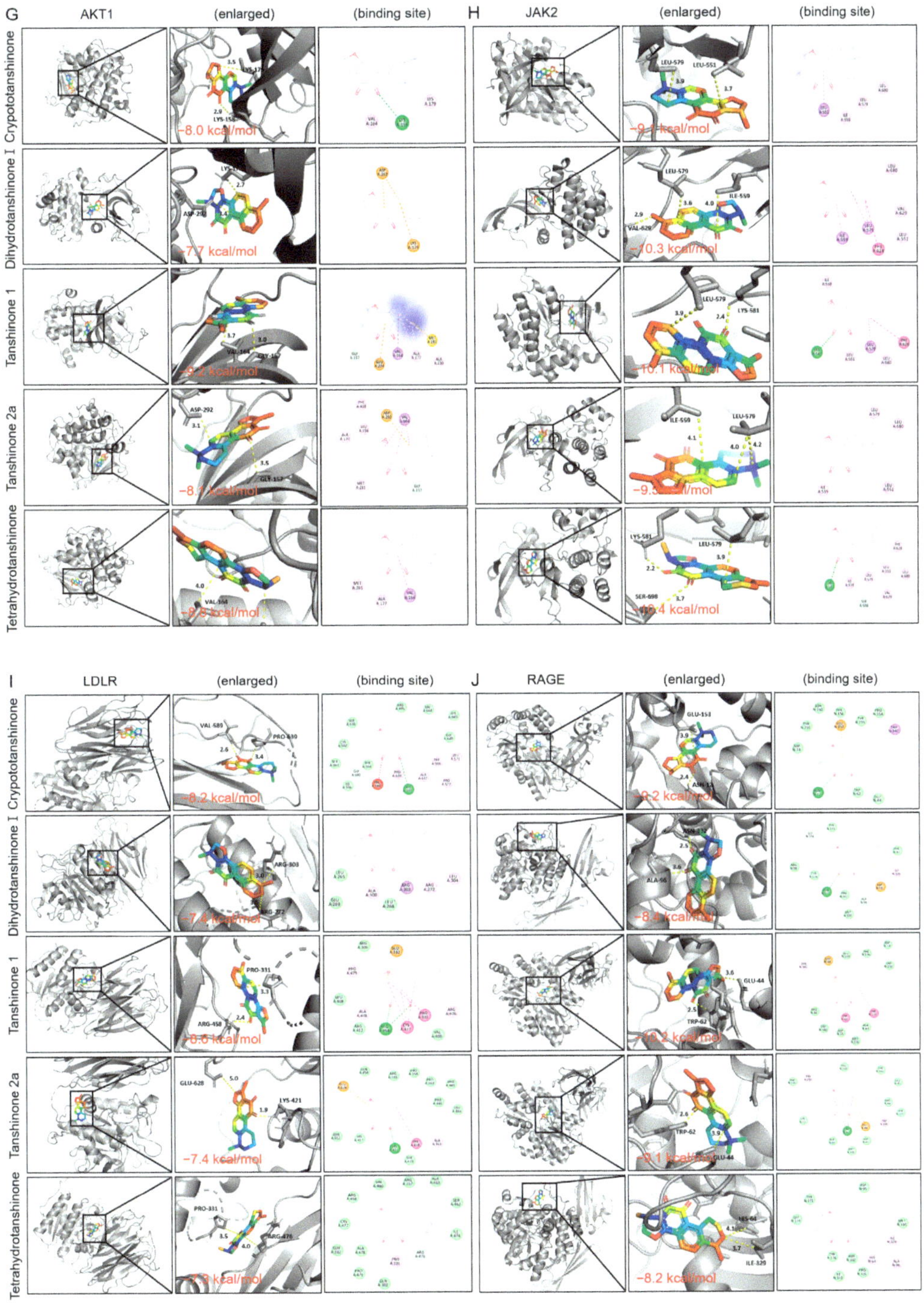

Figure 10. *Cont.*

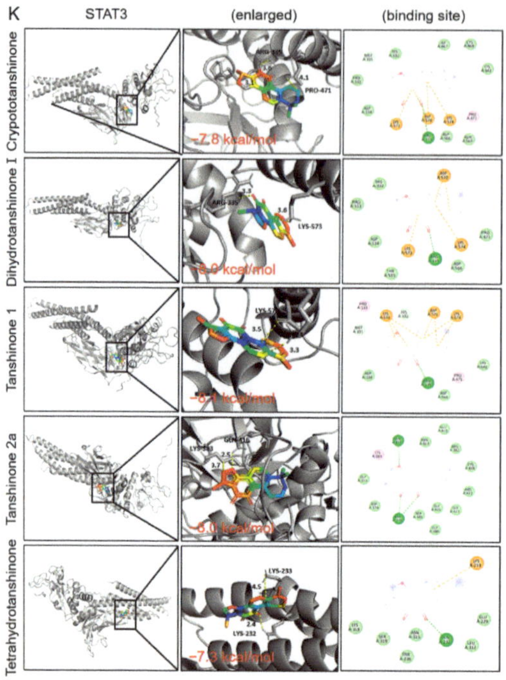

Figure 10. Molecular docking analysis of Danshen compounds with dyslipidemia-related protein targets. (**A**) Compound docking with HMGCR; (**B**) compound docking with class II HMGCR; (**C**) compound docking with LDLR; (**D**) compound docking with PPARA; (**E**) compound docking with JAK2; (**F**) compound docking with RAGE; (**G**) compound docking with STAT3; (**H**) compound docking with macrophage; (**I**) compound docking with TNF; (**J**) compound docking with PIK3CA; (**K**) compound docking with AKT1.

3. Discussion

Dyslipidemia is an eminent risk factor for ASCVD [44], and as it is a leading cause of mortality globally, the management of dyslipidemia is critical. Statin, an HMGCR inhibitor, is a cornerstone for treating dyslipidemia, and other lipid-lowering drugs include ezetimibe, proprotein convertase subtilisin/kexin type 9 (PCSK9) inhibitors, and bile acid sequestrants. While statins are gold standard therapy for dyslipidemia, there are several statin-associated side effects [44]. Most common side effect is statin-associated muscle symptoms that appear as myalgia, and they are reported in 5–20% of the prescribed patients. Moreover, statin use in patients with major risk factors for diabetes showed 28% increased risk for new-onset diabetes [45]. Major risk factors include high fasting glucose, obesity, an HbA1c of over 6% and metabolic syndromes, which are common predispositions in patients with abnormal lipid profiles. While studies show achieving a lower LDL-C level leads to a reduced risk of subsequent cardiovascular events [46], statin usage shows limitation in achieving therapeutic targets. A retrospective cohort study, for example, showed that only 20% of patients with statin treatment achieved the LDL cholesterol target, which is under 70 mg/dL [47]. Thus, the development of novel lipid-lowering agents as an alternative or adjuvant therapy to statin is gaining importance.

Salvia miltiorrhiza Burge, also known as Danshen, has long been used for relieving blood stasis and improving blood circulation in the East Asian medicines. Through scientific research, Danshen has been widely studied for its protective properties in cardiovascular diseases and diabetes, anti-oxidative effects, inhibitory effects on apoptosis, and anti-inflammatory effects [48]. To elucidate this, Tanshinone IIa, the most actively investigated constituents of Danshen, has been reported to possess a wide range of therapeutic effects in cardiovascular diseases [8], and its anti-inflammatory effects include the downregulation of the Toll-like receptor 4 (TLR4)/NF-κB pathway and the inhibition of nitric oxide (NO), interleukin (IL)-1β, IL-6 and TNF-α [49].

While Danshen, and its preparations in pill and injection formulations, is widely used in China for the treatment of dyslipidemia and ASCVD [48] and studies have elucidated the favorable effect of Danshen usage in dyslipidemia, there has been no meta-analysis on the therapeutic efficacy of Danshen against dyslipidemia. Thus, in our study, we conducted a meta-analysis of random clinical trials of Danshen on dyslipidemia through the databases of CNKI and PubMed. Among the 150 screened records, 33 studies were included in the meta-analysis, and the majority of the included studies were statin add-on studies. Danshen treatment with statin showed significant superiority over the statin group in total cholesterol, LDL, HDL, and triglycerides. The results from the meta-analysis suggested the therapeutic efficacy of Danshen as a treatment agent for dyslipidemia.

As natural products normally contain various chemical constituents, the detection of their major bioactive constituents enables us to implement a scientific investigation of their therapeutic effects. With the results from the HPLC fingerprinting of Danshen [43], we selected major constituents with a high possibility of drug utilization among several constituents; Tetrahydrotanshinone, Cryptotanshinone, Dihydrotanshinone I, Tanshinone IIa, and Tanshinone I. These constituents appeared as promising candidate components of Danshen after verification regarding absorption, distribution, metabolism, excretion, drug-likeness, and toxicity.

To further elucidate the possible treatment targets and mechanism responsible for the efficacy of Danshen against dyslipidemia, a network pharmacological analysis was implemented. We obtained overlapping genes between Danshen compounds and dyslipidemia, and through a PPI network analysis, key target genes such as JAK2, STAT3, PI3K, AKT1, and TLR4 were acquired. The KEGG pathway analysis of overlapping genes suggested relevant pathways such as lipid and atherosclerosis, PI3K-AKT, insulin signaling, advanced glycation end product (AGE)-RAGE, TNF, and mitogen-activated protein kinase (MAPK). With overlapping genes associated with the pathways of lipids and atherosclerosis, insulin resistance, hypoxia-inducible factor-1 (HIF-1), and non-alcoholic fatty liver disease (NAFLD), an additional ClueGO analysis revealed lipid and atherosclerosis as the most relevant terms and insulin resistance as the next. Other relevant pathways included AGE-RAGE, TNF, HIF-1, and adipocytokine. Based on the PPI network analysis and pathway analysis, several therapeutic target genes that Danshen might interact with were selected, and we assessed the reliability of binding them with computational molecular docking.

First, we focused on the PI3K/Akt pathway, which plays a key role in the development of atherosclerosis in several stages, such as macrophage polarization, increased intracellular lipid storage, smooth muscle cell proliferation and dysfunction [50]. Oxidized LDL induces PI3K/Akt signaling in macrophages to form foam cells and activate MAPK and the mammalian target of rapamycin (mTOR) downstream, resulting in the proliferation of arterial wall and plaque formation. PI3K/Akt plays a crucial role in regulating macrophage polarization in vascular inflammation as well through NF-κB [51]. Plaque rupture, owing to inflammation, and the following vessel obstruction and thrombosis are related to atherosclerotic progression. Restraining macrophage infiltration in plaque and promoting the autophagy of macrophages enables plaque stability, and PI3K/Akt through mTOR and TLR4 is reported to be a relevant pathway. Also, the inhibition of PI3K/Akt signaling suppresses B-cell lymphoma-extra large (Bcl-xL) and Bcl-2 expressions, which are anti-apoptotic molecules, and thus result in an increase in the apoptosis of foam cells [50,52].

Thus, PI3K/Akt-targeted agents are deemed a novel mechanism of action for treatment for atherosclerosis. All five major constituents of Danshen showed a high binding affinity with PI3K and AKT1 in the molecular docking simulation, and among them, Tanshinone I showed an especially high binding affinity of −9 and −9.2 kcal/mol, respectively.

AGE attenuates RAGE and JAK/STAT cascade expressions [53]. While the JAK/STAT pathway is involved in various processes, its role in inflammation is especially gaining attention as a key factor in inflammation. Cytokines and growth factors initiate the JAK/STAT pathway, and the downstream of JAK/STAT includes diverse genes involved in proliferation and apoptosis in atherosclerosis, such as IL-6, interferon (IFN)-γ, TNF-α and the suppressor of cytokine signaling (SOCS) [54]. The activation of RAGE and the JAK/STAT pathway gives rise to inflammation and oxidative stress, and as a result, accelerations of LDL oxidation, vascular plaque formation, smooth muscle cell proliferation and extracellular matrix production are induced [55–57]. Thus, regulating RAGE and JAK/STAT in atherosclerosis is deemed as an appealing therapeutic target. In our in silico studies, we observed a significantly high binding affinity of Danshen constituents with a JAK2 of over −9.1 kcal/mol. Binding powers with RAGE and STAT3 showed promising results as well.

LDL receptors in the liver plays a critical role in regulating plasma LDL levels as approximately 70% of the circulating LDL is cleared off by hepatic LDL receptors [58]. Because a small number of hepatic LDL receptors lead to elevated plasma LDL levels, the development of drugs that can increase available LDL receptors such as the PCSK9 inhibitor is one of the strategies for dyslipidemia treatment. Thus, our in silico investigation, we performed molecular docking with LDLR and showed a binding power of −7.3~−8.6 kcal/mol.

HMGCR converts HMG-CoA to mevalonic acid, and the mevalonate pathway is responsible for cholesterol biosynthesis [59]. Statin acts as an HMG-CoA reductase inhibitor, thus resulting in decreased intracellular cholesterol activate sterol regulatory element-binding proteins (SREBPs), and its binding to a sterol regulatory element results in an increased transcription of LDL receptor genes and thus an influx of plasma LDL cholesterols into the hepatocyte. While statin is a powerful HMG-CoA reductase inhibitor, HMG-CoA reductases are reported to have two classes. They share the same catalytic mechanism but have a significant difference in their three-dimensional structures and sensitivity to statin [60]. While class I HMG-CoA reductase is sensitive to the inhibition of statin, class II HMG-CoA reductase is reported to be considerably less sensitive. Molecular dockings with two classes of HMG-CoA reductase showed a considerably high binding power. High binding affinities with class II HMG-CoA reductase that showed −7.9 to −8.5 kcal/mol are especially noticeable in that they suggest a novel therapeutic target of Danshen that statin does not interact with. The in silico results suggested a high possibility of Danshen's action on two classes of HMG-CoA reductase and LDL receptors, which is statin's main mechanism of action. These results comply with the results of our meta-analysis which demonstrated the therapeutic efficacy of Danshen in lipid profiles when used as an adjuvant to statins compared to statin therapy alone.

Inflammation plays a crucial role in the development of ASCVD as it is related to stages such as the infiltration of monocytes, foam cell formation, and plaque vulnerability. The importance of inflammation regulation in dyslipidemia is undebatable. TNF-α activates NF-κB, and NF-κB promotes adhesion molecules such as selectin E (SELE) and intercellular adhesion molecule (ICAM)-1 to recruit monocytes into the injured sites in the endothelial cells [61]. Monocyte-derived macrophages and tissue-resident macrophages develop two phenotypes, which are M1 and M2 macrophages, and they have pro-inflammatory and anti-inflammatory characteristics, respectively [50]. Inflammatory changes in ASCVD induce an M1-dominant polarization of macrophages, which leads to plaque pathogenesis [62]. Thus, we assessed the binding affinity of major compounds of Danshen with classical macrophage M1 and major cytokine TNF for the verification of its anti-inflammatory effects, and they showed favorable results.

PPAR-α activates fatty acid beta-oxidation [63]. Due to its triglyceride-lowering effects, fibrate, which is a PPAR-α agonist, has long been utilized for dyslipidemia treatment. PPAR-α suppresses endothelin-1 expression, resulting in an alleviated smooth muscle cell proliferation. Moreover, PPAR-α has anti-inflammatory effects as it inhibits the cytokine-initiated expression of vascular adhesion molecule-1 (VCAM-1), reduces IL-6 production, and suppresses cyclooxygenase (COX)-2 activity by regulating NF-κB. Due to these beneficial effects on metabolic disorders, the PPAR-α agonist is deemed a promising agent for dyslipidemia as well. The binding affinity of Danshen constituents with PPAR-α was remarkably high, showing −8.6 to −9.1 kcal/mol.

Overall, our in silico investigation elucidated that Danshen may alleviate the development of dyslipidemia into ASCVD by suppressing macrophage polarization, apoptosis, and smooth muscle cell proliferation via modulating PI3K/Akt. Also, the modulation of Danshen on RAGE and the JAK/STAT pathway is also suggested, resulting in the inhibition of inflammation and oxidative stress and an acceleration of vascular plaque formation. Danshen may alleviate inflammation by suppressing key inflammatory cytokine TNF and classical macrophage M1. Regarding cholesterol biosynthesis, Danshen may provide an additional lipid-lowering effect as a class II HMG-CoA reductase inhibitor. Also, Danshen may facilitate the beta-oxidation of fatty acids and the elimination of serum LDL by regulating PPAR-α and LDLR. These elucidated potential therapeutic mechanisms suggest further investigational subjects in understanding the therapeutic effect of Danshen in dyslipidemia.

There are a few limitations in our investigation. Our investigation on target genes and pathways was based on computational analysis and thus warrants careful interpretation. Although network pharmacological methods are widely used for drug research, it is important to recognize clear limitations of these techniques, such as data restrictions and data discrepancies [64]. Also, our investigation of Danshen focused on several potential active components of *Salvia miltiorrhiza*, while *Salvia miltiorrhiza* as whole includes various constituents. As many traditional medicines act through multiple bioactives and targets, an investigation on several selected bioactives may not be sufficient to explain the therapeutic effect of *Salvia miltiorrhiza* as a whole [64]. However, through a meta-analysis of randomized clinical trials, we were able to verify the therapeutic effect of Danshen against dyslipidemia to some extent. Our investigation was the first meta-analysis that assessed the therapeutic efficacy of Danshen on dyslipidemia to the best of our knowledge. In addition, we suggested possible therapeutic components, key targets and mechanisms and pathways of Danshen for the treatment of dyslipidemia. We hope our investigation will contribute to furthering the understanding of the mechanism of action of Danshen, and future in vitro or in vivo studies for the verification of the suggested therapeutic targets and pathways of Danshen are warranted.

4. Materials and Methods

4.1. Research Workflow Through Integrated Methodology

This study utilized both in silico modeling and a meta-analysis to assess the effect of *Salvia miltiorrhiza* on dyslipidemia, the latter following these steps: The first step involved identifying randomized controlled trials (RCTs) that explored the effect of *Salvia miltiorrhiza* on dyslipidemia. This initial phase ensured that there was a sufficient pool of data for a thorough analysis. Using the Cochrane's Review Manager program (The Cochrane Collaboration, England and Wales), data from the selected trials were then systematically reviewed and analyzed, focusing on various lipid parameters to determine the effects of

Salvia miltiorrhiza. Lastly, the study used in silico trials to predict and further investigate the estimated effects of *Salvia miltiorrhiza* for dyslipidemia.

4.2. Data Sources and Search Strategy

The systematic review of clinical trials was conducted following the PRISMA 2020 guidelines for systematic reviews and meta-analyses. All researchers actively participated in the data collection. Studies assessing the efficacy and safety of *Salvia miltiorrhiza* for dyslipidemia were identified using two databases, PubMed and the Chinese National Knowledge Infrastructure (CNKI), covering publications up to May 2024.

4.3. Study Selection

4.3.1. Type of Studies

In this study, RCTs evaluating the efficacy and safety of *Salvia miltiorrhiza* for dyslipidemia were only included, with no restrictions on language or publication date. Studies were excluded if they met any of the following criteria: (a) were not randomized control trials; (b) unrelated to dyslipidemia; (c) did not use *Salvia miltiorrhiza* as the primary intervention; (d) involved non-oral administration methods; (e) were case reports or reviews; (f) involved animal studies; (g) did not consider lipid profiles as primary outcomes; and (h) were not published in scientific peer-reviewed journals, including theses and dissertations.

4.3.2. Type of Participants

The review included studies without restrictions based on age, gender, or race but only those with patients diagnosed with dyslipidemia using validated criteria. Studies involving participants with diabetes or acute pancreatitis were excluded in order to maintain focus on dyslipidemia.

4.3.3. Type of Interventions

RCTs that investigated the effects of *Salvia miltiorrhiza* as the main treatment in comparison to a control group were included. Various forms of the intervention, such as decoctions, granules, capsules, and tablets, were accepted. There were no limitations on dosage or treatment duration, but only oral forms of administration were considered. Thus, trials involving injections of *Salvia miltiorrhiza* were excluded. Similarly, trials that combined the use of other therapies, such as non-drug treatments, acupuncture, massage, or additional herbal ingredients, were not included.

4.3.4. Type of Outcome Measures

The primary outcomes analyzed were lipid profiles, specifically total cholesterol (TC), triglycerides (TGs), high-density lipoprotein cholesterol (HDL-C), and low-density lipoprotein cholesterol (LDL-C).

4.4. Methodological Quality Assessment

The methodological quality of each included study was independently assessed by all researchers using the updated version of the Risk of Bias tool (RoB 2.0) for randomized trials. RoB 2.0 focuses on five potential sources of bias: (a) biases related to the randomization process, (b) biases arising from deviations in allocation concealment, (c) biases due to insufficient blinding, (d) biases from incomplete outcome data, and (e) biases in selective outcome reporting. Studies were classified into three levels of methodological quality—"high risk of bias," "low risk of bias," and "uncertain risk of bias"—using color codes of red, green, and yellow, respectively. Disagreements among researchers were resolved through consensus, guided by the corresponding author.

4.5. *Quality of Evidence According to Outcome Measures*

The quality of evidence for each outcome was assessed using the GRADE (Grading of Recommendations Assessment, Development, and Evaluation) system. GRADE categorizes evidence into four levels: very low, low, moderate, and high. The level of evidence can be downgraded based on factors like potential bias, an inconsistency of the results, the lack of direct evidence, imprecision, or publication bias.

4.6. *Statistical Analysis*

Statistical analyses were performed using the RevMan software (ver. 5.4). Trials were categorized based on the type of intervention and control group. The relative risk (RR) and 95% confidence intervals (CIs) were calculated for lipid parameters, including total cholesterol (TC), triglycerides (TGs), HDL-C, and LDL-C. To combine the results measured with different units, the standardized mean difference (SMD) and 95% confidence intervals (CIs) were computed. The analysis detected heterogeneity among the primary outcomes across the randomized controlled trials, and a contour-enhanced funnel plot was created for further analysis.

4.7. *Network Pharmacology Analysis of Anti-Dyslipidemia Mechanisms*

A network pharmacology analysis was conducted to investigate how Danshen may treat dyslipidemia. The key chemical constituents in Danshen were sourced from the TCMSP (Traditional Chinese Medicine Systems Pharmacology database and analysis platform, https://tcmsp-e.com/, accessed on 23 August 2024), and results from the HPLC fingerprinting of Danshen [43]. In this study, the compounds of Danshen were screened based on the following criteria obtained from the TCMSP database, oral bioavailability (OB) $\geq$ 30%, Caco-2 permeability ≥ -0.4, and drug likeness (DL) $\geq$ 0.18, and then only the compounds that also exhibited peaks in the HPLC analysis of Danshen were selected. Additionally, tanshinone I, a known marker compound of Danshen, was included for further analysis. Absorption, distribution, metabolism, and excretion (ADME) predictions for these five compounds were evaluated using the SwissADME platform (http://www.swissadme.ch/, accessed on 23 August 2024). In the SwissADME platform, compounds were evaluated according to the ADME rules of Lipsinski. The SwissTargetPrediction platform (http://www.swisstargetprediction.ch, accessed on 23 August 2024.) (in the "Homo sapiens" mode) was used to predict potential target genes, and the target information was standardized using the Uniprot database (http://www.uniprot.org, accessed on 23 August 2024). Dyslipidemia-related target genes were retrieved from the GeneCards database (http://www.genecards.org, accessed on 23 August 2024) using "dyslipidemia" as the search term.

Using the "Bioinformatics for Genomics and Proteomics site" (https://bioinfogp.cnb.csic.es/tools/venny/, accessed on 23 August 2024), Venn diagrams showing the overlap between key constituents and dyslipidemia were generated. A network illustrating the interrelation between chemicals and targets was constructed using Cytoscape (version 3.10.2; https://cytoscape.org/, accessed on 23 August 2024) to visualize the relationships between the core constituents of Danshen and dyslipidemia targets. The STRING database (12.0; https://string-db.org/, accessed on 23 August 2024) was used to analyze the interaction gene targets of the major constituents and dyslipidemia, filtering by "Homo sapiens". Disconnected nodes in the network were set to "hide", and the confidence score was set to $\geq$ 0.7. The protein–protein interaction (PPI) network was constructed, redundant nodes were removed, and the top 16 genes with the highest connectivity (degree $\geq$ 4) were analyzed in Cytoscape.

To characterize gene targets by function, including biological processes, cellular components, and molecular functions, we used Gene Ontology (GO) functional analysis. A graph was created by selecting the top 4 counts from each category. Kyoto Encyclopedia of Genes and Genomes (KEGG) enrichment analysis was employed to identify shared targets between the core constituents and dyslipidemia within signaling pathways. The gene lists for four pathways—lipid and atherosclerosis, insulin resistance, HIF-1 signaling pathway, and NAFLD—strongly linked to dyslipidemia were analyzed using ClueGO in Cytoscape, with a data selection threshold of $p < 0.05$.

4.8. Docking Interactions of Danshen Compounds and Target Proteins

Based on the results of the KEGG analysis, we examined the detailed mechanisms of the lipid and atherosclerosis pathway, which showed the highest count, using a pathway map (https://www.genome.jp/pathway/map05417, accessed on 23 August 2024). Several key genes involved in the core processes were selected, and molecular docking was performed with the main constituents of Danshen. The three-dimensional structures of Danshen compounds—Cryptotanshinone, Dihydrotanshinone I, Tanshinone I, Tanshinone IIa, and Tetrahydrotanshinone—were downloaded from the PubChem database (http://pubchem.ncbi.nlm.nih.gov/, accessed on 23 August 2024), while the 3D structures of proteins, including HMGCR (PDB ID: 1DQA and 1T02), LDLR (PDB ID: 3P5B), PPARA (PDB ID: 1I7G), JAK2 (PDB ID: 4FVQ), RAGE (PDB ID: 3O3U), STAT3 (PDB ID: 1BG1), macrophage (PDB ID: 1GD0), TNF (PDB ID: 2AZ5), PIK3CA (PDB ID: 8TSA), and AKT1 (PDB ID: 4GV1), were provided by the PDB database (http://rcsb.org), accessed on 8 August 2024. The criteria for selecting PDB entries for each protein are as follows: First, proteins are searched in the UniProt database (https://www.uniprot.org/, accessed on 23 August 2024), ensuring that the organism is specified as Homo sapiens (Human). From the provided PDB list, only entries that utilize the X-ray diffraction method and have a resolution of 3.0 Å or lower are considered. Subsequently, the RCSB PDB links for the qualifying PDB files are accessed to download them in PDB format. Prior to molecular docking, unnecessary protein domains were removed, and hydrogenation was performed using the Biovia Discovery Studio Visualizer. Molecular docking was conducted using PyRx to assess binding affinity. The interaction sites of the ligand and corresponding distances were analyzed using the receptor–ligand interactions tab in Biovia Discovery Studio Visualizer, where a 2D diagram was created. Additionally, the simulated binding interactions were visualized in PyMOL.

5. Conclusions

This study suggests that Danshen has potential as an adjunct therapy for treating dyslipidemia, particularly when combined with statins. The meta-analysis showed that Danshen significantly improves lipid profiles, including reductions in total cholesterol, LDL-C, and triglycerides while increasing HDL-C, compared to statins alone. Mechanistic analyses suggest that Danshen influences key pathways involved in lipid metabolism, inflammation, and atherosclerosis, such as PI3K/Akt, JAK/STAT, and HMGCR pathways. These findings indicate that Danshen may offer additional therapeutic benefits beyond conventional statin treatment, potentially improving cardiovascular outcomes for patients with dyslipidemia. Larger clinical trials with longer follow-up periods are also necessary to assess Danshen's long-term impact on cardiovascular outcomes.

Author Contributions: Conceptualization and methodology, M.-S.L. and B.-C.L.; investigation, data curation, and writing—original draft preparation, M.-S.L., H.-Y.L., S.-H.O., C.-B.K. and S.-Y.L.; writing—review and editing, M.-S.L., J.-H.K., S.-H.Y., Y.-J.Y. and B.-C.L.; funding acquisition and project administration, B.-C.L. All authors have read and agreed to the published version of the manuscript.

Funding: This research was supported by the Korea Health Technology R&D Project through the Korea Health Industry Development Institute (KHIDI), funded by the Ministry of Health and Welfare, Republic of Korea (grant number: HFD20C0022).

Institutional Review Board Statement: Not applicable.

Informed Consent Statement: Not applicable.

Data Availability Statement: Data are contained within the article.

Conflicts of Interest: The authors declare no conflicts of interest.

References

1. Helkin, A.; Stein, J.J.; Lin, S.; Siddiqui, S.; Maier, K.G.; Gahtan, V. Dyslipidemia Part 1—Review of Lipid Metabolism and Vascular Cell Physiology. *Vasc. Endovasc. Surg.* **2016**, *50*, 107–118. [CrossRef] [PubMed]
2. Leong, X.F. Lipid Oxidation Products on Inflammation-Mediated Hypertension and Atherosclerosis: A Mini Review. *Front. Nutr.* **2021**, *8*, 717740. [CrossRef]
3. Du, Z.; Qin, Y. Dyslipidemia and Cardiovascular Disease: Current Knowledge, Existing Challenges, and New Opportunities for Management Strategies. *J. Clin. Med.* **2023**, *12*, 363. [CrossRef]
4. Berberich, A.J.; Hegele, R.A. A Modern Approach to Dyslipidemia. *Endocr. Rev.* **2022**, *43*, 611–653. [CrossRef]
5. Vinci, P.; Panizon, E.; Tosoni, L.M.; Cerrato, C.; Pellicori, F.; Mearelli, F.; Biasinutto, C.; Fiotti, N.; Di Girolamo, F.G.; Biolo, G. Statin-Associated Myopathy: Emphasis on Mechanisms and Targeted Therapy. *Int. J. Mol. Sci.* **2021**, *22*, 11687. [CrossRef]
6. Krahenbuhl, S.; Pavik-Mezzour, I.; von Eckardstein, A. Unmet Needs in LDL-C Lowering: When Statins Won't Do! *Drugs* **2016**, *76*, 1175–1190. [CrossRef] [PubMed]
7. Guo, R.; Li, L.; Su, J.; Li, S.; Duncan, S.E.; Liu, Z.; Fan, G. Pharmacological Activity and Mechanism of Tanshinone IIA in Related Diseases. *Drug Des. Devel Ther.* **2020**, *14*, 4735–4748. [CrossRef]
8. Ren, J.; Fu, L.; Nile, S.H.; Zhang, J.; Kai, G. Salvia miltiorrhiza in Treating Cardiovascular Diseases: A Review on Its Pharmacological and Clinical Applications. *Front. Pharmacol.* **2019**, *10*, 753. [CrossRef]
9. Mahalakshmi, B.; Huang, C.Y.; Lee, S.D.; Maurya, N.; Kiefer, R.; Kumar, V.B. Review of Danshen: From its metabolism to possible mechanisms of its biological activities. *J. Funct. Foods* **2021**, *85*, 104613. [CrossRef]
10. Chao, C. Clinical Efficacy of Compound Danshen Dropping Pills Combined with Atorvastatin in Treatment of Coronary Heart Disease Complicated with Hyperlipidemia and Its Effect on Serum TRAIL. *Fujian J. Tradit. Chin. Med.* **2022**, *53*, 7–9.
11. Chaolan, D. Clinical Efficacy of Coronary Heart Disease Combined Hyperlipidemia CSDP Joint Alto-generation Statin Therapy. *Mod. Diagn. Treat.* **2015**, *26*, 1754–1755.
12. Xingping, G. Clinical Study on Compound Danshen Dropping Pill Combined with Simvastatin Treatment of Hyperlipidemia. *China J. Chin. Med.* **2013**, *28*, 1575–1576.
13. Hongmei, G. Clinical Observation on Compound Danshen Dropping Pills Combined with Simvastatin in the Treatment of Coronary Heart Disease Combined with Hyperlipidemia. *J. North. Pharm.* **2016**, *13*, 47.
14. Ji-bin, J.; Jing-dong, X. Study the Clinical 64 Cases Efficacy of Treatment of Coronary Heartdisease with Hyperlipidemia Using Simvastatin with Compound Danshen Dripping Pills. *Smart Healthc.* **2019**, *5*, 106–108.
15. Shujia, L. Clinical Effiect of Compound Danshen Dropping Pills Combined with Rosuvastatin or Rosuvastatin Only in the Treatment of Coronary Heart Disease Complicated with Hyperlipidemia. *J. Math. Med.* **2017**, *30*, 1663–1664.
16. Si-si, L.; Yi, Z.; Min, Z. Clinical analysis of salvia miltiorrhiza combined with rosuvastatin calcium in the treatment of coronary heart disease complicated with hyperlipidemia. *J. North. Sichuan Med. Coll.* **2019**, *34*, 448–451.
17. Tao, L. A comparative study of combination therapy with Fufang Danshen drop pill and simvastatin and simvastatin alone oncoronary artery disease with hyperlipidemia. *J. Chin. Res. Hosp.* **2017**, *4*, 20–23.
18. Jun, L.; Zhen, W.; Lingyue, K. Clinical assessment on treatment of hyperlipidemia by compositie salviae dropping pill combined with Simvastatin. *Hebei J. Tradit. Chin. Med.* **2015**, *37*, 93–94, 158.
19. Qingan, L. Clinical analysis of 102 cases of coronary heart disease complicated with hyperlipidemia treated with Simvastatin (Shujiangzhi) combined with Compound Danshen Dropping Pills. *J. Med. Theory Pract.* **2012**, *25*, 1583–1584.
20. Yongchun, L. Analysis of the clinical efficacy of Compound Danshen Dropping Pills combined with rosuvastatin calcium in the treatment of coronary heart disease complicated with hyperlipidemia. *China Pract. Med.* **2014**, *9*, 183–184.
21. Chunlan, R. The clinical efficacy of compound Danshen dropping pill treatment on 80 cases with angina pectoris complicated with hyperlipidemia. *China Med. Pharm.* **2014**, *4*, 86–87, 97.
22. Ting, W.; Jia, Z.; Jiang, L.; Kun, F.; Ye, S. Effect of simvastatin combined with Compound Danshen Dropping Pills on blood lipids and lipid ratio in patients with hyperlipidemia. *Shaanxi Med. J.* **2016**, *45*, 1228–1230.

23. Zichun, W. Clinical efficacy of Compound Danshen Dropping Pills combined with rosuvastatin calcium in the treatment of coronary heart disease combined with hyperlipidemia. *Chin. J. Clin. Ration. Drug Use* **2020**, *13*, 32–33.
24. Ling, X. Curative effect of Fufang Danshen Gutta Pills combining rosuvastatin on coronary heart diseasecomplicating hyperlipidemia. *Chin. J. Evid. Based Cardiovasc. Med.* **2014**, *6*, 99–101.
25. Di, Y. Study on the improvement effect of Compound Danshen Dropping Pills combined with atorvastatin on vascular endothelial function and lipid metabolism disorders in patients with coronary heart disease and hyperlipidemia. *Jilin Med. J.* **2022**, *43*, 1266–1268.
26. Sha, Y.; Ai-ping, J.; Jie, G.; Cui-ling, Y.; Qian-rong, Z. Effects of Compound Danshen Dripping Pills and Trimetazidine on the Hemorheology and Blood Lipid of Patients with Coronary Heart Disease Complicated with Hyperlipidemia. *Progress. Mod. Biomed.* **2019**, *19*, 868–871.
27. Hai-long, Z.; Bao-cheng, Z.; Yu-peng, L. Therapeutic effect of compound Danshen dripping pills combined atorvastatin on coronary heart disease complicated hyperlipidemia and its influence on serum TRAIL level. *Chin. J. Cardiovasc. Rehabil. Med.* **2019**, *28*, 360–364.
28. Mingyan, Z.; Xiaogeng, Y.; Fanchao, M. Clinical Observation on Lovastatin Combined with Compound Danshen Dropping Pills in the Treatment of Hyperlipidemia. *Chin. J. Pract. Nerv. Dis.* **2011**, *14*, 75–76.
29. Ming, Z. Clinical efficacy of Compound Danshen Dropping Pills combined with rosuvastatin in the treatment of patients with coronary heart disease and hyperlipidemia. *World Latest Med. Inf.* **2018**, *18*, 147–148.
30. Lewei, Z. Curative Effect Observation of Atorvastatin and Fufang Danshen Diwan in the Treatment of Old-age and High Fat Blood Disease. *Asia-Pac. Tradit. Med.* **2014**, *10*, 102–103.
31. Xiaoyang, L.; Xiaohua, G. Clinical analysis of 25 cases of hyperlipidemia treated with Compound Danshen Dropping Pills. *J. Emerg. Tradit. Chin. Med.* **2003**, *12*, 338.
32. van Poppel, P.C.; Breedveld, P.; Abbink, E.J.; Roelofs, H.; van Heerde, W.; Smits, P.; Lin, W.; Tan, A.H.; Russel, F.G.; Donders, R. Salvia miltiorrhiza root water-extract (danshen) has no beneficial effect on cardiovascular risk factors. A randomized double-blind cross-over trial. *PLoS ONE* **2015**, *10*, e0128695. [CrossRef] [PubMed]
33. Guangming, B.; Yuanlin, L.; Hongru, Z.; Shixiang, H.; Jie, H.; Yuhai, M.; Daohua, G.; Guíe, C. Observation on the long-term efficacy of Salvia miltiorrhiza on urinary microalbumin in patients with refractory hypertension combined with hyperlipidemia. *Chin. J. Exp. Tradit. Med. Formulae* **2010**, *16*, 244–245, 249.
34. Huina, Z. Evaluation of the clinical effect of Compound Danshen Dropping Pills in the treatment of coronary heart disease and angina pectoris combined with hyperlipidemia and its effect on blood lipid levels. *Cardiovasc. Dis. Electron. J. Integr. Tradit. Chin. West. Med.* **2020**, *8*, 86, 89.
35. Shi-jun, Z.; Ze-xiong, C.; You-wu, L.; Jian, Q.; Yong-hua, C.; Su-ling, L. Effection of compositie salviae dropping pill on hyperlipemia patients with phlegm and blood stasis syndrome. *China J. Chin. Mater. Medica* **2007**, *32*, 440–443.
36. Qinghua, H.; Xiaohong, C.; Jiyuan, L. Effects of compound Danshen Dropping Pills on lipoprotein(a) and blood lipids in patients with coronary heart disease and hyperlipidemia. *Chin. J. Integr. Tradit. West. Med.* **2000**, *20*, 514.
37. Yuanyuan, L.; Meng, Y. Observation on the efficacy of Compound Danshen Dropping Pills combined with atorvastatin in the treatment of hyperlipidemia. *Guide China Med.* **2013**, *11*, 517–518.
38. Xiaojian, P. Clinical Observation on Compound Danshen Dropping Pills Combined with Atorvastatin Calcium Tablets in the Treatment of Elderly Patients with Hyperlipidemia. *J. New Chin. Med.* **2015**, *47*, 32–34.
39. Xiangshi, Y.; Lihong, C. Observation on the efficacy of Compound Danshen Dropping Pills in the treatment of 80 cases of coronary heart disease, angina pectoris and hyperlipidemia. *Chin. Community Dr.* **2012**, *14*, 184–185.
40. Jianying, M. Clinical Observation on the Treatment of Hyperlipidemia with Compound Danshen Dropping Pills. *Gianjin J. Tradit. Chin. Med.* **1998**, *15*, 24.
41. Xiangdong, D.; Jingming, S. The efficacy of Compound Danshen Dropping Pills in the treatment of hyperlipidemia and its effect on platelet aggregation. *Chin. J. Hemorheol.* **2002**, *12*, 328–329.
42. Shikui, C.; Guohong, L. 69 cases of hyperlipidemia treated with Compound Danshen Dropping Pills. *Mod. J. Integr. Tradit. Chin. West. Med.* **2002**, *11*, 1029–1030.
43. Hu, P.; Liang, Q.L.; Luo, G.A.; Zhao, Z.Z.; Jiang, Z.H. Multi-component HPLC fingerprinting of Radix Salviae Miltiorrhizae and its LC-MS-MS identification. *Chem. Pharm. Bull.* **2005**, *53*, 677–683. [CrossRef] [PubMed]
44. Grundy, S.M.; Stone, N.J.; Bailey, A.L.; Beam, C.; Birtcher, K.K.; Blumenthal, R.S.; Braun, L.T.; de Ferranti, S.; Faiella-Tommasino, J.; Forman, D.E.; et al. 2018 AHA/ACC/AACVPR/AAPA/ABC/ACPM/ADA/AGS/APhA/ASPC/NLA/PCNA Guideline on the Management of Blood Cholesterol: A Report of the American College of Cardiology/American Heart Association Task Force on Clinical Practice Guidelines. *J. Am. Coll. Cardiol.* **2019**, *73*, e285–e350. [CrossRef]
45. Ridker, P.M.; Pradhan, A.; MacFadyen, J.G.; Libby, P.; Glynn, R.J. Cardiovascular benefits and diabetes risks of statin therapy in primary prevention: An analysis from the JUPITER trial. *Lancet* **2012**, *380*, 565–571. [CrossRef]

46. Amarenco, P.; Kim, J.S.; Labreuche, J.; Charles, H.; Abtan, J.; Bejot, Y.; Cabrejo, L.; Cha, J.K.; Ducrocq, G.; Giroud, M.; et al. A Comparison of Two LDL Cholesterol Targets after Ischemic Stroke. *N. Engl. J. Med.* **2020**, *382*, 9. [CrossRef]
47. Fox, K.M.; Tai, M.H.; Kostev, K.; Hatz, M.; Qian, Y.; Laufs, U. Treatment patterns and low-density lipoprotein cholesterol (LDL-C) goal attainment among patients receiving high- or moderate-intensity statins. *Clin. Res. Cardiol.* **2018**, *107*, 380–388. [CrossRef]
48. Li, Z.M.; Xu, S.W.; Liu, P.Q. Salvia miltiorrhizaBurge (Danshen): A golden herbal medicine in cardiovascular therapeutics. *Acta Pharmacol. Sin.* **2018**, *39*, 802–824. [CrossRef]
49. Shi, M.J.; Dong, B.S.; Yang, W.N.; Su, S.B.; Zhang, H. Preventive and therapeutic role of Tanshinone IIA in hepatology. *Biomed. Pharmacother.* **2019**, *112*, 108676. [CrossRef]
50. Zhao, Y.; Qian, Y.; Sun, Z.; Shen, X.; Cai, Y.; Li, L.; Wang, Z. Role of PI3K in the Progression and Regression of Atherosclerosis. *Front. Pharmacol.* **2021**, *12*, 632378. [CrossRef]
51. Linton, M.F.; Moslehi, J.J.; Babaev, V.R. Akt Signaling in Macrophage Polarization, Survival, and Atherosclerosis. *Int. J. Mol. Sci.* **2019**, *20*, 2703. [CrossRef] [PubMed]
52. Guo, H.; Cui, H.; Peng, X.; Fang, J.; Zuo, Z.; Deng, J.; Wang, X.; Wu, B.; Chen, K.; Deng, J. Modulation of the PI3K/Akt Pathway and Bcl-2 Family Proteins Involved in Chicken's Tubular Apoptosis Induced by Nickel Chloride ($NiCl_2$). *Int. J. Mol. Sci.* **2015**, *16*, 22989–23011. [CrossRef] [PubMed]
53. Huang, J.S.; Guh, J.Y.; Chen, H.C.; Hung, W.C.; Lai, Y.H.; Chuang, L.Y. Role of receptor for advanced glycation end-product (RAGE) and the JAK/STAT-signaling pathway in AGE-induced collagen production in NRK-49F cells. *J. Cell Biochem.* **2001**, *81*, 102–113. [CrossRef] [PubMed]
54. Zhang, X.; Chen, S.; Yin, G.; Liang, P.; Feng, Y.; Yu, W.; Meng, D.; Liu, H.; Zhang, F. The Role of JAK/STAT Signaling Pathway and Its Downstream Influencing Factors in the Treatment of Atherosclerosis. *J. Cardiovasc. Pharmacol. Ther.* **2024**, *29*, 10742484241248046. [CrossRef]
55. Ramasamy, R.; Yan, S.F.; Schmidt, A.M. Receptor for AGE (RAGE): Signaling mechanisms in the pathogenesis of diabetes and its complications. *Ann. N. Y. Acad. Sci.* **2011**, *1243*, 88–102. [CrossRef]
56. El-Far, A.H.; Sroga, G.; Jaouni, S.K.A.; Mousa, S.A. Role and Mechanisms of RAGE-Ligand Complexes and RAGE-Inhibitors in Cancer Progression. *Int. J. Mol. Sci.* **2020**, *21*, 3613. [CrossRef]
57. Baldini, C.; Moriconi, F.R.; Galimberti, S.; Libby, P.; De Caterina, R. The JAK-STAT pathway: An emerging target for cardiovascular disease in rheumatoid arthritis and myeloproliferative neoplasms. *Eur. Heart J.* **2021**, *42*, 4389–4400. [CrossRef]
58. Feingold, K.R. Lipid and Lipoprotein Metabolism. *Endocrinol. Metab. Clin. N. Am.* **2022**, *51*, 437–458. [CrossRef]
59. Lennernas, H.; Fager, G. Pharmacodynamics and pharmacokinetics of the HMG-CoA reductase inhibitors. Similarities and differences. *Clin. Pharmacokinet.* **1997**, *32*, 403–425. [CrossRef]
60. Tabernero, L.; Rodwell, V.W.; Stauffacher, C.V. Crystal structure of a statin bound to a class II hydroxymethylglutaryl-CoA reductase. *J. Biol. Chem.* **2003**, *278*, 19933–19938. [CrossRef]
61. Zhang, F.; Yu, W.; Hargrove, J.L.; Greenspan, P.; Dean, R.G.; Taylor, E.W.; Hartle, D.K. Inhibition of TNF-alpha induced ICAM-1, VCAM-1 and E-selectin expression by selenium. *Atherosclerosis* **2002**, *161*, 381–386. [CrossRef] [PubMed]
62. Mantovani, A.; Garlanda, C.; Locati, M. Macrophage diversity and polarization in atherosclerosis: A question of balance. *Arterioscler. Thromb. Vasc. Biol.* **2009**, *29*, 1419–1423. [CrossRef] [PubMed]
63. van Raalte, D.H.; Li, M.; Pritchard, P.H.; Wasan, K.M. Peroxisome proliferator-activated receptor (PPAR)-alpha: A pharmacological target with a promising future. *Pharm. Res.* **2004**, *21*, 1531–1538. [CrossRef] [PubMed]
64. Chandran, U.; Mehendale, N.; Patil, S.; Chaguturu, R.; Patwardhan, B. *Innovative Approaches in Drug Discovery. Ethnopharmacology, Systems Biology and Holistic Targeting Network Pharmacology*, 1st ed.; Academic Press: Cambridge, MA, USA, 2017; pp. 127–164. [CrossRef]

Disclaimer/Publisher's Note: The statements, opinions and data contained in all publications are solely those of the individual author(s) and contributor(s) and not of MDPI and/or the editor(s). MDPI and/or the editor(s) disclaim responsibility for any injury to people or property resulting from any ideas, methods, instructions or products referred to in the content.

pharmaceuticals

MDPI

Article

Potential Anti-Obesity Effect of Hazel Leaf Extract in Mice and Network Pharmacology of Selected Polyphenols

Jiarui Zhao, Aikebaier Alimu, Yvmo Li, Zhi Lin, Jun Li, Xinhe Wang, Yuchen Wang, Guangfu Lv, He Lin * and Zhe Lin *

College of Pharmacy, Changchun University of Chinese Medicine, Changchun 130117, China; zjr0852@163.com (J.Z.); aikebaieralimu001@163.com (A.A.); 13596272257@163.com (Y.L.); linzhi1@jxutcm.edu.cn (Z.L.); 18224088551@163.com (J.L.); m17704303262_1@163.com (X.W.); wycscar@hotmail.com (Y.W.); message219@163.com (G.L.)

* Correspondence: linhe@ccucm.edu.cn (H.L.); linzhe@ccucm.edu.cn (Z.L.)

Abstract: Background: Obesity is gradually becoming a widespread health problem, and treatment using natural compounds has seen an increasing trend. As a by-product of hazelnut, hazel leaf is usually disposed of as waste, but it is widely used in traditional and folk medicines around the world. **Aim of this study:** Based on previous studies, the effects of the regulation of lipid metabolism and the mechanism of hazel leaf polyphenol extraction obesity were investigated. **Methods:** In this study, a high-fat diet-fed mouse model of obesity and 3T3-L1 preadipocytes were established. The ameliorative effects of the hazel leaf polyphenol extract on obesity and the regulating lipid metabolisms were explored based on network pharmacology, gut microbiota, and molecular docking. **Results:** Network pharmacology showed that hazel leaf polyphenols may play a role by targeting key targets, including PPARγ, and regulating the PPAR signaling pathway. They significantly improved body weight gain, the liver index, and adiposity and lipid levels; regulated the gut microbiota and short-chain fatty acid contents; down-regulated the expression of lipid synthesis proteins SREBP1c, PPARγ, and C/EBP-α; and up-regulated the expression of p-AMPK in obese mice. They inhibited the differentiation of 3T3-L1 cells, and the expression of related proteins is consistent with the results in vivo. The molecular docking results indicated that gallic acid, quercetin-3-O-beta-D-glucopyranoside, quercetin, myricetin, and luteolin-7-O-glucoside in the hazel leaf polyphenol extract had strong binding activities with PPARγ, C/EBP-α, and AMPK. **Conclusions:** The results demonstrate that the hazel leaf polyphenol extract can improve obesity by regulating lipid metabolism, which provides a valuable basis for developing health products made from hazel leaf polyphenols in the future.

Keywords: hazel leaf; polyphenols; obesity; network pharmacology; gut microbiota; molecular docking

1. Introduction

According to the World Health Organization, the worldwide prevalence of obesity more than doubled between 1990 and 2022. In 2022, 2.5 billion adults (18 years and older) were classified as being overweight (43% of men and 44% of women); of these, 890 million were living with obesity. Additionally, 8% of children and adolescents were living with obesity (160 million young people) [1]. Obesity has become a global health problem and a major risk factor for type 2 diabetes, fatty liver, and cancer [2]. The underlying cause of obesity and overweight is an energy imbalance between calorie intake and consumption. Although the exact mechanism has not been elucidated, it is thought that the pathogenesis of obesity-related complications involves the abnormal production of adipokines byadipocytes [3].

Recently, the consumption of herbs, natural crude extracts, and natural compounds has been shown to be a promising, effective, and safe strategy for preventing obesity and related metabolic disorders [4]. *Paeonia lactiflora* root regulated lipolysis and thermogenesis, thus inhibiting obesity [5]. Rosemary inhibited lipid deposition and prevented inflammation and oxidative stress associated with obesity [6].Carambola leaf extracts reduced obesity induced by a high-fat diet (HFD) and improved hyperlipidemia, insulin resistance, and oxidative stress [7]. Mulberry leaf extract was able to inhibit weight gain and fat cell volume increase in HFD-induced obese mice [8]. Lychee polyphenols can reduce the activity of lipase and control the accumulation of adipose tissue in rats [9]. Green tea polyphenols significantly reduced the food intake of female mice, resulting in weight loss [10]. Polyphenols such as catechins contained in fermented green tea can decrease lipid content by inhibiting pancreatic lipase and increasing energy expenditure [11]. Ellagic acid in pomegranate peel can inhibit lipid synthesis by reducing the expression of FAS in 3T3-L1 adipocytes [12]. As one of the important natural products, polyphenols have received more attention in the prevention and treatment of obesity due to their good efficacy and safety.

Hazel leaf, a by-product of hazelnuts, is often disposed of as waste, but it is widely used in traditional and folk medicines around the world. In Iran, it is used as a liver tonic and also for hemorrhoids, varicose veins, phlebitis, and mild edema. In Sweden, hazel leaf and bark are used to treat pain, hemorrhoids, phlebitis, and other conditions. Galenic preparations of hazel leaves are also used to relieve ulcers and oropharyngeal infections and have mild anti-dysentery, anti-fungal, and scarring properties [13]. Studies have shown that hazel leaves are rich in polyphenols [14]. We previously extracted polyphenols from hazel leaves (*Corylus heterophylla* Fisch × *Corylus avellana* L.) and conducted a preliminary characterization of their composition, demonstrating their antioxidant and anti-inflammatory capabilities and preliminarily predicting the treatment of lipid metabolism and atherosclerotic-related diseases [15].

Based on these results, the present study combined network pharmacology and intestinal microbiota to preliminarily investigate the modulatory effects and mechanisms of a hazel leaf polyphenol extract (ZP) on a high-fat diet-induced obesity mouse model and 3T3-L1 cells, and molecular docking predicted the key active components of the hazel leaf polyphenol extract. This may provide a valuable basis for the exploitation and application of hazelnut leaves and their polyphenols as functional ingredients.

2. Results

2.1. Results of Network Pharmacology

The components of hazel leaf polyphenol extract (Table S1) were imported into the Swiss Target Prediction Database to obtain the targets, and a total of 539 targets were obtained after the screening and removal of duplicates (Table S2). The "obesity" targets mined from Gene Cards and other databases were screened and summarized, and duplicates were removed to obtain 1136 disease targets (Table S3). The intersection of the drug and disease targets was mapped, and 144 intersecting targets were obtained (Table S4). Twenty-seven core targets were obtained by Cytoscape analysis, among which the most important targets were ALB, TNF, AKT1, PPARG, and SRC. They are sorted by degree value in Figure 1a–c. GO analysis revealed 4983 biological processes (BPs), 377 cellular components (CCs), and 854 targets (CCs), with the top 10shown in Figure 1d. The KEGG-enriched pathways obtained include insulin resistance, the PPAR signaling pathway, and the apelin signaling pathway (Figure 1e). Among them, the PPAR signaling pathway is closely related to lipid metabolism and fat synthesis. Therefore, in vitro and in vivo experiments were further used to verify the above mechanisms.

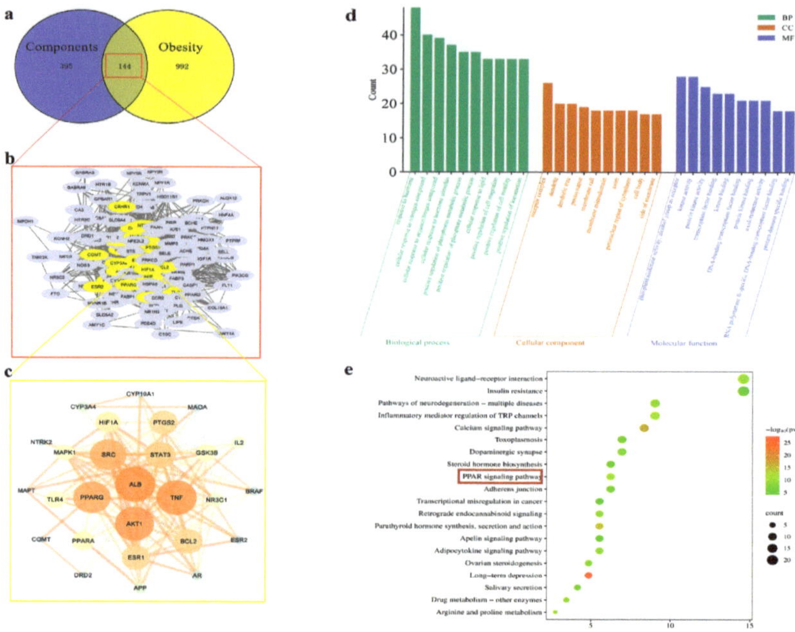

Figure 1. Analysis of network pharmacology. (**a**) Venn diagram analysis of targets of components with obesity. (**b**) A total of 144 intersecting targets. (**c**) The core target obtained from the screening (a larger circle represents a larger degree value of the target). (**d**) GO functional enrichment analysis. (**e**) KEGG pathway analysis.

2.2. Effects of ZP on Weight Gain, Food Intake, Liver Index, and Fat Content in Obese Mice

During the experiment, the weight of mice was recorded weekly, and the weight gain of the control group was stable and slow. Except for the control group, the weight of the other groups increased significantly over the six weeks. However, after the 7th week, the weight gain in the orlistat group and ZP groups began to decrease (Figure 2b). At the end of the experiment (12th week), the weight gain of mice in the orlistat group and ZP groups was significantly lower than that in the model group ($p < 0.01$) (Figure 2a,c). This indicated that hazel leaf polyphenol extract could slow down the mice's weight gain on a high-fat diet. Meanwhile, the food intake of mice was recorded during the experiment. There was no significant difference in the food intake of the orlistat group and ZP groups compared with the model group (Figure 2d), indicating that hazel leaf polyphenol extract did not affect the food intake of mice and may not have exerted weight loss by suppressing appetite.

As shown in Figure 2e–g, compared with the control group, the liver index and fat content of mice in the model group were significantly increased ($p < 0.01$), while the liver index and fat content of mice in the orlistat group and ZP groups (500 and 250 mg/kg) were significantly lower than the model group ($p < 0.05$, $p < 0.01$). These results suggested that the high-fat diet caused the accumulation of lipids in liver tissue and increased body fat content, while hazel leaf polyphenol extract could attenuate the accumulation of lipids.

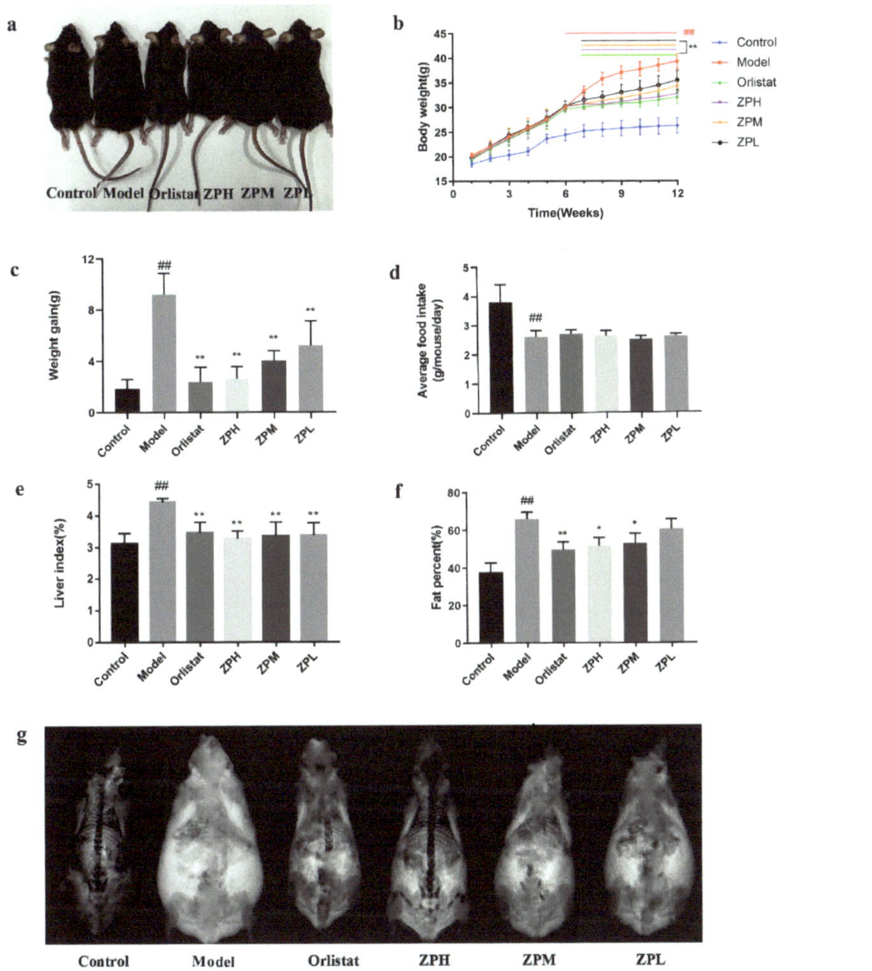

Figure 2. Effects of ZP on obese mice. (**a**) Morphological photographs of mice. (**b**) Mice body weight. (**c**) Weight gain. (**d**) Average food intake of mice. (**e**) Mice liver index. (**f**) Mice fat content. (**g**) Mice fat scan. Data are presented as means ± SD. ## $p < 0.01$ compared to control group. * $p < 0.05$, ** $p < 0.01$ compared to model group.

2.3. Effects of ZP on Blood Lipid Levels in Obese Mice

Blood lipid levels are one of the important indicators of metabolism in the body. As shown in Figure 3a–d, compared with the control group, the serum levels of TG, T-CHO, and LDL-C in the model group were significantly increased ($p < 0.01$), and the level of HDL-C was significantly reduced ($p < 0.01$).However, TG, T-CHO, and LDL-C levels were obviously lower in the ZP groups (500 and 250 mg/kg), and HDL-C levels were obviously higher in the ZPH group (500 mg/kg) compared to the model group ($p < 0.05$, $p < 0.01$).The blood lipid profile of other ZP dose groups was also improved, but there was no significant difference. These results revealed that hazel leaf polyphenol extract ameliorated the abnormal blood lipid levels of obese mice induced by a high-fat diet.

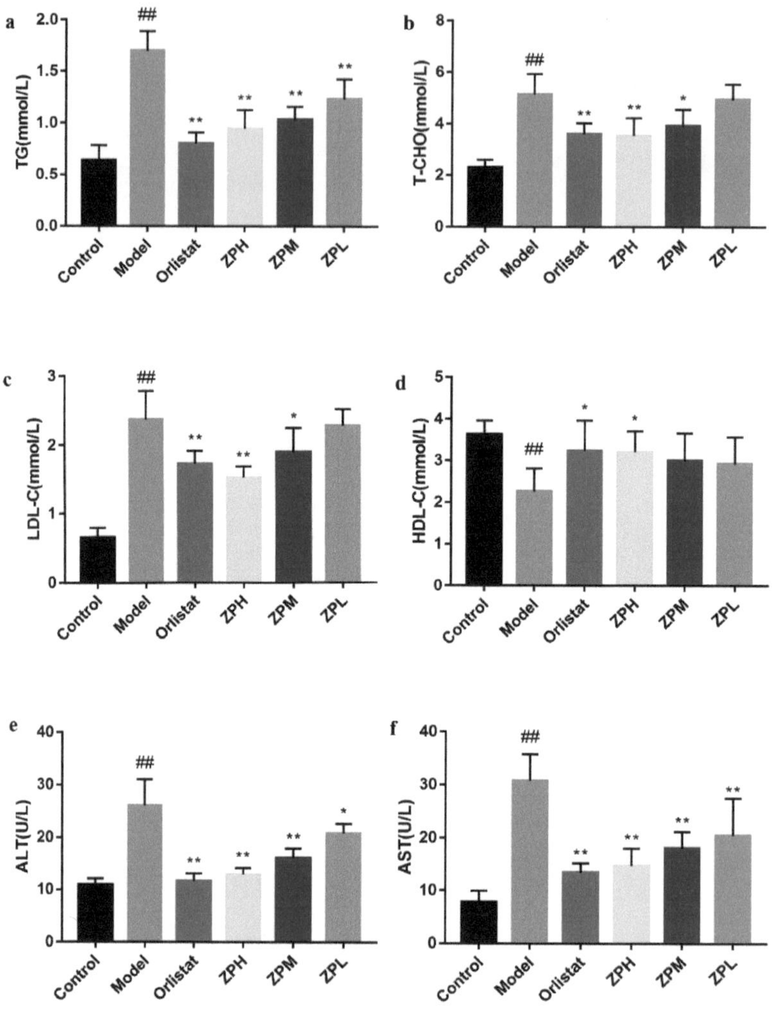

Figure 3. Effects of ZP on blood lipid levels, ALT, and AST in obese mice. (**a**) TG. (**b**) T-CHO. (**c**) LDL-C. (**d**) HDL-C. (**e**) ALT. (**f**) AST. Data are presented as means ± SD. ## $p < 0.01$ compared to control group. * $p < 0.05$, ** $p < 0.01$ compared to model group.

2.4. Effects of ZP on ALT and AST in Obese Mice

ALT and AST are commonly used as markers of liver damage, and serum ALT is the most sensitive indicator of liver fat accumulation. In this experiment, the serum levels of ALT and AST in the model group were significantly elevated compared to the control group ($p < 0.01$), demonstrating that obesity caused liver damage (Figure 3e,f). Compared with the model group, serum ALT and AST levels in ZP groups (500, 250, and 125 mg/kg) were significantly lower in a dose-dependent manner ($p < 0.05$, $p < 0.01$). This suggested that hazel leaf polyphenol extract could reduce liver lipid accumulation and alleviate liver damage in dietary obese mice.

2.5. Effects of ZP on Histological in Obese Mice

Mice liver HE and oil red staining showed that the liver tissue structure of mice in the control group was complete, the morphological size of hepatocytes was normal, the cell arrangement was uniform and compact, no steatosis was observed, and no lipid droplets appeared in liver cells. Compared with the control group, the liver cells in the model group were arranged and scattered, vacuoles of different sizes and numbers appeared, and lipid accumulation occurred in the liver to form lipid droplets. Compared with the model group, the orlistat group, ZPH group (500 mg/kg), ZPM group (250 mg/kg), and ZPL group (125 mg/kg) visibly reduced the number of cell vacuoles and lipid droplets (Figure 4).

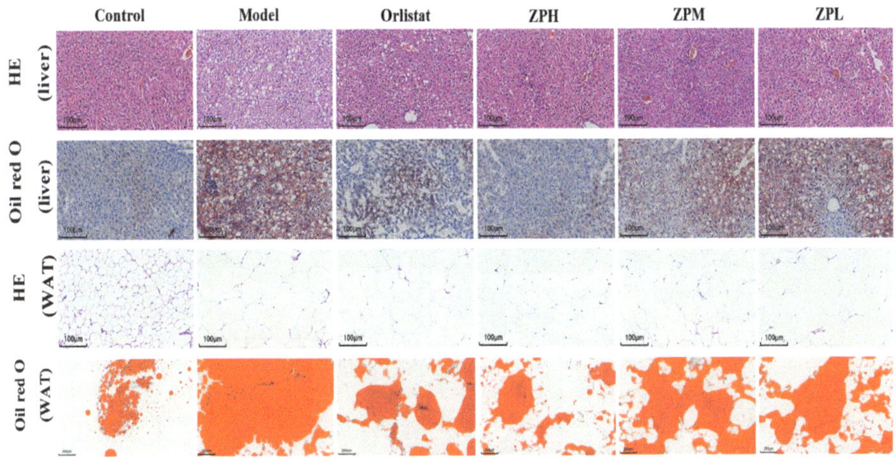

Figure 4. Representative images of liver and perirenal white fat (WAT) H&E staining and oil red O staining in different groups of mice. (HE staining of liver, WAT, and oil red O staining of liver scale bar, 100 μm; oil red O staining of WAT scale bar, 200 μm).

Mice perirenal fat HE and oil red staining showed that the adipocytes in the control group were small in size, clear in morphology, and complete in structure, and the lipid droplet proportion was reduced. Compared with the control group, the fat cell volume in the model group was significantly increased, the cell size was uneven, the cell arrangement was irregular, and the lipid droplet proportion was larger. In the same field of view, the number of fat cells in the model group significantly decreased with the increase in size. Compared with the model group, the volume of adipocytes decreased to varying degrees in the orlistat group, ZPH group (500 mg/kg), ZPM group (250 mg/kg), and ZPL group (125 mg/kg), and the number of cells in the same field of view increased. The area of lipid droplets also clearly decreased. The results indicated that hazel leaf polyphenol extract alleviated the accumulation of lipid droplets in the liver and the increase in adipose tissue in obese mice.

2.6. *Effects of ZP on Gut Microbes in Obese Mice*

Alpha diversity reflects the species richness and species diversity of a single sample. The Ace andChao1 indices measure species richness, which is the number of species. Compared with the control group, the Ace index andChao1 index were decreased significantly in the model group($p < 0.01$), while the indices of ZPH, ZPM, and ZPL groups were significantly higher than those of the model group($p < 0.01$)in a dose-dependent manner(Figure 5a,b).The results showed that the gut microbes were decreased in high-fat diet mice and increased after hazel leaf polyphenol extract intervention. The Shannon index and Simpson index are used to measure species diversity and are influenced by species richness and species uniformity in the sample community. Compared with the control group, the Shannon index and Simpson index were significantly reduced in the model group ($p < 0.01$), while the indices of the ZPH and ZPM groups (500 and 250 mg/kg) were significantly higher than those of the model group($p < 0.05$, $p < 0.01$) (Figure 5c,d). The results showed that the high-fat diet led to a decrease in gut microbes species diversity in mice, which was restored by hazel leaf polyphenol extract.

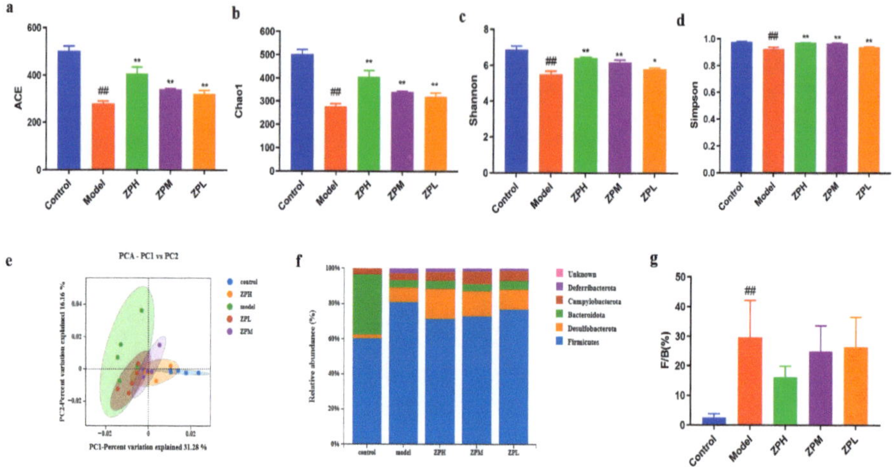

Figure 5. Effects of ZP on gut microbes in obese mice. (**a**) ACE. (**b**) Chao1. (**c**) Shannon. (**d**) Simpson. (**e**) Principal component analysis (PCA). (**f**) Phylum analysis. (**g**) F/B. ## $p < 0.01$ compared to control group. * $p < 0.05$, ** $p < 0.01$ compared to model group.

Principal component analysis (PCA, β diversity analysis) can reflect the similarity of samples between different groups, and the closeness of two samples is positively correlated with the similarity of their composition. As shown in Figure 5e, the samples of the model group and the control group were completely separated, indicating that a high-fat diet caused great changes in the gut microbial composition of mice. However, the gut microbiota of mice with ZP treatment was more similar to that of the control group, and this trend was dose-dependent. These results indicated that hazel leaf polyphenol extracts regulated the gut microbial composition of obese mice to a normal level.

In this study, the phyla with an abundance ratio greater than 1% were analyzed. The dominant phyla were *Bacteroidetes*, *Firmicutes*, *Desulfobacterota*, *Campylobacterota*, and *Deferribacterota*, among which *Firmicutes* and *Bacteroidetes* accounted for the highest proportion [16]. The *Firmicutes* phylum proportion in the model group was higher than that in the control group, and the proportion of *Firmicutes* decreased after ZP intervention (Figure 5f). The *Firmicutes*/*Bacteroides* (F/B) value, as an important indicator used to measure intestinal homeostasis, is closely related to body lipid metabolism [17]. The relative abundance of *Firmicutes* and *Bacteroides* in the gut flora may be biomarkers indicating susceptibility to obesity. Compared to the gut microbiota of healthy individuals, obese individuals have higher F/B values. Studies have shown that an increase in F/B value is beneficial for energy absorption and fat storage, leading to weight gain and obesity [18]. Our results showed that the F/B value of the model group was significantly higher than that of the control group. The F/B values of ZPH and ZPM groups (500 and 250 mg/kg) were decreased significantly compared with the model group (Figure 5g). This suggested that hazel leaf polyphenol extract could affect the structure of the gut flora of obese mice at the phylum level, which is similar to previous studies [19].

At the genus level, compared with the control group, the relative abundance of unclassified_*Lachnospiraceae*, unclassified_*Oscillospiraceae*, and unclassified_*Desulfovibrionaceae* genus in the model group was apparently increased, while the relative abundance of *Ligilactobacillus* was apparently decreased. After the intervention of ZP, the relative abundance of unclassified_*Lachnospiraceae*, unclassified_*Oscillospiraceae*, and unclassified_*Desulfovibrionaceae* was decreased compared to the model group, while the relative abundance of *Ligilactobacillus* was increased (Figure 6a).

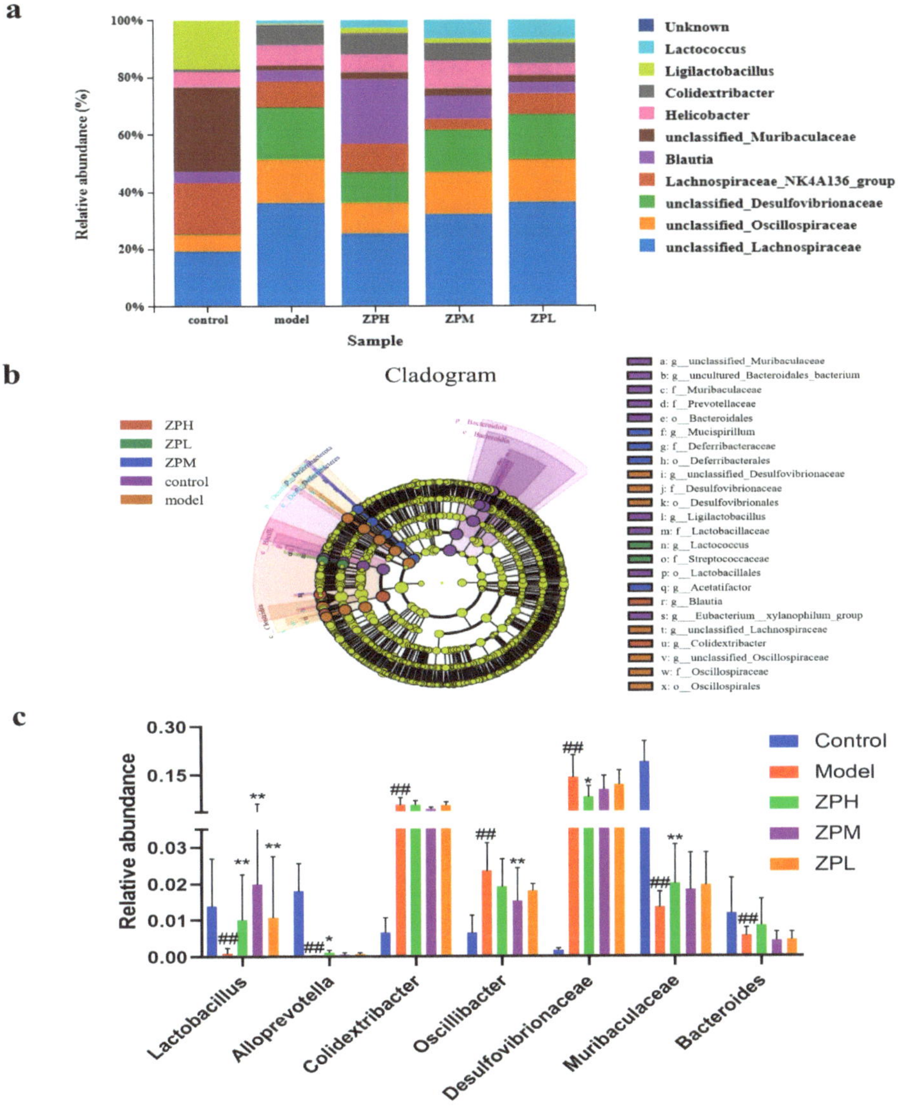

Figure 6. Effects of ZP on gut microbes at genus level in obese mice. (**a**) Genu analysis. (**b**) Cladogram. (**c**) Relative abundance. ## $p < 0.01$ compared to control group. * $p < 0.05$, ** $p < 0.01$ compared to model group.

LEfSe analyses are used to assess the magnitude of colony abundance on differential effects, looking for statistically different species between groups. As can be seen in Figure 6b, different groups have different dominant genera. As shown in Figure 6c, compared with the control group, the relative abundance of *Lactobacillus*, *Alloprevotella*, *Muribacuaceae*, and *Bacteroides* in the model group was decreased significantly ($p < 0.01$), and the relative abundance of *Colidextribacter*, *Oscillibacter*, and *Desulfovibrionaceae* was increased significantly ($p < 0.01$).

Compared with the model group, *Lactobacillus* and *Muribacuaceae* in the ZPH group (500 mg/kg) were significantly increased ($p < 0.01$), and *Desulfovibrionaceae* was significantly lower ($p < 0.05$). This part of the results coincided with the aforementioned changes in the diversity of the flora.

2.7. Effects of ZP on SCFAs Content in Obese Mice

Short-chain fatty acids (SCFAs) are an energy source for gastrointestinal epithelial cells, which affect intestinal mucosal immunity and barrier integrity. They have a variety of biological activities, including improving insulin sensitivity, inducing gluconeogenesis, and weight loss [20]. Compared with the control group, the contents of acetic acid, propionic acid, butyric acid, isovaleric acid, and valeric acid in the model group were remarkedly reduced ($p < 0.05$, $p < 0.01$).Compared with the model group, the contents of acetic acid, butyric acid, valeric acid, and isovaleric acid were remarkedly increased in the ZPH and ZPM groups (500 and 250 mg/kg) ($p < 0.05$, $p < 0.01$) (Figure 7a–g). The results indicated that the hazel leaf polyphenol extract could promote the contents of SCFAs in the obese mice gut.

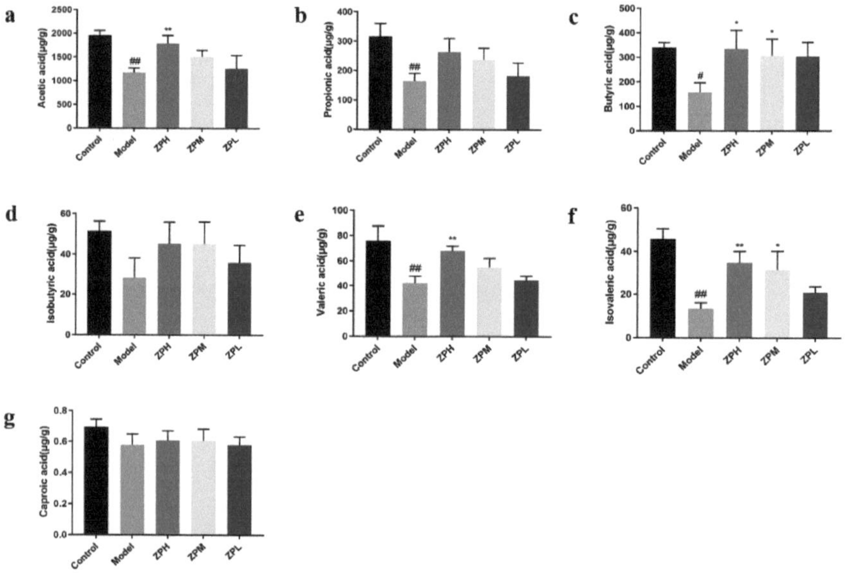

Figure 7. Effects of ZP on short-chain fatty acid content. (**a**) Acetic acid. (**b**) Propionic acid. (**c**) Butyric acid. (**d**) Isobutyric acid. (**e**) Valeric acid. (**f**) Isovaleric acid. (**g**) Caproic acid. # $p < 0.05$, ## $p < 0.01$ compared to control group. * $p < 0.05$, ** $p < 0.01$ compared to model group.

2.8. Effects of ZP on White Adipose Tissue Protein Expression

As shown in Figure 8, the expression levels of lipid synthesis-related proteins SREBP1c (sterol-regulatory element binding protein 1c), PPARγ (peroxisome proliferator-activated receptor gamma), and C/EBPα (CCAAT enhancer binding protein α)in adipose tissue of the model group were significantly higher than those of the control group ($p < 0.01$), while the protein expression levels treated with different ZP concentrations (500, 250, and 125 mg/kg) were significantly lower than those of the model group ($p < 0.01$). AMPK (adenosine 5′-monophosphate AMP-activated protein kinase) is a key target for energy regulation, and the levels of AMPK and its phosphorylated proteins were also detected. The expression of p-AMPK protein in white adipose tissue of ZP treatment groups (500, 250, and 125 mg/kg) was significantly higher than that of the model group ($p < 0.01$), which was close to that of the control group ($p < 0.01$). This suggested that the anti-obesity effect of hazel leaf polyphenol extract might be associated with the inhibition of the expression of lipid synthesis protein and AMPK phosphorylation.

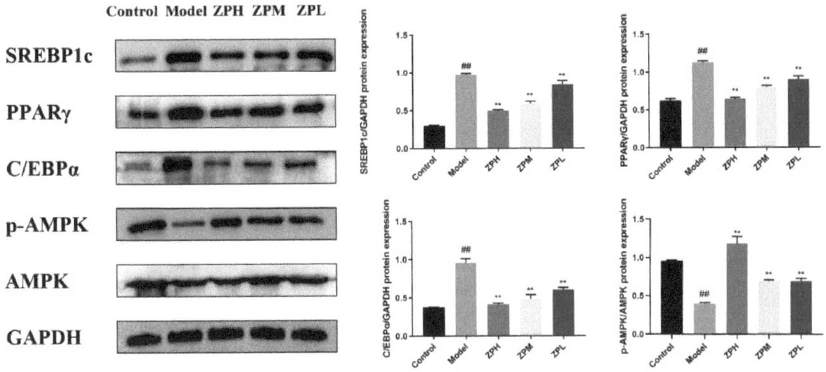

Figure 8. Effects of ZP on white adipose tissue protein expression. ## $p < 0.01$ compared to control group. ** $p < 0.01$ compared to model group.

2.9. Effect of ZP on Lipid Levels in 3T3-L1 Cells

The effects of different concentrations of ZP on 3T3-L1 cell viability were measured by MTT (Figure S2), and subsequent cultures were performed at concentrations of 50, 30, and 10 μg/mL. As shown in Figure 9a, the number of lipid droplets decreased with the increase in the administered dose after the treatment of ZP. This indicated that hazel leaf polyphenol extract could slow down the formation of lipid droplets. Compared with the control group, the content levels of TG and T-CHO in the cells were increased obviously ($p < 0.01$), while the content levels of TG and T-CHO in the cells treated with ZP were obviously decreased compared with the model group ($p < 0.01$) (Figure 9b,c). This suggested that hazel leaf polyphenol extract could inhibit the synthesis of lipids in cells.

Therefore, the expression levels of SREBP1c, PPARγ, and C/EBPα proteins related to lipid synthesis were further detected in 3T3-L1 cells. Compared with the control group, the protein expression levels of SREBP1c, PPARγ, and C/EBPα in the cells were significantly higher in the model group ($p < 0.01$). Compared with the model group, the protein expression levels treated with different ZP concentrations (500, 250, and 125 mg/kg) were significantly reduced ($p < 0.01$) (Figure 9d). In addition, the expression level of p-AMPK in the model group was significantly lower than that in the control group, and the phosphorylation level of AMPK was significantly increased after ZP treatment. These results suggested that hazel leaf polyphenol extract could inhibit the expression of lipogenic genes and promote the activation of AMPK in 3T3-L1 cells. This is consistent with in vivo results.

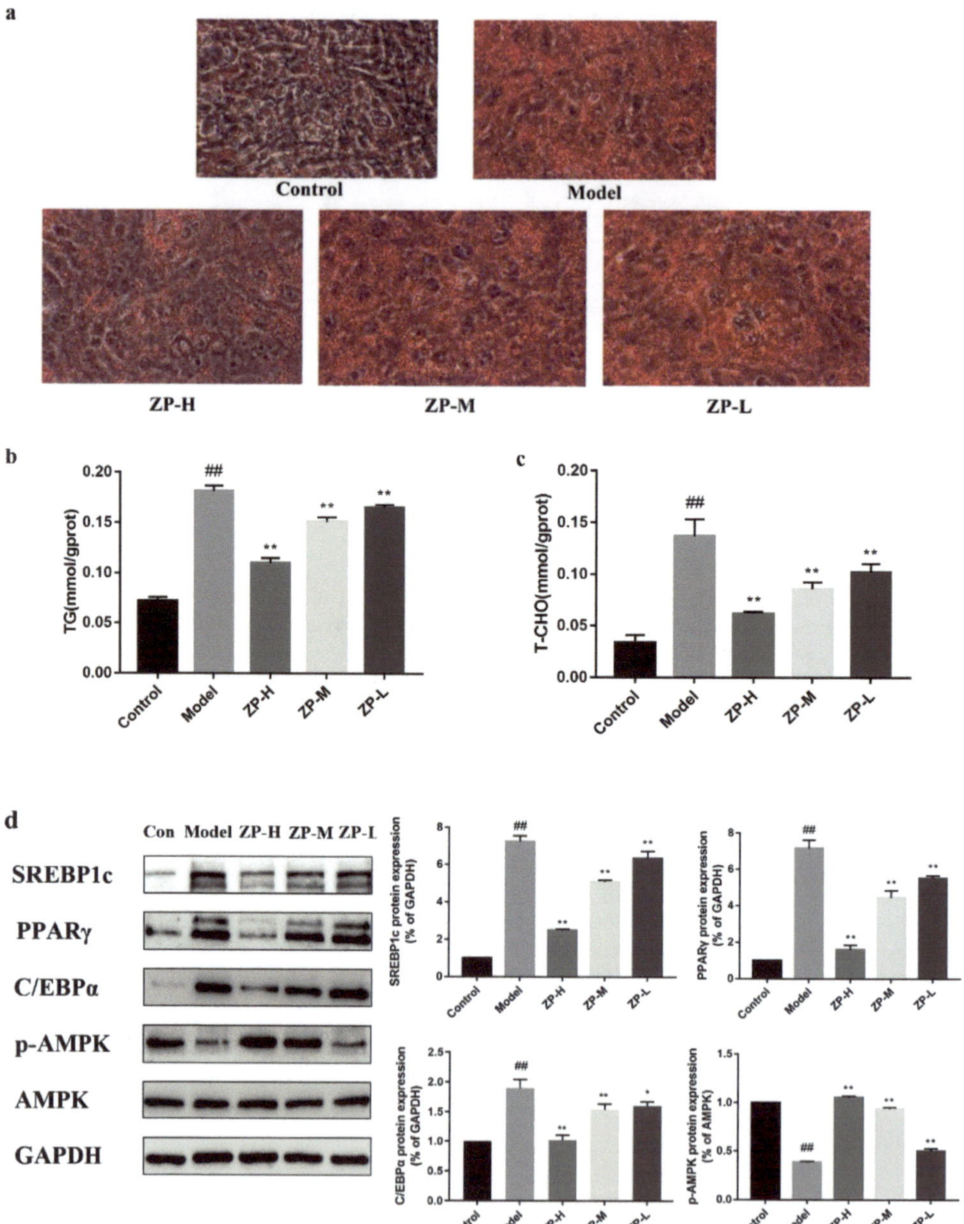

Figure 9. Effects of ZP on 3T3-L1 cells. (**a**) Cell oil red O staining (400×). (**b**) TG. (**c**) T-CHO. (**d**) Western blot. ## $p < 0.01$ compared to control group. * $p < 0.05$, ** $p < 0.01$ compared to model group.

2.10. Molecular Docking Analysis

The molecular docking results showed that gallic acid, quercetin-3-O-beta-D-glucopyranoside, quercetin, myricetin, and luteolin-7-O-glucosidein had good binding with PPARγ, C/EBPα, and AMPK, respectively, according to the binding energy less than or equal to −4 kcal/mol. Their binding sites and binding forms are shown in Table 1. Each of the above compounds binded well to PPARγ, C/EBPα, and AMPK, and the specific binding conditions are shown in Figure 10. These results indicated that gallic acid, quercetin-3-O-beta-D-glucopyranoside, quercetin, myricetin, and luteolin-7-O-glucosidein in hazel leaf polyphenol extract might be the main active components affecting lipid metabolism and anti-obesity.

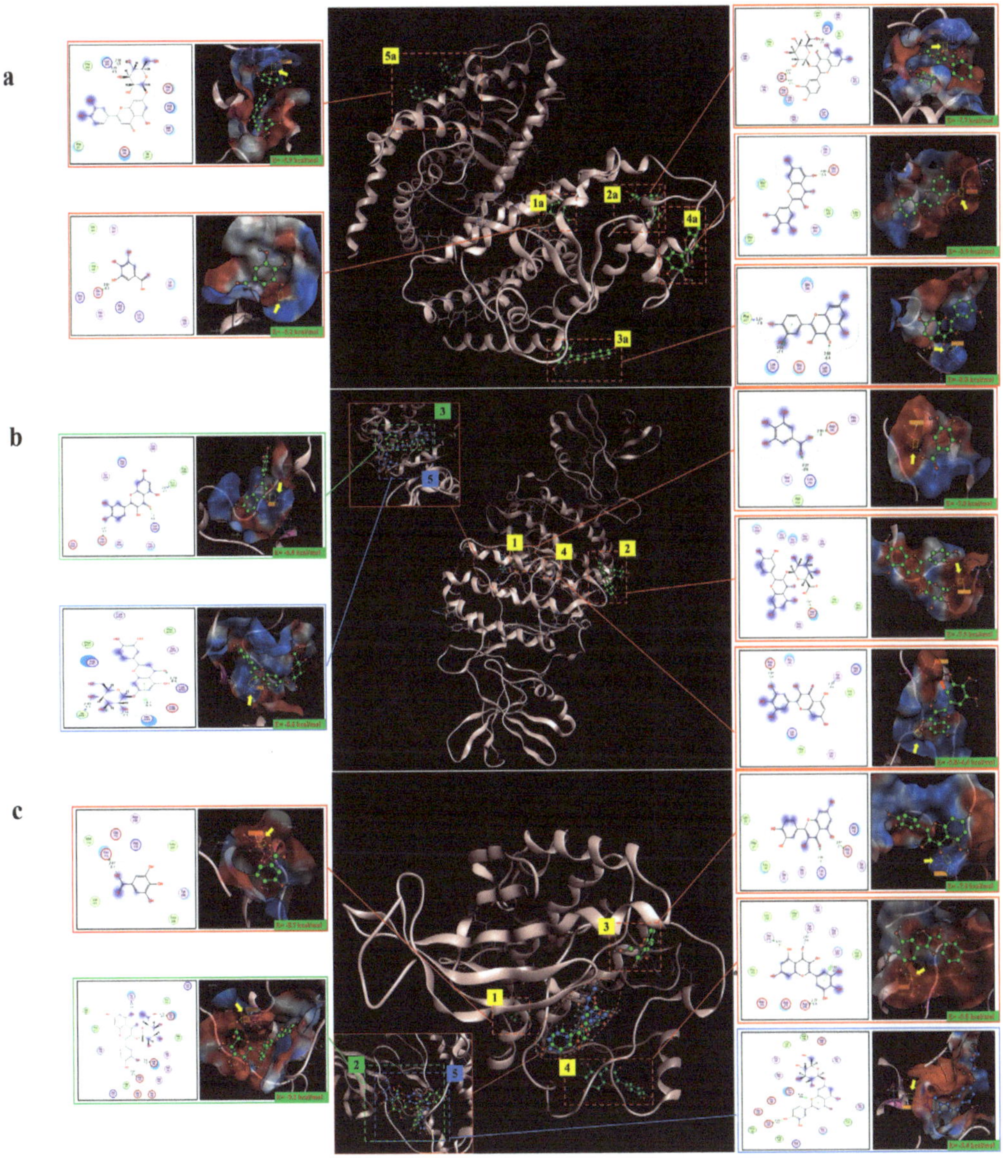

Figure 10. The five compounds and target binding conformation.(**a**) PPARγ. (**b**) C/EBPα. (**c**) AMPK. Label 1 represents gallic acid, label 2 represents quercetin-3-O-beta-D-glucopyranoside, label 3 represents quercetin, label 4 represents myricetin, and label 5 represents luteolin-7-O-glucoside.

Table 1. The binding energy of compounds and targets (kcal/mol).

Target	PDB ID	Target Structure	Compound	Affinity (kcal/mol)	Interaction	Receptor	Distance
PPARγ	2I4J	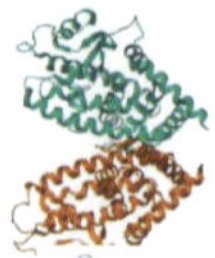	Gallic acid	−5.2	H-donor	Glu 324-OE2	2.81
			Quercetin-3-O-beta-D-glucopyranoside	−7.7	H-acceptor	Lys 457-HZ2	2.82
			Quercetin	−8.8	H-acceptor	Lys 244-HZ2	2.88
			Myricetin	−5.9	H-donor	Glu 272-OE1	2.86
			Luteolin-7-O-glucoside	−5.9	H-acceptor	Lys 263-HZ1	2.98
C/EBPα	6DC0		Gallic acid	−7.0	H-donor	Asp 286-OD2	2.91
			Quercetin-3-O-beta-D-glucopyranoside	−7.5	H-donor	Asp B517-OD1	2.87
			Quercetin	−6.8	H-acceptor	Lys A295-HZ1	3.00
			Myricetin	−5.8	H-donor	Asp 286-OD2	2.91
				−4.4	H-donor	Tyr 282-O	2.76
			Gallic acid	−8.6	H-acceptor	Lys A295-HZ1	4.10
AMPK	3AQV	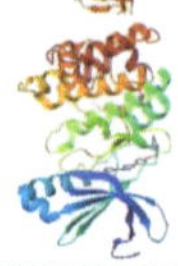	Gallic acid	−8.7	H-donor	Glu 113-OE1	2.87
			Quercetin-3-O-beta-D-glucopyranoside	−9.1	H-donor	Glu 143-OE2	2.72
			Quercetin	−7.0	H-acceptor	Lys 141-HZ3	2.99
			Myricetin	−5.8	H-donor	Asp 216-OD2	2.77
			Luteolin-7-O-glucoside	−5.4	H-donor	Asp 103-OD2	2.91

3. Discussion and Limitation

3.1. Discussion

Obesity is becoming one of the most important metabolic diseases worldwide. The prevalence of overweight or obesity is increasing in both developed and developing countries [21]. The adverse effects of diet pills have prompted the search for safer natural medicines. Natural polyphenols are a class of compounds widely found in plants with antioxidant, anti-inflammatory, and antibacterial properties. In a previous study, we extracted and determined the polyphenols in hazel leaves, characterized their content, and verified their good antioxidant and anti-inflammatory activities. In addition, we predicted their potential therapeutic effects on dyslipidemia, atherosclerosis, and diabetic complications using targeted network pharmacology [15]. Therefore, in the present study, the anti-obesity effects of hazel leaf polyphenol extract were preliminarily investigated using animal and cellular experiments combined with molecular-docking techniques based on network pharmacology.

Based on the results of network pharmacological analysis, the core targets of PPARγ and the KEGG-enriched pathway of PPAR signaling pathway attracted our attention and were further validated by in vivo and in vitro experiments. Adipose tissue in obese mice expands by increasing the size and number of adipocytes. When energy intake exceeds energy expenditure, fat is stored in the adipose tissue, leading to hypertrophy and weight gain [22]. In this study, the body weight and adiposity of mice in the model group were significantly increased after a high-fat diet. HE staining of adipose tissue also showed that the cell volume and number were significantly larger in the model group than those in the control group. The hazel leaf polyphenol extract intervention improved the above indices in obese mice, suggesting that they could inhibit weight gain by reducing the volume and number of adipocytes. Obesity is often associated with dyslipidemia, and elevated TC and TG can promote atherosclerosis [23]. Serum biochemical results showed that high-fat diets resulted in dyslipidemia (TG, T-CHO, LDL-C, and HDL-C) and abnormal liver function (ALT and AST) in mice. Further HE and oil red O-stained pathological sections also confirmed the formation of lipid accumulation in the liver and adipocytes of mice in the model group. Serum lipid levels, as well as lipid accumulation in the liver and adipocytes, were improved in obese mice treated with hazel leaf polyphenol extract, suggesting that hazel leaf polyphenol extract has anti-obesity ability.

Studies have shown that the gut flora is inextricably linked to obesity and related metabolic diseases [24,25]. Gut flora, the microbiota residing in the host's large and small intestines, is essential for maintaining intestinal immune homeostasis and nutrient absorption and also participates in processes related to body energy metabolism by regulating the body's metabolic axis. Clinical data show that the diversity of gut flora is significantly lower in obese individuals than in normal-weight individuals [26]. We examined the intestinal flora of mice and found that supplementation with hazel leaf polyphenol extract ameliorated the changes in the diversity and structure of intestinal microorganisms induced by a high-fat diet, and they reduced the ratio of Firmicutes/Bacillus phylum, an important marker of intestinal homeostasis. In turn, a lower F/B value implies a reduction in fat accumulation, as well as an improvement in lipid metabolism and hepatic steatosis [27,28]. At the genus level, hazel leaf polyphenol extract intervention reduced the abundance of *Oscillibacter* and *Desulfovibrionaceae* and enriched Lactobacillus, *Alloprevotella,* and *Muribaculaceae*. Previous studies have shown that *Alloprevotella* and *Muribaculaceae* are negatively correlated with metabolism-related indices (i.e., body weight, fat mass, and glucose metabolism). *Oscillibacter* is negatively correlated with some indices of intestinal barrier and mucus function [29]. These suggested that hazel leaf polyphenol extract might modulate the gut flora to exert anti-obesity effects.

Gut microbial fermentation of dietary fiber produces short-chain fatty acids, whereas a high-fat, high-protein diet leads to a reduction in short-chain fatty acid production [30]. Further testing of short-chain fatty acid content was found to be consistent with a reduction in short-chain fatty acids in the feces of the obese mouse model group induced by the high-fat diet in this study. Alterations in intestinal flora affect changes in their products' short-chain fatty acids. *Lactobacillus* promotes the production of short-chain fatty acids and maintains the stability of the internal intestinal environment, which is thought to attenuate high-fat diet-induced obesity and related complications [31]. Short-chain fatty acids offer potential therapeutic targets for the treatment of obesity and are related to metabolic diseases by altering host metabolic pathways [32]. Among them, butyric acid affects obesity by modulating the function of the gut barrier-activated peroxisome proliferator γ (PPARγ) [33]. The intervention of hazel leaf polyphenol extract increased the content of various short-chain fatty acids, which may be one of the ways to improve obesity. Meanwhile, short-chain fatty acids directly bind and activate extracellular G protein-coupled receptors [34] and mediate phosphorylation of the AMPK pathway [35]. It is a substrate for energy metabolism in the tricarboxylic acid cycle and regulates protein kinase (AMPK) activation via AMP to maintain homeostasis in the body environment [36]. Short-chain fatty acids promote oxidation and browning of 3T3-L1 adipocytes and inhibit lipid accumulation via the β3-adrenergic receptor/AMP-activated protein kinase α signaling pathway [37].

The PPAR signaling pathway is widely regarded as a potential mechanism for the hypolipidemic effects of Chinese herbal medicines. PPARγ is mainly found in adipose tissue and is an important mediator of energy homeostasis and cell differentiation. It promotes adipose differentiation and lipid synthesis, leading to morphological changes and adipocyte enlargement, and is also an important target for the treatment of type 2 diabetes mellitus [38]. PPARγ is a key transcription factor for adipocyte differentiation, activating the expression of a series of adipocyte phenotypic genes, thus regulating the whole process of preadipocyte to adipocyte differentiation. Western blot analysis of mouse adipose tissue revealed a significant increase in PPARγ protein expression in the model group, which was altered by the administration of hazel leaf polyphenol extract. In general, a high-fat, high-calorie diet modulates the expression of certain lipoproteins, leading to fat accumulation [39]. SREBP-1 is the major transcription factor for lipogenic gene expression in white adipose tissue. Its major isoform, SREBP-1c, is expressed in the liver and white adipocytes and promotes adipogenesis through the activation of triglyceride synthesis genes [40]. C/EBPα is also a key regulator of preadipocyte to mature adipocyte differentiation and is directly involved in cell differentiation by regulating the expression of genes related to intracellular glucose metabolism [41]. Meanwhile, C/EBPα is expressed prior to the initiation of transcription of most adipose-specific genes and is an important regulatory transcription factor for pro-adipocyte differentiation and lipid synthesis [42]. Our results also showed that ZP inhibited the overexpression of SREBP1c and C/EBPα proteins in adipocytes of mice on a high-fat diet, in agreement with previous studies [43], which may be one of the reasons for the reduction in lipid deposition by hazel leaf polyphenol extract.

AMPK is an important target of energy regulation and one of the key pathways in the regulation of lipid metabolism. AMP-dependent protein kinase phosphorylation reduces the incidence of obesity through the down-regulation of its downstream adipose generation-related genes, such as SREBP-1c and C/EBPα, to reduce the incidence of obesity [43,44]. We further examined the protein expression of AMPK and found that hazel leaf polyphenol extract was able to regulate AMPK phosphorylation levels.

In addition, we replicated the lipid differentiation process of 3T3-L1 cells. It was found that the cells were rounded and produced distinct lipid droplets upon maturation, while the administration of hazel leaf polyphenol extract significantly reduced the accumulation of lipid droplets. Meanwhile, the expression of PPARγ, SREBP-1c, C/EBPα, and AMPK proteins in the cells was consistent with the in vivo results, suggesting that hazel leaf polyphenol extract inhibited lipogenesis, which was similar to the results of other studies [45]. Moreover, five compounds of gallic acid, quercetin-3-O-β-D-glucoside, quercetin, myricetin, and lutetin-7-o-glucoside in hazel leaf polyphenol extract were simulated and screened by using the molecular docking technique and were well bound to PPARγ, C/EBP-α, and AMPK. They may be the key effector components.

3.2. Limitation

No anti-obesity experiments so far have been conducted for these components either individually or in combination, and no comparison has been made with the effect of hazel leaf polyphenols extract, which must be validated to support the conclusion of this study.

4. Materials and Methods

4.1. Materials and Reagents

Referring to our previous study [15], after the freeze-dried hazel leaves were crushed, 1.0 g of hazel leaf powder was added to 15 mL of a mixture of 1% hydrochloric acid and 70% methanol (1:1, v/v), vortexed for 30 s, sonicated for 30 min, and then centrifuged at 1800× g for 10 min. The extraction was repeated twice. The lyophilized supernatant and residue were extracted under different acid–base conditions to obtain the hazel leaf polyphenol extract (ZP, 10.5%) and stored at −80°C. The yield of ZP was evaluated by calculating the weight ratio of freeze-dried extracted ZP and leaf powder. The yield of ZP was 10.5%. The ZP was analyzed via chromatographic fingerprint using the UPLC-TOF-MS/MS system (Waters Q-TOF SynaptG2 high-resolution mass spectrometer and Waters ACQUITY UPLC system, Water, Milford, MA, USA) with chromCore 120 C18 column (1.8 μm, 2.1 × 100 mm), equipped with electrospray ionization (ESI). With the automatic matching function and the database UNIFI 1.6 software (Waters, Milford, MA, USA), compounds can be identified quickly. The parameters are set as follows: the analysis time range is 1–25 min. The allowable mass error range is ±10 ppm, the quality detection range is 50–1500 Da, and the negative adduct contains H- and HCOO-. The combined peak area of high-resolution mass spectrometry is used for semi-quantitative analysis [15]. Based on previous results, seventeen compounds were obtained from the identification, and their relative contents are shown in Table S1. Their original chromatograms are shown in Figure S1 [15].

Triglycerides (TGs), total cholesterol (T-CHO), high-density lipoprotein cholesterol (HDL-C), and low-density lipoprotein cholesterol (LDL-C) were obtained from Nanjing Jiancheng Institute of Biological Engineering (Nanjing, China). Insulin, 3-(4,5-Dimethylthiazol-2-yl)-2,5-diphenyltetrazolium bromide (MTT), BCA protein assay reagent kits, and penicillin/streptomycin were purchased from Beijing Solarbio Science & Technology Co., Ltd. (Beijing, China). 3-Isobutyl-1-methyl-2,6 (1H,3H)-purinedione (IBMX) and dexamethasone were obtained from Sigma Aldrich Inc. (St Louis, MO, USA). Orlistat was purchased from Hunan Mingrui Pharmaceutical Co., Ltd. (Liuyang, China).

4.2. Network Pharmacology Analysis

The chemical formulae of seventeen compounds in the ZP obtained by UPLC-TOF-MS/MS analysis were entered into Pubchem [46] (https://pubchem.ncbi.nlm.nih.gov/, accessed on 22 December 2023), and the two-dimensional structural diagrams of the chemical components were downloaded and imported into the Swiss ADME [47] (http://www.swissadme.ch/, accessed on 23 December 2023) system for activity screening (screening criteria: GI absorption, high; "Yes" in the Drug likeness box with a number greater than or equal to one-half of the items). The obtained active ingredients were entered into the Swiss Target Prediction Database [48] (http://swisstargetprediction.ch/, accessed on 23 December 2023) for target prediction, and the targets with a probability ≥ 0 were selected to obtain the gene names of the component targets. The keyword "obesity" was also used for disease target mining in GeneCards database (https://www.genecards.org/, Revelance Score ≥ 0.104990125), OMIM database (https://www.omim.org/, accessed on 25 December 2023), and DisGeNET database (https://www.disgenet.org/, Score ≥ 0.08). The disease targets were compared with the phenolic targets in Venny 2.1.0 (https://bioinfogp.cnb.csic.es/tools/venny/, accessed on 26 December 2023), and it was found that the cross-targets may be the key targets of hazel leaf phenolic compounds against obesity. The key targets were entered into the STRING (https://cn.string-db.org/, accessed on 26 December 2023) online website, and a protein–protein interaction (PPI) network was constructed (human source was selected; confidence level ≥ 0.9). The results were imported into Cytoscape 3.9.0 for visualization and analysis, and the core targets were filtered according to the values of "Closeness", "Betweenness", and "Degree" using the plug-in Centiscape 2.2 [49]. GO and KEGG pathway enrichment analyses were also performed using Metscape Data Platform (https://metascape.org/, accessed on 27 December 2023). Finally, GO enrichment histograms and KEGG enrichment bubble maps were plotted using Microbiology Online Platform [50] (https://www.bioinformatics.com.cn/, accessed on 28 December 2023).

4.3. Animals and the Experimental Design

The maleC57BL/6J mice, weighing 14–16 g, were provided by Liaoning Changsheng Biotechnology Co., Ltd. (Shenyang, China), and the certificate number was SCX (Liao) 2020-0001.This animal study was approved by the Laboratory Animal Ethics Committee of Changchun University of Chinese Medicine (Approval number: YX-2022611) and was carried out under its supervision.

All mice were fed in a constant temperature environment of 23 ± 2 °C and humidity of 55 ± 10% during the entire experiment, with light/dark cycles for 12 h and free drinking water. After one week of acclimation to the environment, the mice were randomly divided into six groups, a control group (n = 8), model group (n = 8), positive control group (Orlistat, 30 mg/kg, n = 8), hazel leaf polyphenol extract high-dose group (ZPH, 500 mg/kg, n = 8), hazel leaf polyphenol extract medium-dose group (ZPM, 250 mg/kg, n = 8), and hazel leaf polyphenol extract low-dose group (ZPL, 125 mg/kg, n = 8) based on our previous studies. All groups were fed a high-fat diet (D12492, Research Diets, New Brunswick, NJ, USA), except the control group, which was fed a regular maintenance diet. After six weeks, each group was given the corresponding dose of the subject via intragastric administration at the same time every day. The control group was given equal distilled water. The body weight and food intake of the mice were recorded every week.

After 6 weeks of administration, the mice were anesthetized with 1% pentobarbital sodium. The fat content of mice was scanned using the UltraFocus high-resolution small animal X-ray imaging system (Faxitron, Marlborough, MA, USA). Then, blood samples were taken, and the mice were euthanized. Liver tissue, perirenal white adipose tissue, and fecal samples were collected. The livers of mice were weighed, and the liver index was calculated.

4.4. Serum Biochemical Analysis

The blood sample was centrifuged at 3000 rpm for 15 min, and the supernatant was collected. TG (triglyceride), T-CHO (total Cholesterol), HDL-C (HDL-C, high-density lipoprotein cholesterol), LDL-C (low-density lipoprotein cholesterol), AST (aspartate aminotransferase), and ALT (alanine aminotransferase) were detected by the commercial kits according to the instructions.

4.5. Histological Analysis

The liver and perirenal white adipose tissue of mice were fixed in 4% (v/v) paraformaldehyde for 24 h.Then, the paraffin-embedded sections were stained with hematoxylin–eosin, or frozen sections were stained with oil red O. The tissue morphology was observed under a microscope (Nikon, Tokyo, Japan).

4.6. Gut Microbial Analysis

The total DNA was extracted from the resulting stool sample, and the gene was sequenced using the method of designing primers in the conserved region and adding them to the sequencing street. The results were then amplified by PCR, purified, quantified, and sequenced using the Illumina HiSeq 2500 platform. The ACE index and Shannon index were analyzed to explore the diversity of α among samples. Principal component analysis (PCA) was performed between the samples. The gut microbial of each group was analyzed at the phylum level and genus level, and the changes of the gut microbial of obese mice were screened using LEfSe analysis.

4.7. Determination of Short-Chain Fatty Acids (SCFAs)

The 50 mg fecal sample was placed into a 2 mL EP tube, in which 1 mL of pure water was added and vortexed for 10 s. Then, steel balls were added to the EP tube, which was treated with a 40 Hz grinder for 4 min, followed by ultrasonic treatment (ice water bath) for 5 min, repeated three times, and then centrifuged at 4 °C and at 5000 rpm for 20 min. After centrifugation, 0.8 mL of supernatant was added into another 2 mL EP tube, along with 0.1 mL of 50% H_2SO_4 and 0.8 mL of extract (containing internal standard 2-methylvaleric acid, 25 mg/L, methyl tert-butyl ether). The EP tube was vortexed for 10 s, oscillated for 10 min, sonicated for 10 min (ice water bath), and centrifuged at 10,000 rpm for 15 min at 4 °C.

After standing at −20°C for 30 min, the supernatant was removed from the injection bottle for GC-MS detection. The Shimadzu GC2030-QP2020 NX gas chromatography–mass spectrometer was equipped with Agilent HP-FFAP capillaries (30 m × 250 μm × 0.25 μm). The chromatographic conditions: injection volume of 1 μL, split ratio of 5:1, solvent delay time of 3.5 min. The mass spectrometry conditions: helium as the carrier gas, column flow rate of 1 mL/min, advance port temperature of 240 °C, transmission line temperature of 240 °C, quadrupole temperature of 150 °C, ion source temperature of 200 °C, and ionization voltage of −70 EV. Data analysis and chromatogram are shown in Table S5 and Figure S3.

4.8. Cell Culture and Treat

Mice embryonic fibroblasts (3T3-L1) were purchased from Wuhan Prosa Life Science Co., Ltd. and were cultured in DMEM culture medium (Hyclone, Logan, UT, USA) containing 10% bovine serum and 1% penicillin/streptomycin, under humidified conditions of 37 °C and 5% CO_2 concentration. The ZP cytotoxicity was detected by MTT method. The cells were cultured in 96-well plates at 5000 per well for 24 h with ZP added and cultured for 24 h or 48 h. A total of 20 μL MTT (5 mg/mL) was added to each well, incubated for 4 h, and the medium was carefully aspirated. Then, 150 μL of dimethyl sulfoxide solution (DMSO) was added to each well and oscillated for 10 min to read the absorbance at 490 nm.

The cells were rolled into six-well plates to induce differentiation. When the cells grew to contact inhibition, differentiation was initiated (0 days). They were then cultured in a fully differentiated medium containing 0.5 mM of 3-isobutyl-1-methylxanthine, 250 nM of dexamethasone, and 10 μg/mL of insulin for 48 h (2 days), continued in a complete culture medium of 10 μg/mL of insulin for 48 h (4 days), and continued with complete culture for 48 h (6 days) [35]. The complete culture medium with or without 10 μg/mL of insulin was replaced every 2 days. Different ZP concentrations were treated on days 0–4, and the cells cultured for six days were collected for relevant index detection. The control group for cell experiments was undifferentiated 3T3-L1 cells.

4.9. Cell Oil Red O Staining

The cells were spread into six-well plates, induced to differentiate for about 8 days, and stained with oil red. The oil red O staining working solution was prepared with the oil red O solution and the oil red O dilution at the ratio of 3:2, mixed evenly, left for 10 min, and filtered by 0.45 μm needle for 2 h. The cell culture medium was slowly pumped, washed with PBS, fixed with 4% paraformaldehyde fixing solution for 10 min, rinsed with PBS, and coated with appropriate amount of dyeing washing solution. Then, the staining wash solution was then removed, washed with PBS for 20 s, observed under a microscope, and photographed.

4.10. Western Blotting

The white adipose tissue or cells were lysed in a lysate containing protease inhibitors and phosphatase inhibitors. After protein quantification, loading buffer was added, vortexed, and mixed. The parafilm was sealed and boiled in a water bath at 100 °C for 5 min to denature the protein. The proteins were isolated by SDS-PAGE and transferred to PVDF membrane, which was incubated overnight at 4 °C with primary antibodies against C/EBPα, PPARγ, SREBP1 c, AMPK, GAPDH (Proteintech Group, Inc., Wuhan, China), and p-AMPK (Cell Signaling Technology, Danvers, MA, USA).The protein expression was analyzed by binding to secondary antibodies using the BeyoECL Plus Enhanced Chemiluminescence Kit. ImageJ 1.8.0 software was used for relative expression analysis.

4.11. Molecular Docking Method

Molecular docking was carried out to evaluate the binding affinity between the compounds and their protein targets. All identified components of hazel leaf polyphenol extract (Table S1) were used as target components for molecular docking with proteins such as C/EBPα, PPARγ, and AMPK (no protein conformation of SREBP1c was found) to simulate the binding state at the molecular level. The structure files of the five molecules (downloaded from Pubchem: https://pubchem.ncbi.nlm.nih.gov/, accessed on 22 December 2023) were imported into MOE 2022.02 software. The ligand molecules were preprocessed using the automatic function of the software "Compute→Molecule→Energy Minimize", energy minimization, optimization, and saved as a ligand library. At the same time, the receptor proteins to be docked (downloaded from RCSB PDB: https://www.rcsb.org/, accessed on 25 December 2023) were imported into the MOE 2022.02 software. The structure of the receptor molecules was protonated using the function "QuickPrep"; the obese key target molecule (receptor macromolecule) was taken as a rigid entity, and the conformation of the active ingredient (ligand macromolecule) was changed within a certain range using the semi-flexible docking method to dock 10 different conformations. The one with a binding energy lower than "−4 kcal/mol" was selected. The binding energies, binding sites, bond lengths, and other data were recorded [51].

4.12. Statistical Analysis

Data were expressed as mean ± standard deviation (SD). The differences between groups were analyzed by one-way ANOVA, Tukey's multi-comparison test, and two-way ANOVA. All data were analyzed using GraphPad Prism 8.0. A p-value of less than 0.05 was considered statistically significant.

5. Conclusions

In conclusion, this study employed network pharmacology, in vivo and in vitro assays, gut microbiota, and molecular docking to investigate the anti-obesity efficacy and possible molecular mechanisms of hazel leaf polyphenol extract. Hazel leaf polyphenol extract plays an anti-obesity role by regulating intestinal flora, short-chain fatty acids, and lipid metabolism. PPAR signaling is one of the potential anti-obesity mechanisms. Gallic acid, quercetin-3-O-β-D-glucopyranoside, quercetin, myricetin, and lutein-7-o-glucoside in the hazel leaf polyphenol extract may be the key active ingredients. However, these conclusions are based on bioinformatics databases and computer analysis and need to be further verified by experimental and clinical pharmacological data.

Supplementary Materials: The following supporting information can be downloaded at: https://www.mdpi.com/article/10.3390/ph17101349/s1, Figure S1: The relative content of the main components in hazel leaf polyphenols; Figure S2: Evaluation of 3T3-L1 viability using MTT assay after exposure to different concentrations of hazel leaf polyphenols for 24 h or 48 h; Figure S3: Total ion flow chromatograms (TIC) of different groups of short-chain fatty acids. (a) Control group. (b) Model group. (c) ZPH group. (d)ZPM group. (e) ZPL group. Table S1: Identification of components in hazel leaf polyphenols; Table S2: The component targets; Table S3: The disease targets; Table S4: The key targets; Table S5: Quality control (RSD) and standard curves for quantitative analysis of short-chain fatty acids.

Author Contributions: J.Z.: Data curation, Writing—original draft. A.A.: Validation. Y.L.: Methodology, Data curation. Z.L. (Zhi Lin): Data curation. J.L.: Data curation. X.W.: Validation. Y.W.: Methodology. G.L.: Methodology. H.L.: Conceptualization, Writing—Review and Editing. Z.L. (Zhe Lin): Supervision, Writing—review and editing. All authors have read and agreed to the published version of the manuscript.

Funding: This research was funded by [Jilin Province Science and Technology Development Project] grant number [20230204006YY].

Institutional Review Board Statement: The animal study protocol was approved by the Laboratory Animal Ethics Committee of Changchun University of Chinese Medicine (protocol code YX-2022611 and date of approval 29 November 2022).

Informed Consent Statement: Not applicable.

Data Availability Statement: Data is contained within the article or Supplementary Materials.

Conflicts of Interest: The authors declare no conflict of interest.

Abbreviations

ZP, hazel leaf polyphenols; TG, triglyceride; T-CHO, total cholesterol; HDL-C, high-density lipoprotein cholesterol; LDL-C, low-density lipoprotein cholesterol; IBMX, 3-Isobutyl-1-methylxanthine; MTT, 3-(4,5-Dimethylthiazolyl-2)-2,5-diphenyl tetrazolium bromide; DMSO, dimethyl sulfoxide solution; HE, hematoxylin eosin; ALT, alanine aminotransferase; AST, aspartate aminotransferase; PCA, principal component analysis; SCFAs, short-chain fatty acids; AMPK, adenosine 5′-monophosphate (AMP)-activated protein kinase; p-AMPK, phosphorylated AMP-activated protein kinase; PPARγ, peroxisome proliferator-activated receptor gamma; C/EBPα, CCAAT enhancer binding protein α; SREBP1c, sterol-regulatory element binding protein 1c.

References

1. WHO. Obesity and Overweight. Available online: https://www.who.int/news-room/fact-sheets/detail/obesity-and-overweight/ (accessed on 14 September 2024).
2. Ginsberg, H.N.; MacCallum, P.R. The obesity, metabolic syndrome, and type 2 diabetes mellitus pandemic: Part I. Increased cardiovascular disease risk and the importance of atherogenic dyslipidemia in persons with the metabolic syndrome and type 2 diabetes mellitus. *J. Cardiometab. Syndr.* **2009**, *4*, 113–119. [CrossRef] [PubMed]

3. Kim, N.H.; Jegal, J.; Kim, Y.N.; Heo, J.D.; Rho, J.R.; Yang, M.H.; Jeong, E.J. Chokeberry Extract and Its Active Polyphenols Suppress Adipogenesis in 3T3-L1 Adipocytes and Modulates Fat Accumulation and Insulin Resistance in Diet-Induced Obese Mice. *Nutrients* **2018**, *10*, 1734. [CrossRef] [PubMed]
4. Song, H.; Zhang, Y.; Huang, Q.; Deng, R.; Zheng, X. *Averrhoa carambola* L. fruit polyphenols ameliorate hyperlipidemia, hepatic steatosis, and hyperglycemia by modulating lipid and glucose metabolism in mice with obesity. *J. Sci. Food Agric.* **2023**, *103*, 6531–6539. [CrossRef] [PubMed]
5. Choi, J.W.; Choi, H.J.; Ryu, G.H.; Lee, J.W.; Beak, J.K.; Koh, E.J.; Jeong, J.B. *Paeonia lactiflora* root decreases lipid accumulation through the induction of lipolysis and thermogenesis via AMPK activation in 3T3-L1 cells. *Int. J. Mol. Med.* **2023**, *52*, 65. [CrossRef] [PubMed]
6. De Oliveira e Silva, A.M.; Pereira, R.O.; Oliveira, A.K.d.S.; Harris, F.S.; de Melo, I.L.P.; Almeida-Souza, T.H.; Yoshime, L.T.; dos Santos Melo, C.; Lopes dos Santos, J.; de Andrade-Wartha, E.R.S.; et al. Ameliorative effects of aqueous extract from rosemary on oxidative stress and inflammation pathways caused by a high-fat diet in C57BL/6 mice. *Appl. Physiol. Nutr. Metab.* **2024**, *49*, 459–472. [CrossRef]
7. Ramadan, N.S.; El-Sayed, N.H.; El-Toumy, S.A.; Mohamed, D.A.; Aziz, Z.A.; Marzouk, M.S.; Esatbeyoglu, T.; Farag, M.A.; Shimizu, K. Anti-Obesity Evaluation of *Averrhoa carambola* L. Leaves and Assessment of Its Polyphenols as Potential alpha-Glucosidase Inhibitors. *Molecules* **2022**, *27*, 5159. [CrossRef]
8. Li, R.; Zhu, Q.; Wang, X.; Wang, H. Mulberry leaf polyphenols alleviated high-fat diet-induced obesity in mice. *Front. Nutr.* **2022**, *9*, 979058. [CrossRef]
9. Samuel Wu, Y.-H.; Chiu, C.-H.; Yang, D.-J.; Lin, Y.-L.; Tseng, J.-K.; Chen, Y.-C. Inhibitory effects of litchi (*Litchi chinensis* Sonn.) flower-water extracts on lipase activity and diet-induced obesity. *J. Funct. Foods* **2013**, *5*, 923–929. [CrossRef]
10. Amorim, R.P.; Daher, I.P.; Arcangelo, F.P.D.; Bellini, M.F.; Orcini, W.A.; Sarria, G.R.; Peruquetti, R.L. Short-term effects of green tea chronotherapy on the metabolic homeostasis of mice on different diets. *Genet. Mol. Res.* **2017**, *16*, gmr16029605. [CrossRef]
11. Seo, D.B.; Jeong, H.W.; Kim, Y.J.; Kim, S.; Kim, J.; Lee, J.H.; Joo, K.; Choi, J.K.; Shin, S.S.; Lee, S.J. Fermented green tea extract exhibits hypolipidaemic effects through the inhibition of pancreatic lipase and promotion of energy expenditure. *Br. J. Nutr.* **2017**, *117*, 177–186. [CrossRef]
12. Wu, D.; Ma, X.; Tian, W. Pomegranate husk extract, punicalagin and ellagic acid inhibit fatty acid synthase and adipogenesis of 3T3-L1 adipocyte. *J. Funct. Foods* **2013**, *5*, 633–641. [CrossRef]
13. Ágnes Alberti, E.R. Szabolcs Béni and Ágnes Kéry, Evaluation of Radical Scavenging Activity of Sempervivum tectorum and Corylus avellana Extracts with Different Phenolic Composition. *Nat. Prod. Commun.* **2016**, *11*, 1934578X1601100412.
14. Oliveira, I.; Sousa, A.; Valentao, P.; Andrade, P.; Ferreira, I.; Ferreres, F.; Bento, A.; Seabra, R.; Estevinho, L.; Pereira, J. Hazel (*Corylus avellana* L.) leaves as source of antimicrobial and antioxidative compounds. *Food Chem.* **2007**, *105*, 1018–1025. [CrossRef]
15. Zhao, J.; Wang, X.; Wang, Y.; Lv, G.; Lin, H.; Lin, Z. UPLC-MS/MS profiling, antioxidant and anti-inflammatory activities, and potential health benefits prediction of phenolic compounds in hazel leaf. *Front. Nutr.* **2023**, *10*, 1092071. [CrossRef]
16. Lin, G.; Liu, X.; Yan, X.; Liu, D.; Yang, C.; Liu, B.; Huang, Y.; Zhao, C. Role of Green Macroalgae Enteromorpha Prolifera Polyphenols in the Modulation of Gene Expression and Intestinal Microflora Profiles in Type 2 Diabetic Mice. *Int. J. Mol. Sci.* **2018**, *20*, 25. [CrossRef]
17. Koliada, A.; Syzenko, G.; Moseiko, V.; Budovska, L.; Puchkov, K.; Perederiy, V.; Gavalko, Y.; Dorofeyev, A.; Romanenko, M.; Tkach, S.; et al. Association between body mass index and Firmicutes/Bacteroidetes ratio in an adult Ukrainian population. *BMC Microbiol.* **2017**, *17*, 120. [CrossRef]
18. Amabebe, E.; Robert, F.O.; Agbalalah, T.; Orubu, E.S.F. Microbial dysbiosis-induced obesity: Role of gut microbiota in homoeostasis of energy metabolism. *Br. J. Nutr.* **2020**, *123*, 1127–1137. [CrossRef]
19. Guo, X.; Tang, R.; Yang, S.; Lu, Y.; Luo, J.; Liu, Z. Rutin and Its Combination With Inulin Attenuate Gut Dysbiosis, the Inflammatory Status and Endoplasmic Reticulum Stress in Paneth Cells of Obese Mice Induced by High-Fat Diet. *Front. Microbiol.* **2018**, *9*, 2651. [CrossRef]
20. Hee, B.v.d.; Wells, J.M. Microbial Regulation of Host Physiology by Short-chain Fatty Acids. *Trends Microbiol.* **2021**, *29*, 700–712. [PubMed]
21. Wang, Y.; Zhao, L.; Gao, L.; Pan, A.; Xue, H. Health policy and public health implications of obesity in China. *Lancet Diabetes Endocrinol.* **2021**, *9*, 446–461. [CrossRef]
22. Kim, K.-H.; Park, Y. Food Components with Anti-Obesity Effect. *Annu. Rev. Food Sci. Technol.* **2011**, *2*, 237–257. [CrossRef]
23. Wang, C.; Chen, Z.; Brennan, M.A.; Wang, J.; Sun, J.; Fang, H.; Kang, M.; Brennan, C.S.; Mu, J. The effect of extruded multigrain powder on metabolism and intestinal flora of high-fat-diet induced C57BL/6J mice. *Food Res. Int.* **2023**, *169*, 112878. [CrossRef]
24. Aron-Wisnewsky, J.; Prifti, E.; Belda, E.; Ichou, F.; Kayser, B.D.; Dao, M.C.; Verger, E.O.; Hedjazi, L.; Bouillot, J.L.; Chevallier, J.M.; et al. Major microbiota dysbiosis in severe obesity: Fate after bariatric surgery. *Gut* **2019**, *68*, 70–82. [CrossRef]
25. Sun, L.; Ma, L.; Ma, Y.; Zhang, F.; Zhao, C.; Nie, Y. Insights into the role of gut microbiota in obesity: Pathogenesis, mechanisms, and therapeutic perspectives. *Protein Cell* **2018**, *9*, 397–403. [CrossRef]
26. Tilg, H.; Zmora, N.; Adolph, T.E.; Elinav, E. The intestinal microbiota fuelling metabolic inflammation. *Nat. Rev. Immunol.* **2020**, *20*, 40–54. [CrossRef]
27. Deng, Y.; Liu, W.; Wang, J.; Yu, J.; Yang, L.Q. Intermittent Fasting Improves Lipid Metabolism Through Changes in Gut Microbiota in Diet-Induced Obese Mice. *Med. Sci. Monit.* **2020**, *26*, e926789. [CrossRef]

28. Lozupone, C.A.; Stombaugh, J.I.; Gordon, J.I.; Jansson, J.K.; Knight, R. Diversity, stability and resilience of the human gut microbiota. *Nature* **2012**, *489*, 220–230. [CrossRef]
29. Paone, P.; Suriano, F.; Jian, C.; Korpela, K.; Delzenne, N.M.; Van Hul, M.; Salonen, A.; Cani, P.D. Prebiotic oligofructose protects against high-fat diet-induced obesity by changing the gut microbiota, intestinal mucus production, glycosylation and secretion. *Gut Microbes* **2022**, *14*, 2152307. [CrossRef]
30. Zhao, Q.; Fu, Y.; Zhang, F.; Wang, C.; Yang, X.; Bai, S.; Xue, Y.; Shen, Q. Heat-Treated Adzuki Bean Protein Hydrolysates Reduce Obesity in Mice Fed a High-Fat Diet via Remodeling Gut Microbiota and Improving Metabolic Function. *Mol. Nutr. Food Res.* **2022**, *66*, 2100907. [CrossRef]
31. Requena, T.; Martínez-Cuesta, M.C.; Peláez, C. Diet and microbiota linked in health and disease. *Food Funct.* **2018**, *9*, 688–704. [CrossRef]
32. Cornejo-Pareja, I.; Munoz-Garach, A.; Clemente-Postigo, M.; Tinahones, F.J. Importance of gut microbiota in obesity. *Eur. J. Clin. Nutr.* **2019**, *72* (Suppl. S1), 26–37. [CrossRef] [PubMed]
33. Coppola, S.; Avagliano, C.; Calignano, A.; Berni Canani, R. The protective role of butyrate against obesity and obesity-related diseases. *Molecules* **2021**, *26*, 682. [CrossRef] [PubMed]
34. Tang, C.; Kong, L.; Shan, M.; Lu, Z.; Lu, Y. Protective and ameliorating effects of probiotics against diet-induced obesity: A review. *Food Res. Int.* **2021**, *147*, 110490. [CrossRef] [PubMed]
35. Zhang, W.; Kong, L.; Zhong, Z.; Lin, L.; Li, J.; Zheng, G. Short chain fatty acids increase fat oxidation and promote browning through β3-adrenergic receptor/AMP-activated protein kinase α signaling pathway in 3T3-L1 adipocytes. *J. Funct. Foods* **2023**, *103*, 105488. [CrossRef]
36. Shimano, H.; Yahagi, N.; Amemiya-Kudo, M.; Hasty, A.H.; Osuga, J.; Tamura, Y.; Shionoiri, F.; Iizuka, Y.; Ohashi, K.; Harada, K.; et al. Sterol regulatory element-binding protein-1 as a key transcription factor for nutritional induction of lipogenic enzyme genes. *J. Biol. Chem.* **1999**, *274*, 35832–35839. [CrossRef]
37. Moseti, D.; Regassa, A.; Kim, W.K. Molecular Regulation of Adipogenesis and Potential Anti-Adipogenic Bioactive Molecules. *Int. J. Mol. Sci.* **2016**, *17*, 124. [CrossRef]
38. Zhang, G.; Li, N.; Tong, Y.; Li, P.; Han, H.; Song, Q.; Yang, B.; Cui, L. Artemisinin derivatives inhibit adipogenic differentiation of 3T3-L1 preadipocytes through upregulation of CHOP. *Biochem. Res Commun.* **2021**, *557*, 309–315. [CrossRef]
39. Liu, S.; Chang, X.; Yu, J.; Xu, W. Cerasus humilis Cherry Polyphenol Reduces High-Fat Diet-Induced Obesity in C57BL/6 Mice by Mitigating Fat Deposition, Inflammation, and Oxidation. *J. Agric. Food Chem.* **2020**, *68*, 4424–4436. [CrossRef]
40. Fang, C.; Kim, H.; Noratto, G.; Sun, Y.; Talcott, S.T.; Mertens-Talcott, S.U. Gallotannin derivatives from mango (*Mangifera indica* L.) suppress adipogenesis and increase thermogenesis in 3T3-L1 adipocytes in part through the AMPK pathway. *J. Funct. Foods* **2018**, *46*, 101–109. [CrossRef]
41. Li, Y.; Li, C.; Wu, J.; Liu, W.; Li, D.; Xu, J. Harmane ameliorates obesity though inhibiting lipid accumulation and inducing adipocyte browning. *RSC Adv.* **2020**, *10*, 4397–4403. [CrossRef]
42. Rao, Y.; Yu, H.; Gao, L.; Lu, Y.T.; Xu, Z.; Liu, H.; Gu, L.Q.; Ye, J.M.; Huang, Z.S. Natural alkaloid bouchardatine ameliorates metabolic disorders in high-fat diet-fed mice by stimulating the sirtuin 1/liver kinase B-1/AMPK axis. *Br. J. Pharmacol.* **2017**, *174*, 2457–2470. [CrossRef] [PubMed]
43. Ney, L.M.; Wipplinger, M.; Grossmann, M.; Engert, N.; Wegner, V.D.; Mosig, A.S. Short chain fatty acids: Key regulators of the local and systemic immune response in inflammatory diseases and infections. *Open Biol.* **2023**, *13*, 230014. [CrossRef] [PubMed]
44. Luo, P.; Lednovich, K.; Xu, K.; Nnyamah, C.; Layden, B.T.; Xu, P. Central and peripheral regulations mediated by short-chain fatty acids on energy homeostasis. *Transl. Res.* **2022**, *248*, 128–150. [CrossRef] [PubMed]
45. Wang, Z.; Liu, J.; Li, F.; Ma, S.; Zhao, L.; Ge, P.; Chen, H. Mechanisms of Qingyi Decoction in Severe Acute Pancreatitis Associated Acute Lung Injury via Gut Microbiota: Targeting the Short-Chain Fatty Acids-Mediated AMPK/NF-kB/NLRP3 Pathway. *Microbiol. Spectr.* **2023**, *11*, e03664-22. [CrossRef]
46. Kim, S.; Chen, J.; Cheng, T.; Gindulyte, A.; He, J.; He, S.; Li, Q.; Shoemaker, B.A.; Thiessen, P.A.; Yu, B.; et al. PubChem in 2021: New data content and improved web interfaces. *Nucleic Acids Res.* **2021**, *49*, D1388–D1395. [CrossRef]
47. Daina, A.; Michielin, O.; Zoete, V. SwissADME: A free web tool to evaluate pharmacokinetics, drug-likeness and medicinal chemistry friendliness of small molecules. *Sci. Rep.* **2017**, *7*, 42717. [CrossRef]
48. Daina, A.; Michielin, O.; Zoete, V. Swiss Target Prediction: Updated data and new features for efficient prediction of protein targets of small molecules. *Nucleic Acids Res.* **2019**, *47*, W357–W364. [CrossRef]
49. Scardoni, G.; Petterlini, M.; Laudanna, C. Analyzing biological network parameters with CentiScaPe. *Bioinformatics* **2009**, *25*, 2857–2859. [CrossRef]
50. Madikyzy, M.; Tilegen, M.; Nazarbek, G.; Mu, C.; Kutzhanova, A.; Li, X.; Ma, C.; Xie, Y. Honghua extract mediated potent inhibition of COVID-19 host cell pathways. *Sci. Rep.* **2022**, *12*, 14296. [CrossRef]
51. Vilar, S.; Cozza, G.; Moro, S. Medicinal Chemistry and the Molecular Operating Environment (MOE): Application of QSAR and Molecular Docking to Drug Discovery. *Curr. Top. Med. Chem.* **2008**, *8*, 1555–1572. [CrossRef]

Disclaimer/Publisher's Note: The statements, opinions and data contained in all publications are solely those of the individual author(s) and contributor(s) and not of MDPI and/or the editor(s). MDPI and/or the editor(s) disclaim responsibility for any injury to people or property resulting from any ideas, methods, instructions or products referred to in the content.

pharmaceuticals

MDPI

Article

Unveiling the Bioactive Efficacy of *Cupressus sempervirens* 'Stricta' Essential Oil: Composition, In Vitro Activities, and In Silico Analyses

Eman Fikry [1], Raha Orfali [2,*], Nora Tawfeek [1], Shagufta Perveen [3,*], Safina Ghafar [2], Maher M. El-Domiaty [1] and Azza M. El-Shafae [1]

[1] Department of Pharmacognosy, Faculty of Pharmacy, Zagazig University, Zagazig 44519, Egypt; efhassan@zu.edu.eg (E.F.); mmeldomyaty@pharmacy.zu.edu.eg (M.M.E.-D.); amelshafaey@pharmacy.zu.edu.eg (A.M.E.-S.)

[2] Department of Pharmacognosy, Collage of Pharmacy, King Saud University, P.O. Box 2457, Riyadh 11451, Saudi Arabia

[3] Department of Bacteriology, University of Wisconsin-Madison, Madison, WI 53706, USA

* Correspondence: rorfali@ksu.edu.sa (R.O.); shagufta792000@yahoo.com or sperveen@wisc.edu (S.P.)

Abstract: Prior studies have extensively investigated the essential oil derived from the Mediterranean cypress, *Cupressus sempervirens*. However, the 'Stricta' variety, known for its ornamental value, has received less attention in terms of its oil composition and potential health benefits. The objective of this research was to comprehensively analyze the chemical components and medicinal properties of the essential oil extracted from *C. sempervirens* 'Stricta' (CSSLEO) grown in Egypt. Utilizing gas chromatography–mass spectrometry (GC–MS), the investigation identified 22 compounds within CSSLEO, with α-pinene and δ-3-carene being predominant, accounting for 96.01% of the oil. In vitro assays evaluated CSSLEO's cytotoxic effects on cancer cell lines, revealing notable anticancer potential. Additionally, the oil displayed antidiabetic properties by impeding crucial enzymes involved in glucose metabolism. Complementary in silico network pharmacology and molecular docking studies provided insights into the possible interactions between CSSLEO's key compounds and essential proteins and pathways in cancer treatment. The results underscored CSSLEO's intricate composition and its promising applications in cancer prevention and diabetes management. The conclusions drawn from this research underscore the need for further investigation to validate CSSLEO's clinical effectiveness and to gain a deeper understanding of its therapeutic mechanisms, with a view to harnessing its potential in oncology and endocrinology.

Keywords: *Cupressus sempervirens* 'Stricta'; MTT assay; antidiabetic; antioxidant; antimicrobial; network pharmacology; gene expression; molecular docking

1. Introduction

Essential oils (EOs) have been utilized for centuries due to their medicinal properties and have played a crucial role in the culinary world by adding unique aromas and flavors, as well as serving as natural preservatives [1]. These oils are intricate combinations of various chemical constituents, each possessing distinct functional groups and molecular structures [2]. As secondary metabolites found in plants, EOs are primarily responsible for the biological activities exhibited by aromatic plants. Numerous in vitro and in vivo studies have confirmed the diverse functions of EOs and their individual components, showcasing antioxidant, antimicrobial, anti-inflammatory, antidiabetic, antiviral, cytotoxic, antihypertensive, antihyperlipidemic, and antispasmodic properties through different mechanisms [3–5].

Gymnosperms are known for their abundance of essential oil (EO) components, such as terpenoids, phenolic compounds, alkanes, and alkenes. These plants have been acknowledged for their numerous health benefits, including their anticancer, antimicrobial, antifungal, and antioxidant activities [6–8]. Among the gymnosperms, conifers, particularly those in the Cupressaceae family, are noteworthy as significant sources of EOs [9].

The Cupressaceae family consists of around 30 genera and 134 species, which are distributed globally. These evergreen, resinous trees and shrubs can be either monoecious or dioecious [10]. EOs play a crucial role as chemotaxonomic markers within this family [11–14], with the genus *Cupressus* being characterized by the predominance of monoterpenes, sesquiterpenes, and diterpenes in their EO profiles [13].

Cypress, known by its classical Latin name *Cupressus*, is native to Asia Minor and Northern Persia and extends westwards into Greece and Crete. The original species, *Cupressus sempervirens*, commonly known as the Mediterranean cypress, has a wide, uneven, spreading crown. This species has a rich history of traditional uses and has been noted for its medicinal properties including cytotoxic, antidiabetic, antioxidant, and antimicrobial activities [13,15–18].

However, through horticultural development, a strain with more upright branching was selectively bred, resulting in the columnar form known as variety 'Stricta' [19]. *Cupressus sempervirens* 'Stricta' is a variety found in the central Mediterranean countries, especially Italy, where it has been used as an ornamental plant in landscapes for centuries. This tree can grow up to 25 m tall, characterized by a single upright trunk and numerous closely-spaced vertical branches covered in dense, dark-green foliage [20,21]. Despite being widely cultivated, there is a lack of research on the chemical composition and biological properties of its essential oil, making it an area that requires further exploration and study.

Essential oils (EOs) derived from the *Cupressus* genus have demonstrated antimicrobial, anticancer, and antioxidant properties. For instance, EOs from *C. arizonica* have shown modest antioxidant potential in DPPH assays, limited antioxidant effects in deoxyribose degradation assays, and moderate effects in non-enzymatic lipid peroxidation assays [13]. The Tunisian *C. sempervirens* EO has exhibited significant antioxidant activities in ABTS and DPPH assays, surpassing the effects of 2, 6-di-tert-butyl-4-hydroxy-boxylic acid (BHT) [22]. Additionally, *Cupressus sempervirens* var. Pyramidalis has demonstrated antiproliferative activity against C32 amelanotic melanoma cells [23]. Given these findings, it was imperative to elucidate the chemical composition of EOs produced by *Cupressus sempervirens* 'Stricta' (CSSLEO) and to investigate its potential biological activities in vitro, including cytotoxic, antioxidant, antidiabetic, and antimicrobial activities. Furthermore, this research aimed to leverage target network analysis and in silico validation to discover novel medicinal applications for EOs, thereby contributing to the global effort to address various health challenges through natural remedies.

2. Results

2.1. Chemical Composition of CSSLEO

A transparent, light yellow essential oil (EO) was obtained through the hydrodistillation of fresh leaves from *Cupressus sempervirens* 'Stricta'. The average yield of the EO was found to be $0.65 \pm 0.03\%$ v/w, based on three separate extractions. Analysis using GC–MS identified a total of 22 components, which accounted for 96.01% of the overall composition. The chemical names and corresponding area percentages of these components can be found in Table 1. Figure 1 displays the GC–MS spectrum and pie chart representing the composition of the EO. The chemical structures of the identified constituents are illustrated in Figure S1.

Table 1. Chemical composition of *Cupressus sempervirens* 'Stricta' leaf essential oil.

Peak	Rt	Compound Name	Molecular Formula	Chemical Class	$RI_{Exp.}$ [a]	RI_{Lit} [b]	Area %	Identification [c]
1	7.205	Tricyclene	$C_{10}H_{16}$	Tricyclic monoterpene hydrocarbon	911	902	0.17	MS, RI
2	7.370	α-Thujene	$C_{10}H_{16}$	Bicyclic monoterpene hydrocarbon	917	908	0.85	MS, RI
3	7.575	α-Pinene	$C_{10}H_{16}$	Bicyclic monoterpene hydrocarbon	923	916	33.51	MS, RI
4	7.855	α-Fenchene	$C_{10}H_{16}$	Bicyclic monoterpene hydrocarbon	933	929	1.50	MS, RI
5	8.615	Sabinene	$C_{10}H_{16}$	Bicyclic monoterpene hydrocarbon	958	958	3.75	MS, RI
6	8.705	β-Pinene	$C_{10}H_{16}$	Bicyclic monoterpene hydrocarbon	961	961	1.64	MS, RI
7	9.210	β-Myrcene	$C_{10}H_{16}$	Acyclic monoterpene hydrocarbon	978	978	3.74	MS, RI
8	9.825	δ-3-Carene	$C_{10}H_{16}$	Bicyclic monoterpene hydrocarbon	998	999	28.73	MS, RI
9	9.945	α-Terpinene	$C_{10}H_{16}$	Monocyclic monoterpene hydrocarbon	1002	1006	0.84	MS, RI
10	10.040	*m*-Cymene	$C_{10}H_{14}$	Aromatic monoterpene hydrocarbon	1005	1012	0.11	MS, RI
11	10.255	β-Phellandrene	$C_{10}H_{16}$	Monocyclic monoterpene hydrocarbon	1012	1013	1.16	MS, RI
12	10.320	Sylvestrene	$C_{10}H_{16}$	Monocyclic monoterpene hydrocarbon	1014	1018	1.99	MS, RI
13	11.200	γ-Terpinene	$C_{10}H_{16}$	Monocyclic monoterpene hydrocarbon	1042	1048	1.24	MS, RI
14	12.115	α-Terpinolene	$C_{10}H_{16}$	Monocyclic monoterpene hydrocarbon	1071	1078	6.77	MS, RI
15	12.945	β-linalool	$C_{10}H_{18}O$	Acyclic monoterpene alcohol	1098	1090	0.13	MS, RI
16	13.725	α-Campholenal	$C_{10}H_{16}O$	Monocyclic monoterpene aldehyde	1123	1125	0.19	MS, RI
17	14.525	Isoborneol	$C_{10}H_{18}O$	Bicyclic monoterpene alcohol	1149	1147	0.26	MS, RI
18	14.610	Terpinen-4-ol	$C_{10}H_{18}O$	Monocyclic monoterpene alcohol	1152	1167	2.36	MS
19	17.810	Bornyl acetate	$C_{12}H_{20}O_2$	Bicyclic monoterpene ester	1259	1276	0.31	MS
20	19.640	α-Terpinyl acetate	$C_{12}H_{20}O_2$	Monocyclic monoterpene ester	1323	1339	5.91	MS

Table 1. *Cont.*

Peak	Rt	Compound Name	Molecular Formula	Chemical Class	RI$_{Exp.}$ [a]	RI$_{Lit}$ [b]	Area %	Identification [c]
21	22.710	α-Humulene	$C_{15}H_{24}$	Sesquiterpene hydrocarbon	1436	1445	0.20	MS
22	23.390	Germacrene D	$C_{15}H_{24}$	Sesquiterpene hydrocarbon	1462	1473	0.65	MS
		Total identified					96.01	
		Monoterpene hydrocarbons					86.00	
		Oxygenated monoterpenes					9.16	
		Sesquiterpene hydrocarbons					0.85	

Compounds are arranged based on their elution sequence on the Rtx-5MS GC column. [a] The retention index was experimentally determined on an Rtx-5MS column in comparison to C_8–C_{28} n-alkanes. [b] Reported retention index. [c] To establish identification, the mass spectral (MS) data and retention index (RI) values of the compounds were compared with those in the Adams library [24], NIST 11 Mass Spectral Library, Wiley Registry of Mass Spectral Data 10th edition, and relevant literature [25–27].

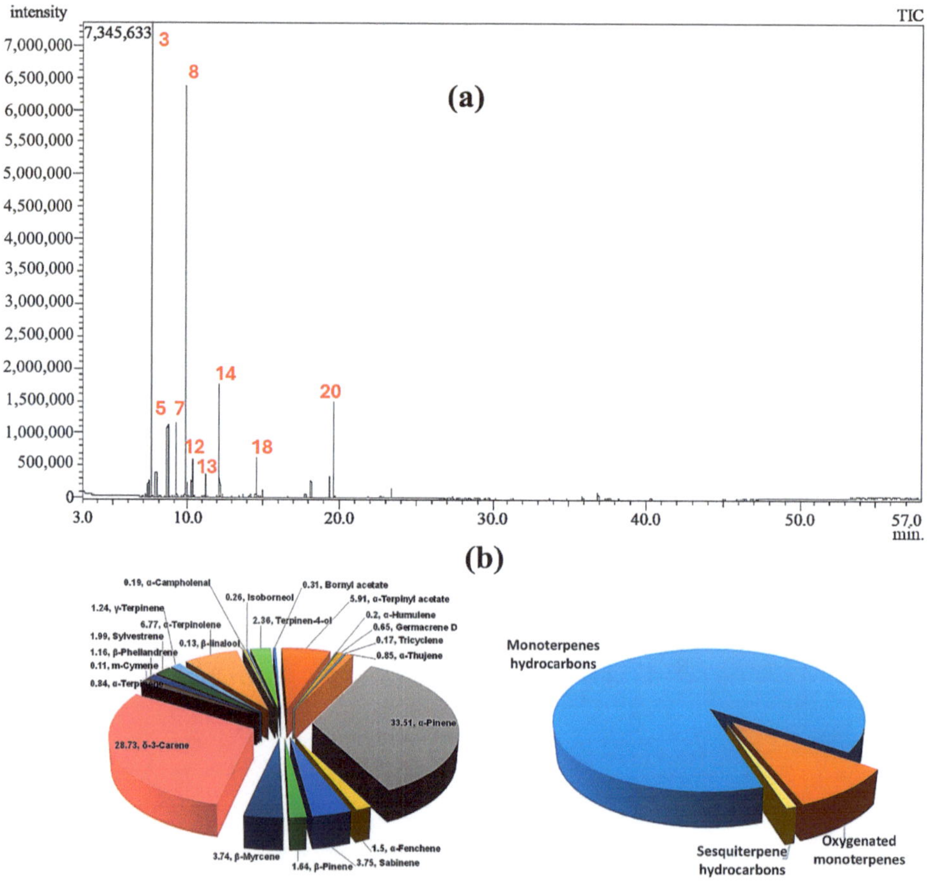

Figure 1. The GC–MS chromatogram (**a**) and the pie charts (**b**) illustrate the relative chemical compositions of *Cupressus sempervirens* 'Stricta' leaf essential oil. The red numbers correspond to Table 1.

2.2. *In Vitro Assessments*

2.2.1. Cytotoxic Potential and Selectivity of CSSLEO

The cytotoxic impact of CSSLEO on different human carcinoma cell lines including breast (MCF-7), colon (HCT-116), hepatocellular (HepG-2), and lung (A-549) carcinoma cell lines, as well as the normal African green monkey kidney cell line (Vero), was evaluated through the MTT assay, as shown in Table 2 and Figure S2. CSSLEO exhibited notable cytotoxicity towards all the cancer cell lines tested, with HepG-2 showing the highest sensitivity (IC_{50} = 4.44 µg/mL) compared to HCT-116, A-549, and MCF-7 (IC_{50} = 6.29, 6.71, and 7.20 µg/mL, respectively). Notably, CSSLEO demonstrated selectivity against the carcinoma cell lines (SI > 3) when compared to Vero, as outlined in Table 2.

Table 2. The cytotoxic effects of CSSLEO and cisplatin on Vero, MCF-7, HCT-116, HepG-2, and A-549 cell lines are presented based on CC_{50}, IC_{50} values in µg/mL (Mean ± SD), and SI.

	Vero	MCF-7		HCT-116		HepG-2		A-549	
	CC_{50}	IC_{50}	SI	IC_{50}	SI	IC_{50}	SI	IC_{50}	SI
CSSLEO	26.95 ± 1.06	7.20 ± 0.32	3.74	6.29 ± 0.29	4.35	4.44 ±0.21	6.07	6.71 ± 0.28	4.02
Cisplatin	40.68 ± 2.75	5.57 ± 0.68	7.30	2.75 ± 0.24	14.28	3.87 ± 0.55	10.52	7.68 ± 0.36	5.29

2.2.2. Antidiabetic Activity

α-Glucosidase Inhibition Activity

The inhibitory effect of CSSLEO on α-glucosidase was illustrated in Table S1, where it was observed that the level of inhibition increased with higher concentrations of CSSLEO. The calculated IC_{50} value of CSSLEO (100.06 ± 2.18 µg/mL) indicated a substantial inhibitory activity, in comparison to the IC_{50} values of the acarbose standard (2.12 ± 0.54 µg/mL).

α-Amylase Inhibition Activity

In Table S2, the potential of CSSLEO to inhibit α-amylase was demonstrated. The extent of enzyme inhibition was directly influenced by the concentration of the tested oil. The IC_{50} value revealed that CSSLEO has half the inhibition potential of acarbose. More specifically, the IC_{50} value for CSSLEO is 26.36 ± 0.91 µg/mL, while acarbose has an IC_{50} value of 12.31 ± 0.87 µg/mL.

2.2.3. Antioxidant Activity

The antioxidant capacity and reducing power of CSSLEO were evaluated using DPPH, ABTS, and FRAP assays, and the results are detailed in Table 3. Unlike ascorbic acid, CSSLEO exhibited lower scavenging abilities against DPPH and ABTS radicals, along with decreased capacity to reduce ferric ions, suggesting a comparatively weaker antioxidant performance.

Table 3. Antioxidant activity of CSSLEO.

	IC_{50} ± SD (µg/mL)		
	DPPH	ABTS	FRAP
CSSLEO	212.08 ± 4.71	138.20 ± 3.69	307.35 ± 6.04
Ascorbic acid	10.19 ± 0.65	10.78 ± 0.65	20.88 ± 0.94

2.2.4. Antimicrobial Activity

The effectiveness of antimicrobial activity against different fungal and bacterial strains was assessed through the well diffusion method, where the diameter of the inhibition zone (IZ) was measured, and the MIC was determined, as outlined in Table 4.

Table 4. The antimicrobial activity of CSSLEO.

Tested Sample	Inhibition Zone (IZ mm) Diameter (Mean ± SD)/Minimum Inhibitory Concentration (MIC mg/mL)											
	Fungi				Gram-Positive Bacteria				Gram-Negative Bacteria			
	Aspergillus fumigatus (RCMB 002008)		*Candida albicans* (RCMB 005003 (1) ATCC 10231)		*Staphylococcus aureus* (ATCC 25923)		*Bacillus subtilis* RCMB 015 (1) NRRL B-543		*Escherichia coli* ATCC 25922		*Proteus vulgaris* RCMB 004 (1) ATCC 13315	
	IZ	MIC	IZ	MIC	IZ	MIC	IZ	MIC	IZ	MIC	IZ	MIC
CSSLEO	NA	-	NA	-	NA	-	8 ± 0.56	25 ± 0.97	13 ± 0.62	6.25 ± 0.83	NA	-
Ketoconazole (100 mg/mL)	17 ± 0.36	-	20 ± 0.27	-	-	-	-	-	-	-	-	-
Gentamycin (4 mg/mL)	-	-	-	-	24 ± 0.58	-	26 ± 0.68	-	30 ± 0.73	-	25 ± 0.81	-

NA: No activity.

2.3. Network Pharmacology-Based Analysis

2.3.1. Drug-Likeness Screening of the Identified Compounds and Acquiring Compound and Disease-Related Targets

Screening pharmacokinetic parameters plays a crucial role in the field of drug discovery, aiding researchers in pinpointing and formulating innovative, potent, and safe pharmaceuticals [28]. The data presented in Table S3 demonstrate that all 22 compounds met the criteria outlined by Lipinski's rules, including a molecular weight below 500 amu, less than 10 H-bond donors, fewer than five H-bond acceptors, and a lipophilicity value under 5. Additionally, they all achieved an Abbott oral bioavailability score of 0.5 or higher.

The molecular targets linked to the bioactive components of CSSLEO were identified using the SwissTargetPrediction and TargetNet databases, resulting in a total of 377 unique targets (Table S4). Cancer-related molecular targets were then identified for breast, colon, liver, and lung cancers using the OMIM, GeneCards, and DisGeNeT databases. This process yielded 13,120 targets for breast cancer (Table S5), 14,801 targets for colon cancer (Table S6), 10,261 targets for liver cancer (Table S7), and 15,226 targets for lung cancer (Table S8). Among these, 294 targets (Table S9) overlapped with the 377 targets associated with CSSLEO bioactive compounds (Figure 2).

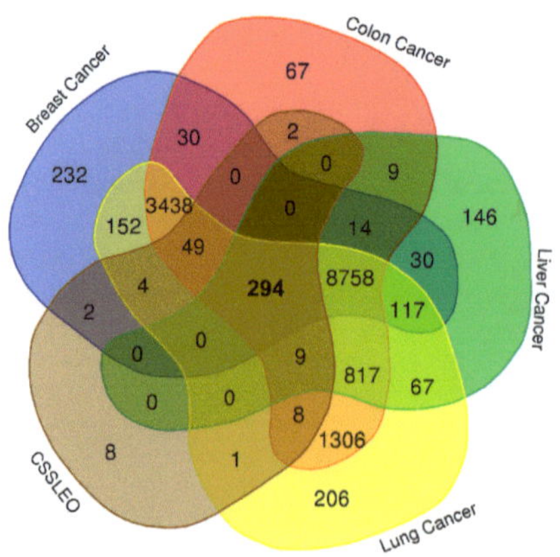

Figure 2. Overlapping molecular targets between breast, colon, liver, and lung cancers and CSSLEO bioactive compounds. The central bold value (294) indicates the number of molecular targets shared across all five categories. CSSLEO: *Cupressus sempervirens* 'Stricta' leaf essential oil.

2.3.2. Examination of Hub Genes Using Protein–Protein Interactions and Topological Analysis

The STRING database was utilized to import the overlapping targets and create a PPI network to investigate the therapeutic mechanism of CSSLEO in combating cancer. To ensure a reliable network, disconnected nodes were eliminated at a high confidence level of 0.9, resulting in a network comprising 228 nodes and 527 edges (Figure 3A). Subsequently, the network was analyzed using Cytoscape 3.10.2. To evaluate the importance of each node, scores were generated using the CytoNCA tool, which employs various network centrality measures including degree centrality (DC), betweenness centrality (BC), closeness centrality (CC), Eigenvector centrality (EC), network centrality (NC), and local average connectivity-based method (LAC). These scores are presented in Table S10. The targets were filtered based on the median values of their scores, with two separate screenings conducted. In the first screening, a threshold of DC $\geq$ 6 was applied, resulting in a subset of 60 nodes and 208 edges, as depicted in Figure 3B and Table S11. Following this, a second screening was carried out to further narrow down the targets. The thresholds used in this screening were DC $\geq$ 9, BC $\geq$ 792.6474792, CC $\geq$ 0.02604854, EC $\geq$ 0.053257737, NC $\geq$ 5.683333333, and LAC $\geq$ 3.393939394. As a result, the second screening yielded seven nodes and nine edges, as shown in Figure 3C. The topological characteristics of these seven core targets are detailed in Table 5.

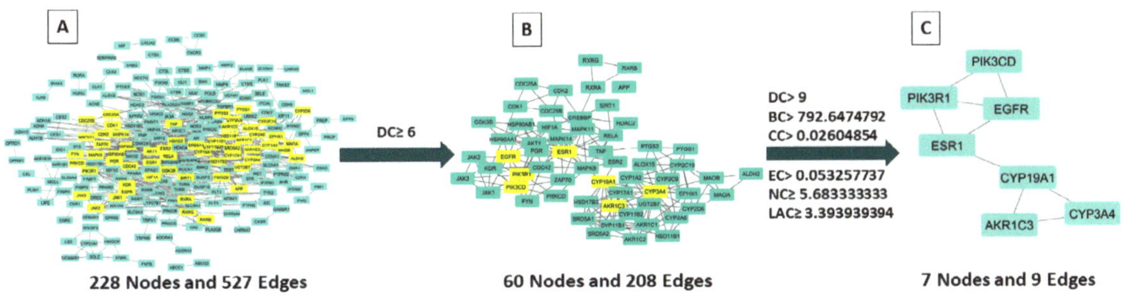

Figure 3. Core targets identified in PPI network analysis: (**A**) A PPI network comprising 294 protein targets; (**B**) initial selection based on exceeding twice the median value of degree centrality across 228 nodes; (**C**) subsequent refinement with centrality values surpassing the median of degree, betweenness, closeness, eigenvector, network, and local average connectivity-based method centralities for 60 nodes.

Table 5. The core protein targets pinpointed through the analysis of PPI and their key topological features (DC = degree centrality, BC = betweenness centrality, CC = closeness centrality, EC = Eigenvector centrality, NC = network centrality, and LAC = local average connectivity-based method).

Target Name	Description	DC	BC	CC	EC	NC	LAC
CYP3A4	Cytochrome P450 3A4	22	5123.8018	0.026203394	0.328590542	14.08333333	4.363636364
PIK3R1	Phosphoinositide-3-kinase regulatory subunit 1	19	4439.497006	0.026478479	0.062474731	11.61941474	3.789473684
PIK3CD	Phosphatidylinositol 4,5-bisphosphate 3-kinase catalytic subunit delta isoform	18	2712.974904	0.026340218	0.054360334	11.06023046	3.888888889
ESR1	Estrogen receptor	18	7909.706627	0.026640066	0.098252818	9.888965283	4.222222222
AKR1C3	Aldo-keto reductase family 1 member C3	16	2055.232626	0.026053024	0.310182363	12.1260101	5.875

Table 5. *Cont.*

Target Name	Description	DC	BC	CC	EC	NC	LAC
EGFR	Epidermal growth factor receptor	15	3023.4256	0.026383078	0.058514997	6.031962482	3.466666667
CYP19A1	Cytochrome P450 19A1 (Human aromatase)	12	6596.306973	0.026364692	0.249898016	7.346969697	5

2.3.3. Key CSSLEO Compounds Associated with Cancer Targets

An investigation was conducted to analyze the relationship between CSSLEO compounds and their targets in cancer treatment, which was then visualized through the creation of a compound-target network using Cytoscape 3.10.2 (Figure 4). The cytoNCA plug-in was utilized to identify the key active constituents among the compounds by applying the degree centrality algorithm to filter out the most significant components. The CSSLEO compounds are organized in Table S12 based on their degree value, with ten key components identified as having degree values exceeding the median (DC > 133).

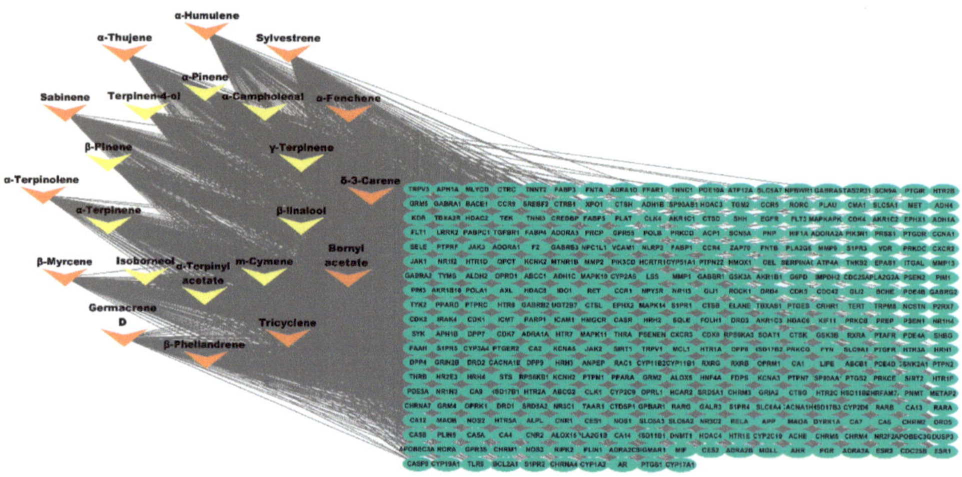

Figure 4. Compound-target network. The key components of CSSLEO with high degree scores are represented by yellow arrows, while the components of CSSLEO with low degree scores are represented by orange arrows. The green ovals symbolize the predictive targets of CSSLEO.

2.3.4. Pathway Enrichment Analysis

In order to uncover the impact of CSSLEO on cancer treatment, gene ontology (GO) enrichment analyses were carried out on 294 common targets, resulting in the identification of a total of 489 terms across GO categories, including 319 BPs, 55 CCs, and 115 MFs. These terms were selected based on an adjusted p-value of less than 0.05, as indicated in Tables S13–S15, with the top ten enriched GO terms visually presented in Figure 5A. Notable BPs enriched in the analysis include pathways such as protein phosphorylation, response to xenobiotic stimulus, phosphorylation, and negative regulation of apoptotic process. Additionally, CCs like the plasma membrane, membrane raft, cytosol, and cell surface were found to be highly enriched in the study. The MFs associated with key therapeutic targets are diverse, encompassing functions like RNA polymerase II transcription factor activity, ligand-activated sequence-specific DNA binding, protein kinase activity, zinc ion binding, and enzyme binding.

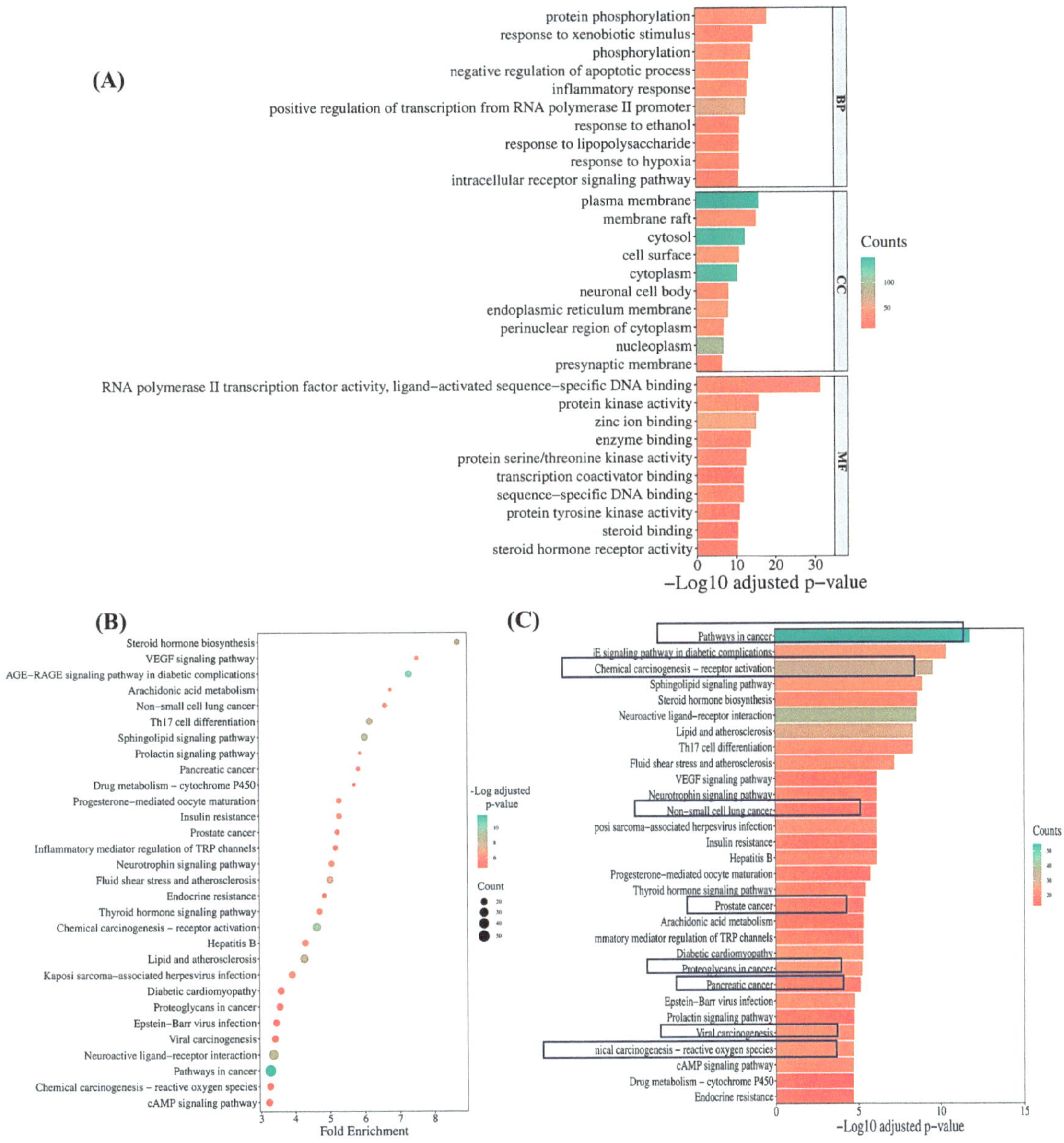

Figure 5. Enrichment analysis. (**A**) Key processes for GO categories. (**B**,**C**) Bubble chart and bar plot of key KEGG pathways. In (**C**), pathways surrounded by boxes indicate those particularly relevant to cancer.

Kyoto Encyclopaedia of Genes and Genomes (KEGG) pathway analysis revealed significant enrichment in 132 signaling pathways among overlapping targets, with a threshold of adjusted p-value < 0.05 (Table S16). The top thirty pathways with the highest negative log10-adjusted p-values were selected, highlighting key pathways relevant to cancer, such as pathways in cancer, chemical carcinogenesis-receptor activation, non-small-cell lung cancer, prostate cancer, proteoglycans in cancer, pancreatic cancer, viral carcinogenesis, and chemical carcinogenesis-reactive oxygen species (Figure 5B,C).

These findings shed light on the multifaceted mechanisms through which CSSLEO can contribute to cancer treatment.

2.3.5. In Silico Analysis of Hub Gene Expression and Prognostic Significance

In silico analysis of hub gene expression and prognostic significance was conducted using publicly available datasets through the Gene Expression Profiling Interactive Analysis webserver (GEPIA2). Box plot analysis of the seven core target genes in breast invasive carcinoma (BRCA), colon adenocarcinoma (COAD), liver hepatocellular carcinoma (LIHC), and lung adenocarcinoma (LUAD), along with their corresponding normal tissues, revealed significant differential expression patterns (Figure 6A). Notably, PIK3R1, AKR1C3, and EGFR mRNAs were significantly downregulated in breast cancer tissue, while ESR1 was significantly upregulated. PIK3R1 also exhibited significant downregulation in colon and lung cancer tissues. In liver cancer tissues, CYP3A4 was significantly downregulated, whereas AKR1C3 was significantly upregulated.

(A)

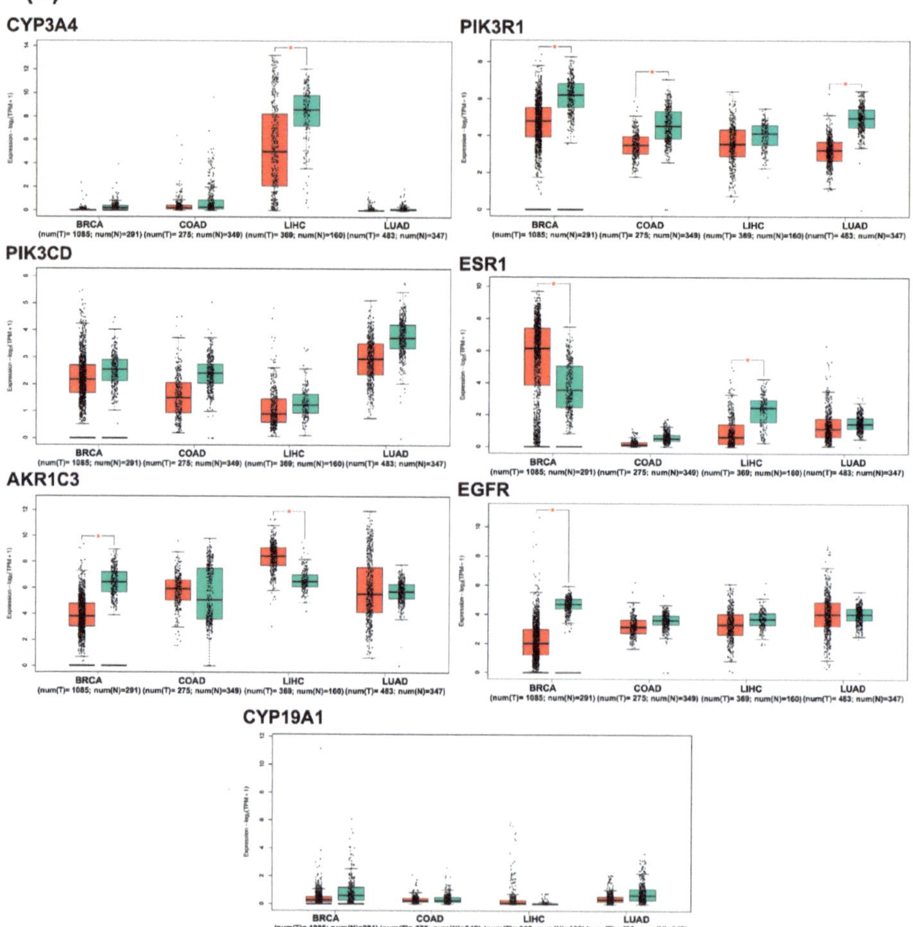

Figure 6. *Cont.*

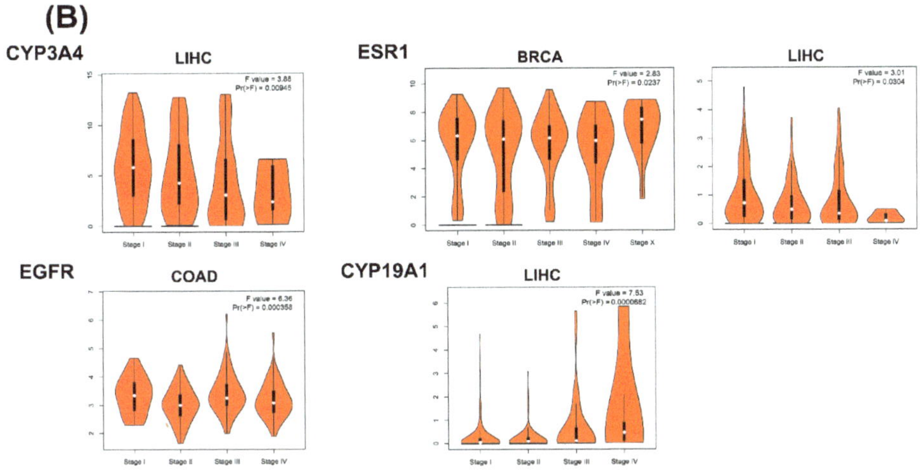

Figure 6. Validation of mRNA expression levels and stage plot analysis of hub genes. (**A**) The box plot illustrates the mRNA expression levels of various hub genes. In the plot, orange represents tumor tissue, while green corresponds to normal tissue (* $p < 0.05$). (**B**) The violin plots display the significant expression of hub genes across different pathological stages of tumors, as analyzed using the GEPIA2 database.

Stage plot analysis, visualized through violin plots (Figure 6B), elucidated the mRNA expression levels of the hub targets across various stages of BRCA, COAD, LIHC, and LUAD. Statistically significant variations ($p < 0.05$) were observed in the expression levels of CYP3A, ESR1, and CYP19A1 across different stages of liver cancer. Additionally, ESR1 demonstrated significant alterations in breast cancer stages, while EGFR showed significant changes in colon cancer stages.

Kaplan–Meier survival analysis (Figure 7) revealed significant correlations between gene expression levels and overall survival (OS) across various cancer types. In BRCA patients, decreased levels of PIK3CD (HR = 0.69, log-rank $p = 0.022$) were associated with poorer prognosis. Elevated expression of CYP19A1 (HR = 2.8, log-rank $p = 0.000074$) in COAD tissues correlated with unfavorable survival outcomes. In LIHC patients, low levels of CYP3A4 (HR = 0.66, log-rank $p = 0.019$) and ESR1 (HR = 0.55, log-rank $p = 0.00069$) were linked to unfavorable OS, while increased expression of AKR1C3 (HR = 1.7, log-rank $p = 0.0042$) and CYP19A1 (HR = 1.8, log-rank $p = 0.0012$) also correlated with poorer survival. Furthermore, reduced expression of PIK3R1 (HR = 0.59, log-rank $p = 0.00053$) in LUAD patients was significantly associated with worse OS. These findings provide insights into hub gene expression patterns and potential prognostic value across multiple cancer types. However, experimental validation is needed to confirm causal relationships and functional significance in cancer progression.

The immunohistochemical findings presented in the Human Protein Atlas (HPA) database (Figure S3) indicated elevated levels of PIK3R1, ESR1, and AKR1C3 in breast cancer tissues as opposed to normal tissues. Conversely, CYP3A4 showed increased expression in normal colon tissues compared to cancerous ones. Liver cancer tissues displayed a higher expression of PIK3R1 in comparison to normal liver tissues. Furthermore, lung cancer tissues exhibited higher expression levels of PIK3R1, AKR1C3, and EGFR when compared to normal lung tissues.

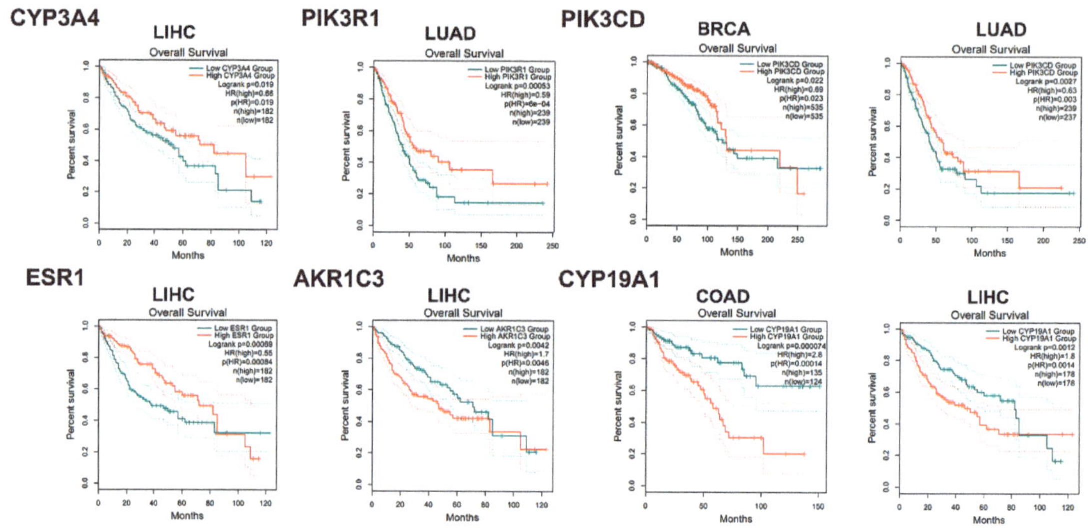

Figure 7. Overall survival associated with hub genes. The solid orange line signifies a group with high expression, while the solid green line represents a group with low expression. The curves were generated using the GEPIA2 database.

2.4. Molecular Docking

A molecular docking analysis was performed using AutoDock Vina software v.1.1.2 to evaluate the binding affinity of CSSLEO compounds to key target proteins associated with cancer pathogenesis. The analysis focused on the top ten CSSLEO compounds, namely α-terpinyl acetate, m-cymene, β-linalool, α-campholenal, γ-terpinene, α-pinene, terpinen-4-ol, β-pinene, α-terpinene, and isoborneol (Table S12), as well as the top seven key targets: CYP3A4, PIK3R1, PIK3CD, ESR1, AKR1C3, EGFR, and CYP19A1 (Table 5). The ligand molecules were docked within a designated grid box generated around the active site of each protein. The results of the docking analysis, including the binding energies between CSSLEO phytochemicals and key proteins, are presented in the heatmap on Figure 8. A lower binding energy score indicates a stronger ligand–receptor association, with a score below −5 kcal/mol indicating a high binding affinity [29]. Figure S4A–F illustrate the top interaction complex for each protein, with the highest docking score below −5 kcal/mol.

The results indicated that the key phytochemicals derived from CSSLEO demonstrated favorable affinities for CYP3A4, with the most significant outcome observed for α-terpinyl acetate. This particular compound exhibited a docking score of −7.302 kcal/mol and participated in a total of six intermolecular interactions (Figure S4A). These interactions consisted of two alkyl bonds with ILE184, three pi-alkyl bonds with PHE271 and PHE447, and a conventional hydrogen bond with CYS442. However, the constituents of CSSLEO displayed weak affinities towards PIK3R1. On the other hand, α-terpinyl acetate, m-cymene, β-linalool, α-campholenal, γ-terpinene, terpinen-4-ol, and α-terpinene exhibited favorable interactions with PIK3CD, with the highest score achieved by α-terpinyl acetate (−6.45 kcal/mol) where it engaged in a total of eight intermolecular interactions, as depicted in Figure S4B. These interactions included four alkyl bonds with ILE825, VAL828, MET900, and ILE910A, along with two pi-alkyl bonds with TYR813 and PHE908, and a conventional hydrogen bond with LYS779. In the case of ESR1, all the key compounds of CSSLEO demonstrated strong bindings, with the best score achieved by α-terpinyl acetate (−6.601 kcal/mol) and a total of five intermolecular interactions, as illustrated in Figure S4C. These interactions comprised three alkyl interactions with LEU387, MET388, and LEU391, as well as one pi-alkyl bond with PHE404.

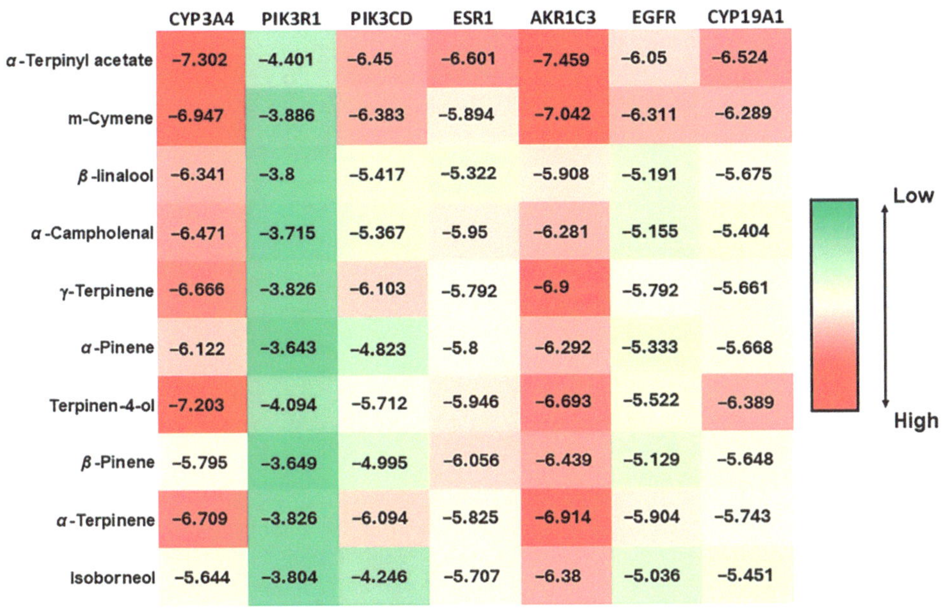

Figure 8. Heatmap visualizing the binding energies between key protein targets and CSSLEO phytochemicals. The Y-axis represents the phytochemicals used as ligands in the molecular docking experiment, while the X-axis represents the proteins.

In addition, AKR1C3 strongly interacted with all the CSSLEO key phytochemicals, with α-terpinyl acetate exhibiting a notable binding energy score of −7.459 kcal/mol. This interaction complex involved seven intermolecular associations, including an alkyl bond with LEU54, four pi-alkyl bonds with TYR24, TYR55, TRP227, and PHE306, a pi-sigma bond with TYR24, and a conventional hydrogen bond with GLN222 (Figure S4D). On the other hand, the CSSLEO components showed good binding scores with EGFR, with m-cymene having the highest interaction score of −6.311 kcal/mol. This interaction involved ten intermolecular bindings, including five alkyl bindings with VAL765, MET766, VAL769, VAL774, and LEU833, three pi-alkyl interactions with MET766, VAL769, and ALA859, and a pi-anion bond with ASP855 (Figure S4E). Furthermore, all the key compounds effectively formed interaction complexes with CYP19A1, with α-terpinyl acetate having the highest binding energy of −6.524 kcal/mol. This interaction involved eight intermolecular associations, including two alkyl bonds with VAL370 and LEU477, three pi-alkyl bonds with PHE221 and TRP224, two conventional hydrogen bonds with ARG115 and MET374, and a carbon–hydrogen bond with VAL373 (Figure S4F).

3. Discussion

The GC–MS analysis of the leaf essential oil extracted from *Cupressus sempervirens* 'Stricta' (CSSLEO) grown in Egypt, for the first time, revealed a unique chemical profile. A total of 22 compounds were identified, constituting 96.01% of the oil. The major compounds were α-pinene (33.51%) and δ-3-carene (28.73%), both bicyclic monoterpene hydrocarbons, making up 62.24% of the composition. Other significant constituents included α-terpinolene (6.77%), α-terpinyl acetate (5.91%), sabinene (3.75%), and β-myrcene (3.74%). This composition exhibits significant variations compared to those previously documented for the essential oils of *C. sempervirens*, the Mediterranean cypress, originating from different geographic regions. For instance, Alexandrian *C. sempervirens* leaf oil primarily contains cedrol (21.29%) and δ-3-carene (17.85%) [30], while *C. sempervirens* L leaf oil from

Cairo is rich in α-pinene (29.21%) and δ-3-carene (18.92%) [31]. In another investigation, it was demonstrated that the essential oil derived from the Saudi Arabian *C. sempervirens* L. aerial parts primarily contains α-pinene (48.6%), δ-3-carene (22.1%), limonene (4.6%), and α-terpinolene (4.5%) [32]. Additionally, the CSSLEO under investigation primarily comprised monoterpene hydrocarbons, contrasting with the chemically diverse *C. sempervirens* oils from other regions, which typically contain abundant monoterpene and sesquiterpene compounds [14,33]. These compositional disparities underscore the potential impact of geographical and environmental factors on essential oil biosynthesis, emphasizing the importance of comparative studies across diverse sources.

Cancer, a major contributor to worldwide mortality rates, is distinguished by unregulated cell growth and the spread of cancer cells to other parts of the body [34]. Despite advances in surgical, radiation, and chemotherapeutic treatments, the search for novel therapeutic strategies with reduced adverse effects persists. Plant-derived compounds, including EOs, are gaining attention in cancer research. Natural substances like vincristine and Taxol, as well as EOs and their components, show promise as potential anticancer agents. Studies in vitro and in vivo have demonstrated their efficacy against various tumor types. This research aims to develop more effective and less toxic cancer treatments compared to traditional chemotherapy [35].

The MTT assay revealed promising preliminary cytotoxic effects of CSSLEO on various carcinoma cell lines, particularly HepG-2 (IC_{50}: 4.44 μg/mL), followed by HCT-116, A-549, and MCF-7 (IC_{50}: 6.29, 6.71, and 7.20 μg/mL, respectively). The selectivity index (SI) exceeding 3 for cancer cell lines compared to normal VERO cells suggests potential cancer-specific targeting [36]. However, these results should be interpreted cautiously as initial findings requiring further investigation.

The observed effects align with previous studies on related EOs. For instance, *Juniperus phoenicea* EO demonstrated selective toxicity towards breast and colon cancer cells, attributed to high concentrations of monoterpenes, particularly α-pinene and β-pinene [37]. Similar antiproliferative effects were noted for EOs from other *Juniperus* species and *Abies* species against various cancer cell lines [38,39]. Notably, EO from the parent plant *Cupressus sempervirens* needles, with α-pinene as the primary component, showed significant cytotoxicity against NB4, HL-60, and EACC cells [31]. Additionally, α-terpinyl acetate from *Levisticum officinale* EO and cymene from *Thymus carmanicus* EO exhibited antiproliferative properties against specific cancer types [40,41].

While these results are encouraging, it is crucial to emphasize that the MTT assay provides only a preliminary assessment of cytotoxicity. To further explore the potential anticancer effects of CSSLEO, this research employed in silico studies to provide initial insights into the properties and potential targets of CSSLEO compounds in relation to cancer pathways. These computational analyses will offer valuable preliminary data to guide future research. However, to establish a definitive anticancer effect and fully elucidate the underlying mechanisms, additional studies will be necessary in future investigations. These may include advanced in vitro assays (e.g., apoptosis detection, cell cycle analysis), experimental investigation of molecular pathways affected by CSSLEO, evaluation of synergistic effects with established chemotherapeutic agents, and in vivo studies to assess efficacy and safety in animal models. Such comprehensive investigations, building upon the current in silico and in vitro findings, would provide a more robust foundation for potential therapeutic applications of CSSLEO in cancer treatment.

Diabetes mellitus presents a significant public health challenge, necessitating the development of safe, cost-effective, and natural therapeutic interventions. Plant-based remedies have shown promise in diabetes management, with recent research elucidating their therapeutic properties and mechanisms of action. Plant metabolites can modulate various pathways implicated in diabetes pathogenesis, including insulin secretion, glucose production inhibition, and reduction of oxidative stress and inflammation [42,43]. An essential approach in managing diabetes is focusing on digestive enzymes that play a role in breaking down carbohydrates. α-Amylase, which is mainly present in saliva and pancreatic juice, facilitates the breakdown of polysaccharides. α-Glucosidase, situated in the mucosal brush border of the small intestine, aids in breaking down complex carbohydrates into forms that can be absorbed. By inhibiting these enzymes, especially α-glucosidase, it is possible to reduce the absorption of glucose after meals, thereby potentially reducing spikes in blood glucose levels [44]. Evaluation of plant metabolites for their inhibitory effects on α-glucosidase and α-amylase provides valuable insights into their potential as natural antidiabetic agents. This approach offers a promising avenue for identifying novel, plant-derived therapeutic strategies for diabetes management [43].

CSSLEO demonstrated dose-dependent in vitro inhibitory effects on α-glucosidase and α-amylase, with more pronounced inhibition of α-amylase. This differential inhibition profile suggests a selective mechanism of action, potentially influencing postprandial glucose absorption. The effects are hypothesized to result from monoterpenes and sesquiterpenes interacting with enzyme active sites [43]. While these findings align with interest in natural products for metabolic disorder management, it is crucial to emphasize that they represent only in vitro enzymatic inhibition. Further research, particularly in vivo studies, is necessary to elucidate the physiological implications and determine potential therapeutic applications. Future investigations should focus on molecular interactions, effects on glucose metabolism in animal models, potential synergies with established treatments, and clinical safety and efficacy assessments.

Oxidative stress, an imbalance between reactive oxygen/nitrogen species and antioxidant defenses, contributes significantly to cancer and diabetes pathogenesis. In cancer, it can induce DNA damage and mutations, promoting carcinogenesis [45]. In diabetes, it disrupts enzymatic systems and enhances lipid peroxidation, leading to β-cell dysfunction [46]. These mechanisms highlight the potential therapeutic role of antioxidants in both diseases.

The antioxidant capacity of CSSLEO was significantly lower than that of ascorbic acid across DPPH, ABTS, and FRAP assays. Its antioxidant profile aligns with certain Cupressaceae species, notably *Juniperus phoenicea* EO. The Jordanian *J. phoenicea* EO demonstrated weak antioxidant activity in DPPH scavenging assays, consistent with CSSLEO's observed properties [37].

On the other hand, the observed weak antibacterial efficacy of CSSLEO against *Bacillus subtilis* (MIC 25 mg/mL) and *Escherichia coli* (MIC 6.25 mg/mL) suggests that while CSSLEO possesses antibacterial properties, its potency is limited compared to essential oils from other *Cupressus* species, which have been reported to exhibit significant antimicrobial activities [47,48]. The absence of antibacterial effects against *Staphylococcus aureus* and *Proteus vulgaris*, as well as the lack of antifungal activity against *Aspergillus fumigatus* and *Candida albicans*, could be attributed to several factors including chemical composition, environmental factors, extraction methods, microbial resistance, concentration, and solubility [49,50].

In silico methodologies, encompassing network pharmacology and molecular docking, were employed to elucidate the properties and potential targets of CSSLEO compounds in cancer pathways. This computational approach facilitated the generation of mechanistic hypotheses and prioritization of compounds for subsequent experimental validation, providing a rational framework for investigating CSSLEO's anticancer potential.

An ADME study of the 22 identified CSSLEO components demonstrated adherence to Lipinski's rules of five, indicating favorable drug-likeness and potential for development as anticancer agents. The compounds achieved Abbott oral bioavailability scores $\geq$0.5, suggesting effective oral administration potential. While the ADME study suggests favorable properties, the challenge of CSSLEO's water-insolubility is acknowledged as a crucial aspect affecting bioavailability. Future studies will investigate advanced techniques, including encapsulation and targeted delivery mechanisms, to optimize bioavailability and pharmacokinetic profiles of CSSLEO. These studies will focus on enhancing ADME parameters during in vivo evaluations, while prioritizing safety considerations. This approach aims to address a critical aspect in the potential therapeutic applications of CSSLEO.

The network pharmacology analysis identified 377 unique molecular targets associated with CSSLEO compounds, with 294 targets overlapping with those related to breast, colon, liver, and lung cancers. This multi-targeted approach aligns with the complex nature of cancer pathogenesis and the need for multi-faceted intervention strategies.

Topological analysis of the PPI network, using the STRING database and Cytoscape 3.10.2, identified seven core targets crucial in cancer pathways: CYP3A4, PIK3R1, PIK3CD, ESR1, AKR1C3, EGFR, and CYP19A1. These hub genes are widely recognized for their significant contributions to cancer biology.

Recent research has illuminated the critical roles of these seven core targets in cancer biology and their potential for therapeutic intervention. CYP3A4, involved in cancer cell metabolism and chemotherapy efficacy, has been identified as a potential target for enhancing treatment effectiveness through its inhibition [51]. PIK3R1, frequently mutated in triple-negative breast cancer, has been associated with other genetic alterations and higher tumor mutational burden, indicating its significance for targeted therapies [52,53]. PIK3CD has been implicated in the progression of various solid tumors, including colorectal, breast, and hepatocellular carcinoma, with its role in tumor growth and treatment resistance offering avenues for therapeutic interventions [54]. ESR1 mutations contribute to endocrine therapy resistance in metastatic hormone-receptor-positive breast cancer, emphasizing the need for a deeper understanding to develop more effective treatment strategies [55]. AKR1C3 has emerged as a potential biomarker and therapeutic target across multiple cancer types, impacting cancer cell proliferation and metastasis, and potentially enhancing immunotherapy response [56]. EGFR, despite being a longstanding target for cancer therapies, continues to present challenges due to acquired resistance, necessitating ongoing research into its signaling mechanisms and resistance pathways to improve treatment outcomes [57]. CYP19A1 has been identified as a key player in lipid metabolism and immune microenvironment modulation in colon cancer, with targeting strategies showing promise for enhancing immunotherapy efficacy and serving as a prognostic predictor [58].

The compound-target network analysis underscores the importance of the identified core targets, with the degree centrality algorithm highlighting key active constituents within the CSSLEO compounds. These widespread compounds are presumed to target CYP3A4, PIK3R1, PIK3CD, ESR1, AKR1C3, EGFR, and CYP19A1, which are known to mediate numerous signaling pathways involved in various cellular processes. This suggests that the anticancer effects of CSSLEO compounds may be attributed to their ability to modulate these core targets, thereby influencing multiple signaling cascades and eliciting diverse biological responses in vitro. However, further research is necessary to elucidate the precise mechanisms and specificity of these interactions in complex biological systems.

The comprehensive pathway enrichment analysis of CSSLEO targets has revealed valuable insights into potential molecular mechanisms in cancer treatment. Identification of 489 terms across GO categories, including BPs, CCs, and MFs, indicates CSSLEO's multifaceted impact on cellular pathways. Key BPs, including protein phosphorylation and xenobiotic response, align with recent findings on kinase activity in cancer progression and drug resistance [59,60]. Negative regulation of apoptosis suggests CSSLEO's potential influence on cell survival pathways [61–63]. CC enrichment in plasma membrane and membrane rafts corroborates current research on cell surface markers in cancer signaling and metastasis [64], while cytosolic enrichment implicates intracellular processes like lysosomal activity [65]. Diverse MFs, particularly RNA polymerase II transcription factor activity and ligand-activated sequence-specific DNA binding, reflect the complexity of gene regulation in cancer, consistent with recent studies on non-coding RNAs and chromatin remodeling in oncogenesis [66,67].

KEGG pathway analysis reinforces these findings, highlighting significant enrichment in cancer-related pathways and suggesting intricate interactions between extracellular matrix components and cell surface receptors in tumor progression [68,69]. Overall, this enrichment study indicates CSSLEO's broad spectrum of cellular targets, providing a foundation for future investigations into its molecular mechanisms in cancer treatment.

The differential expression of PIK3R1, AKR1C3, EGFR, ESR1, PIK3CD, CYP3A4, and CYP19A1 across various cancer types and stages illuminates their potential as biomarkers for cancer progression and prognosis. GEPIA2 database analysis revealed significant downregulation of PIK3R1, AKR1C3, and EGFR mRNAs in breast cancer tissue, with ESR1 upregulation, pointing to a complex interplay in tumorigenesis and treatment response [53]. The stage plot analysis further highlights the dynamic nature of these genes across different cancer stages, particularly in liver cancer, where CYP3A4, ESR1, and CYP19A1 showed significant changes. This suggests that these genes may be involved in the progression of liver cancer and could serve as stage-specific markers [70,71]. Kaplan–Meier survival analysis correlated gene expression levels with overall survival, associating decreased PIK3CD levels with worse prognosis in breast cancer and elevated CYP19A1 expression with poorer survival in colon cancer. These findings are consistent with recent studies that have identified PIK3CD and CYP19A1 as key players in cancer progression and potential therapeutic targets [58,72,73]. Immunohistochemical data from the HPA database supported these findings, showing elevated PIK3R1, ESR1, and AKR1C3 levels in breast cancer tissues, aligning with their proposed roles in cancer cell proliferation and hormone regulation [74–76]. Conversely, increased CYP3A4 expression in normal colon tissues compared to cancerous ones suggests a protective role against tumorigenesis [77]. These findings collectively provide valuable insights into the differential expression patterns, stage-specific changes, prognostic significance, and protein-level alterations of the identified hub genes across multiple cancer types, laying the groundwork for further functional studies and potential therapeutic interventions.

The molecular docking analysis using AutoDock Vina has provided preliminary insights into the binding affinities of CSSLEO compounds to selected cancer-related proteins. While the observed binding energies for compounds such as α-terpinyl acetate suggest potential interactions with targets involved in drug metabolism, cell signaling, and hormone regulation, it is crucial to interpret these results with critical scrutiny.

The computational analysis indicated interactions between CSSLEO compounds and proteins involved in various cellular processes, including drug metabolism (CYP3A4) [51], cell signaling (PIK3CD, AKR1C3) [56,78], hormone regulation (ESR1) [79], and growth factor signaling (EGFR) [80]. However, the widespread nature of these phytochemicals and their potential interactions with multiple proteins necessitate a thorough evaluation of their specificity and potential off-target effects.

It is noteworthy that compounds like α-pinene, prevalent in many essential oils including CSSLEO, are classified as both pan-assay interference compounds (PAINS) and invalid metabolic panaceas (IMPS). This classification indicates such a compound's tendency for non-specific interactions across various biological assays, potentially leading to false positive results and overestimated therapeutic potential. Such promiscuous binding behavior underscores the need for careful validation of any observed effects in molecular docking studies [81–83].

Ultimately, the molecular docking results suggest potential interactions between CSSLEO compounds and cancer-related proteins, but require experimental validation to confirm their pharmacological relevance. These computational findings serve as hypotheses rather than definitive evidence of therapeutic efficacy. Transitioning from in silico observations to clinically viable anticancer agents necessitates extensive in vitro and in vivo studies to evaluate pharmacokinetics, pharmacodynamics, and safety profiles.

Overall, this preliminary investigation into CSSLEO's chemical composition and biological activities provides a foundation for understanding its therapeutic potential. The integration of GC–MS analysis, in vitro assays, and in silico studies offers initial insights into CSSLEO's bioactive properties. These findings serve as a crucial precursor to more advanced research, bridging traditional knowledge with modern analytical techniques. Further investigation is essential to fully elucidate CSSLEO's therapeutic potential and advance natural product research in this domain.

4. Materials and Methods

4.1. Chemicals and Reagents

4.1.1. For Cytotoxic Assay

The experimental design incorporated several mammalian cell lines procured from the American Type Culture Collection (ATCC, Rockville, MD, USA). These included four carcinoma cell lines: MCF-7, HCT-116, HepG-2, and A-549, along with the Vero cell line.

For the investigation procedures, various chemicals and reagents were employed. Sigma (St. Louis, MO, USA) supplied dimethyl sulfoxide (DMSO), MTT reagent, and trypan blue dye. Lonza Bioscience (Basel, Switzerland) provided 0.25% Trypsin-EDTA solution, fetal bovine serum (FBS), HEPES buffer solution, RPMI-1640 culture medium, gentamycin, and L-glutamine.

4.1.2. For Antioxidant and Antidiabetic Assay

The enzymes porcine α-amylase and β-glucosidase, along with the compounds 2,2-diphenyl-1-picrylhydrazyl (DPPH), 2,2′-azino-bis(3-ethylbenzothiazoline-6-sulfonic acid) diammonium salt (ABTS), *p*-nitrophenyl-alpha-D-glucopyranoside (pNPG), and acarbose were obtained from Sigma-Aldrich Corporation (St. Louis, MO, USA). All additional chemicals and reagents utilized in this study were of analytical grade. Double-distilled water was employed for all experimental procedures.

4.2. Plant Material and Sample Preparation

In February 2023, fresh leaves of *Cupressus sempervirens* 'Stricta' were gathered from the El-Orman Botanical Garden in Giza, Egypt (Figure 9). Taxonomic verification was performed by Eng. Therese Labib, a consultant specializing in plant taxonomy at the Egyptian Ministry of Agriculture and previously the director of the El-Orman Botanical Garden. A plant reference material (ZU-Ph-Cog-0307) was archived within the botanical repository of the Pharmacognosy Department, Zagazig University Faculty of Pharmacy.

Figure 9. *Cupressus sempervirens* 'Stricta'.

The essential oil of *C. sempervirens* 'Stricta' was extracted through a hydrodistillation process. Fresh leaves (500 g) were subjected to atmospheric pressure distillation in a

Clevenger-type apparatus for 5 h at approximately 100 °C. The resulting volatile oil was subsequently dehydrated using anhydrous sodium sulphate. For preservation and to facilitate further chemical and biological analyses, the extracted oil was stored in light-resistant containers at 4 °C.

4.3. GC–MS Analysis of CSSLEO

The chromatographic analysis was conducted using a Shimadzu GCMS-QP2010 system (Kyoto, Japan) along with an with an Rtx-5MS fused silica capillary column (Restek, Bellefonte, PA, USA) measuring 30 m in length, with an internal diameter of 0.25 mm and a film thickness of 0.25 µm. The instrument employed a split–splitless injector. The temperature program for the column began with an isothermal phase at 45 °C for 2 min, followed by a linear increase to 300 °C at a rate of 5 °C/min, concluding with a 5 min isothermal period at 300 °C. The injector temperature was maintained at 250 °C throughout the analysis. Helium was employed as the carrier gas, flowing at a rate of 1.41 mL/min. The mass spectrometer scanned a mass range of 35 to 500 m/z, utilizing an ionization voltage of 70 eV and a filament emission current of 60 mA. The ion source temperature was set at 200 °C. Samples were diluted to 1% v/v, and 1 µL aliquots were injected using a split ratio of 1:15. Component identification was achieved by comparing the retention indices (RI) and mass spectra (MS) of the analytes with those found in various reference sources. These included the Adams library [24], NIST11/2011/EPA/NIH database, Wiley library database (10th edition), and published literature data [25–27]. Retention indices were calculated relative to a series of n-alkane standards (C_8–C_{28}) analyzed under identical conditions. The identified components and their relative percentages were compiled and presented in Table 1.

4.4. In Vitro Studies

4.4.1. Assessment of Cytotoxic Activity

Cell Line Propagation

The cytotoxic effects of CSSLEO were assessed in vitro against a panel of cancer cell lines, including MCF-7 (breast), HCT-116 (colorectal), HepG-2 (hepatocellular), and A-549 (lung). These neoplastic cells, representing various types of carcinomas, were maintained in RPMI-1640 growth medium supplemented with 10% heat-inactivated fetal calf serum and 50 µg/mL gentamycin. The culture conditions for these cancer cell lines were standardized at 37 °C in a humidified atmosphere containing 5% CO_2, with subculturing performed two to three times weekly. As a non-cancerous counterpart, Vero cells served as a control and were cultivated in Dulbecco's modified Eagle's medium (DMEM) enriched with 10% heat-inactivated fetal bovine serum, 1% L-glutamine, HEPES buffer, and 50 µg/mL gentamycin. The Vero cells were maintained under similar environmental conditions as the cancer cell lines (37 °C, humidified atmosphere, 5% CO_2), but with a reduced subculturing frequency of once every two weeks.

Assessment of Cytotoxicity Using MTT-Based Cell Viability Assay

Tumor and normal Vero cells were plated at 5×10^4 cells/well in 96-well tissue culture plates and incubated for 24 h. CSSLEO was then added at concentrations ranging from 0.25 to 500 µg/mL in triplicate. For each 96-well microplate, a subset of six wells was allocated as vehicle controls, containing 0.5% (v/v) DMSO. Plates were incubated at 37 °C with 5% CO_2 for 24 h. Untreated cells served as negative controls, while cisplatin-treated cells were used as positive controls. Post-incubation, media was replaced with phenol red-free RPMI 1640, and 10 µL of 12 mM MTT solution was added to each well. After 4 h, 85 µL of media was removed, and 50 µL DMSO was added. Following thorough mixing and a 10 min incubation at 37 °C, absorbance was measured at 590 nm using a microplate reader. Viability percentage was calculated as: Viability (%) = (ODt/ODc) × 100, where ODt and ODc represent the average optical densities of treated and untreated cells, respectively. Here is a rephrased version that aims to be consistent, academic, and professional while avoiding repetitive meanings: The relationship between cell viability and drug concentration is illustrated through survival curves for each tumor cell line post-treatment. The 50% inhibitory concentration (IC_{50}), defined as the drug concentration at which half of the cell population exhibits cytotoxic effects, is derived from the dose–response curve analysis. This analysis is conducted for each concentration using Graphpad Prism software (San Diego, CA, USA) [84]. For the normal cell line (Vero), the equivalent metric is denoted as CC_{50} (50% cytotoxic concentration), representing the concentration at which half of the normal cells are affected.

Data Analysis and Statistical Procedures

Statistical analyses were performed using Statistica software version 8.0 (StatSoft Inc., Tulsa, OK, USA). Data are presented as mean values ± standard deviation (SD) based on triplicate measurements. Significant differences among groups were determined using one-way analysis of variance (ANOVA) followed by Tukey's post hoc test. A p-value < 0.05 was considered statistically significant.

Quantifying CSSLEO's Cytotoxic Selectivity

The selectivity index (SI) was determined by computing the ratio of the average CC_{50} value obtained from the Vero cell line to the IC_{50} values of individual cancer cell lines, specifically MCF-7, A-549, Hep-G2, and HCT-116. This calculation provides a quantitative measure of the compound's selective cytotoxicity towards cancer cells relative to normal cells.

4.4.2. Assessment of Antioxidant Activity

The antioxidant efficacy of CSSLEO was investigated through a triad of established assays: DPPH, $ABTS^{\bullet+}$, and FRAP. These methodologies were implemented in accordance with protocols described by Elkomy et al. [85], Ling et al. [86], and Elaasser et al. [87], respectively. To ensure reliability, each experiment was conducted in triplicate, and the mean values were computed. Ascorbic acid was employed as a reference standard throughout the antioxidant capacity assessment.

4.4.3. Assessment of Antidiabetic Activity

α-Glucosidase Inhibition Activity

The α-glucosidase inhibitory activity was evaluated using a spectrophotometric method employing yeast α-glucosidase and p-nitrophenyl-α-D-glucopyranoside (pNPG) as the substrate [88]. The assay procedure was as follows:

The reaction mixture comprised 0.1 mL of α-glucosidase solution (1 U/mL), 0.2 mL of CSSLEO at various concentrations (0.25–500 µg/mL, prepared in 10% DMSO), and 0.5 mL of phosphate buffer (100 mM, pH 6.8). This mixture was pre-incubated at 37 °C for 15 min. Subsequently, 0.2 mL of 5 mM pNPG was added to initiate the enzymatic reaction, followed by an additional 20 min incubation at 37 °C. The reaction was terminated by adding 0.1 M sodium carbonate (Na_2CO_3). Absorbance readings were taken at 405 nm using a spectrophotometer. Acarbose was used as a positive control, tested at concentrations identical to those of the plant extracts. The percentage inhibition of α-glucosidase activity was calculated using the following equation: % Inhibition = [(A405blank − A405test)/A405blank] × 100, where A405blank represents the absorbance of the control reaction (without inhibitor) at 405 nm, and A405test denotes the absorbance of the sample containing the potential inhibitor at 405 nm.

α-Amylase Inhibition Activity

CSSLEO solutions were prepared at concentrations ranging from 0.25 to 500 µg/mL. Initially, 100 mg of the oil was solubilized in 5 mL of 10% DMSO, and the volume was adjusted to 100 mL using a buffer solution (0.02 M Na_2HPO_4/NaH_2PO_4, 0.006 M NaCl, pH 6.9), following a previously established method [89]. A stock solution of porcine pancreatic α-amylase was prepared at 2 units/mL by dissolving 12.5 mg of the enzyme in a minimum volume of 10% DMSO and diluting to 100 mL with the buffer solution. The starch substrate solution was prepared by dissolving 1 g of corn starch in 100 mL of distilled water. The assay was conducted by combining 200 μL of the CSSLEO solution with 200 μL of the α-amylase stock solution, followed by incubation at 30 °C for 10 min. Subsequently, 200 μL of the starch solution was added, and the mixture was further incubated at 30 °C for 3 min. The reaction was terminated by the addition of 3,5-dinitro salicylic acid, and the mixture was heated in a water bath at 85–90 °C for 10 min. After cooling to room temperature, 5 mL of distilled water was added. A blank solution was prepared by substituting the CSSLEO with 200 μL of buffer solution. Acarbose was employed as a positive control. The absorbance of the samples was measured at 540 nm using a UV–Vis spectrophotometer. The α-amylase inhibitory potential was calculated using the following equation: % Inhibition = [(A540blank − A540test)/A540blank] × 100%, where A540blank and A540test represent the absorbance of the blank and test samples, respectively.

Statistical Analysis

The inhibitory concentration at 50% (IC_{50}) values were determined through nonlinear regression analysis of the dose–response curves. This analysis was performed using GraphPad Prism software version 9.4.1 (GraphPad Software Inc., San Diego, CA, USA). The software generated dose–response curves for each tested concentration, from which IC50 values were interpolated. All experiments were conducted in triplicate to ensure reproducibility. Results are presented as mean values ± standard deviation (SD).

4.4.4. Assessment of Antimicrobial Activity

Antifungal and Antibacterial Activities (Well Diffusion Method)

The antifungal and antibacterial properties of CSSLEO were evaluated using the well diffusion method [90]. The study assessed its efficacy against two fungal strains: *Aspergillus fumigatus* (RCMB 002008) and *Candida albicans* (RCMB 005003). Additionally, the compound's antibacterial activity was tested against two Gram-positive bacteria, *Staphylococcus aureus* (ATCC 25923) and *Bacillus subtilis* (RCMB 015 NRRL B-543), as well as two Gram-negative bacteria, *Escherichia coli* (ATCC 25922) and *Proteus vulgaris* (RCMB 004 ATCC 13315). All microbial strains were sourced from the Regional Center for Mycology and Biotechnology at Al-Azhar University, Egypt. For fungal assays, potato dextrose agar (PDA) was used as the culture medium. The test samples were prepared by dissolving CSSLEO in dimethyl sulfoxide (DMSO) to achieve a concentration of 50 mg/mL. Ketoconazole (100 μg/mL) and Gentamycin (4 mg/mL) served as positive controls for fungi and bacteria, respectively. The assay was conducted by creating 6 mm-diameter wells in the agar plates, which were then filled with 100 μL of the sample solution. DMSO was used as a negative control. The inoculated plates were incubated at 37 °C for 24–48 h. Each assay was performed in triplicate to ensure reliability. The antimicrobial activity was quantified by measuring the diameter of the inhibition zones. Results were reported as the mean inhibition zone diameter (in mm) ± standard deviation (SD). This methodology aligns with established protocols in antimicrobial susceptibility testing [91].

Determination of Minimum Inhibitory Concentration (MIC)

The MIC values of the investigated oil sample were determined using the agar dilution method. The sample was initially dissolved in DMSO and subsequently subjected to serial dilution to achieve a concentration range of 0.39 to 50 mg/mL. Bacterial isolates were cultured on PDA plates, from which individual colonies were selected and suspended in saline solution. The turbidity of the bacterial suspension was adjusted to match the 0.5 McFarland standard, corresponding to approximately 5×10^6 colony-forming units per milliliter (cfu/mL). This suspension was further diluted 1:10 with saline. PDA plates containing various concentrations of the serially diluted sample were prepared. Each plate was inoculated with 2 μL of the prepared bacterial suspension, resulting in a final inoculum concentration of 10^3 cfu per spot. The inoculated plates were then incubated at 30 °C for a period of 24 to 48 h. Throughout the incubation period, the plates were monitored for signs of bacterial growth to determine the MIC values [92,93].

4.5. Network Pharmacology

4.5.1. Evaluating the Pharmacokinetic Properties of CSSLEO Phytoconstituents

The canonical SMILES formulas of the 22 phytoconstituents that were identified through GC–MS analysis were gathered from the PubChem database (https://pubchem.ncbi.nlm.nih.gov/, accessed on 20 March 2024) or generated using ChemDraw v22.0.0.22 (PerkinElmer Informatics, Inc., Buckinghamshire, UK). Subsequently, these SMILES formulas were inputted into the SwissADME web server for drug-like property assessment, where specific screening criteria such as Lipinski's rule [94] and a minimum Abbott oral bioavailability score above 0.5 were applied to evaluate their pharmaceutical potential.

4.5.2. Identification of the Intersection Genes between Breast, Colon, Liver, and Lung Cancers and CSSLEO Bioactive Compounds

Putative targets associated with the compounds were extracted from two predictive databases: Swiss Target Prediction (STP, http://www.swisstargetprediction.ch/) [95] and TargetNet (http://targetnet.scbdd.com/) [96], accessed on 22 March 2024. To compile cancer-related targets, three comprehensive databases were utilized: Online Mendelian Inheritance in Man (OMIM, https://www.omim.org/) [97], DisGeNet (https://www.disgenet.org/search), and GeneCards (https://www.genecards.org/) [98,99], accessed on 23 March 2024. The search process for cancer-related targets was conducted separately for each cancer type, employing specific search terms such as "Breast carcinoma", "Colon Carcinoma", "Hepatocellular Carcinoma", and "Lung Carcinoma".

Following target identification, UniProt IDs and gene symbols were obtained from the UniProt database on 25 March 2024 (https://www.uniprot.org/) [100]. To maintain data integrity, duplicate targets were eliminated from the dataset.

To visualize and analyze the intersection between targets associated with the bioactive compounds and those linked to the four cancer types, a Venn diagram was generated using an online bioinformatics tool (https://bioinformatics.psb.ugent.be/webtools/Venn/, accessed on 25 March 2024).

4.5.3. Constructing the Protein–Protein Interaction (PPI) Network

Following the acquisition of overlapping targets, the STRING database v12.0 (https://string-db.org/, accessed on 25 March 2024) [101] was employed to examine the protein–protein interactions (PPI). The analysis focused on the interactions with a confidence level greater than 0.9, specifically within the species *Homo sapiens*.

The resulting interaction data were then imported into Cytoscape software version 3.10.2 (NIGMS, Bethesda, MD, USA) for the purpose of visualizing the networks [102]. Within Cytoscape, the CytoNCA tool was employed to analyze the network topology parameters of the targets, including DC, BC, CC, EC, NC, and LAC [103]. The DC represents the number of edges connecting to a node, indicating its significance. The BC signifies a target's bridging role within the network. The CC measures the proximity between a target and all other targets in the network. The EC depicts a target's connectivity with other pivotal targets in the network. Lastly, the overall significance of a target in the network is indicated by the NC, while the level of clustering of a target in the network is quantified by the LAC [104]. This analysis aimed to identify hub networks and core targets through a two-step filtration process. The data were first processed to ensure that DC was at least 2 times the median value. A secondary screening was then done to identify core targets with DC, BC, EC, CC, NC, and LAC above or equal to their respective median values [105]. This filtering methodology allowed us to key nodes in the network with the highest importance and influence, which could potentially play a crucial role in in the progression of the disease under investigation or become potential targets for therapeutic interventions.

4.5.4. Establishing the Compound-Target Interaction Network

A compound-target network was created in Cytoscape 3.10.2 to visualize the connection between the bioactive components of CSSLEO and their potential targets for cancer therapy. Within the network, candidate compounds were depicted as nodes, and the interactions between compounds and targets were represented by edges. Using the Cyto-Hubba plugin in Cytoscape [106], essential active constituents of CSSLEO were identified by screening them based on their degree value, highlighting their significance in potential cancer treatment. Furthermore, the key components were determined by selecting the components with a degree value higher than the median.

4.5.5. Enrichment Analysis of GO and KEGG Pathways

The critical therapeutic targets within the biological system and signalling pathways underwent analysis through GO and KEGG enrichment analysis. The overlapping genes were inputted into the Database for Annotation, Visualization, and Integrated Discovery (DAVID, https://david.ncifcrf.gov/, accessed on 28 March 2024) [107] to anticipate the three GO categories (molecular functions-MF, biological processes-BP, cellular components-CC) and KEGG pathways associated with overlapping genes. *Homo sapiens* (Human) was chosen as the organism for this study, and the statistical significance was determined using the Benjamini–Hochberg procedure for multiple testing correction, with a threshold of $p < 0.05$. Subsequently, the visualization analysis was carried out using the Bioinformatics Platform (http://www.bioinformatics.com.cn/en, accessed on 31 March 2024) to generate bar plots and bubble charts for further examination and interpretation.

4.5.6. Computational Analysis of Gene Expression in Cancer Datasets

To further investigate the potential role of the identified core target genes in various cancers, their expression patterns and clinical significance were explored using publicly available databases. The GEPIA2 resource (http://gepia.cancer-pku.cn/, accessed on 3 April 2024) [108,109] was employed to analyze the mRNA expression levels of the seven core target genes. This analysis focused on four cancer types: BRCA, COAD, LIHC, and LUAD. Additionally, expression levels in corresponding normal tissues from the Genotype-Tissue Expression (GTEx) project were evaluated. GEPIA2 was utilized to assess the association between gene expression and clinicopathological features, specifically focusing on the pathological stage of the cancers. Data from The Cancer Genome Atlas (TCGA) project were used for this analysis.

Furthermore, the potential association between gene expression and OS was investigated using Kaplan–Meier survival analysis in GEPIA2. For each gene, patients were stratified into high and low expression groups based on the median expression level. The log-rank test was used to compare survival curves, and hazard ratios (HRs) with 95% confidence intervals were calculated. The HR value indicates the relative risk of death in the high expression group compared to the low expression group. An HR > 1 suggests that higher gene expression is associated with increased risk, while an HR < 1 suggests that higher gene expression is associated with decreased risk. Statistical significance was set at $p < 0.05$ for the log-rank test. To examine protein expression patterns of these hub genes, immunohistochemistry data from the HPA (https://www.proteinatlas.org/, accessed on 5 April 2024) were analyzed [110]. Staining intensity and patterns in cancer tissues were compared to those in corresponding normal tissues.

4.6. Molecular Docking Study

This study investigated the molecular interactions between 10 key active ingredients and 7 core target proteins using computational docking analysis. Three-dimensional crystal structures of the target proteins, obtained from the Protein Data Bank in 10 April 2024 (http://www.rcsb.org) [111], included CYP3A4 (PDB ID: 1TQN/2.05 Å) [112], PIK3R1 (PDB ID: 4JPS/2.20 Å) [113], PIK3CD (PDB ID: 6PYR/2.21 Å) [114], ESR1 (PDB ID: 1SJ0/1.90 Å) [115], AKR1C3 (PDB ID: 4H7C/1.97 Å) [116], EGFR (PDB ID: 5U8L/1.60

Å) [117], and CYP19A1 (PDB ID: 3S79/2.75 Å) [118]. The protein structures were prepared using Biovia Discovery Studio v21.1.0.20298 [119] and USCF Chimera 1.17.3 software [120] as previously described [121]. For CYP3A4, PIK3R1, and CYP19A1, which lacked co-crystallized ligands, the Computed Atlas for Surface Topography of Proteins server (CASTp; http://sts.bioe.uic.edu/castp/index.html) [122] was accessed on 11 April 2024, to predict active pockets and determine coordinates. The active compounds were obtained from the PubChem database and converted to dockable formats using OpenBabel 2.4.1 [123]. Molecular docking was performed using AutoDock Vina 1.1.2, with grid coordinates for each protein's active site established through the AutoDock Vina suite in UCSF Chimera. The docking process generated 10 unique poses for each ligand–protein interaction. The details of the grid boxes, including their centers, sizes, and the amino acid residues of the active sites, can be found in Table S17. Binding affinities were assessed using docking scores (kcal/mol), with lower scores indicating stronger interactions. The best-scoring pose with minimal root mean square deviation (RMSD) value was selected for each ligand. To ensure accuracy, co-crystallized ligands were re-docked where available to validate the active sites.

The resulting molecular interactions were visualized and analyzed using Biovia Discovery Studio Visualizer v21.1.0.20298 (BIOVIA Dassault Systemes, San Diego, CA, USA).

5. Conclusions

In conclusion, this research integrates chemical analysis, in vitro assays, and in silico studies to investigate the diverse bioactive profile of *Cupressus sempervirens* 'Stricta' leaf essential oil (CSSLEO) from Egypt, and its potential therapeutic properties. While preliminary results show promise in cytotoxicity against cancer cell lines and inhibition of diabetes-related enzymes, these findings warrant further rigorous validation and critical evaluation within a broader experimental context. The identification of key molecular targets and potential mechanisms of action through computational methods offers a foundation for future research. However, extensive further investigation, including advanced in vitro and in vivo studies, is necessary to validate these initial findings and fully understand CSSLEO's therapeutic potential. This study serves as a starting point for more targeted research into the possible applications of CSSLEO in cancer and diabetes management.

Supplementary Materials: The following supporting information can be downloaded at: https://www.mdpi.com/article/10.3390/ph17081019/s1, Figure S1: Chemical structures of the identified components in Cupressus sempervirens 'stricta' leaf essential oil; Figure S2: The in vitro cytotoxic effects of Cupressus sempervirens 'Stricta' leaf essential oil and cisplatin on various carcinoma cell lines; Figure S3: The immunohistochemical images of the protein expression of the core targets in breast, colon, liver, lung cancers and their respective normal tissues obtained from the Human Protein Atlas (HPA) database; Figure S4: Illustrations of molecular docking results presented in both three-dimensional and two-dimensional formats; Table S1: Inhibition effects of CSSLEO on α-Glucosidase enzyme activity; Table S2: Inhibition effects of CSSLEO on α-amylase enzyme activity; Table S3: Pharmacokinetics and the drug-likeness properties of CSSLEO constituents; Table S4: Molecular targets of CSSLEO compounds; Table S5: Molecular targets associated with breast cancer; Table S6: Molecular targets associated with colon cancer; Table S7: Molecular targets associated with liver cancer; Table S8: Molecular targets associated with lung cancer; Table S9: Overlapping targets between CSSLEO and breast, colon, liver and lung cancers; Table S10: The topological features of the PPI nodes; Table S11: The protein targets obtained through the first filtration step and their topological features; Table S12: The identified compounds of CSSLEO ranked by the degree scores; Table S13: Details of GO analysis pertaining to biological processes; Table S14: Details of GO analysis pertaining to cellular components; Table S15: Details of GO analysis pertaining to molecular functions; Table S16: Details of KEGGs pathway analysis; Table S17: Target proteins, the corresponding grid coordinates, and amino acid residues of the active sites.

Author Contributions: Conceptualization, E.F., R.O., N.T., S.P., S.G., M.M.E.-D. and A.M.E.-S.; methodology, E.F. and N.T.; software, E.F.; investigation, E.F., R.O., N.T., S.P. and S.G.; validation, E.F. and N.T.; formal analysis, E.F., R.O., N.T., and S.P.; resources, R.O., S.P., M.M.E.-D. and A.M.E.-S.; data curation, E.F., R.O. and N.T.; writing—original draft preparation, E.F. and N.T.; writing—review and editing, R.O, S.P., S.G., M.M.E.-D. and A.M.E.-S.; supervision, A.M.E.-S. and M.M.E.-D.; project administration, E.F., N.T., M.M.E.-D. and A.M.E.-S.; funding acquisition, R.O., S.P. and S.G. All authors have read and agreed to the published version of the manuscript.

Funding: The present research has been financially supported by the Researchers Supporting Project (number RSP2024R431) at King Saud University, Riyadh, Saudi Arabia.

Institutional Review Board Statement: Not applicable.

Informed Consent Statement: Not applicable.

Data Availability Statement: All data and materials used are available in the manuscript.

Conflicts of Interest: The authors declare no conflicts of interest.

References

1. Ali, B.; Al-Wabel, N.A.; Shams, S.; Ahamad, A.; Khan, S.A.; Anwar, F. Essential oils used in aromatherapy: A systemic review. *Asian Pac. J. Trop. Biomed.* **2015**, *5*, 601–611. [CrossRef]
2. Eslahi, H.; Fahimi, N.; Sardarian, A.R. Chemical composition of essential oils. *Essent. Oils Food Process. Chem. Saf. Appl.* **2017**, 119–171. [CrossRef]
3. Bhavaniramya, S.; Vishnupriya, S.; Al-Aboody, M.S.; Vijayakumar, R.; Baskaran, D. Role of essential oils in food safety: Antimicrobial and antioxidant applications. *Grain Oil Sci. Technol.* **2019**, *2*, 49–55. [CrossRef]
4. Alves-Silva, J.M.; Zuzarte, M.; Girão, H.; Salgueiro, L. The role of essential oils and their main compounds in the management of cardiovascular disease risk factors. *Molecules* **2021**, *26*, 3506. [CrossRef] [PubMed]
5. Aljaafari, M.N.; AlAli, A.O.; Baqais, L.; Alqubaisy, M.; AlAli, M.; Molouki, A.; Ong-Abdullah, J.; Abushelaibi, A.; Lai, K.-S.; Lim, S.-H.E. An overview of the potential therapeutic applications of essential oils. *Molecules* **2021**, *26*, 628. [CrossRef] [PubMed]
6. Ghaffari, T.; Hong, J.-H.; Asnaashari, S.; Farajnia, S.; Delazar, A.; Hamishehkar, H.; Kim, K.H. Natural phytochemicals derived from gymnosperms in the prevention and treatment of cancers. *Int. J. Mol. Sci.* **2021**, *22*, 6636. [CrossRef] [PubMed]
7. Joshi, S.; Sati, S. Antifungal potential of gymnosperms: A review. *Microbes Divers. Biotechnol.* **2012**, 333–345.
8. Ancuceanu, R.; Anghel, A.I.; Hovaneț, M.V.; Ciobanu, A.-M.; Lascu, B.E.; Dinu, M. Antioxidant Activity of Essential Oils from Pinaceae Species. *Antioxidants* **2024**, *13*, 286. [CrossRef] [PubMed]
9. Swor, K.; Satyal, P.; Poudel, A.; Setzer, W.N. Gymnosperms of Idaho: Chemical compositions and enantiomeric distributions of essential oils of *Abies lasiocarpa, Picea engelmannii, Pinus contorta, Pseudotsuga menziesii,* and *Thuja plicata. Molecules* **2023**, *28*, 2477. [CrossRef]
10. Schulz, C.; Knopf, P.; Stützel, T. Identification key to the Cypress family (Cupressaceae). *Feddes Repert. Z. Für Bot. Taxon. Und Geobot.* **2005**, *116*, 96–146. [CrossRef]
11. Park, C.; Woo, H.; Park, M.-J. Development of Pinaceae and Cupressaceae Essential Oils from Forest Waste in South Korea. *Plants* **2023**, *12*, 3409. [CrossRef] [PubMed]
12. Santos Filho, F.C.; da Silva Amaral, L.; Rodrigues-Filho, E. Composition of essential oils from *Cupressus lusitanica* and a *Xylariaceous fungus* found on its leaves. *Biochem. Syst. Ecol.* **2011**, *39*, 485–490. [CrossRef]
13. Frezza, C.; De Vita, D.; Sciubba, F.; Toniolo, C.; Tomassini, L.; Nicoletti, M.; Franceschin, M.; Guiso, M.; Bianco, A.; Serafini, M.; et al. There Is Not Only *Cupressus sempervirens* L.: A Review on the Phytochemistry and Bioactivities of the Other *Cupressus* L. Species. *Appl. Sci.* **2022**, *12*, 7353. [CrossRef]
14. Patgar, B.; Satish, S.; Shabaraya, A. Essential Oil of *Cupressus sempervirens*: A Brief Review. *World J. Pharm. Pharm. Sci.* **2021**, *10*, 864–872. [CrossRef]
15. Hussain, K.M.; Saquib, M.; Ahamad, T.; Khatoon, S.; Khan, F.M. *Mediterranean cypress "Cupressus sempervirens"*: A Review on Phytochemical and Pharmacological Properties. *Curr. Tradit. Med.* **2019**, *5*, 278–297. [CrossRef]
16. Orhan, I.E.; Tumen, I. Potential of *Cupressus sempervirens* (*Mediterranean cypress*) in health. In *The Mediterranean Diet*; Elsevier: Amsterdam, The Netherlands, 2015; pp. 639–647. [CrossRef]
17. Al-Rajhi, A.M.H.; Bakri, M.M.; Qanash, H.; Alzahrani, H.Y.; Halawani, H.; Algaydi, M.A.; Abdelghany, T.M. Antimicrobial, Antidiabetic, Antioxidant, and Anticoagulant Activities of *Cupressus sempervirens* In Vitro and In Silico. *Molecules* **2023**, *28*, 7402. [CrossRef] [PubMed]
18. Weick, C.W.; Aamir, N.; Reichart, J. The Ethnobotanical Evolution of the *Mediterranean cypress* (*Cupressus sempervirens*). *Econ. Bot.* **2023**, *77*, 203–221. [CrossRef]
19. Skenfield, M.W. *Historic Trees of the California Gold Rush Towns*; Lulu Press: Morrisville, NC, USA, 2019.
20. Rowell, R.J. *Ornamental Conifers for Australian Gardens*; UNSW Press: Sydney, Australia, 1996.
21. Dirr, M.A. *Dirr's Encyclopedia of Trees and Shrubs*; Timber Press: Portland, OR, USA, 2016.

22. Boukhris, M.; Regane, G.; Yangui, T.; Sayadi, S.; Bouaziz, M. Chemical composition and biological potential of essential oil from Tunisian *Cupressus sempervirens* L. *J. Arid Land Stud.* **2012**, *22*, 329–332.
23. Batiha, G.E.-S.; Teibo, J.O.; Shaheen, H.M.; Akinfe, O.A.; Awad, A.A.; Teibo, T.K.A.; Alexiou, A.; Papadakis, M. Bioactive compounds, pharmacological actions and pharmacokinetics of *Cupressus sempervirens*. *Naunyn-Schmiedeberg's Arch. Pharmacol.* **2023**, *396*, 389–403. [CrossRef]
24. Adams, R.P. *Identification of Essential Oil Components by Gas Chromatography/Mass Spectrometry*, 5th ed.; Texensis Publishing: Gruver, TX, USA, 2017.
25. Hammoda, H.M.; Harraz, F.M.; Farag, M.A.; El-Aswad, A.F.; El-Hawiet, A.; Eid, A.M. Volatiles profiling and bioactivities of *Cupressus* spp. leaf and cone essential oils as analyzed via chemometrics tools. *J. Essent. Oil Res.* **2019**, *31*, 53–62. [CrossRef]
26. Marongiu, B.; Porcedda, S.; Piras, A.; Sanna, G.; Murreddu, M.; Loddo, R. Extraction of *Juniperus communis* L. ssp. nana Willd. essential oil by supercritical carbon dioxide. *Flavour Fragr. J.* **2006**, *21*, 148–154. [CrossRef]
27. Kamal, R.M.; Sabry, M.M.; El-Halawany, A.M.; Rabie, M.A.; El Sayed, N.S.; Hifnawy, M.S.; Younis, I.Y. GC-MS analysis and the effect of topical application of essential oils of *Pinus canariensis* C. Sm., *Cupressus lusitanica* Mill. and *Cupressus arizonica* Greene aerial parts in Imiquimod–Induced Psoriasis in Mice. *J. Ethnopharmacol.* **2024**, *318*, 116947. [CrossRef] [PubMed]
28. Mohanty, D.; Padhee, S.; Priyadarshini, A.; Champati, B.B.; Das, P.K.; Jena, S.; Sahoo, A.; Panda, P.C.; Nayak, S.; Ray, A. Elucidating the anti-cancer potential of *Cinnamomum tamala* essential oil against non-small cell lung cancer: A multifaceted approach involving GC-MS profiling, network pharmacology, and molecular dynamics simulations. *Heliyon* **2024**, *10*, e28026. [CrossRef]
29. Lan, T.; Wang, J.; Zeng, R.; Gao, C.; Liu, X.; Luo, L.; Liang, Y.; Guo, Z.; Wang, W.; Hong, M. Therapeutic targets and molecular mechanisms of Huangqin decoction in liver cancer: A network pharmacology and molecular docking approach. *J. Herb. Med.* **2024**, *43*, 100822. [CrossRef]
30. Elansary, H.O.; Salem, M.Z.; Ashmawy, N.A.; Yacout, M.M. Chemical composition, antibacterial and antioxidant activities of leaves essential oils from *Syzygium cumini* L., *Cupressus sempervirens* L. and *Lantana camara* L. from Egypt. *J. Agric. Sci.* **2012**, *4*, 144–152. [CrossRef]
31. Fayed, S.A. Chemical composition, antioxidant, anticancer properties and toxicity evaluation of leaf essential oil of *Cupressus sempervirens*. *Not. Bot. Horti Agrobot. Cluj-Napoca* **2015**, *43*, 320–326. [CrossRef]
32. Selim, S.A.; Adam, M.E.; Hassan, S.M.; Albalawi, A.R. Chemical composition, antimicrobial and antibiofilm activity of the essential oil and methanol extract of the *Mediterranean cypress* (*Cupressus sempervirens* L.). *BMC Complement. Altern. Med.* **2014**, *14*, 179. [CrossRef]
33. Al-Snafi, A.E. Medical importance of *Cupressus sempervirens*—A review. *IOSR J. Pharm.* **2016**, *6*, 66–76.
34. Alorabi, M.; Elghazawy, H.; Alorabi, M.; Elghazawy, H. Cancer control in Egypt: Investing in health. *ASCO Post* **2021**.
35. Mohamed Abdoul-Latif, F.; Ainane, A.; Houmed Aboubaker, I.; Mohamed, J.; Ainane, T. Exploring the potent anticancer activity of essential oils and their bioactive compounds: Mechanisms and prospects for future cancer therapy. *Pharmaceuticals* **2023**, *16*, 1086. [CrossRef]
36. Fikry, E.; Orfali, R.; Elbaramawi, S.S.; Perveen, S.; El-Shafae, A.M.; El-Domiaty, M.M.; Tawfeek, N. *Chamaecyparis lawsoniana* leaf essential oil as a potential anticancer agent: Experimental and computational studies. *Plants* **2023**, *12*, 2475. [CrossRef] [PubMed]
37. Bajes, H.R.; Oran, S.A.; Bustanji, Y.K. Chemical composition and antiproliferative and antioxidant activities of essential oil from *Juniperus phoenicea* L. cupressaceae. *Res. J. Pharm. Technol.* **2022**, *15*, 153–159. [CrossRef]
38. Sahin Yaglioglu, A.; Eser, F.; Yaglioglu, M.S.; Demirtas, I. The antiproliferative and antioxidant activities of the essential oils of *Juniperus* species from Turkey. *Flavour Fragr. J.* **2020**, *35*, 511–523. [CrossRef]
39. Wajs-Bonikowska, A.; Sienkiewicz, M.; Stobiecka, A.; Maciąg, A.; Szoka, Ł.; Karna, E. Chemical composition and biological activity of Abies alba and A. koreana seed and cone essential oils and characterization of their seed hydrolates. *Chem. Biodivers.* **2015**, *12*, 407–418. [CrossRef] [PubMed]
40. Sertel, S.; Eichhorn, T.; Plinkert, P.K.; Efferth, T. Chemical composition and antiproliferative activity of essential oil from the leaves of a medicinal herb, *Levisticum officinale*, against UMSCC1 head and neck squamous carcinoma cells. *Anticancer Res.* **2011**, *31*, 185–191. [PubMed]
41. Fekrazad, R.; Afzali, M.; Pasban-Aliabadi, H.; Esmaeili-Mahani, S.; Aminizadeh, M.; Mostafavi, A. Cytotoxic effect of *Thymus caramanicus* Jalas on human oral epidermoid carcinoma KB cells. *Braz. Dent. J.* **2017**, *28*, 72–77. [CrossRef]
42. Ojah, E.O.; Moronkola, D.O.; Akintunde, A.-A.M. α-amylase and α-glucosidase antidiabetic potential of ten essential oils from *Calophyllum inophyllum* Linn. *Iberoam. J. Med.* **2020**, *2*, 253–260. [CrossRef]
43. Adefegha, S.A.; Olasehinde, T.A.; Oboh, G. Essential oil composition, antioxidant, antidiabetic and antihypertensive properties of two *Afromomum* species. *J. Oleo Sci.* **2017**, *66*, 51–63. [CrossRef] [PubMed]
44. Mechchate, H.; Es-Safi, I.; Louba, A.; Alqahtani, A.S.; Nasr, F.A.; Noman, O.M.; Farooq, M.; Alharbi, M.S.; Alqahtani, A.; Bari, A. In vitro alpha-amylase and alpha-glucosidase inhibitory activity and in vivo antidiabetic activity of *Withania frutescens* L. Foliar extract. *Molecules* **2021**, *26*, 293. [CrossRef]
45. Rana, S.; Dixit, S.; Mittal, A. Anticancer effects of chemotherapy and nature products. *J. Med. Discov.* **2017**, *2*, 1–8. [CrossRef]
46. Asmat, U.; Abad, K.; Ismail, K. Diabetes mellitus and oxidative stress—A concise review. *Saudi Pharm. J.* **2016**, *24*, 547–553. [CrossRef] [PubMed]

47. Galovičová, L.; Čmiková, N.; Schwarzová, M.; Vukic, M.D.; Vukovic, N.L.; Kowalczewski, P.Ł.; Bakay, L.; Kluz, M.I.; Puchalski, C.; Obradovic, A.D.; et al. Biological Activity of *Cupressus sempervirens* Essential Oil. *Plants* **2023**, *12*, 1097. [CrossRef] [PubMed]
48. Akermi, S.; Smaoui, S.; Elhadef, K.; Fourati, M.; Louhichi, N.; Chaari, M.; Chakchouk Mtibaa, A.; Baanannou, A.; Masmoudi, S.; Mellouli, L. *Cupressus sempervirens* Essential Oil: Exploring the Antibacterial Multitarget Mechanisms, Chemcomputational Toxicity Prediction, and Safety Assessment in Zebrafish Embryos. *Molecules* **2022**, *27*, 2630. [CrossRef] [PubMed]
49. Reyes-Jurado, F.; Franco-Vega, A.; Ramírez-Corona, N.; Palou, E.; López-Malo, A. Essential Oils: Antimicrobial Activities, Extraction Methods, and Their Modeling. *Food Eng. Rev.* **2015**, *7*, 275–297. [CrossRef]
50. Abers, M.; Schroeder, S.; Goelz, L.; Sulser, A.; St. Rose, T.; Puchalski, K.; Langland, J. Antimicrobial activity of the volatile substances from essential oils. *BMC Complement. Med. Ther.* **2021**, *21*, 124. [CrossRef] [PubMed]
51. Wang, F.; Zhang, X.; Wang, Y.; Chen, Y.; Lu, H.; Meng, X.; Ye, X.; Chen, W. Activation/inactivation of anticancer drugs by CYP3A4: Influencing factors for personalized cancer therapy. *Drug Metab. Dispos.* **2023**, *51*, 543–559. [CrossRef] [PubMed]
52. Cobleigh, M.A.; Layng, K.V.; Mauer, E.; Mahon, B.; Hockenberry, A.J.; Abukhdeir, A.M. Comparative genomic analysis of PIK3R1-mutated and wild-type breast cancers. *Breast Cancer Res. Treat.* **2024**, *204*, 407–414. [CrossRef] [PubMed]
53. Liu, Y.; Wang, D.; Li, Z.; Li, X.; Jin, M.; Jia, N.; Cui, X.; Hu, G.; Tang, T.; Yu, Q. Pan-cancer analysis on the role of PIK3R1 and PIK3R2 in human tumors. *Sci. Rep.* **2022**, *12*, 5924. [CrossRef] [PubMed]
54. He, W.; Zhang, H.; Cheng, H.; Wen, J.; Li, D. PIK3CD correlates with prognosis, epithelial–mesenchymal transition and tumor immune infiltration in breast carcinoma. *Discov. Oncol.* **2023**, *14*, 187. [CrossRef]
55. Zundelevich, A.; Dadiani, M.; Kahana-Edwin, S.; Itay, A.; Sella, T.; Gadot, M.; Cesarkas, K.; Farage-Barhom, S.; Saar, E.G.; Eyal, E.; et al. ESR1 mutations are frequent in newly diagnosed metastatic and loco-regional recurrence of endocrine-treated breast cancer and carry worse prognosis. *Breast Cancer Res.* **2020**, *22*, 16. [CrossRef]
56. Li, M.; Zhang, L.; Yu, J.; Wang, X.; Cheng, L.; Ma, Z.; Chen, X.; Wang, L.; Goh, B.C. AKR1C3 in carcinomas: From multifaceted roles to therapeutic strategies. *Front. Pharmacol.* **2024**, *15*, 1378292. [CrossRef]
57. Uribe, M.L.; Marrocco, I.; Yarden, Y. EGFR in cancer: Signaling mechanisms, drugs, and acquired resistance. *Cancers* **2021**, *13*, 2748. [CrossRef] [PubMed]
58. Liu, L.; Mo, M.; Chen, X.; Chao, D.; Zhang, Y.; Chen, X.; Wang, Y.; Zhang, N.; He, N.; Yuan, X.; et al. Targeting inhibition of prognosis-related lipid metabolism genes including CYP19A1 enhances immunotherapeutic response in colon cancer. *J. Exp. Clin. Cancer Res.* **2023**, *42*, 85. [CrossRef]
59. Bhullar, K.S.; Lagarón, N.O.; McGowan, E.M.; Parmar, I.; Jha, A.; Hubbard, B.P.; Rupasinghe, H.P.V. Kinase-targeted cancer therapies: Progress, challenges and future directions. *Mol. Cancer* **2018**, *17*, 48. [CrossRef]
60. Riegel, K.; Vijayarangakannan, P.; Kechagioglou, P.; Bogucka, K.; Rajalingam, K. Recent advances in targeting protein kinases and pseudokinases in cancer biology. *Front. Cell Dev. Biol.* **2022**, *10*, 942500. [CrossRef]
61. Singh, P.; Lim, B. Targeting Apoptosis in Cancer. *Curr. Oncol. Rep.* **2022**, *24*, 273–284. [CrossRef]
62. Tuli, H.S.; Kaur, J.; Vashishth, K.; Sak, K.; Sharma, U.; Choudhary, R.; Behl, T.; Singh, T.; Sharma, S.; Saini, A.K.; et al. Molecular mechanisms behind ROS regulation in cancer: A balancing act between augmented tumorigenesis and cell apoptosis. *Arch. Toxicol.* **2023**, *97*, 103–120. [CrossRef] [PubMed]
63. Razaghi, A.; Heimann, K.; Schaeffer, P.M.; Gibson, S.B. Negative regulators of cell death pathways in cancer: Perspective on biomarkers and targeted therapies. *Apoptosis* **2018**, *23*, 93–112. [CrossRef]
64. Li, B.; Qin, Y.; Yu, X.; Xu, X.; Yu, W. Lipid raft involvement in signal transduction in cancer cell survival, cell death and metastasis. *Cell Prolif.* **2022**, *55*, e13167. [CrossRef] [PubMed]
65. Zhang, Z.; Yue, P.; Lu, T.; Wang, Y.; Wei, Y.; Wei, X. Role of lysosomes in physiological activities, diseases, and therapy. *J. Hematol. Oncol.* **2021**, *14*, 79. [CrossRef]
66. Goodall, G.J.; Wickramasinghe, V.O. RNA in cancer. *Nat. Rev. Cancer* **2021**, *21*, 22–36. [CrossRef] [PubMed]
67. Linzer, N.; Trumbull, A.; Nar, R.; Gibbons, M.D.; Yu, D.T.; Strouboulis, J.; Bungert, J. Regulation of RNA polymerase II transcription initiation and elongation by transcription factor TFII-I. *Front. Mol. Biosci.* **2021**, *8*, 681550. [CrossRef] [PubMed]
68. Greenlee, J.D.; Subramanian, T.; Liu, K.; King, M.R. Rafting down the metastatic cascade: The role of lipid rafts in cancer metastasis, cell death, and clinical outcomes. *Cancer Res.* **2021**, *81*, 5–17. [CrossRef]
69. Skotland, T.; Kavaliauskiene, S.; Sandvig, K. The role of lipid species in membranes and cancer-related changes. *Cancer Metastasis Rev.* **2020**, *39*, 343–360. [CrossRef] [PubMed]
70. Aouiche, C.; Chen, B.; Shang, X. Predicting stage-specific cancer related genes and their dynamic modules by integrating multiple datasets. *BMC Bioinform.* **2019**, *20*, 194. [CrossRef] [PubMed]
71. Sarathi, A.; Palaniappan, A. Novel significant stage-specific differentially expressed genes in hepatocellular carcinoma. *BMC Cancer* **2019**, *19*, 663. [CrossRef] [PubMed]
72. Shao, W.; Azam, Z.; Guo, J.; To, S.S.T. Oncogenic potential of PIK3CD in glioblastoma is exerted through cytoskeletal proteins PAK3 and PLEK2. *Lab. Investig.* **2022**, *102*, 1314–1322. [CrossRef] [PubMed]
73. Barros-Oliveira, M.d.C.; Costa-Silva, D.R.; Dos Santos, A.R.; Pereira, R.O.; Soares-Junior, J.M.; Silva, B.B.d. Influence of CYP19A1 gene expression levels in women with breast cancer: A systematic review of the literature. *Clinics* **2021**, *76*, e2846. [CrossRef]
74. Brett, J.O.; Spring, L.M.; Bardia, A.; Wander, S.A. ESR1 mutation as an emerging clinical biomarker in metastatic hormone receptor-positive breast cancer. *Breast Cancer Res.* **2021**, *23*, 85. [CrossRef]

75. Li, Z.; Wu, Y.; Yates, M.E.; Tasdemir, N.; Bahreini, A.; Chen, J.; Levine, K.M.; Priedigkeit, N.M.; Nasrazadani, A.; Ali, S. Hotspot ESR1 mutations are multimodal and contextual modulators of breast cancer metastasis. *Cancer Res.* **2022**, *82*, 1321–1339. [CrossRef]
76. Reinert, T.; Gonçalves, R.; Bines, J. Implications of ESR1 Mutations in Hormone Receptor-Positive Breast Cancer. *Curr. Treat. Options Oncol.* **2018**, *19*, 24. [CrossRef] [PubMed]
77. Sadeghi, H.; Hashemnia, V.; Nazemalhosseini-Mojarad, E.; Ghasemi, M.R.; Mirfakhraie, R. Correlated downregulation of VDR and CYP3A4 in colorectal cancer. *Mol. Biol. Rep.* **2023**, *50*, 1385–1391. [CrossRef] [PubMed]
78. Chen, J.S.; Huang, J.Q.; Luo, B.; Dong, S.H.; Wang, R.C.; Jiang, Z.k.; Xie, Y.K.; Yi, W.; Wen, G.M.; Zhong, J.F. PIK 3 CD induces cell growth and invasion by activating AKT/GSK-3β/β-catenin signaling in colorectal cancer. *Cancer Sci.* **2019**, *110*, 997–1011. [CrossRef] [PubMed]
79. Lung, D.K.; Reese, R.M.; Alarid, E.T. Intrinsic and extrinsic factors governing the transcriptional regulation of ESR1. *Horm. Cancer* **2020**, *11*, 129–147. [CrossRef]
80. Ali, R.; Wendt, M.K. The paradoxical functions of EGFR during breast cancer progression. *Signal Transduct. Target. Ther.* **2017**, *2*, 1–7. [CrossRef]
81. Bisson, J.; McAlpine, J.B.; Friesen, J.B.; Chen, S.-N.; Graham, J.; Pauli, G.F. Can invalid bioactives undermine natural product-based drug discovery? *J. Med. Chem.* **2016**, *59*, 1671–1690. [CrossRef]
82. Baell, J.B. Feeling nature's PAINS: Natural products, natural product drugs, and pan assay interference compounds (PAINS). *J. Nat. Prod.* **2016**, *79*, 616–628. [CrossRef]
83. Baell, J.; Walters, M.A. Chemistry: Chemical con artists foil drug discovery. *Nature* **2014**, *513*, 481–483. [CrossRef]
84. Gomha, S.M.; Riyadh, S.M.; Mahmmoud, E.A.; Elaasser, M.M. Synthesis and anticancer activity of arylazothiazoles and 1,3,4-thiadiazoles using chitosan-grafted-poly(4-vinylpyridine) as a novel copolymer basic catalyst. *Chem. Heterocycl. Compd.* **2015**, *51*, 1030–1038. [CrossRef]
85. Elkomy, N.M.; El-Shaibany, A.; Elnagar, G.M.; Abdelkhalek, A.S.; Al-Mahbashi, H.; Elaasser, M.M.; Raweh, S.M.; Aldiyarbi, M.A.; Raslan, A.E. Evaluation of acute oral toxicity, anti-diabetic and antioxidant effects of *Aloe vera* flowers extract. *J. Ethnopharmacol.* **2023**, *309*, 116310. [CrossRef]
86. Ling, L.T.; Yap, S.-A.; Radhakrishnan, A.K.; Subramaniam, T.; Cheng, H.M.; Palanisamy, U.D. Standardised *Mangifera indica* extract is an ideal antioxidant. *Food Chem.* **2009**, *113*, 1154–1159. [CrossRef]
87. Elaasser, M.; Morsi, M.; Galal, S.; Abd El-Rahman, M.; Katry, M. Antioxidant, anti-inflammatory and cytotoxic activities of the unsaponifiable fraction of extra virgin olive oil. *Grasas Y Aceites* **2020**, *71*, e386. [CrossRef]
88. Jaradat, N.; Dacca, H.; Hawash, M.; Abualhasan, M.N. *Ephedra alata* fruit extracts: Phytochemical screening, anti-proliferative activity and inhibition of DPPH, α-amylase, α-glucosidase, and lipase enzymes. *BMC Chem.* **2021**, *15*, 41. [CrossRef] [PubMed]
89. Al-Maharik, N.; Jaradat, N.; Al-Hajj, N.; Jaber, S. *Myrtus communis* L.: Essential oil chemical composition, total phenols and flavonoids contents, antimicrobial, antioxidant, anticancer, and α-amylase inhibitory activity. *Chem. Biol. Technol. Agric.* **2023**, *10*, 41. [CrossRef]
90. Hindler, J.; Howard, B.; Keiser, J. Antimicrobial agents and antimicrobial susceptibility testing. In *Howard BJ. Clinical and Pathogenic Microbiology*, 2nd ed.; Mosby: St Louis, MO, USA, 1994.
91. Choudhary, M.I.; Thomsen, W.J. *Bioassay Techniques for Drug Development*; CRC Press: Boca Raton, FL, USA, 2001.
92. Yoshida, T.; Jono, K.; Okonogi, K. Modified agar dilution susceptibility testing method for determining in vitro activities of antifungal agents, including azole compounds. *Antimicrob. Agents Chemother.* **1997**, *41*, 1349–1351. [CrossRef]
93. Wayne, P. *Reference method for broth dilution antifungal susceptibility testing of yeasts, approved standard. CLSI Doc. M27-A2*; Clinical and Laboratory Standards Institute: Wayne, PA, USA, 2002.
94. Lipinski, C.A. Lead-and drug-like compounds: The rule-of-five revolution. *Drug Discov. Today Technol.* **2004**, *1*, 337–341. [CrossRef]
95. Daina, A.; Michielin, O.; Zoete, V. SwissTargetPrediction: Updated data and new features for efficient prediction of protein targets of small molecules. *Nucleic Acids Res.* **2019**, *47*, W357–W364. [CrossRef]
96. Yao, Z.J.; Dong, J.; Che, Y.J.; Zhu, M.F.; Wen, M.; Wang, N.N.; Wang, S.; Lu, A.P.; Cao, D.S. TargetNet: A web service for predicting potential drug-target interaction profiling via multi-target SAR models. *J. Comput. Aided Mol. Des.* **2016**, *30*, 413–424. [CrossRef] [PubMed]
97. Hamosh, A.; Scott, A.F.; Amberger, J.S.; Bocchini, C.A.; McKusick, V.A. Online Mendelian Inheritance in Man (OMIM), a knowledgebase of human genes and genetic disorders. *Nucleic Acids Res.* **2005**, *33*, D514–D517. [CrossRef]
98. Rebhan, M.; Chalifa-Caspi, V.; Prilusky, J.; Lancet, D. GeneCards: Integrating information about genes, proteins and diseases. *Trends Genet. TIG* **1997**, *13*, 163. [CrossRef]
99. Safran, M.; Dalah, I.; Alexander, J.; Rosen, N.; Iny Stein, T.; Shmoish, M.; Nativ, N.; Bahir, I.; Doniger, T.; Krug, H. GeneCards Version 3: The human gene integrator. *Database* **2010**, *2010*, baq020. [CrossRef] [PubMed]
100. Zaru, R.; Orchard, S.; Consortium, U. Uniprot tools: bLAST, align, peptide search, and ID mapping. *Curr. Protoc.* **2023**, *3*, e697. [CrossRef]
101. Szklarczyk, D.; Kirsch, R.; Koutrouli, M.; Nastou, K.; Mehryary, F.; Hachilif, R.; Gable, A.L.; Fang, T.; Doncheva, N.T.; Pyysalo, S. The STRING database in 2023: Protein–protein association networks and functional enrichment analyses for any sequenced genome of interest. *Nucleic Acids Res.* **2023**, *51*, D638–D646. [CrossRef] [PubMed]

102. Shannon, P.; Markiel, A.; Ozier, O.; Baliga, N.S.; Wang, J.T.; Ramage, D.; Amin, N.; Schwikowski, B.; Ideker, T. Cytoscape: A software environment for integrated models of biomolecular interaction networks. *Genome Res.* **2003**, *13*, 2498–2504. [CrossRef] [PubMed]
103. Tang, Y.; Li, M.; Wang, J.; Pan, Y.; Wu, F.X. CytoNCA: A cytoscape plugin for centrality analysis and evaluation of protein interaction networks. *Biosystems* **2015**, *127*, 67–72. [CrossRef] [PubMed]
104. Hu, H.; Wang, H.; Yang, X.; Li, Z.; Zhan, W.; Zhu, H.; Zhang, T. Network pharmacology analysis reveals potential targets and mechanisms of proton pump inhibitors in breast cancer with diabetes. *Sci. Rep.* **2023**, *13*, 7623. [CrossRef] [PubMed]
105. Li, F.; Duan, J.; Zhao, M.; Huang, S.; Mu, F.; Su, J.; Liu, K.; Pan, Y.; Lu, X.; Li, J. A network pharmacology approach to reveal the protective mechanism of *Salvia miltiorrhiza-Dalbergia odorifera* coupled-herbs on coronary heart disease. *Sci. Rep.* **2019**, *9*, 19343. [CrossRef]
106. Chin, C.-H.; Chen, S.-H.; Wu, H.-H.; Ho, C.-W.; Ko, M.-T.; Lin, C.-Y. cytoHubba: Identifying hub objects and sub-networks from complex interactome. *BMC Syst. Biol.* **2014**, *8*, S11. [CrossRef]
107. Dennis, G.; Sherman, B.T.; Hosack, D.A.; Yang, J.; Gao, W.; Lane, H.C.; Lempicki, R.A. DAVID: Database for annotation, visualization, and integrated discovery. *Genome Biol.* **2003**, *4*, 1–11. [CrossRef]
108. Tang, Z.; Kang, B.; Li, C.; Chen, T.; Zhang, Z. GEPIA2: An enhanced web server for large-scale expression profiling and interactive analysis. *Nucleic Acids Res.* **2019**, *47*, W556–W560. [CrossRef]
109. Tang, Z.; Li, C.; Kang, B.; Gao, G.; Li, C.; Zhang, Z. GEPIA: A web server for cancer and normal gene expression profiling and interactive analyses. *Nucleic Acids Res.* **2017**, *45*, W98–W102. [CrossRef] [PubMed]
110. Digre, A.; Lindskog, C. The Human Protein Atlas—Spatial localization of the human proteome in health and disease. *Protein Sci.* **2021**, *30*, 218–233. [CrossRef] [PubMed]
111. Burley, S.K.; Berman, H.M.; Kleywegt, G.J.; Markley, J.L.; Nakamura, H.; Velankar, S. Protein Data Bank (PDB): The Single Global Macromolecular Structure Archive. In *Protein Crystallography: Methods and Protocols*; Wlodawer, A., Dauter, Z., Jaskolski, M., Eds.; Springer: New York, NY, USA, 2017; pp. 627–641.
112. Yano, J.K.; Wester, M.R.; Schoch, G.A.; Griffin, K.J.; Stout, C.D.; Johnson, E.F. The structure of human microsomal cytochrome P450 3A4 determined by X-ray crystallography to 2.05-Å resolution. *J. Biol. Chem.* **2004**, *279*, 38091–38094. [CrossRef] [PubMed]
113. Furet, P.; Guagnano, V.; Fairhurst, R.A.; Imbach-Weese, P.; Bruce, I.; Knapp, M.; Fritsch, C.; Blasco, F.; Blanz, J.; Aichholz, R. Discovery of NVP-BYL719 a potent and selective phosphatidylinositol-3 kinase alpha inhibitor selected for clinical evaluation. *Bioorganic Med. Chem. Lett.* **2013**, *23*, 3741–3748. [CrossRef] [PubMed]
114. Fradera, X.; Methot, J.L.; Achab, A.; Christopher, M.; Altman, M.D.; Zhou, H.; McGowan, M.A.; Kattar, S.D.; Wilson, K.; Garcia, Y. Design of selective PI3Kδ inhibitors using an iterative scaffold-hopping workflow. *Bioorganic Med. Chem. Lett.* **2019**, *29*, 2575–2580. [CrossRef] [PubMed]
115. Kim, S.; Wu, J.Y.; Birzin, E.T.; Frisch, K.; Chan, W.; Pai, L.-Y.; Yang, Y.T.; Mosley, R.T.; Fitzgerald, P.M.; Sharma, N. Estrogen receptor ligands. II. Discovery of benzoxathiins as potent, selective estrogen receptor α modulators. *J. Med. Chem.* **2004**, *47*, 2171–2175. [CrossRef] [PubMed]
116. Heinrich, D.M.; Flanagan, J.U.; Jamieson, S.M.; Silva, S.; Rigoreau, L.J.; Trivier, E.; Raynham, T.; Turnbull, A.P.; Denny, W.A. Synthesis and structure–activity relationships for 1-(4-(piperidin-1-ylsulfonyl) phenyl) pyrrolidin-2-ones as novel non-carboxylate inhibitors of the aldo-keto reductase enzyme AKR1C3. *Eur. J. Med. Chem.* **2013**, *62*, 738–744. [CrossRef] [PubMed]
117. Zhao, Q.; Ouyang, X.; Wan, X.; Gajiwala, K.S.; Kath, J.C.; Jones, L.H.; Burlingame, A.L.; Taunton, J. Broad-Spectrum Kinase Profiling in Live Cells with Lysine-Targeted Sulfonyl Fluoride Probes. *J. Am. Chem. Soc.* **2017**, *139*, 680–685. [CrossRef] [PubMed]
118. Ghosh, D.; Lo, J.; Morton, D.; Valette, D.; Xi, J.; Griswold, J.; Hubbell, S.; Egbuta, C.; Jiang, W.; An, J.; et al. Novel Aromatase Inhibitors by Structure-Guided Design. *J. Med. Chem.* **2012**, *55*, 8464–8476. [CrossRef] [PubMed]
119. Biovia, D.S. *Discovery Studio Visualizer v21. 1.0. 20298*; Dassault Systèmes: San Diego, CA, USA, 2021.
120. Pettersen, E.F.; Goddard, T.D.; Huang, C.C.; Couch, G.S.; Greenblatt, D.M.; Meng, E.C.; Ferrin, T.E. UCSF Chimera--a visualization system for exploratory research and analysis. *J. Comput. Chem.* **2004**, *25*, 1605–1612. [CrossRef]
121. Fikry, E.; Orfali, R.; El-Sayed, S.S.; Perveen, S.; Ghafar, S.; El-Shafae, A.M.; El-Domiaty, M.M.; Tawfeek, N. Potential Hepatoprotective Effects of *Chamaecyparis lawsoniana* against Methotrexate-Induced Liver Injury: Integrated Phytochemical Profiling, Target Network Analysis, and Experimental Validation. *Antioxidants* **2023**, *12*, 2118. [CrossRef] [PubMed]
122. Tian, W.; Chen, C.; Lei, X.; Zhao, J.; Liang, J. CASTp 3.0: Computed atlas of surface topography of proteins. *Nucleic Acids Res.* **2018**, *46*, W363–W367. [CrossRef] [PubMed]
123. O'Boyle, N.M.; Banck, M.; James, C.A.; Morley, C.; Vandermeersch, T.; Hutchison, G.R. Open Babel: An open chemical toolbox. *J. Cheminform.* **2011**, *3*, 33. [CrossRef] [PubMed]

Disclaimer/Publisher's Note: The statements, opinions and data contained in all publications are solely those of the individual author(s) and contributor(s) and not of MDPI and/or the editor(s). MDPI and/or the editor(s) disclaim responsibility for any injury to people or property resulting from any ideas, methods, instructions or products referred to in the content.

MDPI AG
Grosspeteranlage 5
4052 Basel
Switzerland
Tel.: +41 61 683 77 34

Pharmaceuticals Editorial Office
E-mail: pharmaceuticals@mdpi.com
www.mdpi.com/journal/pharmaceuticals

Disclaimer/Publisher's Note: The title and front matter of this reprint are at the discretion of the Guest Editor. The publisher is not responsible for their content or any associated concerns. The statements, opinions and data contained in all individual articles are solely those of the individual Editor and contributors and not of MDPI. MDPI disclaims responsibility for any injury to people or property resulting from any ideas, methods, instructions or products referred to in the content.

www.ingramcontent.com/pod-product-compliance
Lightning Source LLC
Chambersburg PA
CBHW041423210926
51850CB00013B/3345